ExxonMobil
Lubrication Fundamentals
Third Edition, Revised and Expanded

ExxonMobil
Lubrication Fundamentals
Third Edition, Revised and Expanded

Don M. Pirro
ExxonMobil

Martin Webster
ExxonMobil

Ekkehard Daschner
ExxonMobil

CRC Press
Taylor & Francis Group
Boca Raton London New York

CRC Press is an imprint of the
Taylor & Francis Group, an **informa** business

CRC Press
Taylor & Francis Group
6000 Broken Sound Parkway NW, Suite 300
Boca Raton, FL 33487-2742

© 2016 by Taylor & Francis Group, LLC
CRC Press is an imprint of Taylor & Francis Group, an Informa business

No claim to original U.S. Government works

Printed on acid-free paper
Version Date: 20160106

Printed and bound in India by Replika Press Pvt. Ltd.

International Standard Book Number-13: 978-1-4987-5290-9 (Hardback)

This book contains information obtained from authentic and highly regarded sources. Reasonable efforts have been made to publish reliable data and information, but the author and publisher cannot assume responsibility for the validity of all materials or the consequences of their use. The authors and publishers have attempted to trace the copyright holders of all material reproduced in this publication and apologize to copyright holders if permission to publish in this form has not been obtained. If any copyright material has not been acknowledged please write and let us know so we may rectify in any future reprint.

Except as permitted under U.S. Copyright Law, no part of this book may be reprinted, reproduced, transmitted, or utilized in any form by any electronic, mechanical, or other means, now known or hereafter invented, including photocopying, microfilming, and recording, or in any information storage or retrieval system, without written permission from the publishers.

For permission to photocopy or use material electronically from this work, please access www.copyright.com (http://www.copyright.com/) or contact the Copyright Clearance Center, Inc. (CCC), 222 Rosewood Drive, Danvers, MA 01923, 978-750-8400. CCC is a not-for-profit organization that provides licenses and registration for a variety of users. For organizations that have been granted a photocopy license by the CCC, a separate system of payment has been arranged.

Trademark Notice: Product or corporate names may be trademarks or registered trademarks, and are used only for identification and explanation without intent to infringe.

Library of Congress Cataloging-in-Publication Data

Names: Pirro, Don M., 1955- author. | Webster, Martin (Martin N.), author. | Daschner, Ekkehard, author.
Title: Lubrication fundamentals / authors, Don M. Pirro, Martin Webster, Ekkehard Daschner.
Description: Third edition, revised and expanded. | Boca Raton : Taylor & Francis, CRC Press, 2016. | Previous edition: Lubrication fundamentals / D.M. Pirro, A.A. Wessol. 2001. Originally by J. George Wills. | Includes bibliographical references and index.
Identifiers: LCCN 2015050128 | ISBN 9781498752909
Subjects: LCSH: Lubrication and lubricants.
Classification: LCC TJ1075 .W57 2016 | DDC 621.8/9--dc23
LC record available at http://lccn.loc.gov/2015050128

Visit the Taylor & Francis Web site at
http://www.taylorandfrancis.com

and the CRC Press Web site at
http://www.crcpress.com

Contents

Preface xxiii
Acknowledgments xxv
Authors xxvii
ExxonMobil Contributors to the Third Edition xxix

Chapter 1 Introduction 1

    I. Premodern History of Petroleum 1
    II. Petroleum in North America 2
    III. Development of Lubricants 3
    IV. History of Synthetic Lubricants 4
    V. Future Prospects 6
Bibliography 7

Chapter 2 Lubricant Base Stock Production and Application 9

    I. Lubricant Base Stocks and Their Application 9
        A. American Petroleum Institute Group I, II, III, IV, and V Base Stocks 9
            1. Group I Base Stocks 10
            2. Group II and Group III Base Stocks 11
            3. Group II+ and Group III+ Base Stocks 11
            4. Group IV Base Stocks 11
            5. Group V Base Stocks 11
        B. Base Stock Selection 12
        C. Product Applications 14
        D. Base Oil Slate 15
    II. Role of Crude Oil in the Manufacture of Base Stock 18
        A. Chemistry of Crude Oil 18
        B. Crude Selection 20
    III. Refinery Processing—Separation versus Conversion 22
        A. Atmospheric Distillation 22
        B. Vacuum Distillation 24
        C. Propane Deasphalting 25
    IV. Conventional Solvent Processing 25
        A. Solvent Extraction 26
        B. Solvent Dewaxing 27
        C. Hydrofinishing 28
    V. Conversion Processing 29
        A. Hydrocracking 30
        B. Catalytic Dewaxing 30
        C. Alternate Processing for Group III+ Quality 31
        D. Gas-to-Liquids via Fischer Tropsch Synthesis 32
    VI. Base Stock Composition 32
Bibliography 33

Chapter 3 Lubricating Oils 35

- I. Additives 35
  - A. Pour Point Depressants 35
  - B. VI Improvers 36
  - C. Defoamants 36
  - D. Oxidation Inhibitors 37
  - E. Rust and Corrosion Inhibitors 38
  - F. Detergents and Dispersants 40
  - G. Antiwear Additives 41
  - H. Extreme Pressure Additives 41
- II. Physical and Chemical Characteristics 42
  - A. Carbon Residue 43
  - B. Color 43
  - C. Density and Gravity 43
  - D. Flash and Fire Points 44
  - E. Neutralization Number 45
  - F. Total Acid Number 46
  - G. TBN 46
  - H. Pour Point 46
  - I. Sulfated Ash 47
  - J. Viscosity 47
    1. Engine Oil Viscosity Classification 50
    2. Axle and Manual Transmission Lubricant Viscosity Classification 50
    3. Viscosity System for Industrial Fluid Lubricants 52
  - K. VI 54
- III. Evaluation and Performance Tests 54
  - A. Oxidation Tests 54
  - B. Thermal Stability 56
  - C. Rust Protection Tests 56
  - D. Foam Tests 57
  - E. EP and Antiwear Tests 57
    1. Abrasive Wear 57
    2. Corrosive or Chemical Wear 57
    3. Adhesive Wear 57
    4. Fatigue Wear 58
  - F. Emulsion and Demulsibility Tests 59
- IV. Engine Tests for Oil Performance 60
  - A. Oxidation Stability and Bearing Corrosion Protection 62
  - B. Single Cylinder High Temperature Tests 62
  - C. Multicylinder High Temperature Engine Tests 62
  - D. Multicylinder Low Temperature Tests 63
  - E. Rust and Corrosion Protection Tests 64
  - F. Oil Consumption Rates and Volatility 64
  - G. Emissions and Protection of Emission Control Systems 64
  - H. Fuel Economy 65
- V. Automotive Gear Lubricants 65
- VI. ATFs 67

Bibliography 67

Chapter 4 Lubricating Greases 69
  I. Why Greases Are Used 69
  II. Composition of Grease 69
    A. Fluid Components 69
    B. Thickeners 70
    C. Additives 71
  III. Manufacture of Grease 74
  IV. Grease Characteristics 75
    A. Consistency 75
      1. Cone Penetration 75
      2. NLGI Grease Grade Numbers 77
    B. Dropping Point 77
  V. Evaluation and Performance Tests 78
    A. Mechanical or Structural Stability Tests 78
    B. Static Oxidation Test 79
    C. Dynamic Oxidation Tests 80
    D. Oil Separation Tests 80
    E. Water Resistance Tests 81
    F. Rust Protection Tests 82
    G. Extreme Pressure and Wear Prevention Tests 82
    H. Grease Compatibility 83
    I. Apparent Viscosity 83
Bibliography 85

Chapter 5 Synthetic Lubricants 87
  I. SHFs 90
    A. PAOs (Olefin Oligomers) 91
    B. Alkylated Aromatics 92
      1. Application 93
    C. Polybutenes 93
      1. Application 93
    D. Cycloaliphatics 93
      1. Applications 94
  II. Organic Esters 94
    A. Dibasic Acid Esters 94
      1. Application 95
    B. Polyol Ester 95
      1. Application 95
  III. Polyglycols 96
    A. Application 97
  IV. Phosphate Esters 97
    A. Application 97
  V. Other Synthetic Lubricating Fluids 98
    A. Silicones 98
      1. Application 98
    B. Silicate Esters 98
      1. Application 98
    C. Polyphenyl Ethers 98
      1. Application 99

       D. Halogenated Fluids 99
          1. Application 99
Bibliography 99

Chapter 6   Environmental Lubricants 101

   I. Environmental Considerations 101
  II. Definitions and Test Procedures 102
      A. Toxicity 102
      B. Biodegradability 102
      C. Bioaccumulation 103
 III. Environmental Criteria 105
      A. National Labeling Programs 105
          1. Blue Angel 105
          2. Swedish Standard 105
          3. U.S. Department of Agriculture BioPreferred® 106
          4. U.S. VGP (U.S. VGP 2013) 106
      B. International Labeling Programs 106
          1. Nordic Swan 106
          2. European Ecolabel 107
 IV. Environmental Characteristics of Various Base Stocks 107
      A. Overview of Base Stock Options 108
  V. Product Availability and Performance 110
      A. Vegetable Oil–Based EAL Performance Concerns 111
          1. Oxidation Stability 111
          2. Low Temperature Performance 111
          3. Hydrolytic Stability 112
 VI. Product Selection Process 113
      A. Environmental Acceptability 113
      B. Specifications 113
      C. Equipment Builder Approvals 113
      D. Proven Field Performance 114
      E. Supplier Reliability 114
      F. Operating and Maintenance Conditions 114
 VII. Converting to EALs 114
Bibliography 115

Chapter 7   Hydraulics 117

   I. Basic Principles 117
      A. Hydromechanics 117
      B. Fundamental Hydraulic Systems 118
  II. System Components 119
      A. Hydraulic Pumps 119
          1. Gear Pumps 119
          2. Vane Pumps 120
          3. Piston Pumps 120
          4. Radial Piston Pumps 120
          5. Axial Piston Pumps 122
      B. Pump Selection Criteria 122

Contents

- III. Controlling Pressure and Flow 122
  - A. Relief Valves 125
  - B. Directional Control Valves 125
  - C. Unloading Valves 125
  - D. Sequence Control Valve 127
  - E. Flow Control Valves 127
  - F. Accumulators 127
- IV. Actuators 128
  - A. Hydraulic Cylinders 128
  - B. Rotary Fluid Motors 129
- V. Hydraulic Drives 129
  - A. Hydrostatic Drives 130
- VI. Oil Reservoirs 130
- VII. Oil Qualities Required by Hydraulic Systems 132
  - A. Viscosity 133
  - B. Viscosity Index 133
  - C. Antiwear (Wear Protection) 133
  - D. Oxidation Stability 133
  - E. Air Entrainment 134
  - F. Antifoam 134
  - G. Demulsibility (Water Separating Ability) 134
  - H. Rust Protection 134
  - I. Compatibility 134
- VIII. Hydraulic Fluid Types 135
  - A. Industry Standards and OEM Approvals 135
- IX. Hydraulic System Maintenance 135
  - A. Filtration 136
  - B. Controlling Temperatures 138
  - C. Maintaining Proper Reservoir Oil Levels 138
  - D. Periodic Oil Analysis 139
  - E. Routine Inspections 139
- Bibliography 140

Chapter 8 Lubricating Films and Machine Elements: Bearings, Slides, Guides, Ways, Gears, Cylinders, Couplings, Chains, Wire Ropes 141

- I. Types of Lubricating Films 141
  - A. Fluid Films 141
    - 1. Thick Hydrodynamic Films 142
    - 2. Thin Elastohydrodynamic (EHL) Films 146
    - 3. Hydrostatic Films 148
    - 4. Squeeze Films 149
  - B. Thin Surface Films 149
    - 1. Nature of Surfaces 150
    - 2. Surface Contact 150
  - C. Solid or Dry Films 151
- II. Plain Bearings 152
  - A. Hydrodynamic Lubrication 152
    - 1. Grease Lubrication 154

- B. Hydrostatic Lubrication 155
  - 1. Constant Volume System 155
  - 2. Constant Pressure System with Flow Restrictor 155
  - 3. Constant Pressure with Flow Control Valve 156
  - 4. Hydrostatic Bearing Applications 156
- C. Thin Film Lubrication 157
  - 1. Wearing In of Thin Film Bearings 158
- D. Mechanical Factors 158
  - 1. Length/Diameter Ratio 159
  - 2. Projected Area 159
  - 3. Clearance 159
  - 4. Bearing Materials 160
  - 5. Surface Finish 161
  - 6. Grooving of Bearings 161
- E. Lubricant Selection 169
  - 1. Oil Selection 169
  - 2. Grease Selection 171
III. Rolling Element Bearings 171
- A. Need for Lubrication 173
- B. Factors Affecting Lubrication 175
  - 1. Effect of Speed 175
  - 2. Effect of Load 176
  - 3. Effect of Temperature 176
  - 4. Contamination 177
- C. Lubricant Selection 177
  - 1. Oil Selection 177
  - 2. Grease Selection 179
IV. Slides, Guides, and Ways 180
- A. Film Formation 180
  - 1. Grease Lubrication 180
  - 2. Lubricant Characteristics 181
V. Linear Motion Guides 181
- A. Lubricant Selection 182
  - 1. Oil Lubrication 183
  - 2. Grease Lubrication 183
VI. Gears 183
- A. Action between Gear Teeth 186
- B. Film Formation 189
- C. Factors Affecting Lubrication of Enclosed Gears 189
  - 1. Gear Type 191
  - 2. Gear Speed 192
  - 3. Reduction Ratio 192
  - 4. Operating and Startup Temperatures 192
  - 5. Transmitted Power and Load 193
  - 6. Surface Finish 193
  - 7. Drive Type 194
  - 8. Application Method 194
  - 9. Water Contamination 194
  - 10. Lubricant Leakage 194
VII. Lubricant Characteristics for Enclosed Gears 195
- A. AGMA Standard for Industrial Gear Lubricants 195

VIII.	Lubrication of Open Gears	196
	A. Factors Affecting Lubrication of Open Gears	198
	1. Temperature	198
	2. Dust and Dirt	198
	3. Water	198
	4. Method of Application	198
	5. Load Characteristics	199
	B. AGMA Specifications for Lubricants for Open Gearing	199
IX.	Cylinders	199
X.	Flexible Couplings	200
	A. Lubrication of Flexible Couplings	202
	1. Gear Couplings	203
	2. Chain Couplings	203
	3. Spring-Laced Couplings	204
	4. Sliding Couplings	204
	5. Slipper Joint Couplings	205
	6. Flexible Member Couplings	205
	7. Lubrication Techniques	205
XI.	Drive Chains	205
	A. Silent and Roller Chains	205
	B. Cast or Forged Link Chains	206
	C. Viscosity Selection	206
XII.	Cams and Cam Followers	207
XIII.	Wire Ropes	207
	A. Need for Lubrication	208
	1. Wear	208
	2. Fatigue	208
	3. Corrosion	208
	4. Core Protection	208
	B. Lubrication during Manufacture	208
	C. Lubrication in Service	208
	D. Lubricant Characteristics	209
Bibliography		209

Chapter 9   Application of Lubricants 211

I.	All-Loss Methods	211
	A. Oiling Devices	211
	1. Drop Feed and Wick Feed Cups	211
	2. Bottle Oilers	212
	3. Wick and Pad Oilers	213
	4. Mechanical Force Feed Lubricators	214
	5. Air Line Oilers	215
	6. Air Spray Application	216
	B. Grease Application	216
II.	Reuse Methods	218
	A. Circulation Systems	219
	1. Oil Reservoirs	221
	2. Pump Suction	221
	3. Bearing Housings	221
	4. Return Oil Piping	222

| | 5. Metallurgy Composition of Circulation Systems 222
| | 6. Oil Filtration 222
| | 7. Oil Coolers 223
| | 8. Oil Heating 223
| | 9. Monitoring Parameters 223
| | III. Other Reuse Methods 224
| | A. Splash Oiling 224
| | B. Bath Oiling 225
| | C. Ring, Chain, and Collar Oiling 225
| | IV. Centralized Application Systems 226
| | A. Central Lubrication Systems 226
| | 1. Two-Line System 227
| | 2. Single-Line Spring Return 227
| | 3. Series Manifolded System 228
| | 4. Series System, Reversing Flow 228
| | B. Mist Oiling Systems 229
Bibliography 231

Chapter 10 Internal Combustion Engines 233
   I. Design and Construction Considerations 233
       A. Combustion Cycle 233
            1. Four-Stroke Cycle 233
            2. Two-Stroke Cycle 234
       B. Mechanical Construction 236
       C. Supercharging 237
       D. Methods of Lubricant Application 237
   II. Fuel and Combustion Considerations 238
       A. Gasoline Engines 238
       B. Diesel Engines 239
       C. Gaseous Fueled Engines 240
  III. Operating Considerations 240
       A. Wear 240
       B. Cooling 241
       C. Sealing 241
       D. Deposits 242
  IV. Maintenance Considerations 243
   V. Engine Oil Characteristics 244
       A. Viscosity, Viscosity Index 244
       B. Low Temperature Fluidity 245
       C. Oxidation Stability (Chemical Stability) 245
       D. Thermal Stability 246
       E. Detergency and Dispersancy 246
       F. Alkalinity 247
       G. Antiwear 247
       H. Rust and Corrosion Protection 247
       I. Foam Resistance 248
       J. Effect on Gasoline Engine Octane Number Requirement 248
       K. Engine Oil Identification and Classification Systems 249
            1. API Engine Oil Service Categories 249
            2. API Classification System 249

	3. ILSAC Performance Specifications	252
	4. ACEA European Oil Sequences	254
	5. U.S. Military Specifications	254
	6. Manufacturer Specifications	257
VI.	Oil Recommendations by Application	258
	A. Passenger Car	258
	B. Truck and Bus	259
	C. Farm Machinery	260
	D. Aviation	260
	E. Diesel Engines Used in Industrial and Marine Applications	261
	1. High-Speed Engines	261
	2. Medium-Speed Engines	261
	3. Low-Speed Engines	262
	F. Natural Gas Fired Engines	263
	1. Two-Stroke Gas Engines	264
	2. Four-Stroke Low- to Medium-Speed Gas Engines	264
	3. Four-Stroke High-Speed Gas Engines	265
	4. Gas Engine Oil Selection	265
	5. Dual Fuel Engines	265
	G. Outboard Marine Engines	266
	H. Railroad Engines	266
	I. Motorcycles	267
	1. Two-Stroke Motorcycle Oils	267
	2. Four-Stroke Motorcycle Oils	267
Bibliography		268

Chapter 11 Automotive Transmissions and Drive Trains 269

I.	Clutches	269
II.	Transmissions	271
	A. Mechanical Transmissions	271
	B. Automatic Transmissions	273
	1. Torque Converters	273
	2. Planetary Gears	274
	3. Transmission	275
	C. CVT	276
	D. Semiautomatic Transmissions	276
	E. Hydrostatic Transmissions	277
	F. Factors Affecting Lubrication	279
	1. Mechanical Transmissions	279
	2. Automatic Transmissions	280
	3. Semiautomatic Transmissions	281
	4. Hydrostatic Transmissions	281
III.	Drive Shafts and Universal Joints	281
	A. Lubrication	282
	B. Drive Axles	282
	C. Differential Action	283
	D. Limited-Slip Differential	284
	E. Factors Affecting Lubrication	285
IV.	Transaxles	286
	A. Factors Affecting Lubrication	286

- V. Other Gear Cases 287
  - A. Auxiliary Transmissions 287
  - B. Transfer Cases 287
  - C. Overdrives 287
  - D. Final Drives 288
  - E. Factors Affecting Lubrication 288
- VI. Automotive Gear Lubricants 288
  - A. Load-Carrying Capacity 288
  - B. API Lubricant Service Designations 289
  - C. Viscosity 290
  - D. Channeling Characteristics 290
  - E. Storage Stability 291
  - F. Oxidation Resistance 291
  - G. Foaming 291
  - H. Chemical Activity or Corrosion 291
  - I. Rust Protection 291
  - J. Seal Compatibility 291
  - K. Frictional Properties 292
  - L. Identification 292
  - M. U.S. Military Specifications 292
- VII. Torque Converter and ATFs 292
  - A. Torque Converter Fluids 292
  - B. ATFs 293
    1. General Motors 293
    2. Ford ATFs 294
    3. Chrysler ATFs 295
    4. ATFs for Japanese Vehicles 295
    5. ATFs for European Vehicles 295
    6. Multipurpose ATFs 295
    7. Transmission Fluids for CVTs and DCTs 295
    8. Allison Transmission 295
    9. Caterpillar 296
- VIII. Multipurpose Tractor Fluids 296
  - A. Tractor Fluid Characteristics 297
    1. Viscosity and VI 297
    2. Foam and Air Entrainment Control 297
    3. Rust and Corrosion Protection 298
    4. Oxidation and Thermal Stability 298
    5. Frictional Characteristics 298
    6. EP and Antiwear Properties 299
- Bibliography 299

Chapter 12 Automotive Chassis Components 301
- I. Suspension and Steering Linkages 301
  - A. Front Wheel Suspension Systems 301
  - B. Rear Suspension Systems 303
  - C. Active Suspension Systems 303
  - D. Steering Systems 304
  - E. Factors Affecting Lubrication 305
  - F. Lubricant Characteristics 305

Contents   xv

    II. Steering Gear 306
        A. Factors Affecting Lubrication 308
            1. Electric Power Steering 308
        B. Lubricant Characteristics 308
   III. Wheel Bearings 309
        A. Lubricant Characteristics 309
   IV. Brake Systems 309
        A. ABSs 311
        B. Other Braking Systems 311
        C. Fluid Characteristics 311
    V. Medium- and Heavy-Duty Truck Chassis 313
        A. Factors Affecting Lubrication 314
        B. Lubricant Characteristics 315
Bibliography 315

Chapter 13 Stationary Gas Turbines 317

     I. Principles of Gas Turbines 317
        A. The Simple Cycle, Open System 318
        B. Regenerative Cycle, Open System 320
        C. Intercooling, Reheating 320
        D. Essential Gas Turbine Components 321
            1. Compressor 321
            2. Combustor 322
            3. Turbine 323
    II. Jet Engines for Industrial and Marine Propulsion Use 324
        A. Small Gas Turbine Features 325
   III. Gas Turbine Applications 326
        A. Electric Power Generation 326
        B. Pipeline Transmission 327
        C. Process Operations 327
        D. Combined-Cycle Operation 327
        E. Total Energy 328
        F. Marine Propulsion 328
        G. Microturbines 328
   IV. Lubrication of Gas Turbines 329
        A. Large Industrial Gas Turbines 329
        B. Aeroderivative Gas Turbines 331
        C. Small Gas Turbines 333
Bibliography 333

Chapter 14 Steam Turbines 335

     I. Steam Turbine Operation 336
        A. Single-Cylinder Turbines 337
    II. Turbine Control Systems 339
        A. Speed Governors 340
   III. Lubricated Components 342
        A. Lubricated Parts 342
            1. Journal Bearings 342
            2. Thrust Bearings 343

   3. Hydraulic Control Systems 345
   4. Oil Shaft Seals for Hydrogen-Cooled Generators 346
   5. Gear Drives...347
   6. Flexible Couplings 347
   7. Turning Gear 348
  B. Lubricant Application 348
  C. Factors Affecting Lubrication 348
   1. Circulation and Heating in the Presence of Air 349
   2. Contamination 349
  D. Steam Turbine Oil Additives and Characteristics 350
   1. Additives..350
   2. Viscosity 351
   3. Load-Carrying Ability 351
  E. Oxidation Stability...352
   1. Protection against Rusting...352
   2. Water-Separating Ability 353
   3. Foam Resistance 353
  F. Entrained Air Release..353
  G. Turbine Oil Compatibility Testing 353
  H. Less Flammable Fluids..353
  I. Maintenance Strategies..353
Bibliography...354

Chapter 15 Hydraulic Turbines 355

 I. Turbine Types..356
  A. Impulse Turbines 356
  B. Reaction Turbines 356
   1. Francis Turbines 357
   2. Diagonal Flow Turbines 359
   3. Fixed Blade Propeller Turbines 360
   4. Kaplan Turbines...361
 II. Lubricated Parts 364
  A. Turbine and Generator Bearings..364
   1. Journal and Guide Bearings 365
   2. Thrust Bearings 365
  B. Methods of Lubricant Application..367
  C. Governor and Control Systems 368
  D. Guide Vanes...368
  E. Control Valves 368
  F. Compressors 369
 III. Lubricant Recommendations 369
Bibliography...370

Chapter 16 Wind Turbines 371

 I. Wind Turbine Overview..371
  A. Wind Turbine Design..371
  B. Wind Turbine Blades 373
  C. Wind Turbine Generator 374

II.	General Considerations for Wind Turbine Lubrication	375
	A. Main Gearbox Lubrication	375
	1. Industry Standards and Builder Specifications for Wind Turbine Gear Oils	375
	2. Viscosity and Low Temperature Requirements of Wind Turbine Gear Oils	376
	3. Antiwear Performance	376
	4. Interfacial Properties of Gear Oil	380
	5. Material Compatibility (Seals and Paint)	382
	6. Gear Oil Condition Monitoring	382
	B. Rotor Blade Lubrication (Oil and Grease)	383
	C. Generator Lubrication	385
	D. Main Shaft Bearing Lubrication in Wind Turbines and Direct Drive Turbines	385
Bibliography		386

Chapter 17 Nuclear Power Generation 387

I.	Reactor Types	387
	A. Basic Reactor Systems	387
	1. PWR	389
	2. BWR	389
	3. LWGR	389
	4. FBR	389
	5. GCR	390
	6. HTGR	390
II.	Radiation Effects on Petroleum Products	390
	A. Mechanism of Radiation Damage	391
	B. Chemical Changes in Irradiated Materials	392
	1. Turbine Oil Irradiation	396
	2. Grease Irradiation	396
	3. Radiation Stability of Grease Thickeners	396
III.	Lubrication Recommendations	399
	A. General Requirements	399
	B. Selection of Lubricants	401
Bibliography		401

Chapter 18 Compressors 403

I.	Reciprocating Air and Gas Compressors	404
	A. Methods of Lubricant Application	406
	1. Cylinder Lubrication	406
	2. Bearing (Running Gear) Lubrication	408
	B. Single- and Two-Stage Compressors	408
	1. Factors Affecting Cylinder Lubrication	408
	2. Factors Affecting Running Gear Lubrication	414
	C. Multistage Reciprocating Compressors	414
	1. Factors Affecting Lubrication	414
	2. Lubricating Oil Recommendations	415
II.	Rotary Compressors	418
	A. Straight Lobe Compressors	418
	1. Lubricated Parts	420
	2. Lubricating Oil Recommendations	420

- B. Rotary Lobe Compressors 420
  - 1. Lubricated Parts 420
  - 2. Lubricating Oil Recommendations 421
- C. Rotary Screw Compressors 421
  - 1. Lubricated Parts 424
  - 2. Lubricant Recommendations 424
- D. Sliding Vane Compressors 425
  - 1. Lubricated Parts 426
  - 2. Lubricant Recommendations 426
- E. Liquid Piston Compressors 426
  - 1. Lubricated Parts 427
  - 2. Lubricant Recommendations 427
- F. Diaphragm Compressors 427
  - 1. Lubricated Parts 427
  - 2. Lubricant Recommendations 427
- III. Dynamic Compressors 427
  - A. Centrifugal Compressors 428
    - 1. Lubricated Parts 429
    - 2. Lubricant Recommendations 430
  - B. Axial Flow Compressors 430
    - 1. Lubricated Parts 431
    - 2. Lubricant Recommendations 431
- IV. Refrigeration and Air Conditioning Compressors 431
  - A. Factors in the Compressor Affecting Lubrication 433
    - 1. Cylinder Conditions 433
    - 2. Oxidation 434
    - 3. Bearing System Conditions 434
  - B. Refrigeration System Factors Affecting Lubrication 434
    - 1. Fluorocarbons 435
    - 2. Ammonia (R-717) and Carbon Dioxide (R-744) 435
    - 3. Hydrocarbon Refrigerants 436
    - 4. Sulfur Dioxide (R-764) 436
    - 5. Lubricating Oil Recommendations 436
- Bibliography 437

Chapter 19 Lubricant Contribution to Energy Efficiency 439
- I. Friction Loss Mechanisms 440
  - A. Viscosity—Shear Losses 441
  - B. Boundary Friction 441
  - C. Friction Modifiers Used in Lubricants 445
    - 1. Surface Adsorbing Friction Modifiers 445
    - 2. Chemically Reactive Friction Modifiers 446
    - 3. Solid Dispersion Friction Modifiers 447
  - D. Friction in Concentrated Contacts 447
- II. Hydrodynamic Fluid Films 452
  - A. Hydrodynamic Friction in Bearings in Industrial Applications 452
  - B. Measuring Bearing Friction 454
  - C. Friction in Rolling Element Bearings 456
    - 1. EHL in Rolling Element Bearings 458
    - 2. Other Sources of Friction and Churning Losses in Rolling Element Bearings 459

Contents

		D. Friction in Gears and Gearboxes	461
		1. Friction between Gear Teeth	461
		2. Churning Losses in Gears	463
III.	Friction Losses in Hydraulic Systems		464
	A. Sources of Friction in Hydraulic Systems		464
		1. Hydraulic Fluid Selection	465
IV.	Vehicle and Internal Combustion Engine Efficiency		466
	A. Energy Use in Vehicles		466
	B. Trends in Automotive Design Impacting Efficiency		467
		1. Engine Trends	467
		2. Automotive Transmission and Powertrain Trends	468
		3. Other Vehicle Trends	469
	C. Engine Friction Reduction		469
	D. Measuring Fuel Economy		471
	E. Use of Materials and Surfaces to Improve Efficiency		472
Bibliography			473

Chapter 20 Handling, Storing, and Dispensing Lubricants 475

I.	Hazardous Chemical Labeling for Lubricants		476
II.	Handling		478
	A. Packaged Products		478
		1. Moving to Storage	478
	B. Bulk Products		480
		1. Unloading	483
		2. Tank Cars and Tank Wagons	483
		3. Special Bulk Grease Vehicles	483
III.	Storing		483
	A. Packaged Products		484
		1. Outdoor Storage	484
		2. Warehouse Storage	487
	B. Oil Houses		488
		1. Function	488
		2. Facilities	490
		3. Size and Arrangement	490
		4. Optimum Utilization of Manpower	491
		5. Housekeeping	492
		6. Safety and Fire Prevention	493
		7. Security	496
	C. Lubricant Deterioration in Storage		496
		1. Water Contamination	497
		2. High-Temperature Deterioration	497
		3. Low-Temperature Deterioration	498
		4. Long-Term Storage	498
IV.	Dispensing		499
	A. In the Oil House		500
		1. Faucets	500
		2. Transfer Pumps	500
		3. Grease Gun Fillers	502
		4. Highboys	503

    B. From Oil House to Machine 504
        1. Oil Dispensing Containers 504
        2. Portable Oil Dispensers 505
        3. Portable Grease Equipment 507
        4. Lubrication Carts and Wagons 509
    C. Closed System Dispensing 510
    D. Central Dispensing Systems 510
    E. Maintenance and Service 511
Bibliography 511

Chapter 21 Practices for Lubricant Conservation and Machinery Reliability 513

    I. Overview of In-Plant Lubricant Handling 514
   II. Product Selection 515
     A. Long Service Life 515
     B. Compatibility with Other Products 516
     C. Value as By-Product 516
     D. Ease of Disposal 516
  III. In-Service Handling 517
     A. Reuse versus All-Loss Systems 517
     B. Prevention of Leaks, Spills, and Drips 517
     C. Elimination of Contamination 519
        1. Central Reservoir Maintenance 519
        2. Cross Contamination 521
        3. Proper System Operation 521
  IV. In-Service Lubricant Purification 521
     A. Continuous Bypass Purification 522
     B. Continuous Full-Flow Purification 522
     C. Continuous Independent Purification 523
     D. Periodic Batch Purification 523
     E. Full-Flow and Bypass 523
   V. Purification Methods 524
     A. Settling 525
     B. Filtration 525
     C. Size Filtration 526
        1. Depth-Type Filters 526
        2. Dense Media Filters 528
        3. Clay Filtration 528
     D. Multipurpose Purifiers 529
     E. Centrifugation 530
     F. Centrifugal Oil Filters (Separators) 530
     G. Sludge and Varnish Removal 531
        1. Electrostatic Precipitation Filtration 532
        2. Resin Filtration 532
        3. High-Velocity Chemical Flushes 533
  VI. Reclamation of Lubricating Oils 533
     A. Reclamation Units (Oil Conditioners) 533
 VII. Waste Oil Collection and Routing 534
VIII. Equipment Commissioning and Flushing 536
  IX. Final Disposal 536
Bibliography 537

Chapter 22 In-Service Lubricant Analysis 539
  I. Establishing an In-Service Lubricant Analysis Program 539
     A. Elements of a Successful In-Service Lubricant Analysis Program 540
     B. Selecting an In-Service Lubricant Analysis Provider 540
  II. Used Oil Analysis Program Startup Recommendations.................................... 541
     A. Sample Point Selection 542
     B. Oil Sample Intervals 542
     C. Taking a Representative Used Oil Sample 543
        1. Recommended Sampling Procedure 543
        2. Visual Inspection of the Sample 544
  III. In-Service Lubricant Analysis Testing... 546
     A. Common Used Oil Analysis Tests 546
        1. Grease Analysis 546
     B. Application Specific Used Oil Analysis Test Slates 546
     C. Test Precision 549
  IV. Interpretation of Used Oil Analysis Results 551
     A. Overall Sample Result Condition 551
     B. Condemning Limits Used to Evaluate Sample Result Condition.............................. 551
     C. Diagnosing the Cause of Abnormal Oil Analysis Results... 553
     D. Corrective Action Recommendation Related to Abnormal Oil Analysis Results......... 555
     E. Corrective Action Recommendations Correlating to Other Predictive
        Maintenance Techniques 555
     F. Sustaining a Program through Documented Procedures, Metrics, and Assessments.... 556
        1. Procedures 556
        2. Regular Key Performance Indicators or Metrics 557
        3. Assessments.. 557

Index... 559

Preface

Lubrication and the knowledge of lubricants are not only subjects of interest to all of us, but they are also critical to the cost-effective operation and reliability of machinery that is, directly or indirectly, part of our daily lives. Our world, and exploration of regions beyond our world, depend on mechanical devices that require lubricating films. Whether in our homes or at work, whether knowingly or unknowingly, we all need lubricants and some knowledge of lubrication. Home appliances, lawnmowers, bicycles, and fishing reels are a small sampling of devices that have moving parts that require lubrication. The millions of automobiles, trucks, buses, motorcycles, airplanes, ships, and trains depend on lubrication for operation, and it must be effective lubrication for dependability and safety, and to reduce the environmental impact.

Many changes in the field of lubrication have occurred since the second edition of *Lubrication Fundamentals* was published in 2001. Today, intricate and complex machines are used to make paper products; huge rolling mills turn out metal ingots, bars, and sheets; metalworking machines produce very high precision parts; and special machinery is used to manufacture cement, rubber, and plastic products. Emission regulations have had a major impact on all transportation and power generating equipment and the lubricants that protect them. The use of computers, robotics, and higher technology has led to advanced machine designs that result in faster machine speeds, greater load-carrying capability, more compact equipment, smaller capacity lubricant reservoirs, higher machine temperatures, various material compatibility challenges for lubricants, restrictions in lubricant additive content, and less frequent lubrication application up to and including fill-for-life lubrication. As a result, there continues to be an explosion in higher performance and specialty application oils and greases. The impact of these lubricants on our natural environment is also a driver for new lubricant technology.

The third edition of *Lubrication Fundamentals* builds on the machinery basics discussed in the first two editions, much of which is still very applicable today. The third edition also addresses many of the new applications, and new lubricant and base stock technologies that were introduced or improved upon in the past 15 years to meet the needs of modern machinery and sustainability. The lubricants industry will continue to be faced with many challenges going forward, and innovative technologies will be needed to meet these challenges. Critical activities along the lubricant value chain that are impacted by technology include new lubrication requirements, crude oil composition and selection, base stock manufacture, product formulation and evaluation, lube oil blend plant capabilities, lubricant application, and environmental stewardship. These will be exciting times for industry, especially those participating in the quest to develop the new lubricant molecule for the future.

Don M. Pirro
Martin Webster
Ekkehard Daschner

Acknowledgments

Lubrication Fundamentals, Third Edition Revised and Expanded, like many technical publications of this magnitude, is not the work of one or two people. It is the combined work of hundreds of engineers, chemists, scientists, physicists, technologists, designers, writers, and artists—a compendium of a broad spectrum of skills and talent over a long period. The study of lubrication fundamentals starts with the scientists who research the interaction of oil films with moving components under various stresses and loads. It then takes the unique cooperation that exists between the machine designer and equipment builders, on one side, and the lubricant formulators and suppliers on the other. Additionally, collaboration often takes place with many industry associations such as the International Organization for Standardization, American Society for Testing and Materials, American Petroleum Institute, Association des Constructeurs Européens d'Automobiles, Society of Automotive Engineers, Society of Tribologists and Lubrication Engineers, Deutsches Institut für Normung, National Lubricating Grease Institute, and American Gear Manufacturers Association. This frequently culminates in the mating of the right lubricant properly applied to meet the requirements of the most efficient and demanding machines operating today.

The lubricants industry is most grateful to lubrication pioneers such as J. George Wills, the author of the first edition who identified the need for a practical resource on lubrication. He developed a vision, secured the support and resources to undertake such a monumental task, and then dedicated the effort to turn his vision into reality. We are also most appreciative for the efforts of A.A. (Al) Wessol, who was the coauthor of the second edition. Al was a true lubrication expert in many applications and was willing to share his knowledge and years of practical experience. We are privileged to be able to build on their efforts and share the many technological advances in industry.

It would be impossible to list the host of people who have helped to put this third edition together. The book compiles the many technical publications of ExxonMobil and the cooperative offerings of the foremost international equipment builders and associations. We are most appreciative to the many original equipment manufacturers and industry associations, with whom we have worked over many years, for sharing their knowledge and technology.

We thank the following senior managers at ExxonMobil Research and Engineering for their acceptance, support, and encouragement for this project: Grant Karsner, Michele Touvelle, Nick Hilder, and Bill Buck. Additionally, we recognize Jane Walter of ExxonMobil for her assistance with the book cover design and graphics.

Authors

Ekkehard Daschner, Don M. Pirro, and Martin Webster

Don M. Pirro is the Global Alliance Technical Manager at ExxonMobil Research and Engineering, Paulsboro, New Jersey. He started his career in 1978 as a test engineer for Ingersoll Rand's Turbo Compressor Division. He has more than 36 years of lubrication experience with ExxonMobil in technical positions such as Lubrication Engineer, Chief Engineer, Equipment Builder and Application Engineering Manager, U.S. Technical Support Manager, Americas EB and OEM Manager, Global Used Oil Analysis Manager, and Marine Lubricants Technical Manager. He is the author or contributing editor of several scholarly articles on synthetic lubes, environmental awareness applications, grease technology, lubricant interchangeability, used oil analysis, and marine lubricants. Pirro is a member of the Society of Tribologists and Lubrication Engineers (STLE) and the Association of Manufacturing Technology. He is a coauthor of the second edition of *Lubrication Fundamentals*. He graduated from Rutgers University, New Brunswick, New Jersey, with degrees in mechanical engineering and business administration.

Ekkehard Daschner is the Industrial Lubricant Section Head and Greases Technology Program Lead at ExxonMobil Research and Engineering, Paulsboro, New Jersey. Daschner earned his mechanical engineering degree in Germany and started his career in an aluminum company. In 1982, he joined the Mobil Oil AG R&D laboratory in Germany supporting product development projects. After holding various positions in technical support, he was appointed Area Sales Director and Brand Manager in Germany and the European headquarters in London and Brussels, and later served as European and later as Global Industrial and Marine Equipment Builder Services Manager. He is recognized as an expert in lubricant applications and in particular in bearings and gearboxes.

Martin Webster is a senior research associate at ExxonMobil Research and Engineering, Annandale, New Jersey. He earned his BS and MS in aeronautical engineering and his PhD in tribology from Imperial College, London, UK. In 1986, he was awarded the Bronze Medal in tribology from the Institute of Mechanical Engineers in the UK. After spending 4 years working in the UK, he joined ExxonMobil in 1989. At ExxonMobil, Martin has held various positions in research and product development and is currently involved in the study of fundamental lubrication mechanisms at the company's Corporate Strategic Research Laboratory. His publications include peer-reviewed papers, book chapters, and patents in elastohydrodynamic lubrication (EHL), rolling contact fatigue, hydrodynamics, contact mechanics, and wear. His industry activities have included membership in the American Society of Mechanical Engineers (ASME) bearings committee and the American Gear Manufacturers Association (AGMA) gear rating committee. He has been a member of the STLE since 1989. In 2006, he was elected a director of the STLE and in 2012 he joined the executive committee. He is currently serving a 1-year term as the STLE president in 2015–2016.

ExxonMobil Contributors to the Third Edition

We would like to acknowledge contributions to the third edition from the following ExxonMobil engineers, chemists, lubricant formulators, tribologists, scientists, and technologists. Collectively, these people represent over 600 years of lubrication experience. They are not only leading experts within ExxonMobil, but many of them are recognized as among the best lubrication experts in industry:

David Baillargeon	**David G. L. Holt**
Rob Banas	**Percy Kanga**
Neil Briffett	**Sandra Legay**
Jim Carey	**Wojciech Leszek**
Barb Carfolite	**Roger Liao**
Kevin Crouthamel	**Kevin McKenna**
Michael Douglass	**Rob Meldrum**
Oscar Farng	**Ricardo Orta**
Mark Hagemeister	**Dave Scheetz**
Heather Haigh	**Tom Schiff**
Doug Hakala	**Jamie Spagnoli**
Jim Hannon	**Kathy Tellier**
Camden Henderson	**Andrea Wardlow**
John Hermann	**Beth Winsett**
Larry Hoch	**Virginia Wiszniewski**

1 Introduction

Petroleum is one of the naturally occurring hydrocarbons that frequently include natural gas, natural bitumen, and natural wax. The name "petroleum" is derived from the Latin *petra* (rock) and *oleum* (oil). According to the most generally accepted theory today, petroleum was formed by the decomposition of organic refuse, aided by high temperatures and pressures, over a vast period of geologic time.

I. PREMODERN HISTORY OF PETROLEUM

Although petroleum occurs, as its name indicates, among rocks in the earth, it sometimes seeps to the surface through fissures or is exposed by erosion. The existence of petroleum was known to primitive man, as surface seepage, often sticky and thick, was obvious to anyone passing by. Prehistoric animals were sometimes mired in it, but few human bones have been recovered from these tar pits. Early man evidently knew enough about the danger of surface seepage to avoid it.

The first actual use of petroleum seems to have been in Egypt, which imported bitumen, probably from Greece, for use in embalming. The Egyptians believed that the spirit remained immortal if the body was preserved.

Around the year 450 B.C., Herodotus, widely referred to as the father of history, described the pits of Kir A'b near Susa in present-day Iraq as follows:

> At Ardericca is a well which produces three different substances, for asphalt, salt and oil are drawn up from it in the following manner. It is pumped up by means of a swipe; and, instead of a bucket, half a wine skin is attached to it. Having dipped down with this, a man draws it up and then pours the contents into a reservoir, and being poured from this into another, it assumes these different forms: the asphalt and salt immediately become solid, and the liquid oil is collected. The Persians call it Phadinance; it is black and emits a strong odor.

Pliny, the historian, and Dioscorides Pedanius, the Greek botanist, both mention "Sicilian oil," from the island of Sicily, which was burned for illumination as early as the beginning of the Christian Era.

The Scriptures contain many references to petroleum, in addition to the well-known story of Moses, who was set afloat on the river as an infant in a little boat of reeds waterproofed with pitch, and was found by Pharaoh's daughter. Some of these references include the following:

> Make thee an ark of gopher wood; rooms shalt thou make in the ark, and shalt pitch it within and without with pitch. (Genesis VI.14)
> And they had brick for stone, and slime (bitumen) had they for mortar. (Building the Tower of Babel, Genesis XI.3)
> And the Vale of Siddim was full of slime (bitumen) pits; and the kings of Sodom and Gomorrah fled, and fell there.... (Genesis XIV.10)

Other references are found in Strabo, Josephus, Diodorus Siculus, and Plutarch, and since then there is substantial evidence that petroleum was known in almost every part of the world.

Marco Polo, the Venetian traveler and merchant, visited the lands of the Caspian in the thirteenth century. In an account of this visit, he stated:

> To the north lies Zorzania, near the confines of which there is a fountain of oil which discharges so great a quantity as to furnish loading for many camels. The use made of it is not for the purpose of food, but

as an unguent for the cure of cutaneous distempers in men and cattle, as well as other complaints; and it is also good for burning. In the neighboring country, no other is used in their lamps, and people come from distant parts to procure it.

Sir Walter Raleigh, while visiting the island of Trinidad off the coast of Venezuela, inspected the great deposit of bitumen there. The following is taken from *The Discoveries of Guiana* (1596):

At this point called Tierra de Brea, or Piche, there is that abundance of stone pitch that all the ships of the world may be therewith loden from thence, and wee made triall of it in trimming our ships to be most excellent good, and melteth not with the sunne as the pitch of Norway, and therefore for ships trading the south partes very profitable.

II. PETROLEUM IN NORTH AMERICA

In the North American continent, petroleum seepages were undoubtedly known to the aborigines, but the first known record of the substance was made by the Franciscan friar, Joseph de la Roche D'Allion, who in 1629 crossed the Niagara River from Canada and visited an area later known as Cuba, New York. At this place, petroleum was collected by the Indians, who used it medicinally and to bind pigments used in body adornments.

In 1721, Charlevois, the French historian and missionary who descended the Mississippi River to its mouth, quotes a Capitan de Joncaire as follows:

There is a fountain at the head of a branch of the Ohio River (probably the Allegheny) the waters of which like oil, has a taste of iron and serves to appease all manner of pain.

In the *Massachusetts Magazine*, Volume 1, July 1789, there is an account under the heading "American Natural Curiosities," as follows:

In the northern parts of Pennsylvania, there is a creek called Oil Creek, which empties into the Allegheny River. It issues from a spring, on the top of which floats an oil similar to that called Barbadoes tar; and from which one man may gather several gallons in a day. The troops sent to guard the western posts halted at this spring, collected some of the oil, and bathed their joints with it. This gave them great relief from the rheumatic complaints with which they were affected. The waters, of which the troops drank freely, operated as a gentle purge.

Although the practice of deriving useful oils by the distillation of bituminous shales and various organic substances was generally known, it was not until the nineteenth century that distillation processes were widely used for a number of useful substances, including tars for waterproofing; gas for illumination; and various chemicals, pharmaceuticals, and oils.

In 1833, Dr. Benjamin Silliman contributed an article to the *American Journal of Science* that contained the following:

The petroleum, sold in the Eastern states under the name of Seneca Oil, is a dark brown color, between that of tar and molasses, and its degree of consistency is not dissimilar, according to temperature; its odor is strong and too well known to need description. I have frequently distilled it in a glass retort, and the naphtha which collects in the receiver is of a light straw color, and much lighter, more odorous and inflammable than the petroleum; in the first distillation, a little water usually rests in the receiver, at the bottom of the naphtha; from this it is easily decanted, and a second distillation prepares it perfectly for preserving potassium and sodium, the object which led me to distil it, and these metals I have kept under it (as others have done) for years; eventually they acquire some oxygen, from or through the naphtha, and the exterior portion of the metal returns, slowly, to the condition of alkali—more rapidly if the stopper is not tight.

> The petroleum remaining from the distillation is thick like pitch; if the distillation has been pushed far, the residuum will flow only languidly in the retort, and in cold weather it becomes a soft solid, much resembling the maltha or mineral pitch.

Along the banks of the Kanawha River in West Virginia, petroleum was proving a constant source of annoyance in the brine wells; and one of these wells, in 1814, discharged petroleum at periods of from 1 to 4 days, in quantities ranging from 30 to 60 gallons at each eruption. A Pittsburgh druggist named Samuel M. Kier began bottling the petroleum from these brine wells about 1846 and selling the oil for medicinal purposes. He claimed it was remarkably effective for most ills and advertised this widely. In those days, many people believed that the worse a nostrum tasted, the more powerful it was. People died young then, and often did not know what killed them. In the light of today's knowledge, we would certainly not recommend drinking such products. Sales boomed for awhile, but in 1852, there was a falling-off in trade. Therefore, the enterprising Mr. Kier began to distill the substance for its illuminating oil content. His experiment was successful and was a forerunner, in part, of future commercial refining methods.

In 1853, a bottle of petroleum at the office of Professor Crosby of Dartmouth College was noticed by Mr. George Bissel, a good businessman. Bissel soon visited Titusville, Pennsylvania, where the oil had originated, and purchased 100 acres of land in an area known as Watsons Flats and leased a similar tract for the total sum of $5000. Bissel and an associate, J.D. Eveleth, then organized the first oil company in the United States, the Pennsylvania Rock Oil Company. The incorporation papers were filed in Albany, New York, on December 30, 1854. Bissel had pits dug in his land in the hope of obtaining commercial quantities of petroleum, but was unsuccessful with this method. A new company was formed, which was called the Pennsylvania Rock Oil Company of Connecticut, with New Haven as headquarters. The property of the New York corporation was transferred to the new company, and Bissel began again.

In 1856, Bissel read one of Samuel Kier's advertisements on which was shown a drilling rig for brine wells. Suddenly it occurred to him to have wells drilled, as was being done in some places for brine. A new company, the Seneca Oil Company, succeeded the older company, and an acquaintance of some of its partners, E.L. Drake, was selected to conduct field experiments in Titusville. Drake found that, to reach hard rock in which to try the drilling method, some unusual form of shoring was needed to prevent a cave-in. It occurred to him to drive a pipe through the loose sand and shale—a plan afterward adopted in oil well and artesian well drilling.

Drilling then began under the direction of W.A. Smith, a blacksmith and brine well driller, and went down 69 1/2 ft. On Saturday, August 27, 1859, the drill dropped into a crevice about 6 inches deep, and the tools were pulled out and set aside for the work to be resumed on Monday. However, Smith decided to visit the well that Sunday to check on it, and upon peering into the pipe saw petroleum within a few feet of the top. On the following day, the well produced the incredible quantity of 20 barrels of oil a day.

III. DEVELOPMENT OF LUBRICANTS

James Watt's development of the practical steam engine, circa 1769, introduced the industrial age. The coming of new, more demanding machines, in turn, marked the beginning of the search for improved lubricants. They were needed to meet the requirements of devices operating at constantly increasing speeds and loads. The first substances to be used were mineral oils, which were obtained from naturally occurring surface pools and extended with the addition of the already known plant and animal oils.

During the period from 1850 to 1875, many men experimented with the products of petroleum distillation then available, attempting to find uses for them, in addition to providing illumination. Some of the viscous materials were investigated as substitutes for the vegetable and animal oils previously used for lubrication, mainly those derived from olives, rape seed, whale, tallow, lard, and other fixed oils.

As early as 1400 B.C., greases, made of a combination of calcium and fats, were used to lubricate chariot wheels. Traces of this grease were found on chariots excavated from the tombs of Yuaa and Thuiu. During the third quarter of the nineteenth century, greases were made with petroleum oils combined with potassium, calcium, and sodium soaps and placed on the market in limited quantities.

Gradually, as distillation and refining processes were improved, a wider range of petroleum oils were produced to take the place of the fatty oils. These mineral oils could be controlled more accurately in manufacture and were not subject to the rapid deterioration of the fatty oils.

The latter half of the nineteenth century saw the development of many new techniques for the most efficient production of lubricants from petroleum. Not the least of these was vacuum distillation, first put to commercial use by Vacuum Oil Company of Rochester, New York, successor to Mobil Oil Corporation and present-day ExxonMobil Corporation. This process permitted the distillation of crude residuum without the use of destructively high temperatures. What set vacuum-distilled lubricants apart was reliability as machines started easier, lasted longer, and had fewer problems.

In the early 1900s, gasoline became important, and this resulted in ever-increasing demands for crude oil. Until this time, only Pennsylvania crudes were used for lubricating oil but now new processes were developed to make lube stocks from other crude sources. Some of the fatty oils continued to be used in special services as late as the early part of the twentieth century. Tallow was fairly effective in steam cylinders as a lubricant. However, it was not always pleasant to handle, because maggots developed in the tallow particularly in hot weather. Lard oil was used for cutting of metals, and castor oil was used to lubricate the aircraft engines of World War I. Even today, some fatty oils are still used as compounding in small percentages with mineral oils, but chemical additives have taken their place for the majority of users.

During the period 1910–1918, improved mechanical equipment—e.g., high-pressure steam turbines, new electrical machinery, and automobiles, placed greater demands on lubricants. Thus began the first real growth in lubricants research. This covered both refining and the new field of additive technology. It carried on, with a high level of intensity, through the 1920s and 1930s, reaching its peak during World War II. Always the impetus was the same: new innovations in machine design required new innovations in lubricant composition.

IV. HISTORY OF SYNTHETIC LUBRICANTS

A historical survey of the development of synthetic lubricants would be incomplete if it did not include mention of those oils and fats that occur in nature. In the grave of the Egyptian Tehut-Hetep (circa 1650 B.C.), there is a description of how olive oil was applied to wooden planks so that heavy stones might be moved more easily. Plinius (23–79 A.D.) compiled a list of lubricants made from plants and animals that were well known at that time and are still used today. Bearings and slow-moving machinery were lubricated from antiquity up to the beginning of the nineteenth century with oils made from olives, rape seed, castor plant seeds, and other naturally occurring friction reducers. These were the first lubricants.

The Second World War provided the primary stimulus for the development and use of synthetic lubricants. During this time, nations faced a real possibility of running short of crude oil. This challenged America's oil companies to seek ways of manufacturing lubricants and fuels from natural gas and other noncrude resources. Additionally, new devices, such as jet engines, came into being, creating new performance criteria especially in terms of high-temperature degradation resistance. Existing equipment during the war was exposed to environmental extremes—arctic and subarctic regions, for example—that led to requirements that mineral lubricants could not fulfill.

Such demands during World War II resulted in the first truly commercial use of synthetic lubricants. These are defined as products made from base fluids manufactured by chemical synthesis. Polyalkylene glycols (polyglycols) were used for lubrication in the military for the first time in 1940, although these substances had been known for about a hundred years. They were used more extensively in industry after 1945, with the end of military secrecy. Today, polyglycols are well known for use as brake fluids, high-temperature lubricants, and energy-efficient oils for gearboxes (Mobil Oil Corporation, 1999).

The chemical history of o-phosphoric acid esters is as old as that of polyglycols. Triphenyl phosphate was synthesized for the first time in 1854. Phosphoric acid esters attracted industrial attention for the first time around 1920 as softening agents for cellulose nitrate. In recent years, tertiary phosphoric acid esters have been further developed as defoamants, wear-resistant additives, and especially as fire-resistant hydraulic fluids.

Although research work on dibasic acid esters began long before, these materials—in their various chemical forms—assumed an important role as lubricants with the advent of World War II. On both sides of the conflict, researchers were working to develop new lubricants and fluids that would be less sensitive to temperature change—i.e., remaining more fluid at very low temperatures, and being less volatile and more stable at very high temperatures. A great deal of this effort centered on dibasic acid esters, resulting in a variety of specialty products for instruments, fire control apparatus, gyroscopes, and special aviation and naval ordnance equipment. Simple esters, although used as lubricants in early jet engines, lacked thermal stability, which limited flight time between engine overhauls to only a few hundred hours. In Germany, esters were used as automotive engine lubricants. In the early 1950s, British and U.S. researchers produced diester oils for the engines of jet aircraft, first for the military and then for commercial uses including engine oil and industrial applications. More recently, polyol esters have replaced diesters in many jet applications and specific polyol esters, which are readily biodegradable, are finding application in a wide range of equipment used in environmentally sensitive areas. Today, diester and polyol ester-based products make up the second largest synthetic market segment.

Silicone polymers were used during the Second World War in greases and instrument oils. In spite of their good physical characteristics and their high-temperature stability, they have captured only a small share of the market, largely because of their high cost. However, they are indispensable for use in certain types of precision machinery such as synchronous motors and shock absorbers, as well as very high temperature hydraulic systems. They are also used as defoamants. Other synthetic lubricants developed in more recent years include fluoride esters, polyphenyl ethers, tetra-alkylsilanes, ferrocene derivatives, heterocyclenes, aromatic amines, and hexafluorobenzenes, which also have come to find use in very specialized applications.

Early work on the synthesis of hydrocarbons took place in the late 1920s and 1930s. It was based on alphaolefin technology, similar to the polyalphaolefins (PAOs) used today. These early efforts were little more than academic exercises, however, for several reasons:

1. Conventional, mineral-based lubricating oil quality was improving at a rapid rate, availability was seemingly limitless, and cost was low.
2. The raw material from which the alphaolefins were derived was in short supply.
3. Viscosity index (VI) improvers were developed, reducing the need for high VI synthetic base oils.

A fateful discovery was made by one Mobil Oil Corporation researcher on May 18, 1949. This researcher was attempting to synthesize antioxidant additive compounds by adding phosphorous or sulfur to polymerized decene alpha olefin molecules. Following good practices, the researcher included in his experiments a so-called "blank" or control sample for measuring the effects of the added elements. The experiment was a failure as it did not give the desired results, but the curiosity of the researcher persisted as the "blank" seemed like a very interesting material that looked very much like mineral oil. The new material was very unlike mineral oil in two ways. It was an excellent lubricant at high temperatures (>300°F or 150°C), and it flowed at extremely low temperatures (–70°F or –57°C). That day, the first synthetic PAO was invented, marking the second revolution in lubrication science.

Immediately after that discovery, interest in synthetics waned as petroleum supplies were cheap and plentiful with the development of the Middle East oil fields. It was not until the late 1950s that the potential of synthesized hydrocarbon PAO as a high-performance synthetic lubricant was rediscovered and a more intensive research program at Mobil Oil Corporation commenced. The real driving force for the widespread use of synthesized hydrocarbon PAOs began with the invention and successful introduction of Mobil 1™ synthetic engine oil back in 1974. What started out as an engine oil designed for high-performance vehicles and to help the United States and others conserve energy during the energy crisis of the time, Mobil 1 essentially created a "superior" category of lubricants that is still widely recognized throughout the world (Mobil Oil Corporation, 1985).

Today, synthesized hydrocarbons are used in a wide range of industrial and automotive applications and are, by far, the synthetic lubricant segment with the largest market demand. Their greatest use is for automobile engine oils. Recent advances in PAO technology have led to even greater performance attributes in the area of high and low temperature performance. Synthetic lubricants are being used in the most demanding areas of application, particularly where equipment performance and oil life is of paramount concern. In coming decades, their significance will continue to increase as equipment designs with higher operating temperatures and stresses gain an even greater share of their markets, and as their overall cost/performance attributes compared to mineral products are fully valued. Also, most importantly, synthetic lubricants will be chosen increasingly because of their ability to improve application productivity by their ability to contribute to energy conservation, extend equipment life, and reduce maintenance and downtime. Another quickly evolving area for synthetic lubricants is the development of a full range of high-performance lubricants that would be less damaging to the environment where there is potential for spills or leakage.

V. FUTURE PROSPECTS

The twenty-first century continues to see advancements in equipment technology as society places new demands on output and environmental performance. Equipment is being designed to achieve higher production levels with increased energy efficiency and lower total cost of ownership. This will result in new and different materials used in equipment, smaller equipment footprints with greater power density, advances in seal technology, finer oil filtration, smaller lubricant capacity reservoirs, higher operating speeds, increased temperatures, and higher system pressures that will all place greater demands on lubricants. These demands—coupled with the trends of reduced or maintenance-free operation, fill-for-life lubrication, the quest for more energy efficient lubricants, increased environmental awareness and regulation, sustainability, and greater attention to safety issues—will continue to challenge lubricant technology and drive associated research and development activities.

Use of alternative materials is driven by many factors such as weight reduction, increased strength, resistance to fatigue, and improvement in the surface properties that influence friction and wear. Over the past 10 years, there has been a rapid increase in the development of new wear-resistant and friction-reducing coatings. Many of these are now routinely used in commercial equipment. Because the surface now exhibits different chemical and physical properties, the interactions with the lubricant are also modified. More recently, there has been an increased focus on trying to understand how to better optimize these interactions, which may potentially lead to the development of new lubricant components and finished products aimed at taking full advantage of these new coatings. One significant challenge is the increasing number and complexity of materials used in a single piece of equipment. For example, a typical internal combustion engine contains hundreds of different interfaces that require lubrication. A new coating for a cam may not work in the piston area for which a different solution may be sought. Thus, future engines may contain a greater number of different materials for which an overall optimum lubricant needs to be found. This adds further complexity to the already challenging problem of optimizing a lubrication solution that provides an overall performance benefit across the entire engine.

The quest to develop energy sources that are more efficient and increasingly benign to the environment is already having a dramatic impact on lubrication practices and products. A recent example is the rapid growth of wind power. At first glance these systems, which are typically constructed using well-known mechanical components such as rolling element bearings, gears, and electrical generators, would appear to present no major challenges. However, what has transpired is that many of the components can suffer from a wide variety of durability issues including fretting and surface fatigue (e.g., micropitting). Traditional approaches to design, materials, finishing, and lubrication have had to be updated in order to alleviate these problems. It has also lead to the development of lubricant products specifically designed to meet the new challenges. Looking ahead, we can only imagine what new or adapted technologies will be needed to produce the energy of the future. What is known, however, is that there will be moving parts that will require lubrication.

BIBLIOGRAPHY

Mobil Oil Corporation. 1999. History of Mobil Synthetic Lubricant Technology.
Mobil Oil Corporation. 1985. Synthetic Lubricants.

2 Lubricant Base Stock Production and Application

Base stock technology has undergone significant changes as refining moves through the twenty-first century. Two major factors appear to be responsible—demands for higher performance from finished lubricants and process improvements with the implementation of modern refining technology. Crude oil is still the major raw material used for manufacturing mineral oil base stocks and other associated petroleum products such as fuel, asphalt, and wax. However, new quantities of high-quality base stocks are catalytically synthesized from gaseous hydrocarbons. The gas-to-liquids (GTL) process is becoming a significant contributor to the base stock pool by converting natural gas into liquid petroleum hydrocarbons. Moreover, to satisfy lubrication performance demands under very severe operating conditions, it is necessary to replace conventional mineral base stocks with synthetic fluids such as polyalphaolefins (PAOs), alkyl aromatics, esters, polyglycols, and other special materials.

Conventional processing involving separation by distillation and solvent extraction is still widely used to produce base stocks formulated into a wide range of lubricants. As might be expected, different crude feeds and refining processes produce base oils that can vary significantly in composition and performance. It is therefore essential that crude selection and processing must be done carefully to control base stock quality. Additional processing steps, such as hydrotreating, are used to improve and maintain base stock quality. Unfortunately, even with the process step of hydrofinishing, conventional oils often fail to satisfy the performance needs when used in many modern lubricant applications. Compared to more modern conversion techniques, conventional processing yields lower quantities of comparatively low-quality base stocks. Production costs are generally higher for conventional processing, particularly for small operations without the advantage of economy of scale.

To meet many current finished lubricant performance requirements, higher quality base stocks are needed. Therefore, more severe, multiple-step catalytic processes are used. Instead of extracting existing lube molecules already present in the crude, a catalyst is used to make significant changes to the molecular structure of the crude fractions being processed. Lower value molecules are converted into more desirable lube molecules under the influence of high-pressure hydrogen and isomerization catalysts. The process is very efficient and produces high-quality base stocks at high yield off-take.

I. LUBRICANT BASE STOCKS AND THEIR APPLICATION

A. American Petroleum Institute Group I, II, III, IV, and V Base Stocks

The American Petroleum Institute (API) developed a simple classification system for lubricant base stocks as identified in API Publication 1509 Annex E. Although originally intended to help regulators interpret performance data for the licensing and certification of gasoline and diesel engine oils, the classification system is now used as a guide for base stock selection across a broad range of lubricant products. The definitions have evolved to the present five-group standard set in 1995. The API classification of these base stocks is shown in Table 2.1.

TABLE 2.1
API Base Stock Classification

	Saturates (%)		Sulfur (%)	Viscosity Index
Group I (solvent-refined)	<90	and/or	>0.03	80 to <120
Group II (hydroprocessed)	≥90	and	≤0.03	80 to <120
Group III (waxy feeds)	≥90	and	≤0.03	≥120

Group IV all polyalphaolefins (PAO)
Group V naphthenic stocks, synthetics, and other non-PAO base stocks not in Groups I, II, III, or IV

Groups I, II, and III are generally derived from crude, and they are most often thought of as mineral oil based.

Group IV is reserved solely for PAOs. In contrast to the separation processes used for mineral base stocks, PAOs are synthetic in origin, being built up from gaseous hydrocarbons such as ethylene.

Group V is a "catch-all" category for all base stocks not included in the first four groups. It includes both mineral-based naphthenic oils and nonconventional synthesized base stocks. Synthetic Group V base stocks are chemically created to meet severe performance requirements. Synthetic base stocks are used in significant volumes for manufacturing high-performance finished lubricants where other base stocks are likely to fail. This group includes synthetic base stocks such as organic esters, polyglycols, silicones, polybutenes, phosphate esters, and alkylated aromatics. (See Chapter 5 for more information on synthetic base stocks.)

1. Group I Base Stocks

Group I base stocks come from traditional solvent refining techniques. Group I, with higher sulfur and/or lower paraffin content, is distinctively different from mineral base stocks made by catalysis. Because solvent processing is a selective separation process, there are limits as to what can be achieved in terms of producing high viscosity index (VI) and high total saturates content. The process essentially isolates components that are present in the crude with essentially no formation of new molecules. Although crude that is considered good for lubricant base stocks may contain desirable lube molecules, these molecules are not necessarily present in significant concentrations.

Solvent refining makes use of differences in solubility to separate or remove undesirable aromatic and wax molecules from vacuum distillate lube fractions. Hydrocarbon types that are preferentially soluble in a solvent are selectively removed from the feed stream. First, to remove undesirable aromatics, polar solvents such as furfural, n-methyl-pyrrolidone (NMP), or phenol are used to attract and remove heavy aromatics from the vacuum distillate. Next, wax is crystallized from the feed by addition of a solvent such as ketone at refrigerated temperatures. The ketone serves as an antisolvent for the wax so it can be recovered by filtration. As a final step to making Group I base stocks, both color and stability may be improved through hydrofinishing.

Ultimately, 10–40% aromatic components remain after solvent processing. These are mostly single-ring aromatics with some sulfur-containing compounds. These components are unaffected by hydrofinishing and remain in the oil. Group I stocks generally have good solvency and natural antioxidant properties, and are desirable for better additive solubility and natural long-term oxidation control. The nature and composition of the crude oil feed can often determine the final properties of the base stock. Although Group I production is declining, there are some products where they are preferred. Moreover, there is one area where Group I has a strong presence—high viscosity base stocks that are difficult to produce by catalytic processing. Therefore, there is no anticipated end to Group I base stock production.

2. Group II and Group III Base Stocks

Group II and Group III base stocks are produced from feeds by hydroprocessing and catalytic dewaxing. Both groups share the same low sulfur and high saturates characteristics. The distinction between the two is the VI range. The VI range for Group II is the same as that for solvent processed Group I and includes base stocks with a VI between 80 and less than 120. In contrast, the minimum VI for the more severely processed Group III is 120 and above.

There are various hybrid processing routes to manufacture Group II and Group III base stocks involving combinations of both old and new technology. However, in the strict conversion process, distillate or deasphalted oil (DAO) is hydrocracked, saturating nearly all of the aromatic content of the feed. This is followed by catalytic dewaxing to convert straight-chain waxy paraffins to iso-paraffins. To make Group III quality, more severe hydroprocessing is used. Alternately, GTL wax also makes good feed for catalytic isomerization to high VI base stocks. GTL base stocks are similar in composition and performance to Group III base stocks. Although synthesized from natural gas, GTL base stocks are similar to and identified as Group III base stocks.

These high-quality Group II and Group III mineral base stocks are not available in as broad a viscosity range as are conventionally refined Group I or synthetic PAOs. Commercial Group II base stocks are presently available at viscosities as high as 600 Saybolt Universal seconds (SUS) (12 cSt at 100°C), whereas Group III are limited to a maximum viscosity of 300 SUS (8 cSt at 100°C). However, with improvements in both conversion processing and GTL synthesis, it is very likely that the upper viscosity ranges of these high-quality base stocks will be extended to meet lubricant demand.

3. Group II+ and Group III+ Base Stocks

Although not officially part of the API classification, individual companies manufacture Group II+ and Group III+ stocks. These "plus" base stocks all have VIs on the high end of the API guideline. Sulfur and saturates specifications remain unchanged from the official group classification. However, individual marketers have set their own VI expectations for these unofficial grades. Among "plus" producers, the minimum VI for a Group II+ begins somewhere between 110 and 115. For Group III+, the minimum VI falls somewhere between 130 and 140. In addition to the high VI benefit, these stocks generally have lower volatility and lower pour points. The common way to reach Group III+ very high VI quality is to start by refining slack wax or GTL wax feeds.

4. Group IV Base Stocks

PAOs are paraffin-like liquid hydrocarbons with a unique combination of high-temperature viscosity retention, low volatility, very low pour point, and a high degree of oxidation resistance. The VI of these base stocks can range from 125 to more than 200 with pour points down to –85°F (–65°C). These characteristics result from the wax-free combination of relatively unbranched molecules of predetermined chain length. These properties made them ideal base stocks for high-performance lubricants. The most widely known consumer product that uses PAO base stock is Mobil 1™ synthetic motor oil.

5. Group V Base Stocks

Group V naphthenic base stocks are produced from naphthenic crude. The refining process is similar to that of their paraffinic Group I counterparts except that naphthenic crude contains essentially no paraffins, and therefore solvent dewaxing is not necessary. However, many naphthenic stocks are finished by hydrotreating. Naphthenic base stocks are characterized by having very low pour points (down to –80°F or –62°C) and low VI (typically in the range of 20–85) compared to Group I base stocks. These base stocks make them suitable for a number of specialized applications such as metalworking, process oils, refrigeration compressor oils, and automatic transmission oils.

The properties and manufacture of nonconventional synthetic Group V base stocks vary widely. These cover a broad range of materials, each having relatively unique properties that make them suitable for a number of specialized industrial, automotive, and aviation applications.

B. Base Stock Selection

Proper base stock selection is essential to the manufacture of high-quality finished lubricants. Different product lines make use of the individual performance features demonstrated by various stocks. The base stock makes up a significant portion of the finished lubricant. It contributes to finished lubricant performance features in a broad range of areas including viscosity, volatility, thermal stability, additive/contaminant solubility, oxidation stability, air release, foaming resistance, and demulsibility. The balance of lubricant formulations consists of the addition of performance-improving additives, selected according to the needs of the product being manufactured.

Formulators have a wide array of base stocks to pick from, with significant performance differences between various stocks. From a high level perspective, API base stock group classifications are used as a basis for selecting materials for lubricant manufacture. From a cost/performance assessment, the selection of the optimum base stock for a particular application is essential. With greater performance demands being made on lubricants, the need for higher quality base oils is increasing. As shown in Figure 2.1, considerable change in the marketplace is occurring as API Group II base stock is displacing Group I demand. As smaller Group I manufacturing plants are shut down, market penetration of Group II is expected to continue to dominate the demand for paraffinic base stock.

As shown in the global paraffinic base stock demand forecast, market forces are driving a slow but consistent move toward higher quality base stocks. This is primarily being led by new and changing engine oil requirements. However, marine engines and some industrial applications continue to benefit from the use of Group I base stocks, which are available in a broad range of viscosities and have greater solvency. Although many applications can use either Group I or II base stocks, Group I is in steady decline. In contrast, premium passenger vehicle and heavy-duty commercial lubricants need to use Group II/II+ and higher quality oils to satisfy industry performance requirements and meet market claims. Finally, when low temperature performance coupled with evaporative volatility control is essential, Group III+ is the one mineral oil base stock that can fulfill these performance requirements.

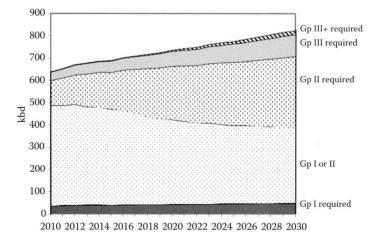

FIGURE 2.1 Global paraffinic base stock demand forecast by API classification. (From ExxonMobil assessment of publicly available information.)

Lubricant Base Stock Production and Application

Each API group has its own set of general properties that make it preferred for certain applications. Molecular composition of the hydrocarbons in base oil, along with the presence of impurities such as sulfur and aromatics, do have a significant influence on performance. As shown in Table 2.2, base stocks within a group share common characteristics. Even though base stocks may be in the same group, quality within a group can vary by individual manufacturer. As a result, it may be necessary to customize product formulations to account for those differences when changing between base oil suppliers.

Group I, also known as conventional base stock, is produced by physical extraction. Solvents are used to remove high concentrations of wax and multiring aromatic components. Group I is valued for its broad viscosity range. Unlike other mineral-based stocks, Group I is abundantly available in very high viscosities. The presence of sulfur compounds provides natural antioxidant properties. In addition, good solvency enables the oil to hold additives, keep dirt and soot in solution, and provide excellent deposit control. As Group I plants close down owing to the decline in demand, and because higher quality base stocks are available in both greater supply and at declining cost, lubricants that can be blended with Group I will continue to be reformulated and displaced with higher quality Group II or III base stocks.

Group II and III refining is typically done by hydroprocessing and catalytic dewaxing. This treatment removes impurities that affect oxidation stability. Compared to Group I, these base stocks have better thermal stability, volatility properties, and low temperature performance. Greater processing severity given to Group III produces higher VI base stocks. The higher VI is attributable to the presence of greater levels of isoparaffins. Higher quality Group II and III base stocks are necessary to meet the greater performance demands of modern engine oils.

Group III+ base stocks fill the performance gap between the best mineral-based stocks and Groups IV and V, highly synthesized stocks that are reacted to a precise chemical structure. Wax, the feed stock for manufacturing Group III+, is obtained either as a by-product from solvent refining of Group I base stocks, or it is built by linking individual molecules in the GTL manufacturing process. An isomerization catalyst is used to add branching to the molecules to produce base oil similar in quality to PAOs. Group III+ is extensively used to meet the increasingly demanding performance requirement of top-tier lubricants.

Evaporation control and good low temperature performance are important properties required to satisfy product claims. Figure 2.2 displays relative volatility and low temperature performance for typical base stocks within each API Group. The Noack test method (American Society for Testing and Materials [ASTM] D5800) measures evaporative loss of a sample subjected to heating and airflow. The Cold Crank Simulator test method (ASTM D5293) measures the apparent viscosity of engine oil, simulating the response of an average engine during a cold start. For both tests, lower results near the lower left hand corner of the graph are preferred.

TABLE 2.2
Base Stock Property Comparison

Group	Conventional	Hydroprocessed		PAO, Esters
	I	II	III	IV, V
SUS viscosity	100–2500	100–600	100–250	100–7000
Oxidation stability	Good	Good	Very good	Excellent
Volatility	Fair	Good	Very good	Excellent
Solvency	Very good	Poor/good	Poor	Poor–Excellent
Low-temperature characteristics	Fair	Good	Very good	Excellent

Note: PAO, polyalphaolefins; SUS, Saybolt Universal seconds.

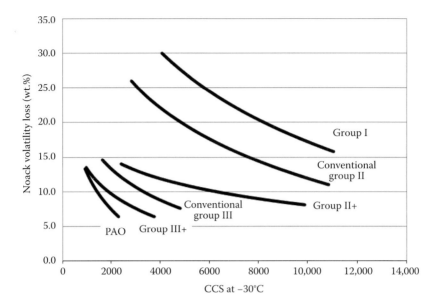

FIGURE 2.2 Blend properties of typical 4 and 6 cSt base stocks. (Note: Group II+ portrayed in the figure is optimized for both VI and volatility.)

C. Product Applications

Lube base stocks make up a significant portion of the finished lubricants, typically ranging from 70% of automotive and diesel engine oils to 99% of some industrial oils. Base stocks contribute significant performance characteristics to finished lubricants in such areas as thermal stability, viscosity, volatility, ability to dissolve both additives and contaminants (oil degradation materials, combustion by-products, etc.), low temperature fluidity, demulsibility, air release/resistance to foaming, and oxidation stability. Table 2.3 identifies whether the base stock or the additive has the

TABLE 2.3
Base Stock and Additive Impact on Lubricant Properties

Lubricant Property	Base Stock Impact	Additives Impact
Viscosity	Primary	Secondary
Viscosity stability (VI)	Primary	Secondary
Thermal stability	Primary	–
Solvency	Primary	–
Air release	Primary	–
Volatility	Primary	–
Low temperature flow	Primary	Secondary
Oxidation stability	Primary	Primary
Deposit control	Secondary	Primary
Demulsibility	Secondary	Primary
Foam prevention	Secondary	Primary
Antiwear/extreme pressure	Secondary	Primary
Color	Secondary	Secondary
Emission control	Secondary	Primary

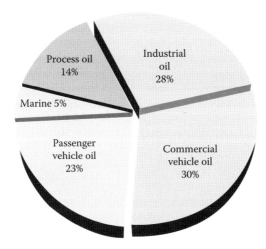

FIGURE 2.3 Estimated lubricant demand by finished product sector. (From ExxonMobil assessment of publicly available information.)

greatest influence on key properties of a lubricant. It clearly demonstrates that both the base stock and additive share responsibility for overall finished lube performance.

Modern lubricants are used in a great number of applications. No one single base stock type is able to satisfy all product requirements. An estimate of lubricant demand by the various finished product sector is shown in Figure 2.3. Demands made on the base oil vary considerably not only between but also within various sectors. For example, some heavy-duty diesel engines require some higher viscosity base stocks available only in Group I. In contrast, to satisfy low-temperature Society of Automotive Engineers (SAE) 0W passenger car engine oils, Group III or IV base stocks are preferred. Moreover, some products under severe service require Group IV or V synthetic oils to withstand extreme stresses encountered in service. Table 2.4 offers a qualitative description of the API Group classifications and their typical application.

Although often overlooked, a considerable volume of base stock is marketed as "process oil." Process oils are used throughout industry to both enhance manufacturing processes and improve "end product" performance and quality. The base stock used in a particular application is selected based on factors such as purity, solubility, viscosity, and safety. Solvent-refined materials, such as Group I base oils or Group V naphthenic base oils, are used extensively in the rubber and plastics industry to make products softer, more flexible, and easier to process. Other common uses include dust suppressants, anticaking agents, mold release oils, and carriers for ink oils, polishes, and additives. Oils refined to "white oil quality" are used where the highest purity ingredients are necessary. For example, they are used in cosmetics, food, pharmaceuticals, and lubricants that may contact food. The formulation of white oils into these products is generally covered by strict government regulations.

D. Base Oil Slate

The API base oil group classification is a part of the Engine Oil Licensing and Certification System (API Publication 1509). This is a voluntary program created by original equipment manufacturers, lube oil marketers, and base stock refiners as a guide to the selection and interchange of base stocks. The API defines a base stock slate as "A product line of base stocks that have different viscosities but are in the same base stock grouping and marketed by the same supplier." A somewhat similar system used in Europe is the Association Technique de l'Industrie Européenne des Lubrifiants (ATIEL) Code of Practice. To comply, base oil manufacturers declare that their base stocks of

TABLE 2.4
Typical Base Stock Applications by API Group

API Group	Typical Applications
Group I—Wax and multiring aromatics removed by solvent processing. Some aromatics and/or sulfur remain. Natural antioxidants, good solvency, and compatible with seals. Unlike Groups II and III, available in very high viscosities to 2500 SUS (30 cSt at 100°C) and above.	Marine and diesel engine oils, hydraulic oils, heat transfer oils, industrial gear oils, conventional greases, paper machine oils, compressor oils, machine tool way and slide lubricants.
Group II—VI performance similar to Group I except wax, aromatics, and sulfur removed catalytically. Available in viscosities up to 600 SUS (12 cSt at 100°C).	Passenger car and commercial engine oils, natural gas engine oils, automatic transmission fluid, turbine oils and automotive gear oils.
Group III—Similar to Group II but higher VI (lower viscosity change with temperature change). High severity hydro-processing eliminates ring structures beyond Group II. Has poor solvency. Viscosities up to 300 SUS (8 cSt at 100°C).	Premium passenger car motor oils (in particular, SAE 0W and 5W viscosity grades), automatic transmission fluids, food grade lubricants, white oil quality lubricants.
Group IV—Polyalphaolefins (PAO) synthetic, man-made, uniform, and free from impurities. Excellent performance at high temperature with superior oxidation stability.	High-performance engine, gear, compressor, hydraulic and circulating oils. High-performance greases, heavy duty transmissions, industrial bearing lubricants.
Group V Naphthenics—Multiring aromatics removed by solvent processing from naphthenic crude feeds. Very good low temperature performance and high solubility. Available in a wide viscosity range.	Products that operate in a narrow temperature range, at low temperatures and/or require high solubility. This includes transformer and process oils, grease, metalworking fluids, refrigeration compressor oils.
Group V Synthetics—Broad range of materials, each having relatively unique properties including alkylated aromatics, polybutenes, organic esters, polyglycols, phosphate esters, silicones.	Depending on the synthetic used: aviation turbine engines, fire-resistant hydraulic oils, heat transfer oils, brake fluids, high-temperature gear oils, refrigeration and compressor oils, chain oils and more.

Note: API, American Petroleum Institute; SAE, Society of Automotive Engineers; SUS, Saybolt Universal seconds.

different viscosities are of the same group or "slate." The intention is that, within a slate, base stocks of various viscosities must be interchangeable without affecting demonstrated performance.

For a new engine oil formulation, laboratory and engine testing to support quality claims approaching $1 million or greater may be necessary. A new formulation must pass a schedule of tests that are included in licensing guidelines. However, once a formulation is proven, there are rules that permit controlled interchange of base stocks, often with limited or no engine testing. Formulation flexibility is permitted by applying either Base Oil Interchange (BOI) or Viscosity Grade Read Across (VGRA) procedures that are part of the engine oil licensing programs. The API defines these in Publication 1509 Annex E and F. BOI defines testing necessary to make certain that performance is not affected when one base stock is substituted with another. The guidelines for both passenger car and heavy diesel engines are adjusted to be in line with current engine service categories. VGRA guidelines permit a single engine oil formulation to be converted, without further testing, to different viscosity grades after passing certification tests using the most severe grade being licensed.

Consider, for example, the EHC slate of Group II base stocks manufactured by ExxonMobil. A 10W-30 engine oil designed using EHC 50 (a branded medium viscosity base oil) must pass a rigorous testing protocol to support market claims. However, after quality claims have been demonstrated for a base oil and additive combination, BOI rules allow higher and lower viscosities of base stock within the same base oil slate to be substituted into the formula. No further engine testing is necessary for the same grade engine oil. In the BOI example demonstrated in Figure 2.4, once a single 10W-30 oil is certified, other 10W-30 engine oils may be made with the same additive formulation using a different viscosity from the EHC slate of Group II base stocks including EHC

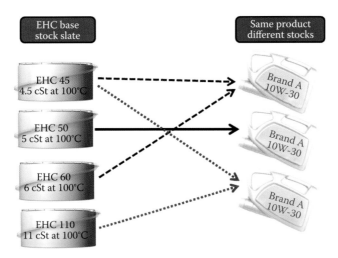

FIGURE 2.4 Example of BOI. Complete approval testing with one base stock in a slate permits blending with other base stock slate members without further certification testing to demonstrate claims.

45, EHC 50, EHC 60, EHC 110, and combinations thereof, without additional certification testing. There are many subtle variations in applying BOI, particularly when substituting one API group for another. BOI is complex and must be carefully applied even when used by a skilled practitioner.

Another practical aspect of the declaration of a base stock slate is the principle of VGRA. As illustrated in Figure 2.5, once a single grade of branded engine oil passes testing to support market claims, no further engine testing is necessary to blend a range of viscosity grades of the same brand oil. For example, the low viscosity (4.5 cSt) EHC 45 could be used to produce a branded 5W-30 engine oil. Then the 4.5, 5.0, 6.0, and 11 cSt stocks in the EHC slate could be used to blend 5W-30, 10W-30, 15W-40, and 20W-50 with essentially similar additive technology with no further engine testing. Generally, however, engine oil testing "reads across" from lighter to heavier viscosity grades.

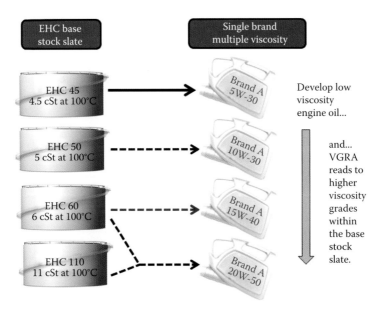

FIGURE 2.5 Viscosity Grade Read Across (VGRA). Once a single grade of branded engine oil passes testing to support market claims, no further testing is necessary to blend a range of viscosity grade oils.

A key API 1509 and ATIEL Code of Practice concept in manufacturing approvals is a declaration of the performance capability of a manufacturer's base stock slate. It is the base stock marketer's responsibility to ensure that the base oils are similar enough to enable this blending. Base stock specifications alone are not sufficient predictors of performance in finished engine oils. In general, single marketers are geographically diversifying their operations, varying refining processes, and using an increasing number of different crude feeds of varying quality. As manufacturing becomes more diverse, better control is needed to ensure consistent base stock quality. Therefore, in addition to relying on standard specifications, more attention must be given to compositional parameters, feedstock quality, and refining processes. A slate requires a single manufacturer to maintain proper control of base stock quality. BOI is best proven and backed by pertinent test data—not just "technical judgment."

II. ROLE OF CRUDE OIL IN THE MANUFACTURE OF BASE STOCK

A. Chemistry of Crude Oil

Crude oil, as it is taken from the earth, is a very complex mixture of hydrocarbons. As might be expected, the term hydrocarbon refers to the hydrogen and carbon atoms linked together and forming a majority of the individual molecules that are present. Having been transformed from living organisms being subject to heat and pressure over millions of years, the crude is termed to be an "organic" material. In addition to hydrogen and carbon, there are small amounts of other atomic elements that are present and mostly incorporated into the hydrocarbon molecules. The most common of these are nitrogen and sulfur. As part of the hydrocarbon, they are referred to as "hetero-" atoms. Heavier crude may also contain some inorganic materials such as nickel, vanadium, and iron, which are removed during refining.

Hydrocarbons form in an extraordinarily large number of configurations. They range in size from the simplest methane molecule, which contains one carbon atom and four hydrogens, to heavy asphaltic resins with structures that have not been fully mapped. Because we are dealing with mineral oil–based lubricants, we are interested in molecules of a size that fit into the range of approximately 15 to 95 carbon atoms long, corresponding roughly to a boiling range of about 300–700°C (572–1292°F). In addition, as the number of carbon atoms in a molecule increase, the number of different ways that a molecule can be structured increases dramatically. Although many different arrangements of molecules may be imagined, for practical purposes, petroleum hydrocarbons are generally classified in a limited number of ways.

Consider the list of hydrocarbon types found in crude and their associated structures as identified in Figure 2.6. The carbon atoms in each of the structures are shown as red dots. Pendent hydrogen

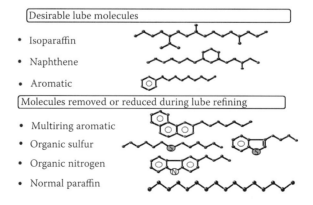

FIGURE 2.6 Typical structures of hydrocarbons found in crude.

Lubricant Base Stock Production and Application

atoms are not displayed. However, each individual carbon atom has four bonds, each of which may be connected to a combination of hydrogens, a heteroatom, or another carbon. With the astronomical number of individual molecules that are possible, only a single representative structure of each is shown. Those molecules that make good lubricants are grouped at the top, whereas less desirable molecules are listed below. Nitrogen compounds promote oxidation and cause the formation of undesirable sludge. During the refining process to remove nitrogen, sulfur compounds are also eliminated. Although sulfur compounds can naturally inhibit oxidation and sludge formation, additives designed to inhibit oxidation generally provide better protection.

Characteristics that differentiate performance between hydrocarbon types are listed along the table heading in Table 2.5. Five important lubricant performance characteristics of a hydrocarbon are assessed. VI measures a lubricant's ability to resist change in fluidity with changes in temperature. Pour point is the temperature where the lubricant will begin to resist flow and cease to function as a lubricant. Oxidation resistance is essential to prevent the formation of harmful sludge and varnish. Solubility is important to not only dissolve the additives used in formulating lubricants but also to suspend contaminants that may find their way into finished oil. Finally, no assessment is complete without concern for toxicity and product safety.

There is a clear relationship between the type of hydrocarbons present in a base stock and its performance as a lubricant. As demonstrated in Table 2.5, although it is evident that some structures are better than others, no molecular type is perfect in every respect. Note that PAO, a synthetic, is included for comparison with the mineral-based materials. It is nearly perfect in every respect except solubility. However, this high-quality PAO is delivered at higher cost, which is typical for a specialty chemical.

Paraffins, having relatively simple structures, can be divided into two classes. Normal paraffins (n-paraffins), with simple linear structures and no side chains, turn waxy at low temperatures and make poor lubricants for products that need to operate in a wide range of temperatures. In contrast, isoparaffins, with side-branching carbon atoms to keep the molecules from crystallizing, remain fluid at low temperatures and are preferred base stock components. In addition, relative to other mineral-based materials, the isoparaffins have excellent VI, oxidation stability, and low toxicity.

TABLE 2.5
Lubricant Performance of Various Hydrocarbons Found in Crude Oil

Molecule	Structure	VI	Pour point	Oxidation	Solubility	Toxicity
n-paraffin		Excellent	Poor	Excellent	Poor	Low
Isoparaffin		Good/excellent	Good	Excellent	Good	Low
Single-ring naphthenic		Good	Good	Good	Good	Low
Multiring naphthenic		Poor	Excellent	Good	Excellent	Low
Alkylbenzene		Good	Excellent	Good	Excellent	Moderate
Polycyclic aromatic		Poor	Poor	Very poor	Good	Very high
Polyalphaolefin		Excellent	Excellent	Excellent	Good/poor	Low

Naphthenics are similar to paraffins except that there is a grouping of carbons that form at least one ring structure generally consisting of six carbon atoms. Although the ring structure of naphthenic molecules appear similar in structure to aromatic molecules, there is one significant difference: the carbons in a naphthenic molecule are all fully saturated or filled with hydrogen. No double bonds exist between two carbon atoms in a naphthenic ring. Naphthenic molecules with one or two rings have relatively well-rounded performance. All considered, this is good, because paraffinic Group I and Group II base stocks often contain a considerable amount of naphthenic material. However, quality drops when the naphthenic content becomes too high, and/or there is a high concentration of multiring naphthenic material. With too much naphthenic material, the VI becomes too low for use in engine oils even though pour point and solubility are excellent.

Aromatic hydrocarbons are present in paraffinic lube crudes. In aromatics, the six carbon atoms in a ring share double bonds and, as a result, aromatics contain fewer hydrogen atoms than other hydrocarbons. All aromatics have poor VI properties, exhibiting significant changes in viscosity with changes in temperature. However, single-ring substituted monoaromatics are considered fair performers in conjunction with isoparaffins. They have a significant presence in the composition of lower quality Group I base stocks, and their primary contribution is to improve solubility and pour point. In contrast, polynuclear aromatics are particularly poor lubricants with respect to VI, pour point, and oxidation stability, and their highly toxic nature make them undesirable components of base stocks.

B. Crude Selection

As mentioned previously, petroleum crude is still the major raw material used for manufacturing base stocks. However, almost 90% of the crude oil that is refined ends up in nonlube products such as gasoline, distillates, and other residuum. The quality and yield of these fuels and other nonlube products can be of overriding importance in the evaluation and selection of a crude.

Historically, the preferred crude to be refined into a lube base stock is called a "paraffinic lube crude." These are sweet (low sulfur) and highly paraffinic crudes. Conventional refining involves use of various solvents to extract unwanted aromatics and separate out waxy materials. Because such processing makes no change in the structure of the crude, the desired molecules that are of lubricant quality must always be present in the crude at the start of processing. In particular, lubricant materials are contained in the atmospheric boiling range of approximately 300–700°C (572–1292°F).

Unfortunately, crude is not a uniform material, and composition can vary greatly, even between wells in the same field. Significant differences are observed from crudes around the world. Coming right from the ground, crude can range from very light condensed hydrocarbons to extremely heavy asphaltic material that needs to be heated to flow. In addition to paraffinic-based crudes, there are also those that are naphthenic in nature. Naphthenic crudes contain mostly naphthenic and aromatic components with little or no waxy paraffin. In contrast to paraffinic base stocks of the same viscosity at 100°C (212°F), naphthenic oils remain liquid at much lower temperatures.

Because of its complexity, identification of good crude for the production of base stocks requires more than a few simple tests. However, API gravity has historically been used as a key test to classify how suitable a particular crude is for manufacturing various products. The gravity of crude typically falls in a broad range, from a very light API gravity of about 50° API to a very heavy 10° API. (API gravity is an inverse scale—the higher the API gravity, the less dense the material.) Heavy asphaltic crude of 10° API is as dense as water and much heavier than a typical lube crude of about 40° API.

Good lube crude, suitable for solvent refining, is relatively light and traditionally falls in a range of 35–50° API. However, with advances in refinery technology, heavier crudes can economically produce high-quality Group II and III base stocks. For example, with hydrocracking and catalytic dewaxing along with hydrofinishing, heavy sour crude such as Hamaca from Venezuela with an API gravity of 26° API yields high-quality base stock. How far down in API gravity a refinery can go, and still make quality lubricant molecules, depends on refining capability and economics. There are practical limits to how heavy a feed can be. Heavy asphaltic crudes in the range of 10–20° API contain catalyst plugging materials that will quickly deactivate catalyst and terminate processing.

Another usual measure of crude quality is sulfur content, being labeled as sweet or sour based on its sulfur content. Sweet is considered to be less than 0.5% sulfur by mass and sour is greater than 1.0% sulfur. Anything between 0.5% and 1.0% is an "intermediate." Table 2.6 lists the properties for a few crudes from around the world. West Texas Intermediate, which is often used as a benchmark against which others are compared, is a good quality sweet lube crude. Because it is easier to make value-added products with lighter sweet crudes, they generally command a premium price over heavier sour crudes.

Identification of good lube crudes go beyond the simple measure of gravity and sulfur content. Analysis of a proposed new crude generally begins by separating it into fractions by distillation. Each fraction is then tested to determine if it has the right components to be able to turn it into a good base stock. Final confirmation on whether quality lubricants can be made from a new crude supply is done by actually putting it into a closely monitored refinery run. This is followed by testing both the base stocks and the products into which they are formulated.

A breakdown of the molecular constituents in a typical lube crude is shown in Figure 2.7. Different classes of molecules are sorted by the number of carbon atoms for the molecules present, starting with the light molecules on the left progressing to the heavier components on the right. The lightest molecules of interest for lubricants begin with a carbon number of about C17, corresponding to an atmospheric boiling point of about 300°C (572°F) (i.e., at a temperature where there are about 17 carbon atoms in each individual molecule). Identification of the constituents by chromatographic separation ends at a carbon count of about C95. This generally corresponds to the largest molecules that are present in high viscosity base stocks as well as the capability of conventional analytical separation instruments.

TABLE 2.6
Crude Oil Properties

Crude Name	°API	Crude Type	Sulfur (Mass %)	Sulfur Level
Bachaquero (Venezuela)	17	Heavy asphaltic	2.4	Sour
Alba (UK)	20	Heavy asphaltic	1.2	Sour
Maya (Mexico)	22		3.4	Sour
Hamaca (Venezuela)	26		1.6	Sour
Arabian Heavy	28		2.9	Sour
Arabian Light	32		1.8	Sour
Alaskan North Slope	32		0.9	Intermediate
Brent (North Sea)	38		0.4	Sweet
Statfjord (Norway)	39		0.2	Sweet
West Texas Intermediate	40	Conventional light	0.3	Sweet
Tapis (Malaysia)	46	Conventional light	0.1	Sweet
Arabian Super Light	51	Conventional light	0.1	Sweet
Condensate	65	–	–	

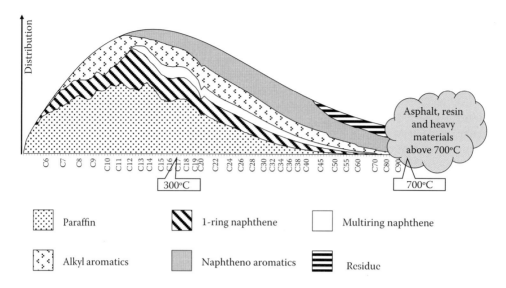

FIGURE 2.7 Hydrocarbon distribution in a typical lube crude versus carbon number.

III. REFINERY PROCESSING—SEPARATION VERSUS CONVERSION

Conventional solvent processing of crude relies on physical *separation* of the preferred lube molecules from the remaining materials in the crude. Separation in refining is often accomplished by distillation and solvent extraction. Therefore, to make Group I base stocks, lubricant quality molecules must be present in concentrations that make processing worthwhile. Aside from some hydrofinishing to remove impurities and partially saturate remaining aromatics, solvent process molecules are essentially unchanged.

Modern lubricant *conversion* technology creates lube molecules through conversion and upgrading of nonlube components in the crude. Instead of simply extracting existing molecules from the crude, it relies on high temperature and high pressure hydrogen to saturate aromatics, break apart, and rearrange common molecules into high quality lube base stocks. Therefore, conversion processing can use a much broader range of crude types than conventional solvent processing, including less expensive crudes. Ultimately, conversion processes create greater volumes of more valuable lube base stocks with less costly processing.

A description of both separation and conversion processing processes follows. There are many exceptions to these schemes. As refineries are upgraded, modern lube refining techniques have been successfully integrated into existing refinery operations. For example, solvent dewaxing can effectively be replaced by catalytic dewaxing, solvent extraction can be used to condition feed going into a conversion process, fractionation of streams can take place at various points in a process. Therefore, in the world of lube processing, hybrid systems are very common. Separate and more in-depth discussion of the separation versus conversion processes will begin following the section on vacuum distillation (Section III.B), a process common to both operations.

A. ATMOSPHERIC DISTILLATION

Before either separation or conversion processing begins, it is necessary to clean the crude and subsequently separate off lighter components by distillation. It is necessary to remove inorganic salts, suspended material, and water. Salt and sludge can break down during processing to form acids that cause severe corrosion of refinery equipment, plug heat exchangers and process equipment, and poison-sensitive catalysts. Cleanup is accomplished by mixing the crude with additional water to dissolve salts and sediment, adding processing chemicals to speed separation, and allowing the contaminants to settle out of the crude.

Atmospheric distillation is next, as shown in Figure 2.8. The desalted crude is pumped through a furnace where it is heated and partially vaporized. The mixture of hot liquid and vapor from the furnace is put into a fractionating column called a tower. The tower operates at a pressure slightly above atmospheric and at a temperature range where the hydrocarbons will boil without cracking. The bottom of the tower operates at a higher temperature than the top. As vapor moves up the column, heavier components condense out and pass down. A multiple series of trays is stacked inside the tower to hold condensed liquid. These multiple trays ensure that as the vapor stream moves up the column, lower boiling vapors are progressively condensed and flow down the column. This fractional distillation separates the feed into multiple narrow boiling range cuts that cannot be duplicated in a simple single pot still.

Different products are drawn from the atmospheric tower at various heights along the column. Naphtha and gaseous hydrocarbons are carried over the top. The naphtha is condensed from the vapor and further processed into gasoline. Kerosene, diesel, and gas oils are withdrawn as side cuts from the successively lower and hotter levels of the tower and used for fuel. A heavy black, atmospheric residuum is collected at the bottom. Because the residuum tends to thermally decompose at about 370°C (700°F), any further distillation of the material from the tower bottoms would cause these bottoms to decompose. Further separation must be done under high vacuum where the components will boil at lower temperatures without cracking.

Boiling point distribution of light products is shown in the center row of Figure 2.9. Distribution is plotted against both atmospheric equivalent boiling point in degrees Celsius and average carbon number of the cut. Similarly, the distribution spectrum for various base stocks by viscosity class is shown in the top row of the graph. (Distillation temperatures for the base stocks have been mathematically converted to theoretical boiling points at atmospheric pressure.) Viscosity grades are given in terms of SUS. Base stock viscosity grades can be manufactured in a range varying from a very low viscosity of 70 SUS (3 cSt at 100°C) to 2500 SUS (30 cSt at 100°C) for bright stock.

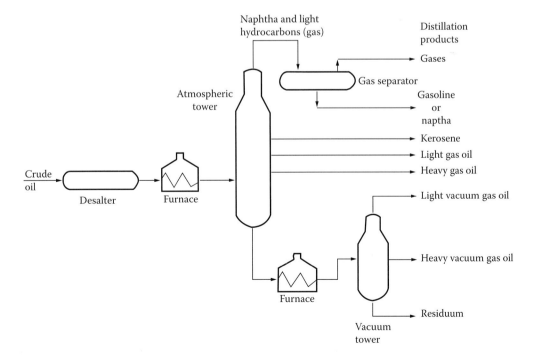

FIGURE 2.8 Crude distillation unit.

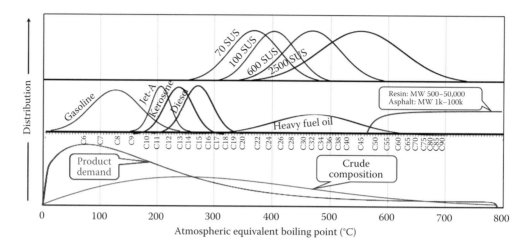

FIGURE 2.9 Products across the distribution spectrum. This graph shows the product volume distribution versus the atmospheric boiling point and the number of carbon atoms in a typical molecule. Lube fractions in the top row are identified by viscosity in SUS, which has been the traditional viscosity measurement for Group I base stocks in commercial practice.

B. Vacuum Distillation

Vacuum distillation is the first process in base stock production for both conventional solvent processing and hydrocarbon conversion processing. The heavy fraction of crude from the atmospheric tower bottoms, a mix of materials boiling above 350°C (662°F), is fed to the bottom of the vacuum tower. Lighter materials rise to the top of the tower, whereas the heavy asphalt, resins, and heavy lube stock remain at the bottom. Notice in Figure 2.9 that there is overlap between the various viscosity grades. This overlap may vary slightly according to the capability of the equipment to separate the extremely complex mixture of molecules present in the feed. Careful operation of the vacuum tower is necessary to prevent too much overlap that would cause processing problems downstream and result in inferior product quality. Base oil properties set by vacuum distillation include viscosity, flash point, and volatility.

Similar to atmospheric distillation, the vacuum tower separates the heavy fraction of the crude into different viscosity classes by boiling range. Unlike the atmospheric unit using multiple trays to effect separation, instead, packing is used to reduce the pressure drop in the vacuum tower. Packing generally consists of small ceramic, metal, or plastic shapes randomly laid down in the column. Alternately open structured monolithic packing can be fitted into the column. Packing permits contact of vapor with liquid without the restriction of having to pass through liquid-filled trays as it rises up the tower. Only a few special trays, called draw trays, are used at various points in the column to enable collection and removal of liquid at the points where product side streams are drawn.

In operation, the pressure at the top of the column is reduced to less than one-tenth of normal atmospheric pressure. The heavy crude fraction is heated in a furnace to about 400°C (752°F) and fed into a flash zone near the bottom of the column. At atmospheric pressure, the feed will not vaporize; however, separation can begin at reduced pressure under a vacuum. As the vapor rises in the column, heavier components condense on the packing and move downward. The lightest components continue to move up the column. Neutral distillate withdrawn from the side streams consists of fractionated liquid with a small amount of lower boiling material. Therefore, each side stream is charged to a stripping unit, where steam is used to strip out low boiling materials to prepare the neutral distillate for further processing. The low boiling material is charged back to the vacuum tower to provide cooling and for further fractionation.

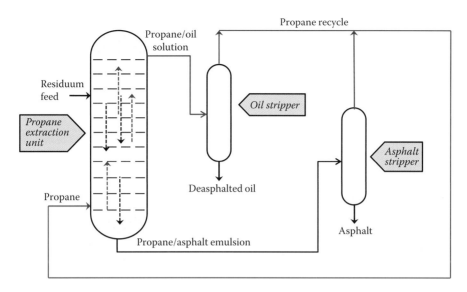

FIGURE 2.10 Continuous propane deasphalting process.

C. Propane Deasphalting

The vacuum residuum collected at the bottom of the tower is typically steam stripped in a stripping section below the flash zone before leaving the tower. The residuum contains the lube fraction in combination with a complex mixture of asphalt and resin. The extremely high viscosity, nonlube components are removed producing a DAO for further lube manufacture. Separation is typically done by extracting the lube fraction with propane solvent in a propane deasphalting unit. Propane is used as an extraction solvent because it will dissolve paraffinic, naphthenic, and aromatic lube fractions from the residuum. After the propane extract is decanted from the heavy asphalt layer, the low boiling propane solvent is evaporated leaving a DAO. Asphalt and resins are not soluble in propane and form a separate liquid phase.

In practice, deasphalting is done continuously in a countercurrent flow extraction unit as diagrammed in Figure 2.10. The residuum, usually diluted with a small amount of propane, is pumped to the middle of an extraction column full of liquid propane. At the same time, propane is injected into the bottom of the column at about 6–8 times the residuum volume. Because the residuum is denser than propane, the residuum flows down the column, with the propane rising up in a counter flow. The two streams are mixed together within the column and the lube fraction dissolves in the propane and continues up the column. The heavy, insoluble asphaltic material passes out of the bottom of the column. The propane and asphalt streams are sent to separate stripping towers, where propane is recovered and recycled. The DAO is next processed in the same manner as the other lower viscosity lube distillates.

IV. CONVENTIONAL SOLVENT PROCESSING

Conventional solvent processing (see Figure 2.11) involves separating existing lube molecules from crude where those molecules already exist. After vacuum distillation, *solvent extraction* is used to remove aromatics, particularly the polynuclear aromatics that have low VI, high pour point, and poor oxidation stability. Next, *solvent dewaxing* removes the waxy straight-chained paraffins that negatively affect pour point and cold flow properties. In the final step, *hydrofinishing* improves the color and stability of the finished Group I base stock.

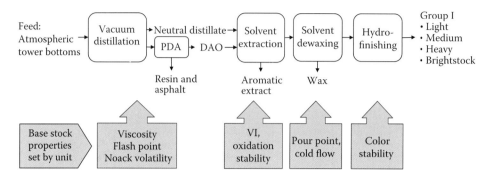

FIGURE 2.11 Conventional solvent refining process to make Group I base stocks.

A. Solvent Extraction

Solvent extraction is applied to both neutral distillates and the heavier propane DAOs. The process works by separating out the poor-quality aromatic molecules from the higher-quality lube components. Out of the extraction unit, the fraction containing desirable lube molecules is called a raffinate, and the extracted low VI phase is called the aromatic extract. For good separation, the choice of solvent is very important. A solvent should have good solvency for the aromatic fraction, while having little or no solubility for the saturated fraction. In addition, the solvent should have a higher density than the raffinate to simplify phase separation, and a relatively low boiling point to aid recovery and recycle after the raffinate and the extract exit the extractor. Several highly polar materials are used extensively as solvents in commercial production: furfural, NMP, and phenol.

In practice, separation takes place by countercurrent extraction in a vertical tower (see Figure 2.12). The various viscosity cuts from the vacuum distillation tower are processed one at a time in block mode to maintain the viscosity grades established by the vacuum tower. The distillate charge is fed into the side of the unit. The extraction solvent enters near the top and flows down the tower, dropping because of its higher density. A series of rotating-disk contactors and baffles drive intimate contact between the rising distillate feed and the descending solvent stream. Aromatics, along with some sulfur/nitrogen compounds, migrate into the solvent phase as it passes down the tower. After exiting the tower bottom, the extract mixture goes to solvent recovery. Recovered solvent is recycled back to the extractor, and the extract is generally cracked into fuel.

As the raffinate stream exits the top of the tower, it contains only a small amount of solvent that is recovered by evaporation and steam stripping. At this point, the raffinate is relatively free of

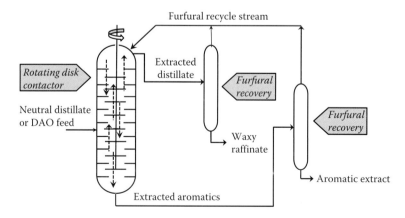

FIGURE 2.12 Counter current extraction of aromatic hydrocarbons in a rotating disk contactor using furfural as the solvent.

Lubricant Base Stock Production and Application

aromatics, particularly those with multiple rings, and is ready for solvent dewaxing. The extent of extraction is controlled by unit operating temperatures, distillate and solvent feed rates, and solvent type. The most evident effects of solvent extraction are increases in VI of the raffinate along with an approximate 50% decrease in sulfur. In addition, there is also a significant improvement in both thermal and oxidation stability. At this point, the raffinate still contains a significant amount of wax that will require removal by solvent dewaxing.

B. Solvent Dewaxing

The purpose of dewaxing is to remove normal paraffins from the waxy raffinate in order to improve its low-temperature performance properties. A mixture of solvents, methyl ethyl ketone (MEK) and toluene, cause the normal paraffins to form crystals at reduced temperatures. The wax crystals are then separated from the oil by filtration. (This is in contrast to phase separation used in the earlier solvent extraction process.) The function of the individual solvents is not entirely clear. Nevertheless, the combination of solvents does adjust the polarity of the solvent to optimize wax precipitation. Other solvent combinations such as MEK/methyl isobutyl ketone are also commonly used.

The first step in MEK dewaxing (see Figure 2.13) is to condition the feed stream by mixing it with solvent and heating it to dissolve the oil and wax. Only wax formed under carefully controlled process conditions can guarantee optimum crystal structure for filtration. After heating, a pair of double-pipe heat exchangers is used to cool the feed to a temperature that is about 5–10°C (9–18°F) below the desired pour point. As the feed passes through the central pipe of the heat exchangers, wax crystals form on the chilled wall surface. To maintain filter efficiency, a rotating blade scrapes the wax off the walls and into the process stream. Additional solvent may be added throughout the process to promote wax crystal formation in the wax slurry.

The wax slurry is collected in a surge tank before it is fed to a rotary vacuum drum filter. Separation of wax from oil takes place on a filter cloth fitted to the outside of a cylindrically framed drum. The drum is suspended on its side and rotated on its central axis through the wax slurry. As shown in the end view of a filter drum in Figure 2.14, the filter drum is segmented into separate chambers to permit variations in vacuum and pressure on the filter cloth as it cycles. A series of

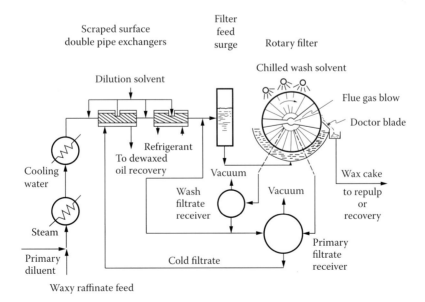

FIGURE 2.13 MEK solvent dewaxing process.

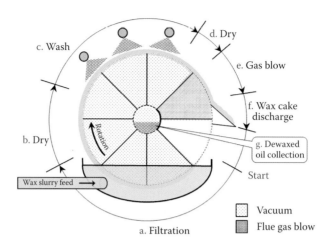

FIGURE 2.14 Wax collection cycle for a rotary drum filter.

valves, located near the central axis of the drum, channels the dewaxed oil passing through the filter to solvent reclamation.

As shown in the wax collection cycle, no vacuum is applied at the "Start" of a cycle before a filter segment passes into the slurry trough. This prevents loss of vacuum through the uncoated filter. Filtration begins (a) as a filter segment enters the wax slurry and a vacuum pulls the filtrate oil through the filter leaving the wax on the filter cloth. The wax continues to build in thickness through the filtration cycle. The vacuum continues as (b) the segment leaves the wax slurry and goes through a short period of drying to remove more oil and solvent from the wax. In the wash cycle, (c) cold solvent is sprayed on the wax to displace more oil from the cake. (d) The wash is followed by another short period of drying. Finally, (e) the vacuum is lifted from the filter with pressure from a flue gas blow and, as the filter drum rotates, (f) the wax is guided from the drum by a doctor blade and is passed by the conveyor to a wax solvent recovery system. At this point, the filter cloth is empty and the cycling continues.

The dewaxed filtrate is removed continuously (g) through a complex set of pipes and valves located in the drum and sent to solvent recovery. Solvent recovered from both the oil and wax phases is recycled back for use in the solvent dewaxing process. After the dewaxed oil is free of solvent, it is often subjected to a light hydrofinishing to improve quality.

The wax that was separated from the oil still contains a relatively high amount of oil. This oily wax is called "slack wax" and, depending on wax disposition, usually contains approximately 15–30% oil. To make a high-quality, fully refined wax, the oil content of the wax is usually reduced to less than 1% through a deoiling process. However, as an alternative to making finished wax, slack wax from conventional solvent processing of middle distillates can be catalytically converted to high-quality Group III+ base stocks. Catalytic isomerization is covered in Section V.

C. Hydrofinishing

Hydrofinishing is often done as the final step in solvent processing. In this process, hydrogen is reacted with an oil to improve base stock color by removing small amounts of sulfur, other heteroatom components, and polar contents. The hydrofinishing reactor consists of a bed of catalyst through which the hot oil contacts a relatively low-pressure hydrogen feed (200–300 psi or 1379–2068 kPa). Essentially no chemical change occurs with the higher performance lube molecules. The process has very little effect on the key physical properties represented in marketing the base stock. However, it does make a significant improvement in demulsibility and both thermal and oxidation stability of the resulting Group I base stock.

Lubricant Base Stock Production and Application

V. CONVERSION PROCESSING

Hydroprocessing is a general term given to using a catalyst with hydrogen to convert less desirable crude fractions into quality lubricants. More specific terms are given these processes based on reaction severity according to temperature, pressure, and the amount of hydrogen present (Table 2.7). Hydrofinishing, at low severity, is typically stabilizing base stock and has very little reaction effect. Hydrocracking is at the other extreme, where high severity operation creates not only high-quality lube oil, but also gasoline, diesel, and lighter hydrocarbons. Hydrotreating severity falls somewhere between finishing and cracking, and is often used to condition a process stream. For example, hydrotreating is used to remove sulfur from a stream before it enters a process unit containing a noble metal catalyst that is very susceptible to poisoning by sulfur.

Hydroprocessing technology changed base oil refining technology from a process involving limited physical separation to an exceptional chemical operation. Conversion processing, in its various forms, is essential in turning crude oil into the high-quality Group II and Group III base stocks needed to meet the increasing performance demands of today. When comparing single viscosity grades, these base stocks have much lower volatility and better cold temperature performance capability than Group I stocks.

A diagram of the steps in conversion processing is shown in Figure 2.15. The first step in the conversion processing of vacuum gas oil and DAO is to pass it through a high severity *hydrocracker*

TABLE 2.7
Hydroprocessing Severity, Purpose, and Effect

Process ⟶	Hydrofinishing	Hydrotreating	Hydrocracking
Process severity	Low	Medium	High
Purpose	Remove impurities, saturate olefins, stabilize color	Improve oxidation stability, reduce heteroatoms	Crack molecules, remove S, N, and aromatics
Processing effects			
Paraffins	No reaction	No reaction	Cracking
Naphthenes	No reaction	No reaction	Ring opening
Aromatics	Partial saturation	Partial saturation	Saturation
Olefins	Saturation	Saturation	Saturation
Sulfur removal	Some	Some	Complete
Nitrogen removal	Little	Some	Complete

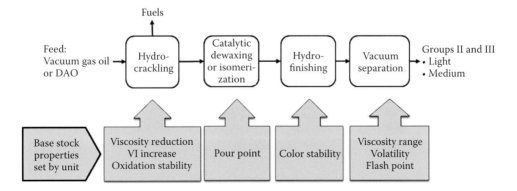

FIGURE 2.15 Conversion process used to make Group II and III base stocks.

to produce a waxy oil. Next, the oil from the hydrocracker is *catalytically dewaxed*. In this process, waxy paraffins are converted into isoparaffins that exhibit good low temperature performance. Finally, this is followed by hydrofinishing and the subsequent distillation into the desired viscosity grades.

A. Hydrocracking

Hydrocracking has clearly replaced solvent extraction as a preferred means for converting crude into paraffinic feeds for base stock production. Vacuum gas oil or DAO is reduced to a lower molecular weight and a higher hydrogen content in a complex process. Significant reactions that take place in hydroprocessing are aromatic ring saturation, naphthenic ring opening, olefin saturation, cracking of paraffins, and removal of sulfur and nitrogen. Refining conditions are carefully controlled because there are optimum conditions for the manufacture of high-quality lubes. Pushing conditions beyond the optimum can actually cause a severe drop in base stock quality.

Hydrocracking is typically done in a reactor with a catalyst that is packed in separate beds arranged along the reactors' length. Because the hydrocracking reaction generates large quantities of heat, quench zones are arranged between the various catalyst beds where hydrogen can be fed to control reaction temperatures to better control the reaction and its severity. Consideration needs to be given to feed composition, catalyst selection, capability of downstream processing, and unit operating conditions of temperature, pressure, and both hydrogen flow and consumption. To convert crude into high-performance base stock, it is necessary to use hydrogen and a catalyst under the severe hydrocracking conditions of high temperature and pressures (as high as 3000 psi or 20,684 kPa). Proper catalyst selection is key to both good yield and long catalyst life.

A single multifunctional catalyst is often used for base stock hydroprocessing, serving both aromatic saturation and cracking. Aromatic saturation occurs at a junction between an active metal and the catalyst base. In turn, cracking is controlled by the acidity of the catalyst base itself. Acid strength varies according to the mixture of materials in the catalyst base. Catalyst acidity has a significant effect on product output. If acidity is too high, it will result in the production of an excess of lower value light products. Hydrocracking catalysts are also sensitive to excessive contaminants in the feed. If necessary, nitrogen and sulfur may be removed by treating the feed with a less sensitive catalyst before it is hydrocracked. Some base stocks may be produced from these two-stage hydrocrackers.

Group II and Group III saturates levels are typically well above 90%, with most of the aromatics undergoing saturation during hydrocracking. Where the goal is to maximize lube production, conversion to light fuel products is limited to about 30% of the feed, leaving a greater portion for base oil. Therefore, in lubes hydrocracking, there is a relatively high content of low VI naphthenic material exiting the hydrocracker in combination with a mix of paraffins. In contrast, where the goal is to maximize fuel production, more severe hydrocracking is necessary. As a result, fuel hydrocracking produces a waxy, paraffinic feed that has a lower naphthenic content, making it ideal for producing high VI Group III base stocks.

B. Catalytic Dewaxing

The lube streams from hydrocracking contain waxy paraffinic material that must be eliminated to improve low temperature properties. In conversion processing, a dewaxing catalyst is used along with high pressure hydrogen. Early catalytic dewaxing processes worked by simply cracking waxy normal paraffins into smaller molecules, resulting in loss of yield. However, by using advanced catalyst technology, dewaxing can take place by converting normal, straight-chain paraffins into lube-quality branched isoparaffins. In contrast to solvent dewaxing, dewaxing by isomerization results in higher performance base stocks at higher product yields.

Lubricant Base Stock Production and Application

ExxonMobil's MSDW lube isomerization dewaxing catalyst converts waxy oil and raffinate to conventional or ultralow pour point base oils and provides a serious boost to the VI of the oil over hydroprocessed feed stocks. The MSDW process uses a simple, fixed bed cascade reactor with shape-selective dewaxing catalyst. Hydrogen and waxy oil are combined, heated, and fed to the top of the dewaxing reactor. Critical to the process is a zeolite catalyst with a noble metal for increased activity. The catalyst is of medium pore size and shaped to permit entry of straight-chain normal paraffins, but is impenetrable to larger bulky lube molecules. Once in intimate contact with the catalyst, normal paraffins undergo isomerization into isoparaffins.

After dewaxing, the lube base oil is separated from the hydrogen-rich recycled gas and by-products. The hydrogen gas is recycled back to processing. The lube base oil is then fractionated in a vacuum column to separate off light products and cut the lube fraction into finished base stocks of various viscosity grades. Because waxy hydrocarbons are converted instead of being extracted, conversion process yields are better than those of conventional solvent processing.

C. Alternate Processing for Group III+ Quality

The highest VI Group III+ base stocks can be created by processing a highly concentrated wax stream through catalytic isomerization. The wax can be sourced as a by-product from the conventional dewaxing operation in the production of Group I base stocks from high wax content crudes, or from wax made in a Fischer–Tropsch (FT)-based GTL process. As shown in Figure 2.16 (see the top process), the wax feed is converted to high-quality base stock by catalytic isomerization. This same process is used to convert hydrocracker bottoms to ordinary Group III base stocks, except that the waxy feeds are essentially free from aromatic and naphthenic hydrocarbons. In fact, FT feeds are approximately 99% straight-chain normal paraffins facilitating the conversion to high VI base stocks. Processing after catalytic isomerization includes stripping of gaseous hydrocarbons, naphtha, and middle distillates followed by vacuum distillation to produce the desired viscosity grades.

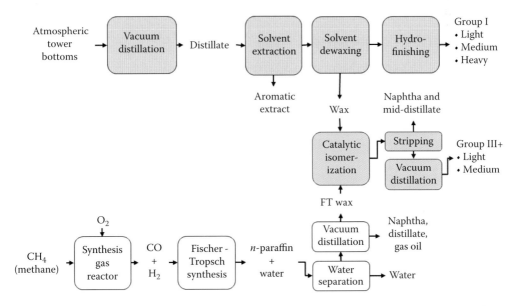

FIGURE 2.16 Group III+ conversion process from wax. The source of wax can vary. It may be a by-product of Group I production (above) or from synthesis by the gas-to-liquid process (below).

D. Gas-to-Liquids via Fischer Tropsch Synthesis

The GTL process, via FT synthesis, converts gaseous hydrocarbons to higher value fuels and lubricants. It works with many types of materials such as coal, biomass, natural gas, or other organic materials that can be transformed into a gas. For lubricant use, the area of most interest is the extraction of methane from natural gas and conversion into high-quality base stock. Methane is a simple hydrocarbon molecule consisting of one carbon with four attached hydrogen atoms. It is available in great quantities around the world, often found in remote locations. Gas at well sites that are "stranded," i.e., far from market locations, can be converted into liquids for easy handling and distribution of products. The resulting GTL base stock contains essentially no aromatics or sulfur and is considered very high quality because the feed has little, if any, contaminants that are present in crude.

Conceptually, the manufacture of FT wax for use in the production of base stocks appears to be straightforward. However, construction to a commercial scale and the ability to efficiently operate are actually quite complicated. A number of companies have put GTL into practice, often producing light products such as fuel or naphtha. For base stocks, higher molecular weight wax is necessary, and one process is clearly leading with commercialization of lubricant base stocks. To begin, methane is separated from heavier hydrocarbons, heterogeneous contaminants such as sulfur, and water (see the bottom process in Figure 2.16). The purified methane and oxygen are fed to the synthesis gas reactor where, at high temperatures, they are converted into synthesis gas. This material, also called syngas, is a combination of carbon monoxide and hydrogen.

Syngas next passes to a fixed bed reactor where it contacts a proprietary cobalt catalyst and begins conversion to hydrocarbon via FT synthesis. In this process, several reactions take place aided by the catalyst. Carbon splits from a carbon dioxide molecule and attaches itself to a growing hydrocarbon chain. As the carbon joins the chain, it picks up several hydrogen atoms. The free oxygen from the carbon dioxide reacts to form water. This process repeats many times over, adding one carbon atom at a time, to form long straight-chain normal paraffins. Depending on reactor conditions, the carbon chain continues to grow until termination by the addition of hydrogen. After exiting the FT reactor, water is separated from the hydrocarbon and lighter materials are recovered by fractionation. Finally, the long-chain FT wax is converted to base stock by catalytic isomerization, the same process that was covered in the previous section on catalytic dewaxing.

VI. BASE STOCK COMPOSITION

This chapter highlighted developments and current trends in base stock technology. Although a great deal of effort has been made to group base stocks into several categories, it is important to note that even the highest quality mineral-based stocks still occupy a fairly broad compositional space. Consider the base stock compositional triangle in Figure 2.17. Each corner is identified with one of

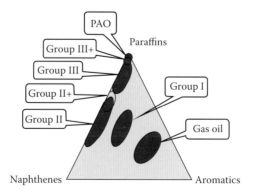

FIGURE 2.17 Base stock compositional triangle.

the three significant hydrocarbon types found in base oil: paraffins, naphthenes, and aromatics. The concentration at each corner is 100% of the type labeled. Any point within the triangle represents a combination of the three materials that adds up to 100%. Typical hydrocarbon compositions associated with the various API groups are identified. As shown, there is considerable variability within the mineral oil base stocks. Experience has shown that oils within a group do not always behave as expected. Ultimately, thorough testing of new lubricant formulations is necessary to demonstrate satisfactory product performance with any new base stock.

BIBLIOGRAPHY

API 1509. Engine Oil Licensing and Certification System, 17th Edition, Addendum 1. American Petroleum Institute, Washington, D.C., 2014, www.api.org.
ASTM D5293-14, Standard Test Method for Apparent Viscosity of Engine Oils and Base Stocks Between –5°C and –35°C Using Cold-Cranking Simulator, ASTM International, West Conshohocken, PA, 2014, www.astm.org.
ASTM D5800-15, Standard Test Method for Evaporation Loss of Lubricating Oils by the Noack Method, ASTM International, West Conshohocken, PA, 2015, www.astm.org.
ATIEL Code of Practice Issue 19. ATIEL, Brussels, 2013, www.atiel.org.

3 Lubricating Oils

I. ADDITIVES

The preceding chapter discussed refining processes and base stock manufacturing. The lube oil base stock is the building block to which appropriate additives are selected and properly blended to achieve a delicate balance in the performance characteristics of the finished lubricant. The various base stock manufacturing processes can all produce base stocks with the necessary characteristics to formulate finished lubricants with the desirable performance levels. The key to achieving the highest levels of performance in finished lubricants is in the understanding of the interactions of base stocks and additives and matching those to the requirements of machinery and operating conditions that they will be subjected to.

Additives are chemical compounds added to lubricating oils to impart specific properties to the finished oils. Some additives impart new and useful properties to the lubricant, some enhance properties already present, whereas some act to reduce the rate at which undesirable changes take place in the product during its service life.

By improving the performance characteristics of lubricating oils, additives have aided in the development of higher technology machinery. Modern passenger car engines, automatic transmissions, hypoid gears, railroad and marine diesel engines, high-speed gas and steam turbines, and industrial processing machinery, as well as many other types of equipment would have been greatly lagging in their development were it not for lubricant additives and the performance benefits they provide.

Additives for lubricating oils were used first during the 1920s, and their use has continued to increase. Today, practically all types of lubricating oils contain at least one additive, and some oils contain many different types of additives. The amount of additives used in a finished lubricant varies from a few hundredths of a percent to 30% or more.

Besides their many beneficial effects, additives can have some detrimental side effects especially if the dosage is excessive or if undesirable interactions with other additives occur. It is the responsibility of the oil formulator to achieve a balance of additives for optimum performance and to ensure this by testing that this combination does not exhibit any unwanted side effects.

The more commonly used lubricant additives are discussed in the following sections. Although some additives are multifunctional, as in the case of certain viscosity index (VI) improvers that also function as pour point, these are discussed in terms of their primary function.

A. Pour Point Depressants

Certain high molecular weight polymers function by inhibiting the formation of a wax crystal structure that tends to impede oil flow at low temperatures. Two general types of pour point depressants are used:

1. Alkylaromatic polymers adsorb on the wax crystals as they form, preventing them from growing and adhering to each other.
2. Polymethacrylates cocrystallize with wax to prevent crystal growth.

The additives do not entirely prevent wax crystal growth, but rather lower the temperature at which a rigid structure is formed. Depending on the type of oil, pour point depression of up to 50°F

(28°C) can be achieved by these additives, although a lowering of the pour point by approximately 20–30°F (11–17°C) is more common.

B. VI Improvers

VI improvers are long-chain, high molecular weight polymers that function by increasing the relative viscosity of an oil more at high temperatures than they do at low temperatures. Generally, this results from the polymer changing its physical configuration with increasing temperature of the mixture. It is postulated that in cold oil the molecules of the polymer adopt a coiled form so that their effect on viscosity is minimized. In hot oil, the molecules tend to straighten out, and the interaction between these long molecules and the oil produces a proportionally greater thickening effect. (Note that although the oil polymer mixture still decreases in viscosity as the temperature increases, it does not decrease as much as the oil would alone.)

Among the principal VI improvers are methacrylate polymers and copolymers, acrylate polymers, olefin polymers and copolymers, and styrene–butadiene/styrene–isoprene copolymers. The degree of VI improvement from these materials is a function of the molecular weight and the molecular weight distribution of the polymer.

In general, the higher the polymer molecular weight, the better the oil thickening power at a fixed polymer treat concentration (e.g., 1 wt.%). Therefore, it is highly desirable to use high molecular weight polymers as VI improvers. On the other hand, the long molecules in VI improvers are subject to degradation due to mechanical shearing in service; therefore, the molecular weight may reach a certain limit before the polymer can be used in a reasonably shear stable service condition. Molecular architecture also plays a role in shear stability as linear polymers are more prone to significant viscosity shear loss than star- and comb-shaped polymers. Shear breakdown occurs by two mechanisms. Temporary shear breakdown occurs under certain conditions of moderate shear stress and results in a temporary loss of viscosity. Apparently, under these conditions the long molecules of the VI improver align themselves in the direction of the stress, so there is less resistance to flow. When the stress is removed, the molecules return to their usual random arrangement and the temporary viscosity loss is recovered. This effect can be beneficial in that it can temporarily reduce oil friction to permit easier starting, as when cranking a cold engine. Permanent shear breakdown occurs when the shear stresses actually rupture the long molecules, converting them into lower molecular weight materials that become less effective VI improvers. This results in a permanent viscosity loss, which can be significant. It is generally the limiting factor controlling the maximum amount of VI improver that can be used in a particular oil blend.

VI improvers are typically used in multiviscosity engine oils, automatic transmission fluids (ATFs), multipurpose tractor fluids, automotive gear lubricants, and some hydraulic fluids. Their use permits the lubricant formulations that provide satisfactory lubrication over a much wider temperature range than is possible with straight mineral oils alone, and they are the foundation of multigrade lubricants.

C. Defoamants

The ability of mineral oils to resist foaming varies considerably depending on the type of crude oil selected, the type of base stock processing and degree of refining, and viscosity. In many applications, the oil is agitated enough to cause foaming. In some applications, even small amounts of foam can be extremely troublesome. In these cases, a defoamant additive with the ability to lower the droplet surface tension may be added to the oil.

Silicone polymers used at a few parts per million are the most widely used defoamants for mineral oils. These materials are essentially insoluble in oil, and the correct choice of polymer size and blending procedures are critical to avoid settling during long-term storage. Under certain conditions, these defoamants may increase air entrainment in the oil. Organic polymers are sometimes used to overcome this difficulty with the silicone defoamants, although much higher concentrations are generally required.

The theory on how defoamants work is that the defoamant droplets attach themselves to the air bubbles. They then can either spread or form unstable bridges between bubbles that then coalesce into larger bubbles, which rise more readily to the surface of the foam layer, where they collapse and release the air.

D. Oxidation Inhibitors

When oil is heated in the presence of air, oxidation occurs. As a result of this oxidation, the oil viscosity and concentration of organic acids in the oil both increase, and varnish and lacquer deposits may form on hot metal surfaces exposed to the oil. In extreme cases, these deposits may be further oxidized to form hard, carbonaceous materials.

The rate at which oxidation proceeds is affected by several factors. As the temperature increases, the rate of oxidation increases exponentially. A general guide is that for each 10°C (18°F) rise in temperature, the oxidation rate will double. Greater exposure to air (and the oxygen it contains), or more intimate mixing with it, will also increase the rate of oxidation. Many materials, such as metals (particularly copper and iron) and organic and mineral acids, may act as catalysts or oxidation promoters.

Although the complete mechanism of oil oxidation is not well defined, it is generally recognized as proceeding by free radical chain reaction. Reaction chain initiators are formed first from unstable oil molecules, and these react with oxygen to form peroxy radicals, which in turn attack the unoxidized oil to form new initiators and hydroperoxides. The hydroperoxides are unstable and divide, forming new initiators to expand the reaction. Any materials that will interrupt this chain reaction will inhibit oxidation. Two general types of oxidation inhibitors are used: those commonly called radical chain terminators that react with the initiators (peroxy radicals and hydroperoxides) to form inactive compounds; and those typically called peroxide decomposers that decompose these materials to form less reactive compounds.

At temperatures below 200°F (93°C), oxidation proceeds slowly and inhibitors of the first type are effective. Examples of this type are hindered (alkylated) phenols such as 2,6-ditertiary-butyl-4-methylphenol (also called 2,6-ditertiary-butylparacresol [DBPC]), and aromatic amines such as N-phenyl-alpha-naphthylamine. These are used in products such as turbine, circulation, and hydraulic oils, which are intended for extended service at moderate temperatures.

When the operating temperature exceeds about 200°F (93°C), the catalytic effects of metals become important factors in promoting oil oxidation. Under these conditions, inhibitors that reduce the catalytic effect of the metals must be used. These materials usually react with the surfaces of the metals to form protective coatings and, for this reason, are sometimes called metal deactivators. Typical of this type of additives are the dithiophosphates, primarily zinc dithiophosphate. The dithiophosphates also act to decompose hydroperoxides at temperatures above 200°F (93°C), so they inhibit oxidation by this mechanism as well.

Oxidation inhibitors may not entirely prevent oil oxidation when conditions of exposure are severe, and some types of oils are inhibited to a much greater degree than others. Oxidation inhibitors are not, therefore, cure-alls, and the formulation of a stable oil requires a suitable base stock combined with careful selection of the type and concentration of oxidation inhibitor. Some other additives can reduce the oxidation stability of an oil while performing their intended functions. Proper lubricant formulation requires the balancing of all of the additive reactions to achieve the desired total performance characteristics.

E. Rust and Corrosion Inhibitors

Various kinds of corrosion can occur in systems supplied with lubricating oils. Probably the two most important types are corrosion from organic acids that develop in the oil itself and corrosion from contaminants that are picked up and carried in the oil (Figure 3.1).

One of the areas where corrosion by organic acids can occur is the bearing inserts used in internal combustion engines. Some of the metals used in these bearing inserts, such as corrosion inhibitor additives used in the oil, form a protective film on the bearing surfaces that prevents the corrosive materials from reaching or attacking the metal (Figure 3.2). The film may be either adsorbed on the metal or chemically bonded to it.

During the combustion process in gasoline or diesel engines, certain materials in the fuel, such as sulfur and antiknock scavengers, can burn to form strong acids. These acids can then condense on the cylinder walls and be carried to other parts of the engine by the lubricant. Corrosive wear of piston rings, cylinder walls, bearings, and other engine components can then occur.

It has been found that the inclusion of highly alkaline materials in the oil will help to neutralize these strong acids as they are formed, greatly reducing this corrosion and corrosive wear. These alkaline materials are also used to provide detergency. See the detailed discussion in Section I.F ("Detergents and Dispersants").

Rust inhibitors are usually compounds having a high polar attraction toward metal surfaces. By physical or chemical interaction at the metal surface, they form a tenacious, continuous film that prevents water from reaching the metal surface. Typical materials used for this purpose are amine succinates and alkaline earth sulfonates. The effectiveness of properly selected rust inhibitors is illustrated in Figure 3.3.

Rust inhibitors can be used in most types of lubricating oils, but the selection must be made carefully in order to avoid problems such as corrosion of nonferrous metals or the formation of

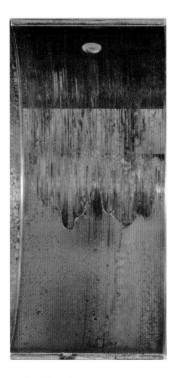

FIGURE 3.1 Heavily corroded copper–lead bearing.

Lubricating Oils

FIGURE 3.2 Satisfactorily protected copper–lead bearing.

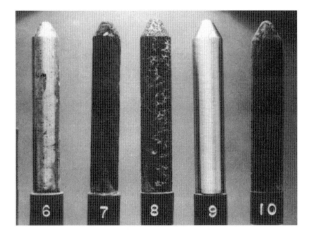

FIGURE 3.3 American Society for Testing and Materials (ASTM) rust test specimens.

troublesome emulsions with water. Because rust inhibitors are adsorbed on metal surfaces, an oil can be depleted of rust inhibitor over time.

F. Detergents and Dispersants

In internal combustion engines, a variety of effects tend to cause oil deterioration with the potential for the formation of harmful deposits. These deposits can interfere with oil circulation, can accumulate behind piston rings to cause ring sticking with rapid ring wear and increased oil consumption, and affect clearances and proper functioning of critical components such as hydraulic valve lifters. Once formed, such deposits are generally difficult to remove except by mechanical cleaning. The use of proper detergents and dispersant additives in the oil can prevent or delay the formation of deposits and reduce the rate at which they deposit on metal surfaces. An essential factor to reducing deposit formation in engines is the regular draining and replacement of the oil so that the contaminants are removed from the engine before they exceed the oil's capacity to hold them.

Detergents are chemical compounds that chemically neutralize deposit precursors that form either under high temperature conditions or as the result of burning fuels with a high sulfur content or other materials that form acidic combustion by-products. Dispersants, on the other hand, are chemical compounds that disperse or suspend potential sludge or varnish forming materials in the oil, particularly those formed during low-temperature operations when condensation and partially burned fuel find their way into the oil. These contaminants are removed from the system when the oil is drained. There is no sharp line of demarcation between detergents and dispersants. Detergents have some ability to disperse and suspend contaminants, whereas dispersants have some ability to prevent the formation of high-temperature deposits.

The principal detergents used today are organic soaps and salts of alkaline earth metals such as magnesium, calcium, and barium. These materials are often referred to as metallo-organic compounds. Barium, calcium, and magnesium sulfonates, calcium and magnesium salicylates, calcium and magnesium carboxylates, and barium and calcium phenates (or phenol sulfides) are widely used, and barium phosphonates are still used in some applications. The sulfonates, salicylates, carboxylates, and phenates may be neutral or overbased—that is, they contain more of the alkaline metal than is required to neutralize the acidic components used in diesel engine oils to neutralize the strong acids formed from combustion of the sulfur in the fuel. This neutralization reduces corrosion and corrosive wear and minimizes the tendency of these acids to cause oil degradation.

Overbased materials are generally used in lower concentrations in gasoline engine oils where fuel sulfur is generally lower. Higher concentrations of overbased detergents are used in lubricants for marine diesel engines that operate on high sulfur (e.g., 3.5% sulfur) residual fuels where cylinder lubricants can have a TBN of 100 or greater. The overbased materials are included in the formulation to help reduce corrosion in low-temperature operations. Both neutral and overbased materials also act to disperse and suspend potential varnish forming materials resulting from oil oxidation, preventing these materials from depositing on engine surfaces.

During combustion, metallo-organic detergents can leave an ashy residue often referred to as sulfated ash. In some cases, this may be detrimental in that the ash can contribute to combustion chamber deposits. In other cases, it may be beneficial in that the ash provides wear-resistant coatings for surfaces such as valve faces and seats.

Typical dispersants (also called polymeric dispersants and ashless dispersants) in use today are described as polymeric succinimides, succinic acid esters/amides, and their borated derivatives. Other dispersant additive chemistries such as olefin/phosphorus pentasulfide (P_2S_5) reaction products, polyesters, polymeric alkylphenol/formaldehyde/polyamine reaction products, and benzylamides have become less popular and are generally used in specific applications. These are based on long-chain hydrocarbons (typically, polyisobutylene), which are acidified via melation with maleic anhydride and then neutralized with a monoamine/diamine or polyamine compound containing basic nitrogen. The hydrocarbon portion provides oil solubility, whereas the nitrogen portion provides an active site that attracts and holds potential deposit-forming materials to keep them suspended in the oil.

Although the primary use of detergent and dispersant additives is for engine oils, they may also be used in some lubricants such as ATFs, hydraulic oils, and circulation oils for high-temperature service. In these applications, the detergents and dispersants help to prevent the deposition of lacquer and varnish resulting from oil oxidation, and in this way supplement the effects of the oxidation inhibitors.

G. Antiwear Additives

Antiwear additives are used in many lubricating oils to reduce friction, wear, and scuffing and scoring under boundary lubrication conditions, that is, when full lubricating films cannot be maintained. As the oil film becomes progressively thinner because of increasing loads or temperatures, contact through the oil film is first made by minute surface irregularities or asperities. As these opposing asperities make contact, friction increases and welding can occur. The welds break immediately as sliding continues, but this can form new roughness through metal transfer, and also form wear particles that can cause scuffing and scoring. Two general classes of materials are used to prevent metallic contact, depending on the severity of the requirements.

Mild antiwear and friction-reducing additives, sometimes called boundary lubrication additives, are polar materials such as fatty oils, acids, and esters. They are long-chain materials that form an adsorbed film on the metal surfaces with the polar ends of the molecules attached to the metal and the molecules projecting more or less normal to the surface. Contact is then between the projecting ends of the layers of molecules on the opposing surfaces. Friction is reduced, and the surfaces move more freely relative to each other. Wear is reduced under mild sliding conditions, but under severe sliding conditions the layers of molecules can be rubbed off so that their wear-reducing effect is lost.

Some of the more effective antiwear additives have been zinc-containing additives or ashless phosphorus-based additives. Zinc dialkyl dithiophosphate (ZDDP) has been successfully used as an antiwear additive for decades in a multitude of applications including engine oils, hydraulic oils, and circulation oils. One ashless phosphorus type of antiwear additive that has been used is tricresyl phosphate, which has been used in transmission fluids, turbine oils, and hydraulic oils. Other ashless antiwear additives are organic compounds containing sulfur and phosphorus.

As vehicle emission regulations become more challenging, increasing restrictions are likely to be placed on other engine oil additives besides phosphorus, which was one of the first identified to have an impact on emission control systems. Sulfur and metals are also under scrutiny as sulfur is suspected as a poison of DeNOx catalysts, and ash (from metals) may plug after-treatment particulate traps. Although modern engine oils still rely heavily on ZDDP to provide antiwear, antioxidation, and anticorrosion protection, ZDDP is an obvious target for emission control because it contains phosphorus, sulfur, and zinc. The future longer trend will likely be toward a continuous reduction or replacement of ZDDP in engine oils providing that the performance integrity can be maintained through the use of alternate additives.

H. Extreme Pressure Additives

At high temperatures or under heavy loads where more severe sliding conditions exist, compounds called extreme pressure (EP) additives are required to reduce friction, control wear, and prevent severe surface damage. They are also referred to as antiscuffing additives. These materials function by chemically reacting with the sliding metal surfaces to form relatively oil insoluble surface films. The kinetics of the reaction are a function of the surface temperatures generated by the localized high temperatures that result from rubbing between opposing surface asperities, and breaking of junctions between these asperities.

Even with EP additives in the lubricant, wear of new surfaces may be high initially. In addition to normal break-in wear, nascent metal (freshly formed, chemically reactive surfaces), time, and temperature are required to form the protective surface films. After the films are formed, relative motion is between the layers of surface films rather than the metals. The sliding process can lead to some film removal, but replacement by further chemical reaction is rapid so that the loss of metal is extremely low. This process gradually depletes the amount of EP additive available in the oil although the rate of depletion is usually very slow. Thus, there will be sufficient additive left to provide adequate protection for the metal surfaces except possibly under severe operating conditions where oil makeup rates are extremely low and normal oil drain intervals are exceeded.

The severity of the sliding condition dictates the reactivity of the EP additives required for maximum effectiveness. The optimum reactivity occurs when the additives minimize the adhesive or metallic wear without leading to appreciable corrosive or chemical wear. Additives that are too reactive lead to the formation of excessively thick surface films, which have less resistance to attrition, so some metal is lost by the sliding action. As a particular EP additive may have different reactivity with different metals, it is important to match additive metal reactivity to the additives not only with the severity of the sliding system but also with the specific metals involved. For example, some additives that are excellent for steel-on-steel systems may not be satisfactory for bronze-on-steel systems operating at similar sliding severity because they are too reactive with the bronze.

Another important function of EP additives is that they contribute to the polishing of the sliding surfaces because the chemical reaction is greatest at the asperities where contact is made and localized temperatures are highest. The load is then distributed more uniformly over a greater contact area, which allows for a reduction in sliding severity, more effective lubrication, and a reduction in wear.

EP agents are usually compounds containing sulfur, phosphorus, chlorine, borate, or metals, either alone or in combination. The EP compounds used today depend on the end use application and the chemical activity that is required. Sulfur compounds, sometimes used in combination with chlorine or phosphorus, are used in many metal cutting fluids. Sulfur and phosphorus combinations are used in most industrial and automotive gear lubricants. These materials provide excellent protection against gear tooth scuffing and have the advantages of better oxidation stability, lower corrosivity, and often lower friction than other combinations that have been used in the past. The use of metallic EP additives is diminishing because of the influence of environmental concerns. Heavy metals are considered pollutants, and their presence is no longer welcomed in the environment. As an example, several decades ago lead was successfully used as an EP additive for gear oils but was banned because of environmental, health, and safety concerns. Based on performance needs and cost, ashless EP additives (dithiocarbamates, dithiophosphates, thiolesters, phosphorothioates, thiadiazoles, benzothiazoles, amine phosphates, phosphites, phosphates, etc.) may be preferred for some lubricant applications.

II. PHYSICAL AND CHEMICAL CHARACTERISTICS

There are a multitude of physical and chemical tests that yield useful information on the characteristics of lubricating oils. However, the quality or the performance features of a lubricant cannot be adequately described on the basis of physical and chemical tests alone. As a result many entities such as the military, equipment builders, industry associations, and some commercial consumers include performance tests as well as physical and chemical tests in their lubricant specifications. Physical and chemical tests are of considerable value in maintaining uniformity of products during manufacture. Moreover, some of these tests may be applied to used oils to determine the changes that have occurred in service, and to indicate possible causes for those changes.

Some of the most commonly used tests for physical or chemical properties of lubricating oils are outlined in the following sections, with brief explanations of the significance of the tests. For detailed information on test methods, publications from the various testing societies or standards organizations should be consulted.

A. Carbon Residue

The carbon residue of a lubricating oil is the amount of deposit, in percentage by weight, left after evaporation and pyrolysis of the oil under prescribed conditions. Oils of naphthenic type usually show lower residues than those of similar viscosity made from paraffinic crudes. The more severe the refining process for an oil—whether by solvent extraction, hydroprocessing, filtration, or acid treatment—the lower the carbon residue value will typically be. Those oils containing residual materials found in base stocks such as bright stocks will typically also have higher residue values. Although many finished lubricating oils now contain additives that may contribute significantly to the amount of residue in the test, their effect on performance is distinctly beneficial.

Originally, the carbon residue test was developed to determine the carbon-forming tendency of steam cylinder oils. In the years that followed, unsuccessful attempts were made to relate carbon residue values to the amount of carbon formed in the combustion chambers and on the pistons of internal combustion engines. The carbon residue values alone have only limited significance because the properties of a lubricating oil are of equal or greater importance to other factors such as fuel composition, engine operation, and the mechanical condition of the engine. The carbon residue determination is now made mainly on base oils used for engine oil manufacture; straight mineral engine oils such as aircraft engine oils; heat transfer oils and some of the cylinder oil type products used for reciprocating air compressors. In these cases, the determination is an indication of the degree of refining to which the base stock has been subjected.

B. Color

The color of lubricating oils, as observed by light transmitted through them, varies from practically clear or transparent to opaque or black. Usually, the various methods of measuring color are based on a visual comparison of the amount of light transmitted through a specified depth of oil with the amount of light transmitted through one of a series of colored glasses. The color is then given as a number corresponding to the number of the colored glass.

Color variations in lubricating oils result from differences in crude oils, viscosity, method and degree of treatment during refining, and in the amount and nature of additives included in them. During processing, color is a useful guide to the refiner to indicate if processes are operating properly. In finished lubricants, color has little significance except in the case of medicinal and industrial white oils, which are often compounded into, or applied to, products where staining or discoloration would be undesirable. Although color changes in used lube oils should not be used as a condemning criteria, a color change accompanied by other physical changes such as odor or oxidation would possibly signal a need for action. Dyes are sometimes used in oils to differentiate the color versus other types of lubricants (e.g., automatic transmission fluids).

C. Density and Gravity

The density of a substance is its mass per unit volume. The specific gravity (relative density) is the ratio of the mass of a given volume of a material at a standard temperature to the mass of an equal volume of water at the same temperature. American Petroleum Institute (API) gravity is a special function of specific gravity that is related to it by the following equation:

$$\text{Gravity API} = (141.5/\text{specific gravity at } 60°F) - 131.5$$

The API gravity value, therefore, increases as the specific gravity decreases. Because both density and gravity change with temperature, determinations are made at a controlled temperature and then corrected to a standard temperature by use of special tables.

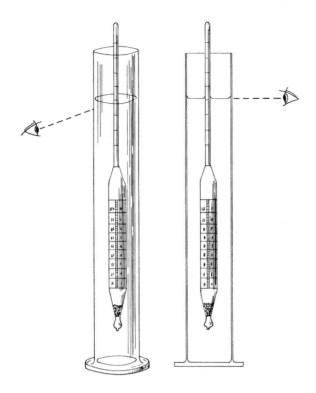

FIGURE 3.4 Use of hydrometer to determine density and gravity.

Density and gravity can be determined by means of hydrometers (see Figure 3.4). The hydrometer can be calibrated to read any of the three properties—density, gravity, or API gravity.

Gravity determinations are quickly and easily made. Because products of a given crude oil—having definite boiling ranges and viscosities—will fall into definite ranges, this property is widely used for control in refinery operations. It is also useful for identifying oils, provided that the distillation range or viscosity of the oils is known. Its primary use, however, is to convert weighed quantities to volume and measured volumes to weight.

In testing used oils, particularly used engine oils, a decrease in specific gravity (increase in API gravity) may indicate fuel dilution, whereas an increase in specific gravity might indicate the presence of contaminants such as fuel soot or oxidized materials. Additional test information is necessary to fully explain changes in gravity because some effects tend to cancel out others. For hydrocarbon-based materials, API gravity can also be used to determine the heating value of the material (BTUs/gal). Tables are readily available to read these values directly once the API gravity is determined.

D. Flash and Fire Points

The flash point of an oil is the temperature at which the oil releases enough vapor at its surface to ignite when an open flame is applied. For example, if a lubricating oil is heated in an open container, ignitable vapors are released in increasing quantities as the temperature rises. When the concentration of vapors at the surface becomes great enough, exposure to an open flame will result in a brief flash as the vapors ignite. When a test of this type is conducted under certain specified conditions, as in the Cleveland Open Cup method (Figure 3.5), the bulk oil temperature at which this happens is reported as the flash point. The release of vapors at this temperature is not sufficiently rapid to sustain combustion, so the flame immediately dies out. However, if heating is continued, a temperature

Lubricating Oils

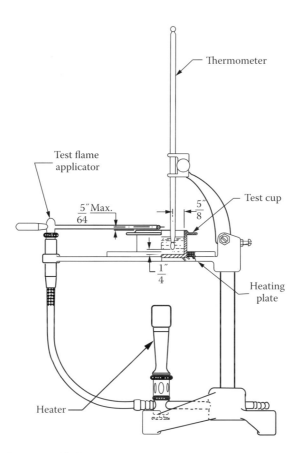

FIGURE 3.5 Cleveland open cup flash tester.

will be reached at which vapors are released rapidly enough to support combustion. This temperature is called the fire point. For any specific product, both flash and fire points will vary depending on the apparatus and the heating rate. Temperatures are raised in 5°F increments for this test.

The flash point of new oils varies with viscosity—higher viscosity oils have higher flash points. For mineral oils, flash points are also affected by the type of crude and refining process. For example, naphthenic oils generally have lower flash points than paraffinic oils of similar viscosity.

Flash and fire point tests are of value to refiners for control purposes and are significant to consumers under certain circumstances for safety considerations. Also, in certain high-temperature applications, use of an oil with a low flash point, indicating higher volatility, may result in higher oil consumption rates. Flash and fire points are of little value in determining whether fire-resistant fluids are safe near possible points of ignition.

During the inspection of a used oil, a significant reduction in the flash point usually indicates contamination with lower flash material such as diesel fuel with a diesel engine oil. An exception to this occurs when certain products, such as heat transfer oils, are used for long periods at high temperatures and undergo thermal cracking with the formation of lighter hydrocarbons and a reduction in the original flash point.

E. Neutralization Number

Tests were developed to provide a quick determination of the amount of acid in an oil by neutralizing it with a base. The amount of acid in the oil was expressed in terms of the amount of a standard base required to neutralize a specified volume of oil. This quantity of base came to be called the neutralization number (NN) of the oil.

Some of the products of oxidation of petroleum hydrocarbons are organic acids. As an oil oxidizes in service, caused by exposure to elevated temperatures, there is a tendency for it to become more acidic. Measuring the acidity of an oil in service was one way of following the progress of oxidation. This technique is still used with a number of products, primarily those intended for extended service, such as steam turbine oils and electrical insulating oils.

Many of the additives now used to improve performance properties may make an oil acidic or basic depending on their composition. In other cases, the additives may contain weak acids and weak bases, which do not react with each other in the oil solution, but react with both the strong acid and the strong base used in the NN tests to give both acidic and basic NNs. There are also other additives that undergo exchange reactions with the base used for neutralization to give false acid values in the test. These effects of the additives tend to obscure any changes in acidity occurring in the oil itself, so NN determinations may have little significance for some oils containing additives.

F. TOTAL ACID NUMBER

The total acid number (TAN) of an oil is synonymous to NN. The TAN of an oil is the weight (in milligrams) of potassium hydroxide required to neutralize 1 g of oil and is a measure of all the materials in an oil that will react with potassium hydroxide under specified test conditions. The usual major components of such materials are organic acids, soaps of heavy metals, intermediate and advanced oxidation products, organic nitrates, nitro compounds, and other compounds that may be present as additives. It is worth mentioning that new and used oil can exhibit both TAN and TBN (total base number) values.

Organic acids may form as a result of progressive oxidation of the oil, and the heavy metal soaps result from the reaction of these acids with metals. Mineral acids (i.e., strong inorganic acids), if present in an oil sample, would be neutralized by potassium hydroxide and would, therefore, affect the TAN determination. However, such acids are seldom present except in internal combustion engines using high sulfur fuels or in cases of contamination.

Because a variety of degradation products contribute to the TAN value, and the organic acids present can vary widely in corrosive properties, the test cannot be used to predict the corrosiveness of an oil.

G. TBN

The TBN of an oil is the quantity of acid, expressed in terms of the equivalent number of milligrams of potassium hydroxide, required to neutralize all basic constituents present in 1 g of oil. This test is normally used with oils that contain alkaline, acid-neutralizing additives. The rate of consumption of these alkaline materials (TBN depletion) can be an indication of the projected serviceable life of the oil. With used oils, it indicates how much acid-neutralizing additive remains in the oil. Typical oils of this nature would be diesel engine oils or other internal combustion engine oils where acid-producing constituents such as sulfur or chlorine exist in the fuel. As long as any significant amount of TBN remains in the oil, there should not be any strong acids present. However, the nature of high alkaline and metallic antioxidant additives sometimes allow for both TAN and TBN values to be obtained on the same sample. This occurs for both new and used oils.

H. POUR POINT

The pour point of a lubricating oil is the lowest temperature at which it will pour or flow when it is chilled and undisturbed under prescribed conditions. Most mineral oils contain some dissolved wax that begins to separate when an oil is chilled. These wax crystals will interlock to form a

Lubricating Oils

rigid structure that traps the oil into small pockets in the structure. When this wax crystal structure becomes sufficiently complete, the oil will no longer flow under the conditions of the test. Mechanical agitation can break up the wax structure so that it is possible to have an oil flow at temperatures considerably below its pour point. Cooling rates also affect wax crystallization; it is possible to cool an oil rapidly to a temperature below its pour point and still have it flow.

Although the pour point of most oils is related to the crystallization of wax, certain oils, such as some synthetics that are essentially wax-free, have viscosity-limited pour points. In these oils, the viscosity becomes progressively higher as the temperature is lowered, until at some temperature no flow can be observed. The pour points of such oils cannot be lowered with pour point depressants, because these agents act by interfering with the growth and interlocking of the wax crystal structure.

Untreated lubricating oils show wide variations in pour points. Distillates from waxy paraffinic or mixed base crudes typically have pour points in the range of 80°F to 120°F (27°C to 49°C), whereas raw distillates from naphthenic crudes may have pour points on the order of 0°F (−18°C) or lower. After solvent dewaxing, the paraffin distillates will have pour points on the order of 20°F to 0°F (−7°C to −18°C). Where lower pour points oils are required, pour point depressant additives are normally used. Some of the more common pour point depressants include polyalkyl methacrylates, styrene maleic copolymers, alkylaromatic polymers, and alkylated napthalenes.

The importance of the pour point of an oil is almost entirely dependent on its intended use. For example, the pour point of a winter-grade engine oil must be low enough so that the oil can be dispensed readily, and will flow to the pump suction in the engine at the lowest anticipated ambient temperatures. On the other hand, there is no particular need for low pour points for oils to be used inside heated plants or in continuous service such as steam turbines, or for many other applications.

I. Sulfated Ash

The sulfated ash of an oil is the residue, in percent by weight, remaining after burning the oil, treating the initial residue with sulfuric acid, and burning the treated residue. It is a measure of the noncombustible constituents (usually metallic materials) contained in the oil.

New, straight mineral lubricating oils contain essentially no ash-forming materials. Many additives in oil, such as detergents that contain metallo-organic components, will form a residue in the sulfated ash test. During lubricant blending, this test can provide a simple method to check that the additives have been incorporated in approximately the correct amounts. Additional testing is usually necessary to determine if the various metallic elements are in the oil in the correct proportions because the sulfated ash test combines all metallic elements into a single residue.

Several original engine manufacturers (OEMs) have included a maximum limit on sulfated ash content in their engine oil specifications. The sulfated ash specification exists owing to the concern that excessive quantities of some of these materials may contribute to such problems as combustion chamber deposits, top ring wear, or port deposits particularly in two-stroke engines.

When analyzing in-service engine oils, an increase in sulfated ash content usually indicates a buildup of contaminants such as dust and dirt, wear debris, or use of a higher ash oil for makeup.

J. Viscosity

Probably the most important single property of a lubricating oil is its viscosity. It is a key factor in the formation of lubricating films under both thick and thin film conditions; it affects heat generation in bearings, cylinders, and gears; it governs the sealing effect of the oil and the rate of consumption or loss; and it determines the ease with which machines may be started under cold conditions. The first step toward having satisfactory equipment performance is to use an oil of the proper viscosity for the expected operating conditions.

The basic concept of viscosity is shown in Figure 3.6, where a plate is shown being drawn at a uniform speed over a film of oil. The oil adheres to both the moving surface and the stationary

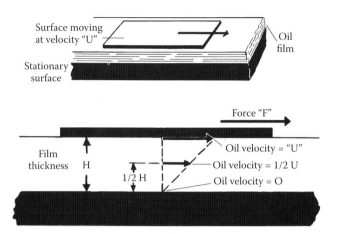

FIGURE 3.6 Concept of dynamic viscosity.

surface. Oil that is in contact with the moving surface travels with the same velocity (U) as that surface, whereas oil that is in contact with the stationary surface is at zero velocity. In between, the oil film may be visualized as being made up of many layers, each being drawn by the layer above it at a fraction of velocity U that is proportional to its distance above the stationary plate (see Figure 3.6, lower view). A force (F) must be applied to the moving plate to overcome the friction between the fluid layers. Because this friction is the result of viscosity, the force is proportional to viscosity. Viscosity can be determined by measuring the force required to overcome fluid friction in a film of known dimensions. Viscosity determined in this manner is called dynamic or absolute viscosity.

Dynamic viscosities are usually reported in poise (P) or centipoise (cP; 1 cP = 0.01P), or in SI units in Pascal seconds (Pa s; 1 Pa s = 10 P). Dynamic viscosity, which is a function only of the internal friction of a fluid, is the quantity used most frequently in bearing design and oil flow calculations. Because it is more convenient to measure viscosity in a manner such that the measurement is affected by the density of the oil, kinematic viscosities normally are used to characterize lubricants.

The kinematic viscosity of a fluid is the quotient of its dynamic viscosity divided by its density, both measured at the same temperature and in consistent units. The most common units for reporting kinematic viscosities now are the Stokes (St) or centistokes (cSt; 1 cSt = 0.01 St), or in SI units, square millimeters per second (mm^2/s; 1 mm^2/s = 1 cSt). Dynamic viscosities, in centipoise, can be converted to kinematic viscosities, in centistokes, by dividing by the density in grams per cubic centimeter (g/cm^3) at the same temperature. Kinematic viscosities, in centistokes, can be converted to dynamic viscosities, in centipoise, by multiplying by the density in grams per cubic centimeter. Kinematic viscosities, in square millimeters per second (mm^2/s), can be converted to dynamic viscosities, in Pascal seconds, by multiplying by the density in grams per cubic centimeter, and dividing the result by 1000.

Other viscosity systems, including the Saybolt, Redwood, and Engler, are in limited use today for lubricating oils. Most actual viscosity determinations are made in centistokes and converted to values in the other viscosity systems by means of published SI conversion tables.

The viscosity of any fluid changes with temperature—increasing as the temperature is decreased, and decreasing as the temperature is increased. Thus, it is necessary to have some method of determining the viscosities of lubricating oils at temperatures other than those at which they are measured. This is usually accomplished by measuring the viscosity at two temperatures, then plotting these points on special viscosity–temperature charts developed by the American Society for Testing and Materials (ASTM). A straight line can then be drawn through the points and viscosities at other temperatures read from it with reasonable accuracy (Figure 3.7). The line should not be extended

Lubricating Oils

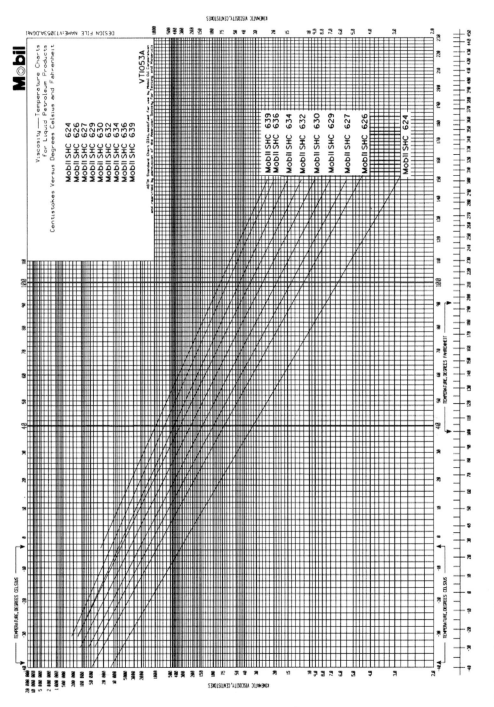

FIGURE 3.7 Example viscosity–temperature chart.

below the pour point or above approximately 150°C (300°F) for most lubricating oils, because viscosity may no longer be linear in these regions.

The two temperatures most often used for reporting kinematic viscosities are 40°C (104°F) and 100°C (212°F).

In selecting the proper oil for a given application, viscosity is a primary consideration. It must be high enough to provide proper lubricating films but not so high that friction losses in the oil will be excessive. Because viscosity varies with temperature, it is necessary to consider the actual operating temperature of the oil in the machine. Other considerations, such as whether a machine must be started at low ambient temperatures, must also be taken into account.

Three viscosity numbering systems are in use to identify oils according to viscosity ranges. Two of these are for automotive lubricants and one is for industrial oils.

1. Engine Oil Viscosity Classification

The Society of Automotive Engineers (SAE) J300 standard classifies oils for use in automotive engines by viscosities determined at low shear rates and high temperature (100°C), high shear rate at high temperature (150°C), and at both low and high shear rates at low temperature (−5°C to −40°C). The ranges for grades in this classification are shown in Table 3.1.

In these classifications, grades with the suffix letter W are intended primarily for use where low ambient temperatures will be encountered, whereas grades without the suffix letter are intended for use where low ambient temperatures will not be encountered. Multigrade oils are used for year-round service in automotive engines. An example of a multigrade oil is as follows: oils can be formulated that will meet the low temperature limits of the 5W grade and the 100°C limits for the 30 grade. It can then be designated an SAE 5W-30 grade and is referred to as a multigrade or multi-viscosity oil. Oils of this type generally require the use of VI improvers in conjunction with mineral or synthetic base stocks.

High temperature/high shear (HT/HS) viscosities are measured at 150°C by ASTM D4683, D4741, D5481, or Coordinating European Council (CEC) L-36-90. The HT/HS is measured under shear stresses similar to those experienced in engine bearings. The HT/HS value indicates the temporary shear stability of the VI improvers used to formulate multigrade engine oils. The SAE J300 classification shows that an SAE 30 grade oil has to have an HT/HS rate viscosity of ≥2.9 cP at 150°C or it will not perform as an SAE 30 grade oil under engine operating conditions.

At low temperature, the high shear rate viscosity is measured by ASTM D5293, a multitemperature cold cranking simulator method. The low temperature, low shear rate viscosity is measured by ASTM D4684. Results of both tests have been shown to correlate with engine starting and engine oil pumpability at low temperatures.

This SAE system is widely used by engine manufacturers to determine suitable oil viscosities for their engines, and by oil marketers to indicate the viscosities of internal combustion engine oils.

2. Axle and Manual Transmission Lubricant Viscosity Classification

SAE Recommended Practice J306 classifies lubricants for use in automotive manual transmissions and drive axles by viscosity measured at 100°C (212°F), and by maximum temperature at which they reach a viscosity of 150,000 cP (150 Pa s) when cooled and measured in accordance with ASTM D2983, Method of Test for Apparent Viscosity at Low Temperature using the Brookfield Viscometer.

The limits for the automotive gear lubricant oil classification are given in Table 3.2. Multigrade oils, such as 80W-90 or 75W-140, can be formulated under this system. This limiting viscosity of 150,000 cP was selected on the basis of test data indicating that lubrication failures of pinion bearings of a specific axle design could be experienced when the lubricant viscosity exceeded this value. Because other axle designs, as well as transmissions, may have higher or lower limiting viscosities,

TABLE 3.1
Engine Oil Viscosity Classification[a,b]—SAE J300 (January 2015)

SAE Viscosity Grade	Low-Temperature (°C) Cranking Viscosity,[c] cP Max	Low-Temperature (°C) Pumping[d] Viscosity (cP) Max with No Yield Stress	Low-Shear-Rate Kinematic Viscosity[e] (cSt) at 100°C Min	Low-Shear-Rate Kinematic Viscosity[e] (cSt) at 100°C Max	High-Shear-Rate Viscosity[f] (cP) at 150°C Max
0W	6200 at −35°C	60,000 at −40°C	3.8	—	—
5W	6600 at −30°C	60,000 at −35°C	3.8	—	—
10W	7000 at −25°C	60,000 at −30°C	4.1	—	—
15W	7000 at −20°C	60,000 at −25°C	5.6	—	—
20W	9500 at −15°C	60,000 at −20°C	5.6	—	—
25W	13,000 at 10°C	60,000 at −15°C	9.3	—	—
8	—	—	4.0	<6.1	1.7
12	—	—	5.0	<7.1	2.0
16	—	—	6.1	<8.2	2.3
20	—	—	6.9	<9.3	2.6
30	—	—	9.3	<12.5	2.9
40	—	—	12.5	<16.3	3.5 (0W-40, 5W-40, 10W-40) 3.7 (15W-40, 20W-40, 25W-40, 40 grades)
50	—	—	16.3	<21.9	3.7
60	—	—	21.9	<26.1	3.7

[a] 1 mPa·s = 1 cP; 1 mm²/s = 1 cSt.
[b] All values, with the exception of the low-temperature cracking viscosity, are critical specifications as defined by ASTM D3244.
[c] ASTM D5293: Cranking viscosity—the noncritical specification protocol in ASTM D3244 shall be applied with a *p* value of 0.95.
[d] ASTM D4684: Note that the presence of any yield stress detectable by this method constitutes a failure regardless of viscosity.
[e] ASTM D445.
[f] ASTM D4683, ASTM D4741, ASTM D5481, or CEC L-36-90.

TABLE 3.2
Automotive Gear Lubricant Viscosity Classification

SAE Viscosity Grade	Maximum Temperature for Viscosity of 150,000 cP, °C	Kinematic Viscosity at 100°C, cSt Minimum	Kinematic Viscosity at 100°C, cSt Maximum
70W	−55	4.1	–
75W	−40	4.1	–
80W	−26	7.0	–
85W	−12	11.0	–
80	–	7.0	<11.0
85	–	11.0	<13.5
90	–	13.5	<18.5
110	–	18.5	<24.0
140	–	24.0	<32.5
190	–	32.5	<41.0
250	–	41.0	–

Note: 1 cP = 1 mPa s; 1 cSt = 1 mm²/s.

it is the responsibility of the gear manufacturer to specify the actual grades that will provide satisfactory service under different ambient conditions.

3. **Viscosity System for Industrial Fluid Lubricants**

This system was jointly developed by the ASTM and the Society of Tribologists and Lubrication Engineers to establish a series of definite viscosity levels that could be used as a common basis

TABLE 3.3
Viscosity System for Industrial Fluid Lubricants

Viscosity System Grade Identification	Midpoint Viscosity (cSt [mm²/s]) at 40.0°C	Kinematic Viscosity Limits (cSt [mm²/s]) at 40.0°C	
		Minimum	Maximum
ISO VG 2	2.2	1.98	2.42
ISO VG 3	3.2	2.88	3.52
ISO VG 5	4.6	4.14	5.06
ISO VG 7	6.8	6.12	7.48
ISO VG 10	10	9.00	11.0
ISO VG 15	15	13.5	16.5
ISO VG 22	22	19.8	24.2
ISO VG 32	32	28.8	35.2
ISO VG 46	46	41.4	50.6
ISO VG 68	68	61.2	74.8
ISO VG 100	100	90.0	110
ISO VG 150	150	135	165
ISO VG 220	220	198	242
ISO VG 320	320	288	352
ISO VG 460	460	414	506
ISO VG 680	680	612	748
ISO VG 1000	1000	900	1100
ISO VG 1500	1500	1350	1650

Lubricating Oils

for specifying or selecting the viscosity of industrial fluid lubricants, and to eliminate unjustified intermediate grades. The system was originally based on viscosities measured at 100°F but was converted to viscosities measured at 40°C in the interests of international standardization. In this form, the system now appears as ASTM D2422 and International Organization for Standardization (ISO) Std. 3448. The ISO viscosity ranges and corresponding grade numbers are shown in Table 3.3. Table 3.4 is a general guide for comparing the various viscosity classification systems.

TABLE 3.4
General Guide for Comparing Viscosity Classifications

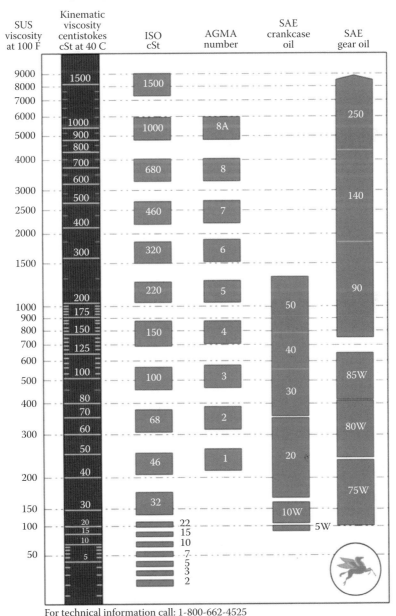

For technical information call: 1-800-662-4525

K. VI

Different oils have different rates of change of viscosity with temperature. For example, an oil derived from a naphthenic-based crude would show a greater rate of change in viscosity with temperature than would an oil derived from a paraffinic crude. VI is a method of applying a numerical value to this rate of change, based on comparison with the relative rates of change of two arbitrarily selected types of oils that differ widely in this characteristic. A high VI indicates a relatively low rate of change in viscosity with temperature; a low VI indicates a relatively high rate of change in viscosity with temperature. If a high VI oil and a low VI oil had the same viscosity at room temperature, the high VI oil would thin out less as the temperature increased and would have a higher VI than the low VI oil at higher temperatures.

The VI of an oil is calculated from viscosities determined at two temperatures by means of tables published by the ASTM. Tables based on viscosities determined at both centigrade (40°C and 100°C) and Fahrenheit (100°F and 212°F) are available.

Mineral oil base stocks made by conventional solvent refining methods range in VI from below 0 to slightly above 100. Mineral oil base stocks refined through special hydroprocessing techniques can have VIs well above 120. Some synthetic lubricating oils have VIs both below and well above this range. Additives called VI improvers can be blended into oils to increase the VI of an oil. Some of these VI improvers are not always stable in lubricating environments exposed to shear or thermal stresses, and the VI of the oil may decrease over time while in service. In many types of service, where the operating temperature remains more or less constant, VI is of little concern as long as the proper oil film thickness is maintained.

III. EVALUATION AND PERFORMANCE TESTS

Field testing under actual service conditions is usually considered the best way to evaluate the performance of a lubricant. In-service testing is not always practical during the early phases of lubricant development because of the time and costs involved. Therefore, shorter and less costly tests are used in the laboratory during development of both commercial and experimental lubricants. The correlation of results from these laboratory bench and rig tests with field performance continues to be a major emphasis of lubricant developers. Many tests have been proposed, but a large proportion of them end up being discarded for various reasons. Most of the tests that have survived have done so because they have shown the ability to predict with a fair degree of reliability some of the aspects of how a lubricant will perform in service.

Selecting lubricants based solely on meeting a set of laboratory bench tests will not always guarantee meeting the requirements of the intended application. Most standard laboratory test procedures were developed for evaluating new oil under laboratory conditions with some expectations of correlation to field application. It is extremely difficult to duplicate operating conditions and contamination for equipment that operates over a period of years in a laboratory test that is completed in minutes to months. This is the reason that extensive field testing of new products should be an integral part of new product development.

Because the process of test development and refinement of these tests is continuous, the following discussion treats the tests in a fairly general manner.

A. Oxidation Tests

Oxidation of lubricating oils depends on the temperature, amount of oxygen contacting the oil, and the catalytic effects of metals. If the oil's service conditions are known, these three variables can be adjusted to provide a test that closely represents actual service. However, oxidation in service is often an extremely slow process, so that using such a test may be time consuming. In order to shorten the test time, the test temperature is usually raised and catalysts added so that more rapid oxidation will result. Unfortunately, this tends to make the test a less reliable indication of expected field performance. As a result, very few oxidation tests have received wide acceptance for all applications, although a considerable number are used by specific laboratories that have developed satisfactory correlations for them.

One bulk oil oxidation test that is widely used is ASTM D943, Oxidation Characteristics of Inhibited Steam-Turbine Oils. This test is commonly known as Turbine Oil Stability Test (TOST). Although it is mainly intended for use on inhibited steam turbine oils, it has been used with mixed results for hydraulic and circulation oils, and for base oils used in the manufacture of turbine, hydraulic, and circulation oils. The test is operated at a moderate temperature (95°C/203°F). Iron and copper catalyst wires are immersed in the oil sample, to which a small amount of water is added. Oxygen is bubbled through the sample at a prescribed rate. The test is either run for a prescribed number of hours, after which the NN of the oil is determined, or until the NN reaches a value of 2.0. The result in the latter case is than reported as the hours to an NN of 2.0.

Objections to the regular use of ASTM D 943 are that extremely long test times (up to 10,000 h) are required for stable oils, and that the only criterion for acceptability is the NN. Severe sludging and deposits on the catalyst wires can occur with some oils without any excessive increase in the NN. A modification of the procedure, called Procedure B, which requires a determination of the sludge content, overcomes some of these latter objections.

Two additional bulk oil oxidation tests that have been used internationally are the Institute of Petroleum (IP) 280 procedure for turbine oils and the PNEUROP IP 48 procedure of the Comite European des Constructeurs de Compresseurs et d'Outillage for compressor oils (PNEUROP is the European Association of Manufacturers of Compressors, Vacuum Pumps, Pneumatic Tools and Air and Condensate Treatment Equipment). In the IP 280 procedure, often referred to as the Conference Internationale des Grandes Reseaux Electriques (CIGRE) test, oxygen is passed through a sample of oil containing soluble iron and copper catalysts. The sample is held at 120°C (248°F) and the test time is 164 h. During the test, volatile acids formed are absorbed in an absorption tube. At the end of the test, the acid numbers of the oil and the absorbent are determined and combined to give the total acidity; the sludge is determined as a weight percent. These may then be further combined to give the total oxidation products. Compared to ASTM D943, the CIGRE test requires a short, fixed test time, and the amount of sludge formed during the test is an important criterion of the evaluation. Where the limits for satisfactory performance in the test are properly set, correlation with performance in modern turbines are felt by some to be good. A concern of the procedure is that soluble catalyst is introduced into the oil that does not recognize the benefits of oils that specifically resist catalyst dissolution in service.

In the IP 48 PNEUROP test, a sample of oil containing iron oxide as a catalyst is aged by being held at 200°C (392°F) for 24 h while air is bubbled through it at 15 L/h. This is a high-temperature bulk oil oxidation test originally designed for the characterization of base oils. At the end of the test, the evaporation loss and the Conradson carbon residue (CCR) value of the remaining sample are determined. The evaporation loss is only significant in that it must not exceed 20%, so the main criterion is the CCR value. The test is believed to correlate to some extent with the tendency of oils to form carbonaceous deposits on compressor valves.

The most widely used oxygen uptake test is the ASTM D2272, Rotary Pressure Vessel Oxidation Test (RPVOT), which is originally known as the rotating bomb oxidation test. This test was designed to monitor the oxidative stability of new and in-service turbine oils. In some cases, it can also be used to characterize hydraulic fluids, circulating fluids, and many other types of industrial lubricants. The test uses a steel pressure vessel where sample oil is initially pressurized to 90 psi with oxygen and thermally stressed to 150°C (300°F) in the presence of copper coil catalyst and water until a pressure drop of 25 psi or more is observed. The test temperature was chosen to promote measurable oil breakdown in a relatively short time. The test is more suitable for the determination of remaining useful life of in-service turbine oils rather than the qualification of new oils.

Similarly, a Thin-Film Oxidation Uptake test (TFOUT) was developed to monitor batch-to-batch variations in the oxidative stability of rerefined lubricating base oils. The test stresses a small amount of oil to 160°C (320°F) in a high-pressure reactor pressurized with oxygen along with a metal catalyst package, a fuel catalyst, and water to partially simulate the high-temperature oxidation conditions in automotive combustion engines. TFOUT can be carried out in an RPVOT apparatus upon proper modification to the sampling accessories.

A widely used thin film oxidation test for engine oils is called the Thermal-Oxidation Engine Oil Simulation Test (TEOST), which was developed to assess the high-temperature deposit formation characteristic of engine oils under turbocharger operating conditions. In the original TEOST test, oil containing ferric naphthenate is in contact with nitrous oxide and moisture air and is cyclically pumped to flow past a tared depositor rod. The rod is heated through 12 temperature cycles, each going from 200°C (392°F) to 480°C (896°F) for 9.5 min (a total of 114 min). After the heating cycle is completed, deposit formed on the depositor rod is determined by differential weighing. The successful use of high-temperature deposition test to characterize engine oils has later led to the development of a TEOST mid-high temperature protocol, a simplified procedure for the assessment of oil deposition tendency in the piston ring zone and under crown areas of fired engines. The depositor assembly was revised to allow the oil flows down the rod in a slow and even manner to obtain a desired thin film. The test runs for 24 h and a constant depositor temperature at 285°C (545°F).

B. Thermal Stability

Thermal stability, as opposed to oxidation stability, is the ability of an oil or additive to resist decomposition under prolonged exposure to high temperatures with minimal oxygen present. Decomposition may result in thickening or thinning of the oil, increased acidity, the formation of sludge, or any combination of these.

Thermal stability tests usually involve static heating or circulation over hot metal surfaces. Exposure to air is usually minimized, but catalyst coupons of various metals may be immersed in the oil sample. Although no tests have received wide acceptance, a number of proprietary tests are used to evaluate the thermal stability of products such as hydraulic fluids, gas engine oils, and oils for large diesel engines.

C. Rust Protection Tests

The rust-protective properties of lubricating oils are difficult to evaluate. Rusting of ferrous metals is a chemical reaction initiated almost immediately when a specimen is exposed to air and moisture. Once initiated, the reaction is difficult to stop. Thus, when specimens are prepared for rust tests, extreme care must be taken to minimize exposure to air and moisture so that rusting will not start before the rust protective agent is applied and the test begun. Even with proper precautions, rust tests do not generally show good repeatability or reproducibility.

Most laboratory rust tests involve polishing or sandblasting a test specimen, coating it with the oil to be tested, then subjecting it to rusting conditions. Testing may be in a humidity cabinet, by atmospheric exposure, or by some form of dynamic test such as the ASTM D665 method, Rust Preventive Characteristics of Steam Turbine Oil in the Presence of Water. In this test, a steel specimen is immersed in a mixture of distilled or synthetic seawater and the oil under test. The oil and water moisture is stirred continuously during the test, which usually lasts for 24 h. The specimen is then examined for rusting.

The IP 220 "Emcor" method is more widely used in Europe for oils and greases. This procedure uses actual ball bearings that are visually rated for rust on the outer races at the end of the test. This is a dynamic test that can be run with a given amount of water flowing into the test bearing housings. Severity may be increased with the use of salt water or acid water in place of the standard distilled water.

Other dynamic tests may involve testing of actual mechanisms lubricated with test oil such as the CEC L-33 (CEC, 1995) test used for automotive gear lubricants. Additional rust tests for automotive engines will be discussed later on in this chapter (Section IV).

D. Foam Tests

The most widely used foam test is ASTM D892, Foaming Characteristics of Lubricating Oils. Air is blown through an oil sample held at a specified temperature for a specified length of time. Immediately after blowing is stopped, the amount of foam is measured and reported as the foaming tendency. After allowing a specified period for the foam to collapse, the volume of foam is again measured and reported as the foam stability.

This test gives a fairly good indication of the foaming characteristics of new, uncontaminated oils, but service results may not correlate well if contaminants such as moisture or finely divided rust, which can aggravate foaming problems, are present in a system. For a number of applications, such as ATFs, special foam tests involving severe agitation of the oil have been developed. These are generally proprietary tests used for specification purposes.

E. EP and Antiwear Tests

One of the main functions of a lubricant is to reduce mechanical wear. Closely related to wear reduction is the ability of EP-type lubricants to prevent scuffing, scoring, and seizure as applied loads are increased. As a result, a considerable number of machines and procedures have been developed to try to evaluate EP and antiwear properties. In a number of cases, the same machines are used for both purposes, although different operating conditions may be used.

Wear can be divided into four classifications based on the cause:

- Abrasive wear
- Corrosive (chemical) wear
- Adhesive wear
- Fatigue wear

1. Abrasive Wear

Abrasive wear is caused by abrasive particles either through external contamination or through wear particles that are generated from within the system. In either case, the oil may not have much direct influence on the amount of wear that occurs other than their ability to carry wear particles away to filtering systems that can remove them.

2. Corrosive or Chemical Wear

Corrosive or chemical wear results from chemical action on the metal surfaces combined with rubbing action that removes corroded metal. A typical example is the wear that may occur on cylinder walls and piston rings of diesel engines burning high sulfur fuels. The strong acids formed by combustion of the sulfur can attack the metal surfaces, forming compounds that can be fairly readily removed as the rings rub against the cylinder walls. Direct measurement of this wear requires many hours of testing, which is often done as the final stage of new formulation evaluation. However, useful indications have been obtained in relatively short periods via sophisticated electronic techniques. Rusting is a chemical action that may occur because of long periods of inoperation under high moisture conditions with the inability of the oil to protect nonwetted surfaces. During equipment startup for this situation, the rust is removed in the contact areas that results in a loss of surface metal, which generates particles that cause abrasive wear.

3. Adhesive Wear

Adhesive wear in lubricated systems occurs when the lubricating film becomes so thin that opposing surface asperities can make contact. This can occur under extreme conditions of load, speed, or temperature. If adequate EP additives are not present, scuffing and scoring can result, and eventually

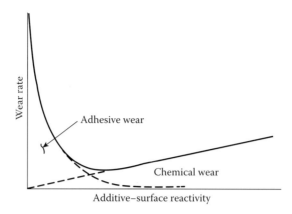

FIGURE 3.8 Balance and correlation of additive–surface reactivity with wear.

seizure may occur. Adhesive wear can also occur with EP lubricants when the reaction kinetics of the additives with the surfaces are such that metal-to-metal contact is not fully controlled.

A number of testing apparatuses, such as the Falex, Forschungsstelle fuer Zahnraeder und Getriebebau (FZG), Four Ball, SAE, Timken, High Frequency Reciprocating Rig, Almen, Alpha LFW-1, and Optimol SRV, are used to determine the loading conditions where seizure, welding, or drastic surface damage to test specimens can occur. Of these, the Alpha LFW-1, Four-Ball, Falex, Timken, and Optimol SRV are also used to measure wear at loads below the failure load. The results obtained with these machines do not necessarily correlate with field performance, but in some cases, the results from certain machines have been reported to provide useful information for specific applications. As shown in Figure 3.8, as additive reactivity increases, adhesive wear decreases and chemical wear increases. To some extent, the machines used to judge additive-surface reactivity also can be used in selecting additives for optimum effectiveness in balancing adhesive and chemical wear for certain metal combinations.

Either actual machines or scale model machines are used in testing the antiwear properties of lubricants. Antiwear type hydraulic fluids are tested for antiwear properties in pump test rigs using commercial piston and vane type hydraulic pumps. Frequently, industrial gear lubricants are tested for scuffing and micropitting protection in the FZG Spur Gear Tester, a scale model machine that reasonably approximates commercial practice as to gear tooth loading and sliding speeds. Correlation of the FZG Test with field performance has been good.

4. Fatigue Wear

Fatigue wear occurs under certain conditions where the lubricating oil film is intact and metallic contact of opposing asperities is either nonexistent or is relatively small. Because of cyclic stressing of the surfaces, fatigue cracks form in the metal and lead to fatigue spalling or pitting. Fatigue wear occurs in rolling element bearings and gears where there is a high degree of rolling and where adhesive wear associated with sliding is negligible. A variation of the fatigue pitting wear is the micropitting wear mechanism that results in small pits in the surfaces of some gears and bearings. Asperity interaction and high loads as well as metallurgy are factors influencing micropitting. An upgraded version of the standard FZG Spur Gear Tester is used to evaluate gear oil micropitting performance according to the Forschungsvereinigung Antriebstechnik (FVA) Method NR 54.

The understanding of the influence of lubricant composition and characteristics continues to be an evolving technology. Some products are currently available that address fatigue-induced wear in antifriction bearings and gears caused by the presence of small amounts of water in the hertzian load zones. In these instances, the fatigue is called water-induced fatigue, and oil chemistry can reduce the negative effects of water in the load zones of antifriction bearings and gears. Pitting of gears can be assessed using the standard FZG Spur Gear Test and the method developed by FVA.

The FZG Four-Square Gear test (FZG A/8.3/90, ASTM D-5182) rig consists of two gearsets, arranged in a four-square configuration, driven by an electric motor. The test gearset is run in the test fluid, while increasing load stages (from 1 to 13) until failure. Each load stage is run for a 15-min period at a fixed speed. Two methods are used for determining the damage load stage. The visual rating method defines the damage load stage as the stage at which more than 20% of the load-carrying flank area of the pinion is damaged by scratches or scuffing. The weight loss method defines the damage load stage as the stage at which the combined weight loss of the drive wheel and pinion exceeds the average of the weight changes in the previous load stages by more than 10 mg. The test is used in developing industrial gear lubricants, ATFs, and hydraulic fluids to meet various manufacturers' specifications.

a. Pitting and Micropitting

Pitting, a form of rolling contact fatigue, is a complex phenomenon, and several factors influence its occurrence, particularly under lubricated conditions. Contact fatigue is commonly observed on rolling contacts such as bearing raceways and pitch lines of gear teeth. It occurs when repeated stresses are concentrated at a microscopic scale on a surface. Damage progresses from microscopic pitting to macroscopic pitting and spalling (chipping). It is often used as a general term that does not specify whether the fatigue is surface induced or stems from subsurface cracking. Subsurface-initiated contact fatigue is caused by flaws at the atomic level below the surface of the metal. Subsurface cracks are formed that propagate through the metal in response to repeated stress caused by the rolling elements and lead to pits or spalling. Unlike surface-induced fatigue, no surface damage is necessary.

Improvements in the manufacturing process for steel have largely decreased these atomic-level flaws, and subsurface fatigue is relatively uncommon. Surface-initiated contact fatigue begins with microscopic surface defects in the metal caused by contaminant particles in the lubricant. Inside the gap area between bearing surfaces, entrapped particles are under very high loads and will either dent the surface or embed themselves in the bearing surfaces. The loaded rolling motion of the surfaces causes extremely high pressure spikes at these surface defects (dents). The lubrication film may be reduced in thickness, and some sliding motion may occur. Microcracks form at the surface leading to macroscopic pitting damage and component failure.

Micropitting occurs on surface-hardened gears and is characterized by extremely small pits. Micropitted metal typically has a frosted or a gray appearance. This condition generally appears on rough surfaces and is exacerbated by use of low-viscosity lubricants. Slow speed gears are also prone to micropitting owing to the presence of thin oil films. Micropitting may be sporadic and may stop when good lubrication conditions are restored following run-in. Therefore, maintaining adequate lubricant film thickness is the most important factor influencing the formation of micropitting. Higher speed operation and smooth gear tooth surfaces also hinder the formation of micropitting. The FZG micropitting test according to FVA 54/7 consists of a load stage test and an endurance test. Test gears type C-GF run at a circumferential speed of 8.3 m/s and a lubricant temperature of 90°C (194°F) or 60°C (140°F). The load and the test periods are varied.

F. Emulsion and Demulsibility Tests

In some cases, lubricating oils are expected to mix with water to form emulsions. Products such as steam engine oils, rock drill oils, soluble cutting oils, and certain types of metal rolling lubricants are examples of products that should form stable emulsions with water. Moisture is usually present in the lubricated areas served by steam engine oils and rock drill oils, and emulsification with this water is necessary to form adherent lubricating films that protect the metal surfaces against wear and corrosion. Soluble cutting oils and some types of rolling oils are designed to be mixed with water to provide the high cooling capability of water combined with some lubricating properties. Various proprietary tests for emulsion stability are in use.

Lubricating oils used in circulating systems should separate readily from water that may enter the system as a result of condensation, leakage, or splash of process water. If the water separates easily, it will settle to the bottom of the reservoir, where it can be periodically drawn off. Steam turbine oils, hydraulic fluids, circulation oils, and industrial gear oils are examples of products where it is particularly important to have good water separation properties. These properties are usually described as having emulsion, demulsibility, or demulsification characteristics.

To determine the emulsion and demulsibility characteristics of lubricating oils, three tests are widely used: ASTM D1401, D2711, and IP 19.

ASTM D1401, Emulsion Characteristics of Petroleum Oils and Synthetic Fluids, was developed specifically for steam turbine oils with viscosities of 32–100 cSt at 40°C but can be used for testing lower and higher viscosity oils and synthetic fluids. The normal test temperature is 130°F (54°C), but for oils more viscous than an ISO 100 viscosity grade, it is recommended that the test temperature be increased to 180°F (82°C). In the test, equal parts of oil (or fluid) and water are stirred together for 5 min. The time for separation of the emulsion is recorded. If complete separation does not occur after standing for 1 h, volumes of oil (or fluid), water, and emulsion remaining are reported.

ASTM D2711, Demulsibility Characteristics of Lubricating Oils, is intended for use in testing medium- and high-viscosity oils such as paper machine oil applications that are prone to water contamination and that may encounter the turbulence of pumping and circulation capable of producing water-in-oil emulsions. A modification of the test is suitable for evaluation of oils containing EP additives. In the test, water and the oil under test are stirred together for 5 min at 180°F (82°C). After 5 h of settling, the percentage of water in the oil and the percentages of water and emulsion separated for the oil are measured and recorded.

IP 19, Demulsification Number, is intended for testing new inhibited and straight mineral turbine oils. A sample of the oil held at about 90°C (194°F) is emulsified with dry steam. The emulsion is then held in a bath at 94°C (201°F), and the time in seconds for the oil to separate is recorded and reported as the demulsification number. If the complete separation has not occurred in 20 min, the demulsification number is reported as 1200+.

Properly used, these three tests have value in describing the emulsion or demulsibility characteristics of new lubricating oils. The IP 19 procedure is sometimes used in describing products that emulsify readily as well as products that easily separate from water. A demulsification number of 1200+ is sometimes specified for such products as rock drill oils where the formation of lubricating emulsions is desired.

IV. ENGINE TESTS FOR OIL PERFORMANCE

A large number of engine tests are currently used to qualify oils for specific categories of classification. This number continues to expand as new or additional engine and oil characteristics are needed to meet not only the requirements of the engines but also cover environmental aspects. Most of the standardized tests are designed to evaluate oils intended for automotive-type engines. Considerable effort has been devoted to standardizing the tests in the United States, Europe, Japan, and other parts of Asia.

The API went from an Engine Service Classification System to an Engine Oil Licensing and Classification System in 1994. Given that the purpose of oil specifications is to prevent in-use performance issues, the historical absence of any major issues with oil when it is used as specified is a strong indicator of the strength of the current specification development system. In addition to API Services categories, the International Lubricant Standardization and Approval Committee (ILSAC), in conjunction with automobile manufacturers, created the gasoline fueled (GF) series and the proposed changes (PC) series in 1992. Currently, there are two new engine oil specifications being developed: PC-11 for heavy-duty diesel engines and GF-6 for passenger automobiles. For each of these specifications, it is very possible there will be two versions: one for current and future engines and another for compatibility with older engines.

Lubricating Oils

There is always a need to develop new tests to evaluate lubricant performance in emerging hardware platforms where engine hardware design changes are being dictated by the need to improve fuel efficiency and reduce emissions in order to meet environmental regulations.

The ASTM has published the "Multicylinder Engine Test Sequences for Evaluation of Automotive Engine Oils" (Sequences III, IV, V, and VI) in Special Technical Publication (STP) 315 and the "Single Cylinder Engine Tests for Evaluating Crankcase Oils" (Sequence VIII, 1-N, 1-K, and 1-P) in STP509. In addition, the major diesel engine manufacturers have developed engine tests that are now part of the API Classification System. These tests were initially developed for qualifying oils for their specific engines. Examples are the Mack EO-L, EO-L Plus, and EO-M using the Mack T-8, T-11, and T-12 engine tests. Cummins Diesel uses the ISB, ISM, and the M-11 tests. All of these engine tests are very high in costs to develop and run. This adds significant costs to engine oil development and testing. All of these tests can be grouped into eight categories of performance measurement:

1. Tests for oxidation stability and bearing corrosion protection
2. Single cylinder high temperature tests
3. Multicylinder high temperature tests
4. Multicylinder low temperature tests
5. Rust and corrosion protection
6. Oil consumption rates and volatility
7. Emissions and protection of emission control systems
8. Fuel economy

A summary of the common engine oil tests used in the United States is shown in Table 3.5. Some additional new engine tests or extension of existing engine tests are used to evaluate the ability

TABLE 3.5
Common Gasoline and Diesel Engine Oil Tests

Engine Test	Principle Parameters Evaluated
Sequence IIIG	Oxidation/Piston Deposits/Cam & Lifter Wear
Sequence IIIH	Oxidation/Piston Deposits
Sequence IVA/B	Cam Wear
Sequence VG/H	Sludge/Varnish formation
Sequence VID/E	Fuel Economy
Sequence VIII	Bearing Corrosion (Copper/Lead)
Mack T-8E	Oxidation/Soot/Filterability
Mack T-10	Ring-Liner Wear/Bearing Wear/Extended Drain Capability
Mack T-11	Soot/Viscosity Control with Exhaust Gas Recirculation (EGR) system
Mack T-12	Soot/Wear Control with Exhaust Gas Recirculation and operating on Ultra-low Sulfur diesel fuel
Mack T-13	Soot/Viscosity Control with Exhaust Gas Recirculation (EGR) system and operating on Ultralow sulfur fuel
Cat 1-K	Piston Cleanliness/Oil Consumption/Ring and Liner Scuffing
Cat 1M-PC	Piston Cleanliness/Varnish/Ring and Liner Wear
Cat 1-N	Piston Cleanliness/Ring and Liner Wear/Oil Consumption
Cat 1-P	Deposit Control/Oil Consumption
Cat C13	Aeration/Deposits/Oil Consumption
Cummins M11-EGR	Bearing Corrosion/Sliding Wear/Filterability/Sludge
Cummins ISB	Cam and Tappet Wear (with EGR system)
Cummins ISM	Ring Wear (with EGR system)
GM 6.2/6.5 l	Roller Cam Follower Wear/Ring Sticking

of oils to provide extended drain capability. The ability to extend drain intervals is important to builders because of the pressure from users to reduce costs associated with maintenance and also conserve nonrenewable resources. Extended drain capability is perceived by users to mean higher quality engines and oils.

A. Oxidation Stability and Bearing Corrosion Protection

Several engine tests are used to evaluate the ability of an oil to resist oxidation and oil thickening under high temperature conditions as well as protect sensitive bearing materials from corrosion. The tests are operated under conditions that promote oil oxidation and the formation of oxyacids that cause bearing corrosion. Copper–lead inserts are used for connecting rod and main bearings in many gasoline and diesel engines. Even the few engines that use aluminum rod and main bearings will use a lead–tin flashing for break-in purposes. Although the aluminum bearings are more resistant to oxyacid corrosion, they can still see some effects of acids in the oil. After the tests are completed, the bearing inserts are examined for surface condition and weight loss to determine the protection afforded by the oil. The engine is also rated for varnish and sludge deposits. The most commonly used bearing corrosion test is Sequence VIII (CEC L-02-A-78).

B. Single Cylinder High Temperature Tests

Single cylinder engines are designed and operated to duplicate longer-term operating conditions in a laboratory. They are specifically designed for oil test purposes. The tests are used primarily for the evaluation of detergency and dispersancy—that is, the ability of the oil to control piston deposits and ring sticking under operating conditions or with fuels that tend to promote the formation of piston deposits. These tests also evaluate oil consumption rates. All of the current single cylinder tests are of the diesel engine design. After completion of the test, the engine is disassembled and rated for piston deposits, top ring groove filling, and wear. The operating conditions for several of these tests are shown in Table 3.6.

C. Multicylinder High Temperature Engine Tests

Although the single cylinder engine tests discussed earlier provide useful information and are extremely valuable in the development of improved oil formulations, there is a trend toward use of full-scale commercial engines for oil testing and development. This trend has been precipitated by the different needs of various engine designs as well as to satisfy demands for reduced emissions

TABLE 3.6
Single Cylinder Engine Oil Test Conditions

Test	1-K	1-N	1-P
Area of use	US	US	US
Fuel injection	DI	DI	DI
Aspiration	SC	SC	SC
Displacement	2.4l	2.4l	2.2l
Load	52 kW	52 kW	55 kW
Speed (rpm)	2100	2100	1800
Oil capacity	6.0 L	6.0 L	6.8 L
Test duration	252 h	252 h	360 h
Oil temp (°C)	107	107	130

Lubricating Oils

TABLE 3.7
Multicylinder Test Conditions for Passenger Vehicle Engine Oils

Test	Sequence IIIG	Sequence IVA	Sequence VG	Sequence VID	Sequence VIII
Engine	GM V-6	Nissan I-4	Ford V-8	GM V-6	LABECO V-1
Fuel	Gasoline	Gasoline	Gasoline	Gasoline	Gasoline
Displacement	3.8 L	2.4 L	4.6 L	3.6 L	0.7 L
Oil capacity	5.5 L	3.0 L	7.5 L	5.4 L	7.0 L
Test duration	100 h	100 h	216 h	100 h	40 h
Speed (rpm)	3600	I 800	I 1200	I 2000	3150
		II 1500	II 2900	II 2000	
			III 700	III 1500	
				IV–VI 695	
Load	250 Nm	25 Nm	Report/18 Nm	20–105 Nm	Based on fuel flow
Oil temp (°C)	155	50/60	68/100/45	35–115	143.5

TABLE 3.8
Multicylinder Test Conditions for Commercial Vehicle Engine Oils

Test	Mack T-8E	Mack T-11	Mack T-13	Cat C13	Cummins ISB
Engine	E7-350	E-Tech	MP-8	Cat C13	Cummins ISB
Fuel	Diesel (Low S)	Diesel (Low S)	Diesel (Ultra Low S)	Diesel (Ultra Low S)	Diesel (Ultra Low S)
Displacement	12 L	12 L	12 L	12.5 L	5.9 L
Test duration	300 h	252 h	360 h	500 h	350 h
Speed (rpm)	1800	1800	1500	1800–2100	1600/Ramped
Oil temp (°C)	100–107	88	130	98	110

and improved fuel economy. The full-scale engine tests are used to evaluate several of the oil's performance characteristics under the different conditions subjected by the various designs. Oxidation stability, deposit control, wear, and scuffing are a few of the oil's characteristics evaluated by these tests. Some of the full-scale engine tests used to evaluate the oil's performance characteristics are the ASTM Sequence IIIH, Mack T-11 and T-13, and the OM 602A (CEC L-51-T-95). Several of the multicylinder test conditions used for passenger vehicles are shown in Table 3.7 and those for commercial vehicles are shown in Table 3.8.

D. Multicylinder Low Temperature Tests

Many engines idle for long periods because of frequent start–stop driving conditions or for cold weather warmup (e.g., diesel engines). In these situations, it is important that the oil has enough detergency and dispersancy to satisfactorily control soot and sludge in the engine. Both sludge and soot will increase the oil's viscosity in addition to reducing antiwear protection, filterability, and fuel economy (note: higher viscosity reduces fuel economy). The oil's detergency and dispersancy characteristics must be balanced with the other performance requirements to handle the negative effects of soot and reduce the buildup of sludge in the engines that can often occur in low temperature operation. Several multicylinder tests are designed to predict the oil's ability to handle the soot and sludging. The Mack T-8, the M111 (CEC L-53-T-95), and the Sequence VG tests are some of the tests used to evaluate low temperature sludge and soot-handling capabilities of oils.

E. Rust and Corrosion Protection Tests

This property has received considerable attention in the United States primarily because of the high proportion of stop-and-go driving in combination with severe winter weather. These conditions tend to promote condensation and accumulation of partially burned fuel in the crankcase oil, both of which promote rust and corrosion in addition to the soot and sludge problems discussed earlier. Two engine tests are currently used to evaluate the rust and corrosion protection properties of oils. These are the Mack T-11 and the Sequence VIII. At the end of the tests, valve train components, oil pump relief valves, and bearings are rated for rust, and any sticking of lifters and relief valves is noted. A bench test called the Ball Rust Test is also used to evaluate rust performance of engine oils.

F. Oil Consumption Rates and Volatility

There is a direct association with an oil's volatility characteristics and oil consumption rates. Although volatility is not the sole reason for oil consumption in any given engine, it provides a measure of the oil's ability to resist vaporization at high temperatures. Typically, distillations were run to determine the volatility characteristics of base stocks used to formulate engine oils. This is still true today. The objective of further defining an oil's volatility led to the introduction of several new nonengine bench tests. The most common of these tests is the NOACK Volatility, Simulated Distillation or "sim-dis" (ASTM D2887). All of these tests measure the percentage of oil that is lost when exposed to high temperatures and is therefore a good measure of the oil's relative potential for increased or decreased oil consumption during severe service. Reduced volatility limits are continuing to place more restraints on base stock processing and selection and the additive levels necessary to achieve the required volatility levels. The trend of using more and more of API Group II, III, and IV base stocks to formulate engine oils continues. In addition to the bench tests discussed above, many of the actual engine tests monitor and report oil consumption rates as part of the test criteria. As an example, OEMs such as Caterpillar, Cummins, and Mack are all concerned about controlling oil consumption for extended service conditions.

G. Emissions and Protection of Emission Control Systems

Control of engine emissions is becoming increasingly important because of health concerns as well as the long-term effects in the earth's atmosphere (greenhouse effects). As a result, legislation is in place for engine builders and users to reduce engine exhaust emissions. These emissions include sulfur dioxide, oxides of nitrogen, carbon monoxide, hydrocarbons, and particulates. There are several ways that an oil can contribute to an engine's emissions. The most common is by providing a seal between the pressure in the combustion chamber at the rings and pistons. Wear or deposits in this area will reduce combustion efficiency and lead to greater emissions. Control of piston land and groove deposits is crucial to maintaining low oil consumption rates and long-term engine performance. Another way that the oil can contribute to increased emissions is by blocking or poisoning catalysts on the engines equipped with catalytic converters. Blocking of catalyst reactions can occur through excessive oil consumption, and the poisoning effects result from chemical components either in the fuel or the oil's additive package. A common additive used in engine oil formulations is ZDDP, which is rich in three nonhydrocarbon and oxygen elements: zinc, sulfur, and phosphorus. Phosphorus is known to be an element that will poison three-way catalysts, sulfur is also known to be poisonous for DeNOx catalyst, and zinc could add ash level that becomes burdensome to keep particulate traps clean. ZDDP still remains a key additive for protecting the long-term oxidation and wear performance of engines. Because calcium and magnesium detergents also contribute a significant amount of ash to the engine oils, another trend has been the reduction in the overall level of detergent that is acceptable when using ultralow sulfur fuel. Since ash, sulfur, and phosphorus cannot currently be completely eliminated from engine oil formulations, the oils are formulated to keep oil consumption rates low, which minimizes the effects of oil additive elements that end up in exhaust gases. The use of low sulfated ash, phosphorus, and sulfur engine oils remains critical with many after-treatment devices.

H. Fuel Economy

Similar to emissions, there is a strong trend to increase the fuel efficiency of both gasoline and diesel engines. The first engine test developed to measure fuel economy was Sequence VI, which was developed in the mid-1980s using a Buick V-6. The Sequence VIA test replaced the Sequence VI test in the mid-1990s using a 1993 Ford 4.6l V-8. Later on, Sequence VIB replaced Sequence VIA with the advent of ILSAC GF-4 in 2004, and the Sequence VID test replaced the Sequence VIB test in year 2010 when ILSAC GF-5 was introduced. All of the fuel economy tests indicate the importance of an oil's viscosity in achieving mandated Corporate Average Fuel Economy (CAFE) requirements. Lower viscosity oils such as 5W-20 and 0W-20 are the principal recommendations of many automobile manufacturers, and they provide measurable economy benefits relative to heavier viscosity grades. The future ILSAC GF-6 even proposes to recognize 0W-16 engine oils where a new Sequence VID test will be used to evaluate the fuel economy effect of all low viscosity oils. Although these tests are used to measure an oil's contribution to fuel economy, the official federal test uses a carbon balance of the tailpipe emissions, and it is this test that is used to establish compliance to CAFE requirements.

V. AUTOMOTIVE GEAR LUBRICANTS

The automotive vehicles manufactured today and in the past have all required gearing of some sort to allow transfer of engine power to the driving wheels. This gearing is composed of a range of gear designs encompassing spur, helical, herringbone, and/or hypoid gears. All these gears require lubrication. As there is a wide range of gearing and application requirements, so is there a range of performance levels to meet the mild to severe operating and application conditions.

Finished automotive gear lubricants will typically be composed of high-quality base stocks (mineral and/or synthetics), and between 5% and 20% additive depending on the desired performance characteristics. Up to 10 different additive materials can be used to formulate these oils, and based on the increasing requirements of extended service intervals and environmental concerns, more may be needed. These additives would include antiwear, EP, oxidation stabilizers, metal deactivators, foam suppressors, corrosion inhibitors, pour point depressants, dispersants, and VI improvers. As with the other high-performance lubricants, these additives compete with each other to perform their functions and must be balanced to provide the required performance requirements.

Three primary technical societies composed of equipment builders, lubricant formulators, additive suppliers, and the users of the equipment have combined efforts to define automotive gear lubricant requirements. These three technical societies are SAE, ASTM, and API. The SAE has established a viscosity classification system (SAE J306) for automotive gear lubricants, as shown in Table 3.2. The ASTM established test methods and criteria for judging performance levels and defining test limits. The API defines performance category language. The MIL-PRF-2105E specification that the U.S. military had established for automotive gear lubricants has been replaced by SAE J2360.

Unlike automotive engine oils, there are no current licensing requirements for gear oils classified by the API. There are however, some major OEMs that offer licenses to use their designations for transmission and axle lubricants.

The API performance categories as defined by API Publication 1560 are as follows:

API GL-1 (not in current use): Lubricants for manual transmissions operating under mild service conditions. These oils do not contain antiwear, EP, or friction modifier additives. They do contain corrosion inhibitors, oxidation inhibitors, pour point depressants, and antifoam agents. This category is not in current use.

API GL-3 (not in current use): This category designates the type of lubricants for manual transmissions, operating under moderate to severe conditions of speed and load, and spiral bevel axles operating in mild to moderate conditions. These service conditions require a lubricant having greater load-carrying capabilities than those that will satisfy API GL-1 service, but below the requirements of lubricants satisfying API GL-4 service.

API GL-4: Designates the type of lubricants typically used for differentials and transmissions operating under moderate to severe conditions of speed and load where spiral bevel gears are used or moderate conditions where hypoid gears are used. These oils may be used in manual transmissions and transaxles where EP oils are acceptable and API MT-1 oils are unsuitable. Limited-slip differentials generally have special lubrication requirements, and the manufacturer or lubricant supplier should be consulted regarding the suitability of this lubricant for such differentials.

API GL-5: Lubricants for differentials containing hypoid gears operating under severe conditions of torque and occasional shock loading. These oils generally contain high levels of antiwear and EP additives.

API MT-1: Lubricants for manual transmissions that do not contain synchronizers. Nonsynchronized manual transmissions are used in buses and heavy-duty trucks. These oils are formulated to provide higher levels of oxidation and thermal stability, component wear, and oil-seal compatibility that may not be provided by API GL-1, GL-4, and GL-5 category oils. See Table 3.9 for a comparison of testing for API GL-5 and MT-1.

Lubricants that satisfy SAE J2360 will satisfy the requirements for API GL-5; however, SAE J2360 has additional performance requirements that exceed API GL-5.

In addition to the automotive gear lubricant tests, various car and other automotive axle and transmission manufacturers have gear tests, many of which are conducted in cars or over-the-road vehicles either on dynamometers or in actual road tests. These tests generally represent special requirements such as the ability of lubricants to provide satisfactory performance in limited slip axles. In general, most laboratory and bench testing has shown good correlation to field performance.

TABLE 3.9
Gear Lubricant Testing for API GL-5 and MT-1

Characteristic	API GL-5	API MT-1
Corrosion resistance	CRC L-33	ASTM D130
Load-carrying capability	CRC L-37	ASTM D5182
Scoring resistance	CRC L-42	ASTM D5182
Oxidation stability	CRC L-60	ASTM D5704
Elastomer compatibility	–	ASTM D5662
Cyclic durability	–	ASTM D5579
Foam resistance	ASST. D892	ASTM D892
Gear oil compatibility	FTM 3430/3440	FTM 3430/3440

VI. ATFs

ATFs are among the most complex lubricants currently available. In the converter section, these fluids are the power transmission and heat transfer medium; in the gear box, they lubricate the gears and bearings, and control the frictional characteristics of the clutches and bands; and in control circuits, they act as hydraulic fluids. All of these functions must be satisfactorily performed over temperatures ranging from the lowest expected ambient temperatures to operating temperatures on the order of 300°F (149°C) or higher, and for extended periods of service. Obviously, very careful evaluation is required before a fluid can be considered acceptable for such service.

Major automotive companies (General Motors, Ford, and Chrysler) in the United States continue to strive for improved ATFs. These improvements are aimed at fill-for-life applications (150,000 miles or more), which means improvements are needed in oxidation stability, antiwear retention, shear stability, low temperature fluidity, material compatibility, and fluid friction stability. Ford Motor Company had already phased out their MERCON® V and replaced it with MERCON® LV (low viscosity for better fuel economy), and GM is pursuing an update of their DEXRON® VI. In anticipation of improved fuel efficiency and extended oil drain intervals, low-viscosity ATFs, with better high-temperature frictional and oxidative stability for six to nine speed automatic transmissions, are being favored by many OEMs.

BIBLIOGRAPHY

American Petroleum Institute. 2013. API 1560: Lubricant Service Designations for Automotive Manual Transmissions, Manual Transaxles, and Axles. Washington, D.C.

ASTM D4683-13, Standard Test Method for Measuring Viscosity of New and Used Engine Oils at High Shear Rate and High Temperature by Tapered Bearing Simulator Viscometer at 150°C, ASTM International, West Conshohocken, PA, 2013, www.astm.org.

ASTM D4741-13, Standard Test Method for Measuring Viscosity at High Temperature and High Shear Rate by Tapered-Plug Viscometer, ASTM International, West Conshohocken, PA, 2013, www.astm.org.

ASTM D5481-13, Standard Test Method for Measuring Apparent Viscosity at High-Temperature and High-Shear Rate by Multicell Capillary Viscometer, ASTM International, West Conshohocken, PA, 2013, www.astm.org.

ASTM D5293-14, Standard Test Method for Apparent Viscosity of Engine Oils and Base Stocks Between −5°C and −35°C Using Cold-Cranking Simulator, ASTM International, West Conshohocken, PA, 2014, www.astm.org.

ASTM D4684-14, Standard Test Method for Determination of Yield Stress and Apparent Viscosity of Engine Oils at Low Temperature, ASTM International, West Conshohocken, PA, 2014, www.astm.org.

ASTM D2983-09, Standard Test Method for Low-Temperature Viscosity of Lubricants Measured by Brookfield Viscometer, ASTM International, West Conshohocken, PA, 2009, www.astm.org.

ASTM D2422-97(2013), Standard Classification of Industrial Fluid Lubricants by Viscosity System, ASTM International, West Conshohocken, PA, 2013, www.astm.org.

ASTM D943-04a(2010)e1, Standard Test Method for Oxidation Characteristics of Inhibited Mineral Oils, ASTM International, West Conshohocken, PA, 2010, www.astm.org.

ASTM D2272-14a, Standard Test Method for Oxidation Stability of Steam Turbine Oils by Rotating Pressure Vessel, ASTM International, West Conshohocken, PA, 2014, www.astm.org.

ASTM D665-14e1, Standard Test Method for Rust-Preventing Characteristics of Inhibited Mineral Oil in the Presence of Water, ASTM International, West Conshohocken, PA, 2014, www.astm.org.

ASTM D892-13, Standard Test Method for Foaming Characteristics of Lubricating Oils, ASTM International, West Conshohocken, PA, 2013, www.astm.org.

ASTM D5182-97(2014), Standard Test Method for Evaluating the Scuffing Load Capacity of Oils (FZG Visual Method), ASTM International, West Conshohocken, PA, 2014, www.astm.org.

ASTM D1401-12, Standard Test Method for Water Separability of Petroleum Oils and Synthetic Fluids, ASTM International, West Conshohocken, PA, 2012, www.astm.org.

ASTM D2711-11, Standard Test Method for Demulsibility Characteristics of Lubricating Oils, ASTM International, West Conshohocken, PA, 2011, www.astm.org.

ASTM D2887-14, Standard Test Method for Boiling Range Distribution of Petroleum Fractions by Gas Chromatography, ASTM International, West Conshohocken, PA, 2014, www.astm.org.

ASTM STP315I, Multicylinder Test Sequences for Evaluating Automotive Engine Oils: Sequence IID, Tenth Edition, 1993, ASTM International, West Conshohocken, PA, 2014, www.astm.org.

ASTM STP509, Single Cylinder Engine Tests for Evaluating the Performance of Crankcase Lubricants (Abridged Procedures), 1972, ASTM International, West Conshohocken, PA, 2014, www.astm.org.

CEC (1995). Biodegradability of two-stroke cycle outboard engine oils in water. Co-ordinating European Council for the Development of Performance Tests for Lubricants and Engine Fuels Method CEC L-33-A-93. The Coordinating European Council. Leicestershire.

CEC Oil oxidation and bearing corrosion. CEC L-02-A-78. The Coordinating European Council. Leicestershire.

CEC Low temperature thickening and wear test (OM602A). CEC L-51-T-95. The Coordinating European Council. Leicestershire.

CEC Evaluation of Sludge Inhibition Qualities of Motor Oils in Gasoline Engines (MB M111 E20). CEC L-53-T-95. The Coordinating European Council. Leicestershire.

FVA 54/7. "Test Procedure for the Investigation of the Micropitting Capacity of Gear Lubricants," FVA Information Sheet, Drive Technology Research Association, 1993.

ISO 3448. Industrial liquid lubricants—ISO viscosity classification. 1992. International Organization for Standardization. Geneva.

IP 48: Determination of oxidation characteristics of lubricating oil. 2012. Energy Institute. London.

IP 280: Petroleum products and lubricants—Inhibited mineral turbine oils—Determination of oxidation stability. 1999. Energy Institute. London.

IP 220: Determination of rust–prevention characteristics of lubricating greases. 2007. Energy Institute. London.

IP 19: Determination of demulsibility characteristics of lubricating oil. 2012. Energy Institute. London.

Mobil Oil Corporation. 1973. Plain Bearings Fluid-Film Lubrication. New York.

Rudnick, R.L. 2003. *Lubricant Additives Chemistry and Applications*. New York: Marcel Dekker, Inc.

SAE J300. 2015. Engine Oil Viscosity Classification. SAE International, Warrendale, PA.

SAE J306. 2005. Automotive Gear Lubricant Viscosity Classification. SAE International, Warrendale, PA.

SAE J2360. 2012. Automotive Gear Lubricants for Commercial and Military Use. SAE International, Warrendale, PA.

4 Lubricating Greases

Lubricating grease can simply be described as a solid to semifluid material composed of a thickening agent with a nonaqueous, friction reducing fluid. Despite this fundamental definition, the manufacture, composition, and performance of grease is a complex matrix of engineering controls and chemical reactions that result in specific performance for unique applications.

End users often choose greases not only based on their performance characteristics, as defined by industry standard specifications or following equipment manufacturer guidance, but also on traits that appeal to the senses such as color, feel, and odor, which have little effect on their performance in the application. Therefore, it is important to develop an elementary understanding of grease composition and performance along with how and why they are used before making a product selection.

I. WHY GREASES ARE USED

The reason greases are used in preference to a fluid lubricant is directly related to their gel-like consistency, which enables them to remain in equipment under a variety of conditions while providing the controlled oil release needed to maintain separation of moving surfaces. Their ability to stay in place makes them particularly beneficial for applications that are sealed for life or difficult to access such as remote gearboxes and bearings. Greases are formulated with a wide variety of base oils enabling them to provide lubrication in a broad array of applications ranging from high-speed spindle bearings operating under very light load to slow-moving, heavily loaded gears in earth-moving equipment. The selection of an appropriate thickener with suitable shear stability for the grease to yield to mechanical stress is equally important, as the grease may also act to seal out contaminants by creating a thickener boundary while releasing oil as needed. Greases are additionally formulated to operate under extreme environments and can provide a protective film that withstands water contamination, prevents corrosion, tolerates high- and/or low-temperature operation, and is compatible with elastomers, polymers, and other greases while providing antiwear and extreme pressure load protection.

II. COMPOSITION OF GREASE

Greases are composed of three main ingredients. The liquid portion may be any fluid, such as mineral or synthetic oil, that has friction-reducing properties. The thickener may be any material that, in combination with the selected fluid, will produce the solid to semifluid structure required to maintain the lubricant at the point of contact. The final ingredients are additives or modifiers that are used to impart special properties or modify existing ones. Greases are made by combining these three components: additive, thickener, and fluid (as shown in Figure 4.1).

A. Fluid Components

Most greases use hydrocarbon-based refined mineral oils as their fluid component. These can range in viscosity from as light as mineral oil up to the heaviest cylinder stocks, with International Organization for Standardization (ISO) viscosity grades 100 through 460 as the most common. In the case of greases requiring a high-viscosity fluid, waxes, petrolatums, resins, or asphalts may also be used. Although these latter materials may not be accurately considered "liquid lubricants," they can perform the same function as the fluid components in conventional greases.

FIGURE 4.1 Grease composition.

Greases made with mineral oils generally provide satisfactory performance in most automotive and industrial applications. In very low or high temperature applications or in applications where temperature may vary over a wide range, greases made with synthetic fluids generally are used. For a detailed discussion on synthetics, see Chapter 5.

B. Thickeners

The National Lubricating Grease Institute (NLGI) survey continues to confirm that the principal thickeners used in greases are metal soaps. These are created using the fundamental reaction of an acid and a base in a stoichiometric balance to create a stable product.

The earliest greases were calcium soaps, made by combining lime with a vegetable oil for which the composition includes a fatty component such as a triglyceride. This created a thickened material suitable for lubrication. As knowledge of chemical reactions expanded, so did the variety of soap-based greases, which has grown to include sodium, aluminum, lithium soap-based greases, followed by clay, polymers, and polyurea thickeners. Greases can also be made with combinations of soaps, such as sodium and calcium, and are referred to as mixed base greases. Soaps made with other metals have been used but have not received general acceptance, because of cost, health, and safety issues, environmental concerns, or performance problems.

The earliest simple forms of calcium-type greases provided a low-cost lubricant with good water resistance and fair shear stability but limited performance over a wide temperature range. This led to the use of sodium-based greases, which had high-temperature performance capability and more desirable texture versus calcium. However, many of these greases are challenged in the presence of water, giving them limited applicability.

Lubricating Greases

A greater understanding of raw materials, reaction, and resulting soap chemistry led to the use of more pure raw materials such as 12-hydroxystearic acid with a basic metal hydroxide to provide more reliable reaction chemistry and resulting soap performance.

The need for higher temperature performance introduced the use of more than one acid compound to the metal soap reaction, and the creation of complex grease soaps. These complex greases are made by using a combination of a conventional metallic soap-forming material with a complexing agent. The complexing agent may be either organic or inorganic and may or may not involve another metallic constituent. Among the most successful of the complex greases are the lithium complex greases. These are made with a combination of conventional lithium soap-forming materials and a low molecular weight organic acid as the complexing agent. Greases of this type are characterized by very high dropping points, usually above 500°F (250°C). Other complex greases, such as aluminum and calcium, are also manufactured for certain applications.

The use of nonsoap thickeners continues to change and grow beyond the special applications for which they were originally intended. Although modified bentonite (clay) and silica aerogel are used to manufacture nonmelting greases for high-temperature applications, these have limited applicability because of other performance drawbacks and the reduced number of additives that are compatible with these polar thickeners. Because oxidation can still cause the oil component of these greases to deteriorate, regular relubrication is required.

Polyurea-based greases are increasing in use as a common thickener technology. Formed by the reaction of an isocyanite and an amine, these compounds can be used in formulations to provide excellent high-temperature lubrication, which are particularly beneficial in bearings with limited access for relubrication. Table 4.1 is an application guide that outlines the typical lubricating grease characteristics classified by thickener type for various major grease soaps.

C. Additives

Additive classes commonly used in lubricating greases are listed below. Most of these materials have much the same function as similar materials added to lubricating oils:

- Oxidation inhibitor
- Antiwear agents
- Extreme pressure additives
- Solid boundary lubrication additives
- Rust and corrosion inhibitors
- Friction modifiers
- Adhesive (tackiness) agents
- Odorants (perfumes)
- Dyes

Grease is unique in its ability to incorporate liquid and solid additives to provide wear protection. Solid additives, such as molybdenum disulfide, graphite, and Teflon®, are used to enhance specific performance characteristics such as load carrying ability. Solid additives are different from extreme pressure additives that react with the lubricated surface to form a chemical film. Solid additives provide a physical separation between moving parts when the mating surfaces cannot maintain a fluid film. Molybdenum disulfide is used in many greases for applications where loads are heavy, surface speeds are low, and restricted or oscillating motion is involved.

TABLE 4.1
Typical Lubricating Grease Characteristics by Thickener Type

Properties	Aluminum	Sodium	Calcium Conventional	Calcium Anhydrous	Lithium	Aluminum Complex	Calcium Complex	Lithium Complex	Polyurea	Organo Clay
Dropping point (°F)	230	325–350	205–220	275–290	350–400	500+	500+	500+	470	500+
Dropping point (°C)	110	163–177	96–104	135–143	177–204	260+	260+	206+	243	260+
Maximum usable temperature (°F)	175	250	200	230	275	350	350	350	350	350
Maximum usable temperature (°C)	79	121	93	110	135	177	177	177	177	177
Water resistance	Good to excellent	Poor to fair	Good to excellent	Excellent	Good	Good to excellent	Fair to excellent	Good to excellent	Good to excellent	Fair to excellent
Work stability	Poor	Fair	Fair to good	Good to excellent	Good to excellent	Good to excellent	Fair to good	Good to excellent	Poor to good	Fair to good
Oxidation stability	Excellent	Poor to good	Poor to excellent	Fair to excellent	Fair to excellent	Fair to excellent	Poor to good	Fair to excellent	Good to excellent	Good
Protection against rust	Good to excellent	Good to excellent	Poor to excellent	Poor to excellent	Poor to excellent	Good to excellent	Fair to excellent	Fair to excellent	Fair to excellent	Poor to excellent

(Continued)

TABLE 4.1 (CONTINUED)
Typical Lubricating Grease Characteristics by Thickener Type

Properties	Aluminum	Sodium	Calcium Conventional	Calcium Anhydrous	Lithium	Aluminum Complex	Calcium Complex	Lithium Complex	Polyurea	Organo Clay
Oil separation	Good	Fair to good	Poor to good	Good	Good to excellent	Good to excellent	Good to excellent	Good to excellent	Good to excellent	Good to excellent
Appearance	Smooth and clear	Smooth to fibrous	Smooth and buttery	Smooth and buttery	Smooth and buttery	Smooth and buttery	Smooth and buttery	Smooth and buttery	Smooth and buttery	Smooth and buttery
Other properties		Adhesive and cohesive	EP grades available	EP grades available	EP grades available, reversible	EP grades available, reversible	EP and antiwear inherent	EP grades available	EP grades available	
Production volume and trend[a]	No change	Declining	Declining	No change	The leader	Increasing	Declining	Increasing	No change	Declining
Principal uses[b]	Thread lubricants	Rolling contact bearings	General uses for economy	Military multiservice	Multiservice and industrial	Multiservice industrial	Multiservice automotive and industrial	Multiservice automotive and industrial	Multiservice automotive and industrial	High temperature (frequent relube)

Source: Courtesy of NLGI.

[a] Lithium grease over 50% of production and all others below 10%.
[b] Multiservice includes rolling contact bearings, plain bearings, and others.

III. MANUFACTURE OF GREASE

The manufacture of grease involves the reaction and stable dispersion of a thickener in a fluid base and the incorporation of performance additives. This is accomplished in a number of ways depending on the thickener type. In some cases, the thickener is purchased by the grease manufacturer in a finished state and then mixed with oil until the desired grease structure is obtained. In most cases with metallic soaps, the thickener is produced through a controlled reaction during the manufacture of the grease.

In the manufacture of simple lithium soap grease, for example, hydrogenated castor oil, fatty acids, and/or glycerides are dissolved in a portion of the oil before they are reacted with an aqueous solution of lithium hydroxide. This reaction is commonly called saponification, and produces a wet lithium soap that is partially dispersed in the mineral oil. Through controlled heating and mixing the material is dehydrated. After dehydration, the soap–oil mixture generally has a dense, grainy, nonuniform appearance and requires further dilution or "cutback" with additional fluid. During the manufacturing process, depending on additive chemistry and soap reactivity, performance additives are incorporated. It is only after the cutback operation that the grease is further processed by kettle milling or homogenization to modify this structure. Once the proper structure and consistency are obtained, the grease is ready for finishing and packaging.

As noted in the preceding discussion, manufacture of a simple soap grease involves all or some of the following four steps:

1. Saponification
2. Dehydration
3. Cutback
4. Milling

The actual manufacture can be done either with only a kettle (sometimes referred to as "open cook") or in a contactor/kettle arrangement. The most critical aspect of either method is to ensure that the temperatures and mixing are sufficient to enable the complete saponification of the reactants, using a controlled heating and cooling schedule. Hot oil or steam usually heats the pressure vessel, whereas the mixing kettle might be heated with oil, steam, or fire.

In the mixing kettle, counterrotating mixing paddles move the grease to first dehydrate it in the heated kettle and then to aid with mixing of additional oil and additives. Circulating an appropriate hot or cold fluid through the jacketed portion of the mixing kettle carries out heating and cooling of the grease. Adding oil to the grease also cools the grease to a temperature appropriate for both including additives and packaging. A typical batch manufacturing process is illustrated in Figure 4.2.

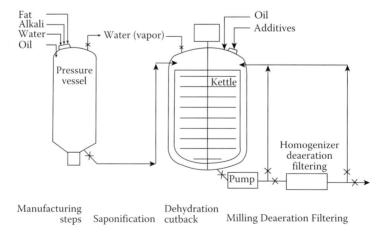

FIGURE 4.2 Typical batch manufacturing of lubricating grease.

Critical to the final structure of the grease is the milling process. This may be done at high or low temperatures using a variety of equipment as fundamental as a high shear rate pump, through which the grease circulates, to the more intense piston homogenizer or colloid mill. The most common purposes of milling are to break a fibrous structure or to improve the dispersion of the lubricating fluid in the thickener. Kettle milling via recirculation will break some fibrous structure, but milling in a homogenizer or other high shear equipment is required to improve dispersion.

Although considerable work has been done on the development of in-line or continuous manufacturing processes for greases, its use is limited to only a few manufacturers. In many ways, in-line grease manufacturing can be thought of as an automation of the batch manufacturing process. Advantages of the in-line process include less labor and a smoother, more uniform final product.

IV. GREASE CHARACTERISTICS

The general description of a grease includes its soap type, base oil viscosity, and consistency in terms of the materials used in its formulation and physical properties (some of which are visual observations). The type and amount of thickener and the viscosity of the fluid lubricant are critical formulation properties. Color and texture, or structure, although observed visually, can also have some correlation to performance and selection. Grease composition can have an impact on the performance attributes of a grease as shown in Table 4.2.

This description normally is supplemented by tests to determine the critical characteristics of grease such as consistency, apparent viscosity, and dropping point. Most of the other tests used to describe greases come under the category of evaluation and performance tests.

A. Consistency

Consistency is defined as the degree to which a plastic material resists deformation under the application of a force. In the case of lubricating greases, it is a measure of the relative hardness or softness and may indicate something of flow and dispensing properties. Consistency is reported in terms of American Society for Testing and Materials (ASTM) D217, Cone Penetration of Lubricating Grease or NLGI Grade. Consistency is measured at a specific temperature, 77°F (25°C), after a specific degree of shear (working).

1. Cone Penetration

The cone penetration of greases is determined with the ASTM D217 Penetrometer (Figure 4.3). The test itself involves filling a standard cup with grease conditioned to an established temperature of 77°F (25°C). The technique used to fill the cup is critical to obtaining consistent results, and therefore clearly established to ensure the top surface is free from defects. At this point, a weighted standard cone is released and allowed to sink into the grease, under its own weight, for 5 s. The

TABLE 4.2
Impact of Grease Composition on Performance

Grease Composition	Performance Consideration
Thickener type	Shear stability and resistance to water
Thickener amount	Flow and distribution in the application
Base oil type	High/low temperature performance
Base oil viscosity	Elastohydrodynamic lubrication/lubricity
Grease color	Visibility in service

FIGURE 4.3 Measuring grease consistency by cone penetrometer. The scale reads the depth that the cone has penetrated into the grease.

depth that the cone has penetrated is then recorded, in tenths of a millimeter, and is reported as the "unworked" penetration of the grease. Because the cone will sink farther into softer greases, higher penetration values indicate softer greases.

The grease is then "worked" or sheared using a standard drilled plate (Figure 4.4) for 60 double strokes in 60 s, after which the sample surface is again struck flush and the cone is released

FIGURE 4.4 Drilled plate and cup used to measure the worked penetration of a grease.

Lubricating Greases

TABLE 4.3
NLGI Grease Classification

NLGI grade	ASTM worked penetration[a]
000	445–475
00	400–430
0	355–385
1	310–340
2	265–295
3	220–250
4	175–205
5	130–160
6	85–115

Note: ASTM, American Society for Testing and Materials; NLGI, National Lubricating Grease Institute.

[a] Ranges are the penetration in tenths of a millimeter after 5 s at 77°F (25°C).

into the grease-filled cup. This measurement is known as the "worked" penetration, and is the basis for the value used to determine the NLGI consistency grade. It is considered to be the most reliable grease consistency test because the amount of disturbance of the sample is controlled and repeatable.

In addition to the standard full-size penetrometer equipment (ASTM D217) shown in Figure 4.3, 1/4- and 1/2-scale cone penetrometer equipment (ASTM D1403) is available for determining the penetrations of small samples. An equation is used to convert the penetrations obtained by ASTM D1403 to equivalent penetrations for the full-scale ASTM D217 test.

2. NLGI Grease Grade Numbers

On the basis of the ASTM worked penetrations, the NLGI has standardized a numerical scale for classifying the consistency of greases. The NLGI grade and corresponding penetration ranges, in order of increasing hardness, are shown in Table 4.3. This system has been well accepted by both manufacturers and consumers. It has proved adequate for specifying the preferred consistency of greases for most applications.

B. DROPPING POINT

The dropping point of a grease is the temperature at which a drop of material falls from the orifice of a test cup under prescribed test conditions (Figure 4.5). Two procedures are used (ASTM D566 and ASTM D2265) that differ in the type of heating units and, therefore, the upper temperature limits. An oil bath is used for ASTM D566 with a measurable dropping point limit of 500°F (260°C); ASTM D2265 uses an aluminum block oven with a dropping point limit of 625°F (330°C). Organo-clay soap thickened greases do not have a true melting point but have a melting range during which they become progressively softer. Some other types of greases may, without change in state, separate oil. In either case, only an arbitrary, controlled test procedure can provide a temperature that can be established as a characteristic of the grease.

The dropping point of a grease is not related to the upper operating temperature to which a grease can successfully provide adequate lubrication. Additional factors must be taken into account in high-temperature lubrication with grease. The dropping point test is most useful as a quality control measure during grease manufacture.

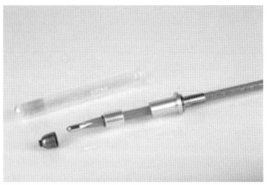

FIGURE 4.5 Dropping point test. This is the complete testing apparatus with viewing window. An enlarged view of the cup and thermometer is shown separately.

V. EVALUATION AND PERFORMANCE TESTS

The tests described in the previous section characterize greases. Most of the other tests for lubricating greases are designed to be useful in predicting performance under certain conditions. It should be noted that the precision of most grease tests is considered to be marginal at best. So, although the comparison of results obtained from the same laboratory should have good agreement when evaluating differences from batch to batch of the same material, the evaluation of greases using different laboratories, equipment, or staff should be done with consideration of this variability.

A. Mechanical or Structural Stability Tests

The ability of a grease to resist changes in consistency during mechanical working is termed its mechanical or structural stability. This is important in most applications because excessive softening as a result of mechanical shearing in service can lead to oil leakage and equipment failure. Hardening or firming as a result of shearing can be equally harmful in that it can prevent the grease from feeding oil properly to the equipment and can also result in its failure.

Generally, two methods are used for determining the structural stability of greases: Extended Worked Penetrations and Roll Stability. Determinations of prolonged worked penetration (see Section IV.A, "Consistency") can be made by extending the period of grease shearing using the ASTM D217 worker apparatus for 1000, 10,000, 50,000 or 100,000 double strokes.

In the standard Roll Stability Test (ASTM D1831), a small sample of grease is subjected to high shear from a weighted cylinder rolling in a chamber for 2 h at room temperature (Figure 4.6). The penetration after shearing is then determined with the 1/2- or 1/4-scale cone equipment. As with the extended worker test, the conditions of the roll stability test can be modified to increase the

FIGURE 4.6 Roll stability test: a heavy cylindrical roller rotates freely inside the tubular chamber on the test apparatus. Results on the left cylinder show that grease is more adhesive than the grease on the right after testing.

duration and resulting shear. In addition, it is possible to run the test at an elevated temperature to more closely simulate an application and evaluate thickener stability.

In both of these tests, the change in consistency with mechanical working is reported as either the absolute change in penetration or the percent change in penetration. The significance of the tests is somewhat limited because of the differences in test shear rates and the actual rates of shearing in a bearing. The shear rates in the tests range between 100 and 1000 reciprocal seconds (s^{-1}), whereas the shear rates in bearings may be as high as, or higher than, 1,000,000 s^{-1}.

B. Static Oxidation Test

Resistance to oxidation is an important characteristic for greases that are intended for use in rolling element bearings. Improvement in this property through the use of oxidation inhibitors has enabled the development of the so-called "sealed for life" bearings.

Both the oil- and the soap-based constituents in a grease oxidize; the higher the temperature, the faster the rate of oxidation, and the relationship of oxidation to temperature increase is exponential. When grease oxidizes, it generally acquires a rancid or oxidized odor and darkens in color. Simultaneously, organic acids usually develop and the lubricant becomes acid in reaction. These acids are not necessarily corrosive but may affect the grease structures, causing hardening or softening.

Laboratory tests have been developed to evaluate oxidation stability under both static and dynamic conditions. The two most frequently used static tests are the ASTM D942 (Oxidation Stability of Lubricating Greases by the Oxygen Pressure Vessel Method) and the ASTM D5483 (Standard Test Method for Oxidation Induction Time of Lubricating Greases by Pressure Differential Scanning Calorimetry).

In the ASTM D942 pressurized vessel test method, the grease is placed in a set of five dishes. The dishes are then placed in a stacked vessel, which is pressurized with oxygen to 110 psi (758 kPa), and placed in a bath held at 210°F (99°C), where it is allowed to remain for a specific time length, usually 100, 200, or 500 h. At the end of this time, the pressure is recorded and the amount of pressure drop reported. For specification purposes, pressure drops of 5 to 25 psi (34–172 kPa) are usually referenced, depending on the test time and the intended use of the grease.

The ASTM D5483 Pressure Differential Scanning Calorimetry (PDSC) method provides greater latitude in establishing a test temperature for relative performance ranking. In this test, a 2-mg sample of grease in an aluminum pan is placed into the front platform of the PDSC cell. An empty pan is placed as a reference on the rear platform. Both pans are subjected to a 500-psi pressure and constant flow (100 ml/min) of oxygen. The desired temperature is maintained at a steady-state value, and the test continues for 2 h. A graph of heat flow versus time is evaluated, and the oxidation induction time is reported.

The results of the ASTM D942 test are probably most indicative of the stability of thin films of a grease in extended storage such as in prelubricated bearings. It is not intended to predict the stability of a grease under dynamic conditions or in bulk storage in the original containers. Although there is no direct correlation between the ASTM D942 and D5483 method, the latter provides a fast and efficient technique to compare greases using a minimal sample size.

C. Dynamic Oxidation Tests

A number of tests are used to evaluate the oxidation stability of a grease under dynamic conditions. Two commonly used tests are the ASTM D3527 (Standard Test Method for Life Performance of Automotive Wheel Bearing Grease) and the ASTM D3336 (Performance Characteristics of Lubricating Greases in Ball Bearings at Elevated Temperatures). The tests differ principally in the type of bearing used and maximum operating temperature—ASTM D3527 provides for operation at 160°C (320°F), whereas the temperature of ASTM D3336 can reach as high as 371°C (700°F). The tests are run until the bearing fails or for a specific number of hours if failure has not occurred. Although these tests are considered to be useful screening methods for determining the projected service life of greased bearings operating at elevated temperatures, the FAG FE9 rolling bearing grease test rig has become an industry standard for evaluating the high-temperature performance of grease.

Run in accordance with the German Institute for Standardization DIN 51 821 procedure, the FAG FE9 test is similar to the ASTM D3336 test in that five bearings are run to failure. However, the equipment has much greater flexibility to run at a wider range of speeds (3000, 6000, and 9000 rpm), temperatures (120–200°C), and loads (1500, 3000, or 4500 N axial). The time at which the bearings have a failure probability of 10% and 50% (denoted by L10 and L50, respectively) are calculated from the data and reported.

D. Oil Separation Tests

The resistance of a grease to separation of oil from the thickener involves certain compromises. When greases are used to lubricate rolling element bearings, a certain amount of bleeding of the oil is necessary in order to perform the lubrication function. On the other hand, if the oil separates too readily from a grease in application devices, a hard, concentrated soap residue may build up, which will clog the devices and prevent or retard the flow of grease to the bearings. In bearings, excessive oil separation may lead to the buildup of a hard soap in bearing recesses, which in time could be troublesome. Furthermore, leakage of separated oil from bearings can damage materials in production or equipment components such as electric motor windings.

In application devices, such as central lubrication systems and spring loaded cups where pressure is applied to the grease on a more or less continuous basis, oil can be separated from greases by a form of pressure filtration. The pressure forces the oil through the clearance spaces around plungers, pistons, or spool valves; but because the soap cannot pass through the small clearances, it is left behind. This may result in blockage of the devices and lubricant application failure.

Some oil release resulting in free oil on the surface of the grease in containers in storage is normal. However, excessive separation is indicative of off-specification product, which should be discussed and possibly returned to the supplier.

Lubricating Greases

Generally, there is no accepted method for evaluating the oil separation properties of a grease in service. Trials in typical dispensing equipment may be conducted, and some of the dispensing equipment manufacturers such as Trabon, Alemite, and SKF-Lincoln have tests for their specific equipment to try to identify oil separation characteristics.

The tendency of oil separation in a grease during storage can be evaluated with ASTM D1742, Oil Separation from Lubricating Grease During Storage. In this test, air pressurized to 0.25 psi (1.72 kPa) is applied to a sample of grease held on a 75-μm (No. 200) mesh screen. After 24 h at 77°F (25°C), the amount of oil that separated is determined and reported. The test correlates directly with oil separation in other sizes of containers. Alternately, IP 121, Determination of Oil Separation from Lubricating Grease, can be used for extended periods at higher temperatures 40°C (102°F).

The tendency of a grease to separate oil at elevated temperatures under static conditions can be evaluated by Method 321.2 of FTM 791b. In this test, a sample of grease is held in a wire mesh cone suspended in a beaker. The beaker is placed in an oven, at approximately 212°F (100°C), for the desired time, usually 30 h. After the test, the oil collected in the beaker is weighed and calculated at a percentage of the original sample. Sometimes the test is used for specification purposes. Another previously used test method (ASTM D6184) was withdrawn in 2014.

E. Water Resistance Tests

The ability of a grease to resist washout under conditions where water may splash or impinge directly on a bearing is an important property in many applications (e.g., paper machine bearings and automobile wheel bearings). Comparative results between different greases can be obtained with ASTM D1264, Water Washout Characteristics of Lubricating Greases.

In this test, a ball bearing with increased clearance shields is rotated with a stream of water impinging on it at a rate of 5 ml/s (Figure 4.7). Resistance to washout is measured by the amount of grease lost from the bearing during the test. This test is considered to be a useful screening test for greases that are to be used wherever water washing may occur.

The ASTM D4049, Standard Test Method for Determining the Resistance of Lubricating Grease to Water Spray, is a more aggressive assessment of grease adhesion under pressurized water. Grease removal is measured when 38°C (100°F) water is sprayed at 40 psi for 5 min on a

FIGURE 4.7 Water washout test.

weighed quantity of grease. The cohesive and adhesive properties of the grease are measured by the percentage of grease removed after the first drying of the test fixture.

The DIN 51 807 test evaluates grease behavior in the presence of water and provides a challenging assessment for a submerged grease. A thin film is applied to a glass strip and placed in a test tube of water for 3 h at 90°C (194°F). At the end of the test, the grease and water appearance are visually rated.

In many cases, direct impingement of water may not be a problem, but a moist atmosphere or water leakage may expose a grease to water contamination. One method of evaluating a grease for use under such conditions is to homogenize water into it. This can be achieved using the ASTM D217 apparatus, and working the grease with various amounts of water to determine if the consistency changes over time. The grease may then be reported on the basis of the amount of water it will absorb without loss of grease structure, or on the amount of hardening or softening resulting from the admixture of a specific proportion of water. Table 4.1 contains some information characterizing the water resistance properties of greases based on thickener type.

F. Rust Protection Tests

In many applications, greases are not only expected to lubricate, but are also expected to provide protection against rust and corrosion. Some types of greases have inherent rust-protective properties, whereas others do not. Rust inhibitors can be incorporated in greases to improve rust-protective properties.

Both static and dynamic tests are used to evaluate the rust-protective properties of greases. Often, the test specimen is a rolling element bearing lubricated with the grease under test and then exposed to conditions designed to promote rusting. One typical static test is ASTM D1743, Standard Test Method for Determining Corrosion Preventive Properties of Lubricating Greases. In this test, tapered roller bearings are packed with the test grease, which is distributed by rotating the bearings for 60 s under a light load. The bearings are then dipped in distilled water and stored for 48 h at 125°F (52°C) and 100% relative humidity. After storage, the bearings are cleaned and examined for rusting or corrosion. Results are judged on a pass/fail basis.

An alternate method to determine corrosion protection performance is the ASTM D6138 Standard Test Method for Determination of Corrosion-Preventive Properties of Lubricating Greases Under Dynamic Wet Conditions (Emcor Test). This method evaluates grease-lubricated ball bearing rust prevention in the presence of water or a variety of water solutions. Double-row ball bearings are packed with grease and are rotated at 80 rpm for 8 h during the first 3 days of the test. Four days of no rotation follow. The test is generally run with the solution placed in the pillow block at the beginning of the test, but can also be run using small pumps to continuously replenish the solution during an 8-h running period.

G. Extreme Pressure and Wear Prevention Tests

Although the results of laboratory extreme pressure and wear prevention tests do not necessarily correlate with service performance, the tests presently provide a means to evaluate these properties at a reasonable cost. The ASTM has standardized the test procedures to determine the extreme pressure properties of greases using the ASTM D2596, Four-Ball Extreme Pressure test (Figure 4.8). The ASTM has also standardized tests for wear prevention properties using the ASTM D2266, Four-Ball Wear Tester Machine.

The extreme pressure test (D2596) is considered to be capable of differentiating between greases having low, medium, or high levels of extreme pressure properties. The wear prevention test (ASTM D2266) is intended to compare only the relative wear preventive characteristics of greases in sliding steel-on-steel applications. It is not intended to predict wear characteristics with other metals.

FIGURE 4.8 Four-ball extreme pressure test.

H. Grease Compatibility

Greases are available with many different types of thickeners, additives, and base oils. The mixing of different greases could result in altering the performance or physical properties (owing to incompatibility), which could lead to a grease mixture that exhibits inferior characteristics to those of either grease before mixing. The mixing of incompatible greases will alter properties such as consistency, pumpability, shear stability, oil separation, and oxidation stability. Generally, when two incompatible greases are mixed, the result is a softening that can lead to increased leakage as well as loss of other performance features.

Equipment performance problems as a result of mixing incompatible greases could manifest themselves after a relatively short period of operation but usually occur over longer periods, sometimes making it difficult to diagnose the problem source back to the mixing of incompatible greases. When it is necessary to use different greases in an application, it is best to mix greases of the same thickener. In all cases where different greases will be mixed, it is recommended to first review grease compatibility charts (such as in Table 4.4) and then consult with grease suppliers to identify whether potential problems could exist. The safest practice is to avoid mixing of greases.

I. Apparent Viscosity

Newtonian fluids, such as normal lubricating oils, are defined as those materials for which the shear rate (or flow rate) is proportional to the applied shear stress (or pressure) at any given temperature. That is, the viscosity, which is defined as the ratio of shear stress to shear rate, is constant at a given temperature. Grease is a non-Newtonian material that does not begin to flow until a shear stress exceeding a yield point is applied. If the shear stress is then increased further, the flow rate increases more proportionally, and the viscosity, as measured by the ratio of shear stress to shear rate, decreases. The observed viscosity of a non-Newtonian material such as grease is called its apparent viscosity. Apparent viscosity varies with both temperature and shear rate; thus, it must always be reported at a specific temperature and flow rate.

TABLE 4.4
Grease Compatibility Guidelines

	Aluminum Complex	Barium	Calcium	Calcium 12-Hydroxystearic Acid	Calcium Complex	Clay	Lithium	Lithium 12-Hydroxystearic Acid	Lithium Complex	Polyurea
Aluminum complex	X	I	I	C	I	I	I	I	C	I
Barium	I	X	I	C	I	I	I	I	I	I
Calcium	I	I	X	C	I	C	C	B	C	I
Calcium 12-hydroxystearic acid	C	C	C	X	B	C	C	C	C	I
Calcium complex	I	I	I	B	X	I	I	I	C	C
Clay	I	I	C	C	I	X	X	I	I	I
Lithium	I	I	C	C	I	X	X	C	C	I
Lithium 12-hydroxystearic acid	I	I	B	C	I	I	C	X	C	I
Lithium complex	C	I	C	C	C	C	C	C	X	I
Polyurea	I	I	I	I	C	I	I	I	I	X

Source: National Lubricating Grease Institute (NLGI) spokesman.

Note: B, borderline compatibility; C, compatible; I, incompatible; X, same grease.

Apparent viscosities of greases are determined in accordance with ASTM D1092. In this test, samples of a grease are forced through a set of capillary tubes, at predetermined flow rates. From the dimensions of the capillaries, the known flow rates, and the pressure required to force the grease through the capillaries at those flow rates, the apparent viscosity of the grease, in poise (p), can be calculated. Results are usually reported graphically as apparent viscosity versus shear rate at a constant temperature, or apparent viscosity versus temperature at a constant shear rate.

Apparent viscosity is used to predict the handling and dispensing properties of a grease. In addition, it can be related to starting and running torque, in grease lubricated mechanisms, and is useful in predicting leakage tendencies.

BIBLIOGRAPHY

Mobil Oil Corporation. 1970. *Mobil Technical Bulletin: Extreme Pressure Lubricant Test Machines*. Internal. Mobil Oil Corporation Publication, New York.

Mobil Oil Corporation. 1988. *Lubricating Greases: Physical and Performance Tests*. Internal. Mobil Oil Corporation Publication. Fairfax, VA.

5 Synthetic Lubricants

Considerable attention has been focused on synthetic lubricants since their introduction into the retail market of synthetic engine oils in the 1970s. The use of synthetic lubricants in aviation and industrial applications also extends back many years. Past interest in synthetic lubricants was primarily attributed to their ability to provide equipment protection advantages under extreme operating conditions. These conditions included very low or very high temperatures and their ability to resist oxidation to a much greater degree than mineral oils. More recent interest still focuses on these performance capabilities, but also on how synthetic lubricants can save money and minimize the impact on the environment.

The terms "synthetic" and "synthesized" are both used to describe the base fluids used in these lubricants. A synthesized material is one that is produced by combining or building individual components into a unified entity. The production of synthetic lubricants starts with synthetic base stocks that are often manufactured from petroleum-based chemicals. The base fluids are made by chemically combining (synthesizing) lower molecular weight compounds into higher molecular weight compounds that have adequate viscosity for use as lubricants. Unlike mineral oils, which are a complex mixture of naturally occurring hydrocarbons, synthetic base fluids are man-made and tailored to have a controlled molecular structure with predictable properties.

As discussed in Chapter 3, the properties of a mineral lubricating oil result from the selection of those crude oil compounds that have the best properties for the intended application. This is accomplished through fractionation, solvent refining, hydrogen processing, solvent dewaxing, and filtration. However, even with extensive treatment, the finished product is a mixture of many compounds. There is no way to select from this mixture only those materials with the best properties, and if there were, the yield would be so low that the process would be uneconomical. Thus, the mineral oils produced have properties that are the average of the mixture, including both the most and the least suited components. This is not to say that lubricants made from mineral-based oils are unsatisfactory. On the contrary, lubricants properly formulated from mineral-based oils will provide good performance in the majority of applications. Synthetic lubricants, on the other hand, possess additional performance advantages.

With synthetic lubricant base stocks, the process of combining individual units can be controlled so that a large proportion of the finished base fluid is either one or only a few compounds. Depending on the starting materials and the combining process, the synthetic base stock can have the properties of the best compounds found in mineral base oil. It can also have unique properties not found in any mineral oil, such as miscibility with water or nonflammable properties just to name a couple.

In either case, the special properties of finished synthetic lubricants often justify any additional cost when used in applications where mineral oil lubricants do not provide adequate performance. Some of the primary applications for synthetic lubricants are listed in Table 5.1.

There are several performance advantages associated with the use of synthetic lubricants. Two of the primary advantages include their extended service life capability and ability to handle a wider range of application temperatures. Their outstanding low-temperature flow characteristics and their stability at high temperatures often make them the preferred choice of lubricant. The comparative operating temperature limits of mineral oil and synthetic lubricants are shown in Figure 5.1.

There are various ways to classify synthetic base fluids. Some of these classification approaches neglect similarities between certain types of materials, whereas others may lead to confusion by grouping materials that have similarities in chemical structure but are totally dissimilar from a

TABLE 5.1
Synthetic Lubricants Application and Recommendations

Equipment: Lubricant Unit	Operating Conditions	Advantages of Synthetic Oils
	Industrial	
Calendars—rubber, plastics, board, tile	High temperature	Extended service; reduced deposits, oxidation, and thermal cracking
Paper machines—dryers, calendars, drive gear units	High temperatures, water contamination	Extended service; reduced deposits, oxidation, and thermal cracking
Nuclear power plants—vertical coolant motors, 6000–9000 hp	Annual oil change 8000 hours min	Extended service, fewer deposits
Gas engines	Low-temperature start-up	Extended drains, cold temperature starting, fuel economy, lower rates of degradation
Gas turbines	High temperature and low ambients	Extended service, broader temperature application range, fewer deposits
Steam turbines—electrohydraulic control, throttle/governor	Near-superheated steam lines	Fire resistance
Tenter frame and high-temperature conveyor chains	150–260°C (302–500°F)	Reduced deposits and improved wear protection, extended service, less consumption, little or no smoke
Hydraulic systems	−40°C to 93°C (−40°F to 200°F)	Better low-temperature pumpability and high-temperature stability
Enclosed gears—parallel, right angle, worm	Moderate to heavy duty, shock-loaded, severe service	Extended service, better oxidation resistance at elevated temperatures, improved efficiency
Refrigeration compressors—SRM license screw and reciprocating compressors	Severe service	Improved refrigerant solubility, compatibility with HFC refrigerants
Machine tool spindles, freezer plants—motors, conveyors, bearings	High speeds, low temperature	Extended service, minimize thermal distortion, low-temperature start-up
Metal die-casting hydraulic systems	Molten metal, source of ignition	Fire resistance
Mining—continuous miners and associated equipment	Fire hazards exist	Fire resistance
Primary metals—slab, continuous casters, rolling mills, shears, ladles, furnace controls	Fire hazards exist	Fire resistance
Air compressors—reciprocating	Severe service	Extended service, fewer deposits
Air compressors—rotary screw	Severe service	Extended service, fewer deposits
Greased/lubricated bearings	Low to high speeds	Wide-temperature service range, longer relubrication intervals
	High to very high speeds	Improved low-temperature starting, lower stabilized temperature in service
Grease/lubricated gears	Heavy-duty severe service	Extended service, low-temperature starting
	Automotive	
Passenger car gasoline engines and front wheel drive manual transmissions	All	Improved fuel economy, low-temperature starting, oil economy, wear protection
On- and off-highway gas and diesel engines	All	Improved low-temperature starting and operation, longer drain intervals, fuel and oil economy
Truck and rear wheel drive cars, drive axles: hypoid spiral bevel, and spur gears	All	Improved low-temperature starting and operation, wear protection

(Continued)

TABLE 5.1 (CONTINUED)
Synthetic Lubricants Application and Recommendations

Equipment: Lubricant Unit	Operating Conditions	Advantages of Synthetic Oils
Automotive		
Passenger car automatic transmissions—Ford, GM	All	Longer life, improved high- and low-temperature operation, greatly improved wear protection
Commercial manual transmissions	All	Improved low-temperature starting and operation, longer drain intervals, fuel and oil economy
Wheel and clutch bearings	All	Longer relubrication intervals, improved low-temperature starting, reduced leakage vs. oils
Aviation—Military and Commercial		
Commercial turbine engines—Pratt & Whitney; Allison: GE, SNECMA, Rolls-Royce Avon, IAE, MIL-L-23699D approved	Temperatures to 220°C (428°F)	Wide-temperature service range, high-temperature stability
Military turbine engines—MIL-L-7808J approved	Temperatures to 190°C (374°F)	High-temperature stability
Aircraft all—wheel bearings, wing flap-screws—MIL-G-81322D approved	Temperature −55°C to 180°C (−67°F to 351°F)	Wide-temperature service range, high-temperature stability
Marine		
High- to medium-speed marine diesel engines (1.5% sulfur fuel)	Very severe	Improved fuel economy, low-temperature starting, oil economy and wear protection, high operating temperature stability

Source: Mobil Oil Corporation, 1994.

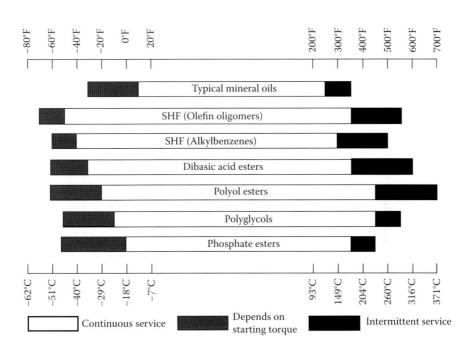

FIGURE 5.1 Comparative temperature limits of mineral oil and synthetic lubricants.

performance or application standpoint. For the purposes of this discussion, synthetic base fluids are classified as follows:

1. Synthesized hydrocarbon fluids (SHFs)
2. Organic esters
3. Polyglycols
4. Phosphate esters
5. Other synthetic lubricating fluids

The first four classes account for more than 90% of the volume of synthetic lubricant base stocks used today. The fifth class includes a number of materials, some of which are quite high in cost, that are generally used in specialized, lower volume applications. The classes of synthetic lubricant base stocks and some of the various types of synthetics associated with each class are listed in Table 5.2. Performance advantages, as well as some limiting properties, of synthetic base stocks are outlined in Table 5.3.

I. SHFs

SHFs represent the largest type of synthetic lubricant base stocks in the market. They are pure hydrocarbons, manufactured from chemical raw materials. Three types are used in considerable volume: polyalphaolefins (PAOs), alkylated aromatics, and polybutenes. A fourth type, cycloaliphatics, is used in small volumes in specialized applications.

TABLE 5.2
Various Classes and Types of Synthetic Base Stocks

Class	Type
Synthesized hydrocarbon Fluids	Polyalphaolefins
	Alkylated aromatics
	Polybutenes
	Cycloaliphatics
Organic esters	Dibasic acid ester
	Polyol ester
Polyglycols	Polyalkylene
	Polyoxyaklylene
	Polyethers
	Glycol
	Polyglycol esters
	Polyalkylene glycol ester
	Polyethylene glycol
Phosphate esters (phosphoric acid esters)	Triaryl phosphate ester
	Triakyl phosphate ester
	Mixed alkylaryl phosphate esters
Other	Silicones
	Silicate esters
	Fluorocarbons
	Polyphenyl ethers

TABLE 5.3
Advantages and Limiting Properties of Synthetic Base Stocks

Synthetic	Advantages versus Mineral Oil	Limiting Properties
SHF	High temperature stability	Solvency/detergency[a]
	Long life	Seal compatibility[a]
	Low temperature fluidity	
	High viscosity index	
	Improved wear protection	
	Low volatility, oil economy	
	Compatibility with mineral oils and paints	
	No wax	
Organic Esters	High temperature stability	Seal compatibility
	Long life	Mineral oil compatibility
	Low temperature fluidity	Antirust[a]
	Solvency/detergency	Antiwear and extreme pressure[a]
		Hydrolytic stability
		Paint compatibility
Phosphate Esters	Fire resistance	Seal compatibility
	Lubricating ability	Low viscosity index
		Paint compatibility
		Metal corrosion[a]
		Hydrolytic stability
Polyglycols	Water versatility	Mineral oil compatibility
	High viscosity index	Paint compatibility
	Low temperature fluidity	Oxidation stability[a]
	Antirust	
	No wax	

[a] Limiting properties of synthetic base fluids which can be overcome by formulation chemistry.

A. PAOs (Olefin Oligomers)

These materials are formed by combining a low molecular weight material, into a specific olefin. Molecules of this olefin are linked together to make a lubricant range material, which is then hydrogen-stabilized. By varying the number of olefin molecules linked together, a wide array of lubricant viscosities can be produced, with a range exceeding that of mineral-based oils. Because they are synthesized from pure chemical feeds, PAOs are generally free of sulfur, nitrogen, and aromatics, resulting in high-performance product formulations. The structure of a typical PAO base oil molecule is shown in Figure 5.2.

PAOs can be considered a special type of paraffinic mineral oil, comparable in properties to the very best components found in petroleum-derived base oils. They have high viscosity indexes (VIs; from 135 to more than 200), excellent low temperature fluidity, and very low pour points. Their shear stability is excellent, as is their hydrolytic stability. Because of the saturated nature of hydrocarbons, both their oxidation and thermal stability are good. Volatility is lower than with comparably viscous mineral oils; thus, evaporation loss at elevated temperatures is lower. In many applications, it is important that PAOs are similar in composition to mineral oils and that they are compatible with mineral oils, additive systems developed for use in mineral oils, and machines designed to operate on mineral oils. PAOs do not cause any significant softening or swelling of typical seal materials, which may be a disadvantage in systems where slight swelling of the seals

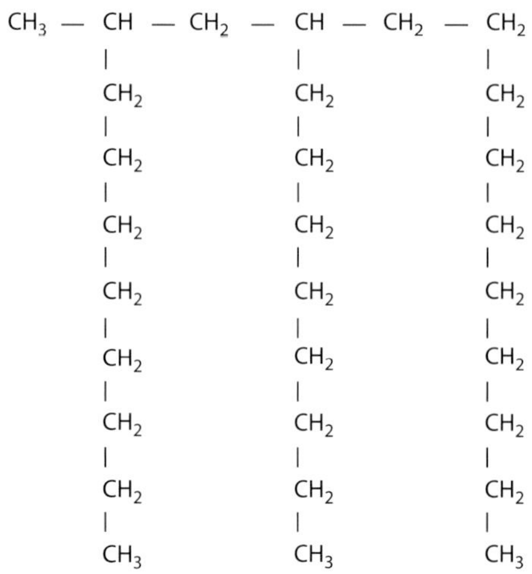

```
CH₃ — CH — CH₂ — CH — CH₂ — CH₂
      |              |              |
      CH₂            CH₂            CH₂
      |              |              |
      CH₂            CH₂            CH₂
      |              |              |
      CH₂            CH₂            CH₂
      |              |              |
      CH₂            CH₂            CH₂
      |              |              |
      CH₂            CH₂            CH₂
      |              |              |
      CH₂            CH₂            CH₂
      |              |              |
      CH₂            CH₂            CH₂
      |              |              |
      CH₃            CH₃            CH₃
```

FIGURE 5.2 Polyalphaolefin. This compound, an oligomer of 1-decene, is a low-viscosity oil suitable for blending low-viscosity engine oils, as well as a variety of other lubricants.

is desirable to keep them tight and pliable to prevent leakage. However, proper formulation of the finished lubricants can overcome this problem.

PAOs are used widely as automotive and industrial lubricants. In aviation applications, greases formulated from PAOs and an inorganic thickener are widely used as general-purpose aircraft greases.

In industrial applications, PAOs may be combined with one of the organic esters or other synthetic compounds as the base fluid in engine oils, gear oils, compressor oils, turbine oils, circulating oils, and hydraulic fluids. SHFs are widely used in gear applications because they possess two unique advantages over mineral oils: lower traction coefficients and higher oxidation stability. Lower traction coefficients translate directly into greater energy efficiency (up to 8%) and their higher oxidation stability enables longer drain intervals. Properly formulated SHF lubricants can yield 6–10 times the oil drain intervals of comparable mineral oils barring any significant contamination of the fluid. They are also formulated as wide temperature range hydraulic fluids, power transmission fluids, and heat transfer fluids. Wide temperature range greases made from PAOs combined with a grease thickener are finding increasing acceptance as long-life rolling element bearing greases for severe duty applications. Occasionally, the acronym SHC, which stands for synthetic hydrocarbon, has been used interchangeably with the acronym SHF. Mobil SHC™ is a registered trademark of ExxonMobil Corporation.

B. Alkylated Aromatics

Alkylated aromatics are classified by the American Petroleum Institute (API) as Group V fluids. The alkylation process involves joining alkyl groups containing 10–18 carbon atoms to the aromatic molecule. The properties of the final product can be altered by changing the structure and position of the alkyl group. Dialkylated benzene, a typical alkylated aromatic in the lubricating oil range, is shown in Figure 5.3.

Alkylated aromatics have excellent low temperature fluidity and low pour points, and exhibit good solubility for additives and degradation materials. Their VIs are about the same as, or slightly higher than, high VI mineral oils. They are less volatile than comparable viscous mineral oils, and more stable to oxidation, high temperatures, and hydrolysis. However, it is more difficult to incorporate inhibitors, and the lubrication properties of specific structures may be poor. As with PAOs, alkylated aromatics are compatible with mineral oils and systems designed for mineral oils.

Synthetic Lubricants

Dialkylated benzene   Alkylated naphthalene

FIGURE 5.3 Dialkylated benzene and alkylated naphthalene are examples of alkylated aromatics. The R group is an alkyl group that usually contains 10 to 18 carbon atoms for products in the lubricating oil range.

1. Application

Alkylated aromatics are used as the base fluid in some engine oils, gear oils, power transmission fluids, gas turbine oils, air and refrigeration compressor lubricant, hydraulic fluids, and greases in subzero applications. They are also used as additives or supplements in the formulation of some specialty lubricants.

C. Polybutenes

Polybutenes are not always considered in discussions of synthesized lubricant bases, although there are a number of lubricant-related applications for the materials. Polybutenes are produced by controlled polymerization of butenes and isobutene (isobutylene). A typical structure is shown in Figure 5.4. The lower molecular weight materials produced using this process have lubricating properties, whereas higher molecular weight materials, usually referred to as polyisobutylenes, are often used as VI improvers and thickeners.

Polybutenes in the lubricating oil range have VIs between 70 and 110, have fair lubricating properties, and can be manufactured to have excellent dielectric properties. An important characteristic in a number of applications is that polybutenes decompose completely to gaseous materials when they exceed their decomposition temperature of approximately 550°F (288°C).

1. Application

One of the major uses of polybutenes is for electrical insulating oils. They are used as cable oils in high-voltage underground cables, as impregnates for insulating paper for cables, as liquid dielectrics, and as impregnates for capacitors. Significant volumes have been used as lubricants for rolling, drawing, and extrusion of aluminum when the aluminum is to be annealed afterward. Other applications include gas compressor lubrication, open gear applications, grease additives, and viscosity modifier additives for lubricants; they are also used as a carrier for solid lubricants.

D. Cycloaliphatics

Cycloaliphatics are a class of synthesized hydrocarbons now used in small quantities because of certain special properties they possess. One typical structure is shown in Figure 5.5. Cycloaliphatics are sometimes referred to as traction fluids. Under high stresses, they develop a glasslike structure

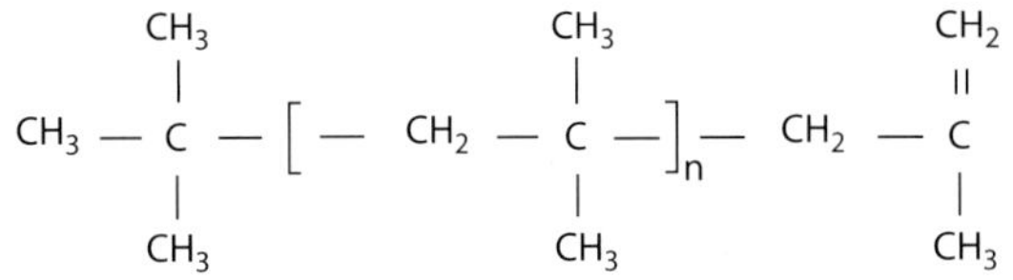

FIGURE 5.4 Typical molecular structure found in polybutenes. Viscosity is a function of how many units, n, of butene compose the oligomer.

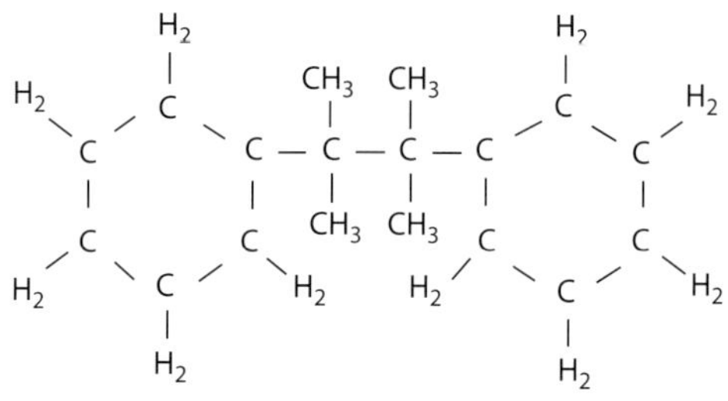

FIGURE 5.5 Typical cycloaliphatic structure.

and can transmit shear forces—that is, they have high traction coefficients. At the same time, they perform somewhat of a lubricating function in that they prevent welding and metal transfer from one surface to the other. Their stability is excellent under these conditions.

1. Applications

The main application for cycloaliphatics at present is in stepless, variable speed drives in which the torque is transmitted from the driving member to the driven member by the resistance to shear of the lubricating fluid. The high traction coefficients of the cycloaliphatics permit higher power ratings than do conventional lubricants.

Cycloaliphatics have also found some application in rolling element bearings where, because of speed and load conditions, skidding of the rolling elements may occur with conventional lubricants.

II. ORGANIC ESTERS

Organic esters have been an important class of synthesized base fluids longer than any of the other materials now in use. Their use dates back to World War II in Germany when they were used in mineral oil blends to improve low-temperature properties and to supplement scarce supplies of mineral oils. They were first used as aircraft jet engine lubricants in the 1950s and are now used as the base fluid for essentially all aircraft jet engine lubricants. They are also used as the base fluid in many wide temperature range aircraft greases.

Organic esters are oxygen-containing compounds that result from the reaction of an alcohol with an organic acid. Two of the commonly used classes of organic esters are dibasic acid esters and polyol esters.

A. Dibasic Acid Esters

The various diesters differ in their acid and alcohol components. For all diesters, acid and alcohol are reacted either thermally or in the presence of a catalyst in an esterification reactor. After the ester has been formed, the water by-product is distilled off, and the unreacted dibasic acid is neutralized and removed by filtration. The base stock is then suitable for final product blending. As shown in Figure 5.6, the backbone of the structure is formed by the acid, with the alcohol radicals joined to its ends.

Dibasic acid esters (diesters) exhibit good metal-wetting ability, high film strength, high oxidation and thermal stability, and good shear stability. Diesters will dissolve system deposits and keep metal surfaces clean. This could be a disadvantage when installing diesters in dirty systems. Their

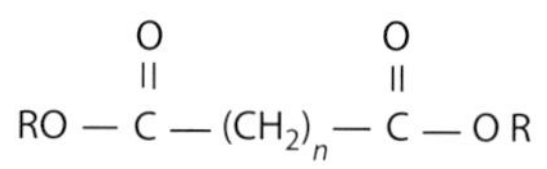

FIGURE 5.6 Dibasic acid ester. Commonly used acids are adipic ($n = 4$), azelaic ($n = 7$), and sebacic ($n = 8$).

hydrolytic stability and antirust properties are fair. When changing over to diesters from another family of products, care should be taken to assure that thorough cleaning and flushing of the system takes place before their installation. A review of compatibility with elastomers and paints that are used in the system should also be conducted. Diesters are compatible with mineral oils.

1. Application

Diesters are used as base oils, or components of the base oil, for automotive engine oils and air compressor lubricants. They are also used as the base fluid in some aircraft greases. Dibasic acid esters have been used as the base fluid for older type I jet engine oils. Generally, the use of these oils has been restricted mainly for older military jet engines, and some very limited use in jet engines for industrial service. These products that were developed originally for use as jet engine oils have been replaced, to a very large extent, by polyol esters.

B. Polyol Ester

Polyol esters are made by reacting a polyhydric alcohol with a monobasic acid to give the desired ester. In contrast to diesters, with polyol esters (Figure 5.7) the polyol forms the backbone of the structure with the acid radical attached to it. As with diesters, the physical properties of polyol esters can be varied by using different polyols or acids. Trimethylol propane and pentaerithritol are two of the polyols that are commonly used. Usually, the acids are obtained from animal or vegetable oils.

Polyol esters have better high temperature stability than diesters. Their low temperature properties and hydrolytic stability are about the same, but their VIs may be lower. Their volatility is equal to or lower than that of diesters. The polyol esters also may have more effect on paints and cause greater swelling of elastomers.

1. Application

Their primary use of polyol esters is in type II jet engine oils. They also are used in air compressor oils and as components in some synthesized hydrocarbon blends. More recently, polyol ester have become widely used as refrigeration lubricants because of their miscibility with

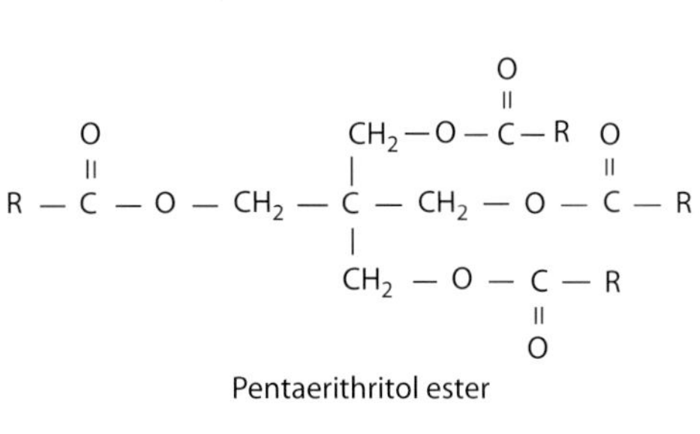

$$\begin{array}{c}
 CH_2 - O - \overset{\overset{O}{\|}}{C} - R \\
 | \overset{O}{\|} \\
CH_3 - CH_2 - C - CH_2 - O - C - R \\
 | \\
 CH_2 - O - C - R \\
 \underset{\|}{}\\
 O
\end{array}$$

Trimethylolpropane ester

$$\begin{array}{c}
 CH_2 - O - \overset{\overset{O}{\|}}{C} - R \\
\overset{O}{\|} | \overset{O}{\|} \\
R - C - O - CH_2 - C - CH_2 - O - C - R \\
 | \\
 CH_2 - O - C - R \\
 \overset{\|}{O}
\end{array}$$

Pentaerithritol ester

FIGURE 5.7 Polyol esters. The esters shown here are made from the alcohols trimethylolpropane and pentaerythritol. Acid radicals (contain R) typically are composed of 5–10 carbons.

hydrofluorocarbon (HFC) refrigerants. Polyol esters are also the lubricant of choice in refrigeration systems where nonchlorine HFC refrigerants and blends are used. They are also used in environmentally sensitive products such as hydraulic fluids and other lubricants where biodegradable materials are required.

III. POLYGLYCOLS

Polyglycols cover a wide range of products and properties. At present, they are the largest single class of synthetic lubricant bases.

Polyglycols are described as polyalkylene glycols (PAGs), polyethers, polyglycol ethers, and polyalkylene glycol ethers. The latter term is the most complete and accurate for the bulk of the materials used in lubricants. Smaller quantities of simple glycols, such as ethylene and polyethylene glycol, are also used as hydraulic brake fluids. Typical structures for the two types are shown in Figure 5.8.

PAGs are polymers made from ethylene oxide (EO), propylene oxide (PO), or their derivatives. The primary raw materials are ethylene or propylene, oxidized to form cyclic ethers (alkylene oxides). Combining ethers derived from EO to PO has a profound effect on the solubility of the product in other fluids, as shown below.

EO:PO = 4:1 Water-soluble, not soluble in hydrocarbons
EO:PO = 1:1 Soluble in cold water, soluble in alcohol and glycol ethers, not soluble in hydrocarbons
EO:PO = 0:1 Not soluble in water, conditionally soluble in hydrocarbons

These comparisons help explain the differences between the various types of, and uses for, polyglycol: automotive antifreeze, brake fluid, water-based hydraulic fluids, hydrocarbon gas compressors, and high-temperature bearing lubricants. In addition to using ethylene and POs, butylene oxide is used to provide some PAGs with specific properties and are oil soluble. Polyglycols made with butylene oxide are more expensive and do not exhibit traction coefficients equal to combinations of EO and PO.

One of the major advantages of PAGs is that they decompose completely to volatile compounds under high temperature conditions. This results in low sludge or varnish buildup under high operating temperatures or could result in complete decomposition, without leaving any deposits, in certain extremely hot applications.

Polyglycols have good viscosity–temperature characteristics, although at low temperatures they tend to become more viscous than some of the other synthesized base stocks. Pour points are relatively low. High temperature stability ranges from fair to good and may be improved with additives. Thermal conductivity is high. Polyglycols are not generally compatible with mineral oils or additives developed for use in mineral oils but are not as compatible with many paints and finishes. They have low solubility for hydrocarbon gases and some refrigerants. Seal swelling is low, but with the water-soluble types some care must be exercised in seal selection to be sure that the seals are compatible with water. Even if the glycol fluid does not initially contain any water, it has a tendency to absorb moisture from the atmosphere.

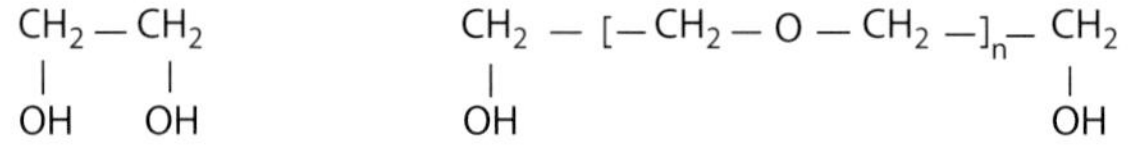

$$CH_2 - CH_2 \qquad CH_2 - [-CH_2 - O - CH_2 -]_n - CH_2$$
$$|\quad\quad |\qquad\qquad\qquad |\qquad\qquad\qquad\qquad\qquad\quad |$$
$$OH \quad\ OH \qquad\qquad\ OH \qquad\qquad\qquad\qquad\qquad\ OH$$
Ethylene glycol   Polyethylene glycol ether

FIGURE 5.8 Polyglycols. This class of fluids is most commonly formed from ethylene oxide to form polyethylene glycol or propylene oxide to form polypropylene glycol.

A. Application

The applications for polyglycols are divided into those that are water soluble and those that are insoluble with water.

The largest volume application of water-soluble polyglycols is in hydraulic brake fluids. Other major applications include metalworking lubricants and fire-resistant hydraulic fluids. In the latter application, polyglycol is mixed with water, which provides the fire resistance. Water-soluble polyglycols are also used in the preparation of water-diluted lubricants for rubber bearings and joints.

Water-insoluble polyglycols are used as heat transfer fluids, in certain types of industrial hydraulic fluids, and as high-temperature gear and bearing oils. They are also used as lubricants for refrigeration compressors operating on some HFC refrigerants, and for compressors handling hydrocarbon gases. In these latter applications, the low solubility of the gases in PAG minimizes the dilution effect, contributing to better high temperature lubrication.

IV. PHOSPHATE ESTERS

Phosphate esters are one of the other commonly used classes of API Group V synthetic base fluids. A typical phosphate ester structure is shown in Figure 5.9.

One of the major features of phosphate esters is their fire resistance, which is superior to that of mineral oils. Their lubricating properties are also generally good. The high temperature stability of phosphate esters is only fair, and their decomposition products can be corrosive. Generally, they have poor viscosity–temperature characteristics (low VIs), although pour points are reasonably low and their volatility is quite low. Phosphate esters do have a considerable effect on many paints and finishes, and may cause swelling of many seal materials. Their compatibility with mineral oils ranges from poor to good, depending on the ester. Their hydrolytic stability is only fair. They have specific gravities greater than 1, which means that any water contamination tends to float rather than settle to the bottom, and pumping losses are higher than lower specific gravity products. Their costs are generally high, and they are limited in viscosity.

A. Application

The major application of phosphate esters is in fire-resistant fluids of various types. Hydraulic fluids for commercial aircraft are phosphate ester based, as are many industrial fire-resistant hydraulic fluids. These latter fluids are used in applications such as the electrohydraulic control systems of steam turbines and industrial hydraulic systems, where hydraulic fluid leakage might contact a source of ignition. In some cases, they may also be used in the turbine bearing lubrication system.

Phosphate esters are also used as lubricants for compressors where discharge temperatures are high, to prevent receiver fires and explosions that might occur with conventional lubricants. Some phosphate esters are used in greases and mineral oil blends as a wear- and friction-reducing additive.

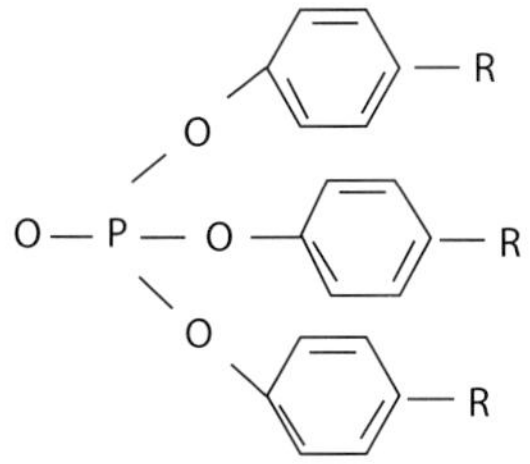

FIGURE 5.9 Phosphate ester. The "R" group can be either an aryl or an alkyl type. If, for example, a tolyl group is used, the ester is tricresyl phosphate.

V. OTHER SYNTHETIC LUBRICATING FLUIDS

Brief descriptions and principal applications of some of the other synthetic base stocks are given in Sections V.A–V.D.

A. SILICONES

These are one of the older types of synthetic fluids. As shown in Figure 5.10, their structure is a polymer type with the carbons in the backbone replaced by silicon.

Silicones have high VIs, some on the order of 300 or more. Their pour points are low and their low temperature fluidity is good. They are chemically inert, nontoxic, fire resistant, and water repellent, and have low volatility. Seal swelling is low. Compressibility is considerably higher compared with that of mineral oils. Thermal and oxidation stability of silicones are good up to quite high temperatures. If oxidation does occur, their oxidation products include silicon oxides, which can be abrasive. A major disadvantage of the common silicones is that they have low surface tension that permits extensive spreading on metal surfaces, especially steel. As a result, effective adherent lubricating films are not formed. Unfortunately, the silicones that exhibit this characteristic also show poor response to wear- and friction-reducing additives.

1. Application

Silicones are used as the base fluid in both wide temperature range and high temperature greases. They are also used in specialty greases designed to lubricate elastomeric materials that would be adversely affected by other types of lubricants. Silicones are also used in specialty hydraulic fluids for such applications as liquid springs and torsion dampers where their high compressibility and minimal change in viscosity with temperature are beneficial. They are also being used as hydraulic brake fluids and as antifoam agents in lubricants. Some newer silicones are also offered as compressor lubricants.

B. SILICATE ESTERS

Silicate esters have excellent thermal stability and with proper inhibitors, show good oxidation stability (see Figure 5.11 for their chemical structure). They have excellent viscosity–temperature characteristics, and their pour points are low. Their volatility is low, and they have fair lubricating properties. A major factor for their limited use is their poor resistance to hydrolysis.

1. Application

Silicate esters are used as heat transfer fluids and dielectric coolants. Some specialty hydraulic fluids are formulated with silicate esters.

C. POLYPHENYL ETHERS

These organic materials typically contain two to six phenyl rings and can contain oxygen or sulfur. They have excellent high temperature properties and outstanding radiation resistance. Polyphenyl

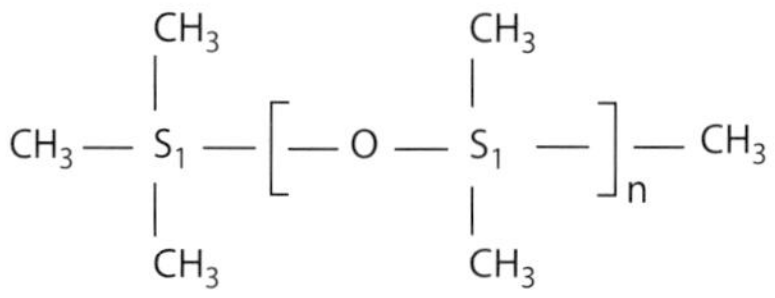

$$CH_3 - \underset{\underset{CH_3}{|}}{\overset{\overset{CH_3}{|}}{Si}} - \left[-O - \underset{\underset{CH_3}{|}}{\overset{\overset{CH_3}{|}}{Si}} - \right]_n - CH_3$$

FIGURE 5.10 Dimethyl polysiloxane; one of the more widely used silicone polymer fluids.

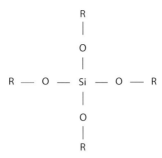

FIGURE 5.11 Silicate ester. These can be made with R groups containing either alkyl or aryl groups. The physical properties are dependent on the nature of these groups.

ethers are thermally stable to above 800°F (450°C) and have excellent resistance to oxidation at elevated temperatures. However, they have high viscosities at normal ambient temperatures, which tend to restrict their use.

1. Application

Small quantities of polyphenyl ethers are used as heat transfer fluids, as lubricants for high vacuum pumps, and as the fluid component of radiation-resistant greases.

D. Halogenated Fluids

Chlorine or fluorine or combinations of the two are used to replace part (or all) of the hydrogen in hydrocarbon or other organic structures to form lubricating fluids. Generally, these fluids are chemically inert, essentially nonflammable, and often show excellent resistance to solvents. Halogenated fluid examples include perfluoroalkylpolyethers and polychlorotrifluoroethylene. Some have outstanding thermal and oxidation stability, being completely unreactive even in liquid oxygen. Their volatility may be also extremely low.

1. Application

Some of the lower-cost halogenated hydrocarbons are used alone or in blends with phosphate esters as fire-resistant hydraulic fluids. Other halogenated fluids are used for such applications as oxygen compressor lubricants, lubricants for vacuum pumps handling corrosive materials, solvent-resistant lubricants, and other lubricant applications where highly corrosive or reactive materials are being handled.

BIBLIOGRAPHY

Mobil Oil Corporation. 1985. Synthetic Lubricants. Mobil Oil Corporation, New York.

6 Environmental Lubricants

Protecting and preserving the environment is a concern in all aspects of our daily lives. It is only a natural progression that advances in lubricant technology would result in the development of oils and greases that would be less harmful to the environment if inadvertently spilled or leaked. The acceleration of research and development in this area has been driven by public demand, industry concerns, and governmental agencies to find better ways to protect the biological balance in nature or at least reduce the negative impact where spills or leakage of lubricants do occur. Many names such as environmentally friendly, ecofriendly, environmentally acceptable, environmentally aware, biodegradable, and nontoxic have traditionally been given to these lubricants. For purposes of discussions in this chapter, we will refer to this class of lubricants as environmentally acceptable lubricants (EALs).

Of primary interest in the selection and use of this class of EALs is defining and measuring the product attributes that could affect the environment. In particular, the importance of providing satisfactory lubrication performance has increased over time. These lubricants must provide performance in key areas such as oxidation stability, viscosity–temperature properties, wear protection, friction reduction, rust and corrosion protection, hydrolytic stability (where water or moisture may be present), and overall oil life. In other words, the EALs must perform equivalent to conventional mineral- or synthetic-based lubricants in the equipment while mitigating potential adverse environmental impact. Performance compromises are no longer accepted.

I. ENVIRONMENTAL CONSIDERATIONS

There have been recent standardization efforts through various international organizations such as the International Organization for Standardization (ISO) and Deutsches Institut für Normung (DIN) (e.g., CEN/TR 16227:2011) or governmental agencies (e.g., European Union [EU] Ecolabel, U.S. Environmental Protection Agency [EPA]) to better define the environmental acceptability of lubricants. This can encompass a broad range of potential environmental benefits. These would include use of renewable resources, resource conservation, pollutant source reduction, recycling, disposability, degradability, and reduced environmental impact. Therefore, any claim of environmental acceptability must be supported by appropriate technical documentation. Most petroleum-based lubricants can be considered environmentally acceptable by various standards. For example, long-life synthetics (discussed in Chapter 5) and other lubricants that provide extended oil drain capability or energy efficiency benefits might be classified as EA materials because they contribute to reducing consumption of resources. Many oils can be reclaimed, recycled, or burned for their heat energy value, again resulting in conservation of resources. All of these efforts to help reduce the environmental impact of lubricants have positive effects and should be an integral part of the planning process to establish an environmental program.

When initially introduced in the 1990s, biodegradability and low toxicity were the main criteria used to define EALs. More recently, the potential for bioaccumulation has been added to the scope. In addition to being biodegradable and exhibiting low toxicity to aquatic organisms, an EAL should also have a low potential for bioaccumulation. The remainder of this chapter will be devoted to discussions of a class of lubricants that exhibit specific characteristics such as biodegradability, low toxicity, and low bioaccumulation potential.

II. DEFINITIONS AND TEST PROCEDURES

The three environmental characteristics most desirable in EALs are speed at which the products will biodegrade if introduced into nature, the toxicity characteristics that might affect bacteria or aquatic life, and the bioaccumulation of the lubricant with the tissues of organisms. Most lubricants are inherently biodegradable, which means that given enough time, they will biodegrade by natural processes. They will not persist in nature. There are, however, applications where much faster rates of biodegradation are desired. These will be referred to as *readily biodegradable* products. All lubricants range in toxicity from low (sometimes called nontoxic) to relatively high. Toxicity has a direct effect on naturally occurring bacteria and aquatic life and therefore needs to be an important part of the development of EALs. Bioaccumulation is an indication of the speed and concentration of the accumulation of chemicals over time within the tissues of an organism.

Unlike traditional lubricant development, where the predominant focus is on product performance in equipment, a major part of developing EALs also involves the understanding and definition of environmental test criteria. This would include assessing the effects of new and used lubricants in actual applications where environmental sensitivity is an issue. Because both base stocks and additive systems impact the environmental characteristics, these tests must evaluate the ecotoxicity of base stocks, additives, and finished lubricants. However, lubricant performance remains critical when the challenge of EAL development is to achieve environmental benefits without compromising the performance of the equipment. This performance concern has been an area of focus for lubricant developers in addition to ensuring that definitions and test methods are well understood by both lubricant suppliers and end users.

A. Toxicity

The impact of lubricants on the aquatic environment is evaluated by conducting acute aquatic toxicity studies on aquatic life forms that are sensitive to changes in their environment (an example is using rainbow trout, a freshwater fish that is sensitive to environmental changes).

There are several test methods used to evaluate the toxicity impact on aquatic life. One such test recognizes that oil is insoluble in water and exposes aquatic specimens to an oil–water dispersion condition, through mechanical dispersion, in increasing concentrations of test materials up to a maximum concentration of 5000 ppm. This procedure, which simulates physical dispersion by wave and current action, is used to evaluate the relative toxicity of lighter-than-water materials. The aquatic specimens are exposed to five concentrations of test material and a control (without test material) during each study. Toxicity is then expressed as the concentration of test material in ppm (wt/vol) required to kill 50% of the aquatic specimens after 96 h of exposure (LC_{50}).

Table 6.1 provides a listing of the most common Organization for Economic Cooperation and Development (OECD) test methods, along with the species tested, that are used in industry to evaluate aquatic toxicity.

B. Biodegradability

Biodegradability measures the extent and speed of degradation of a chemical (or chemical mixture) by microorganisms. Traditionally, there has always been a distinction between primary and ultimate biodegradation. Primary biodegradability is the degradation of one part of the molecule, resulting into the conversion of a toxic material into a less toxic or even nontoxic substance. Ultimate biodegradation refers to the decomposition of the chemical, or the mixture of chemicals, into three basic molecules: carbon dioxide, water, and mineral salts.

Biodegradability can then be defined as either *inherent* or *ready*. At times, this has created confusion in the market because end users are not necessarily familiar with the detailed definitions. *Inherent* biodegradability refers to the material showing some degree of biodegradability

TABLE 6.1
OECD Aquatic Toxicity Tests

Aquatic Toxicity Test	Test Method
Growth inhibition test, algae	OECD 201
Acute immobilization test, *Daphnia*	OECD 202
Acute toxicity test, fish	OECD 203
Prolonged toxicity test (14-day study), fish	OECD 204
Respiration inhibition test, bacteria	OECD 209
Early-life stage toxicity test, fish	OECD 210
Reproduction test, *Daphnia magna*	OECD 212
Short-term toxicity test on embryo and sac-fry stages, fish	OECD 212

in one of several industry tests. This definition can be misleading because most chemicals or chemical mixtures are biodegradable to some extent and could be classified as inherently biodegradable. In comparison, readily biodegradable substances have to demonstrate a certain ratio of ultimate biodegradation within a specific time frame and according to a defined test method.

Common test methods are described in Table 6.2 and can be classified into two main categories:

1. OECD or American Society for Testing and Materials (ASTM) test methods, which measure ready biodegradability defined as the conversion of 60% of the original material into CO_2, after a 10-day window after the start of the biodegradation, which must be completed within a maximum of 28 days after the start of the test.
2. Co-ordinating European Council (CEC) test methods, which measure the overall biodegradation of organic compounds through the infrared absorbance of extractable lipophilic material, and requires 80% or more of biodegradation.

The CEC method does not differentiate between primary and ultimate biodegradation, contrary to the OECD and ASTM methods. As a result, the CEC method is usually considered less severe than the other two methods. The measured biodegradability percentages using the CEC test method can therefore be higher compared with any of the OECD or ASTM methods. This is the reason why any biodegradability claim should always refer to the specific test method that was used to determine the biodegradation ratio of the original material.

C. BIOACCUMULATION

As mentioned previously, bioaccumulation is an indication of the speed and concentration of the accumulation of chemicals over time within the tissues of an organism.

Bioaccumulation is directly related to water solubility. The less water soluble the material is, the more lipophilic, and therefore the more it will have affinity for organic tissues. Water solubility is directly related to the types of atoms present in the molecules. The more oxygen atoms a substance has, the more likely it will be water soluble because water is composed of hydrogen and oxygen atoms. Alkanes, alkenes, and alkynes (all hydrocarbons) are all present in mineral oil base stocks and are composed of carbon and hydrogen only. As a result, mineral oils are very lipophilic and not water soluble. Comparatively, vegetable oils (essentially composed of natural esters), synthetic esters, and polyalkylene glycols (PAGs) do have a number of oxygen atoms in their molecules and are therefore more water soluble. As a result, mineral base stocks have directionally a much higher bioaccumulation potential than synthetic or natural esters or PAGs (see Table 6.3).

TABLE 6.2
Examples of Standardized Test Methods to Measure Biodegradability

Test Type	Test Name	Measured Parameter	Short Description	Minimum Requirement to Claim Biodegradability	Method
Ready Biodegradability: a substance can be considered biodegradable if it shows a minimum of 20% biodegradability within the test duration	Modified Sturm Test	CO_2	Aerobic degradation, ultimate biodegradation	≥60%	OECD 301B
	Modified MITI Test	DOC	Aerobic degradation, measurement of CO_2 consumption, suitable for volatile components	≥70%	OECD 301C
	Closed Bottle Test	BOD/COD	Aerobic degradation, preferred for water soluble products but suitable for nonwater-soluble substances	≥60%	OECD 301D
	MOST	DOC	Manometric respirometry test, suitable for both water soluble and nonwater-soluble	≥70%	OECD 301F
	Sturm Test	CO_2	Aerobic degradation, measurement of CO_2 consumption, specifically designed for water insoluble materials	≥60%	ASTM D5864
	Shake Flake Test	CO_2	Aerobic degradation, measurement of CO_2 consumption, specifically designed for water insoluble materials	≥60%	EPA 560/6-82-003
	BODIS Test	BOD/COD	Two-phase closed bottle test (similar to OECD 301D)	≥60%	ISO 10708
Hydrocarbon degradability	CEC Test	IR spectrum	Hydrocarbon degradability through IR measurement	≥80%	CEC-L33-A-94
Screening test	CO_2 Headspace Test	CO_2	Aerobic degradation (sealed vessels with a headspace of air) suitable for both water soluble and nonwater-soluble (under certain test conditions for the latter)	≥60%	ISO 14593

Note: BOD, biochemical oxygen demand; COD, chemical oxygen demand; DOC, dissolved organic carbon; IR, infrared.

TABLE 6.3
Summary of Bioaccumulation Potential by Base Oil Types

Lubricant Base Oil	Potential for Bioaccumulation
Mineral oil	High
Polyalkelene glycols (PAGs)	Low
Synthetic esters	Low
Vegetable oils	Low

Another factor influencing the potential for bioaccumulation is molecular weight. The higher the molecular weight, the larger the molecules are and the more difficult it is for them to pass through the very thin cell membranes of organic tissues. Therefore, smaller molecules have a lower potential for bioaccumulation than long-chain, higher molecular weight molecules. This is one of the reasons why the use of long-chain polymers, as an additive in lubricants, is very often under scrutiny in environmental requirements such as EU Ecolabel.

III. ENVIRONMENTAL CRITERIA

At present, there is no single universally accepted regulation that defines the criteria for lubricants used in environmentally sensitive areas. However, there have been significant efforts around standardization, and the U.S. EPA Vessel General Permit (VGP) 2013 or European Ecolabel have started to be well recognized and used as international references.

Although the European Ecolabel was intended to replace all European national or regional labels, the latter are still in place and most often are required in the countries that developed them. This section provides a high-level overview of the biodegradability, aquatic toxicity, and bioaccumulation requirements of these different labels.

A. NATIONAL LABELING PROGRAMS

1. Blue Angel

The Blue Angel Ecolabel is the first national environmental labeling scheme that was introduced in Germany in 1988. Criteria were defined by applications, with different requirements and specific standard for hydraulic fluids, lubricating oils, and greases.

Under this labeling scheme, a lubricant can carry a Blue Angel label if it is biodegradable, nontoxic to aquatic species, and nonbioaccumulative. Biodegradability can be assessed either through OECD 301 B to F tests to determine ultimate biodegradation or CEC-L-33-A-93 to measure primary biodegradability.

The secondary criteria include a ban on specific hazardous materials and must also meet aquatic toxicity limits. This labeling scheme has been more stringent than other schemes because of its requirements on ultimate biodegradability. Aquatic toxicity is measured using the typical OECD 201 to 203 methods. There are also performance requirements defined depending on the application, and they are part of the requirements to obtain the certification. However, contrary to other programs, it does not have any requirement with regard to content of renewable materials.

2. Swedish Standard

This Swedish Ecolabel includes standards for hydraulic fluids (SS 155434) and greases (155470). It was jointly developed by industry and the Swedish government and is still very popular, in particular, for hydraulic oils. It includes biodegradability (according to ISO test methods), aquatic toxicity, and also the sensitizing properties of both the finished product and its components. Contrary to the German Blue Angel, it also includes requirements for renewable resource content.

3. U.S. Department of Agriculture BioPreferred®

The focus of the U.S. Department of Agriculture BioPreferred program, like the European Ecolabel, is to increase the purchase of biobased products. The U.S. Department of Agriculture BioPreferred program is applicable not only to lubricants, but it actually defines 97 different categories of products. Each of these categories defines the minimum biobased content for products within the category. Its definition of biobased products for lubricants is directly related to the use of renewable raw materials as an alternative to conventional petroleum-derived raw materials.

4. U.S. VGP (U.S. VGP 2013)

The U.S. VGP started in 1972 with the Clean Water Act, aimed at regulating discharges in U.S. national waters. At that time, the National Pollutant Discharge Elimination System was established, which requires a permit for discharge of pollutant from so-called "point sources." Vessels operating within defined U.S. waterways including 3 nautical miles from the coast and within navigable Great Lake waters are under U.S. jurisdiction and therefore need a permit to operate their discharges.

The latest version of the U.S. VGP was published in 2013 and applies to large vessels exceeding 79 ft (24 m), operating in U.S. national waters, excluding the vessels of the Armed Forces of the United States. All vessels built on or after December 19, 2013, must comply with the 2013 VGP requirements.

A specific subset of requirements was developed for smaller vessels under the EPA's VGP.

One of the key requirements of the 2013 VGP is that all vessels must use EALs in all oil-to-sea interfaces, unless technically unfeasible. EALs are defined as lubricants being biodegradable, "minimally toxic," and not bioaccumulative. Definitions of these criteria are as follows:

- Biodegradable: minimum of 60% biodegradation within 28 days according to OECD 301 B for 90% of the lubricant formulation or 75% of the grease formulation. The finished lubricant may contain up to 10% of components not meeting the 60% threshold of biodegradability and up to 5% of nonbiodegradable (but not bioaccumulative) components. For grease, 25% may be either inherently or nonbiodegradable (but not bioaccumulative).
- Minimally toxic: the finished product must pass the OECD 201, 202, and 203 acute toxicity tests or OECD 210 and OECD 211 for chronic toxicity. As an alternative, an evaluation may be conducted on a constituent basis.
- Not bioaccumulative: each component of the lubricant formulation that is not biodegradable has to be tested using one of the five designated test methods to demonstrate that it is not bioaccumulative.

B. INTERNATIONAL LABELING PROGRAMS

1. Nordic Swan

The Nordic Swan was jointly developed by Norway, Sweden, Finland, Iceland, and Denmark. It is considered to be among the most comprehensive and demanding label in Europe because it includes requirements for biodegradability, aquatic toxicity, renewability, and technical performance. The renewability requirements are the most stringent of all ecolabeling programs (minimum of 65% content of renewable material required for hydraulic, transmission, gear oils, and greases). As a result, there are actually very few products that can claim they meet this ecolabel.

2. European Ecolabel

Issued in 2009, the European Ecolabel for lubricants had the objective to replace all existing national labeling schemes. Although this objective may have not been reached yet, this ecolabel establishes the framework for a more international approach to ecolabeling.

Some of the details of the European Ecolabel include prohibiting the use of some substances that have a specific hazard classification and limiting the concentrations of other substances.

It defines five categories of lubricants for which the program is applicable, along with their associated specific technical and environmental requirements:

Category 1: hydraulic fluids and tractor transmission oils
Category 2: greases and stern tube greases
Category 3: chainsaw oils, concrete release agents, wire rope lubricants, stern tube oils and other total loss lubricants
Category 4: two-stroke oils
Category 5: industrial and marine gear oils

The environmental criteria for the European Ecolabel include biodegradability, aquatic toxicity, and bioaccumulation on both the finished lubricant and the components used to manufacture the lubricants. Requirements for biodegradability and bioaccumulative potential must be fulfilled for each stated substance present above 0.10%. The lubricant should not contain substances that are both nonbiodegradable and (potentially) bioaccumulative.

Renewability requirements are defined for each of the categories as follows:

Category 1: ≥50% (m/m)
Category 2: ≥45% (m/m)
Category 3: ≥70% (m/m)
Category 4: ≥50% (m/m)
Category 5: ≥50% (m/m)

In addition, minimal technical requirements are defined that refer to existing international standards:

- Hydraulic fluids should at least meet the technical performance criteria specified in ISO 15380, Tables 6.2 through 6.5.
- For industrial and marine gear oils, the minimum technical performance requirements of DIN 51517 should be met.
- Chainsaw oils should at least meet the technical performance criteria laid down in the RAL UZ 48 of the Blue Angel.
- Two-stroke oils for terrestrial application should at least meet the technical performance criteria specified in ISO 13738:2000.
- For two-stroke oils for marine applications, at least the technical performance criteria specified in NMMA TC-W3 of the NMMA Certification for Two-Stroke Cycle Gasoline Engine Lubricants' should be met.
- For all other lubricants: fit for purpose.

IV. ENVIRONMENTAL CHARACTERISTICS OF VARIOUS BASE STOCKS

Aquatic toxicity and ready biodegradation studies were conducted on products formulated with mineral oils and nonmineral oil base stocks (Tables 6.4 and 6.5). In general, base stocks, which comprise the major component of most lubricant formulations, are nontoxic. Any aquatic toxicity

TABLE 6.4
Summary of Typical Biodegradability of Various Lubricant Base Oils

Lubricant Base Oil	Biodegradation
Mineral oil	Persistent/inherently[a]
Polyalkylene glycol (PAG)	Readily
Synthetic ester	Readily
Vegetable oils	Readily

[a] Depends on the type of mineral base oil (e.g., paraffinic base stocks are typically more biodegradable than naphthenic base stocks).

TABLE 6.5
Comparison of Fully Formulated EA Lubricants with Various Base Stocks

Properties	Mineral	Vegetable	PAOs	Diesters	Polyol Esters	PAGs
Viscosity/temperature characteristics	Fair	Good	Good	Fair	Very good	Very good
Low temperature properties	Poor	Poor	Very good	Good	Good	Good
Oxidation stability	Fair	Poor	Very good	Good	Good	Good
Compatibility with mineral oils	Excellent	Excellent	Excellent	Good	Fair	Poor
Low volatility	Fair	Good	Excellent	Excellent	Excellent	Good
Varnish and paint compatibility	Excellent	Very good	Excellent	Poor	Poor	Poor
Seal swell (NBR)	Excellent	Excellent	Very good	Fair	Fair	Good
Lubricating properties	Good	Very good	Good	Very good	Very good	Good
Hydrolytic stability	Excellent	Poor	Excellent	Fair	Fair	Very good
Thermal stability	Fair	Fair	Fair	Good	Good	Good
Additive solubility	Excellent	Excellent	Fair	Very good	Very good	Fair

Note: These ratings are generalizations. Specific manufacturers of products should be consulted for current data. Properties of PAG will differ depending on whether they are water soluble or water insoluble.

observed after exposure to the formulated lubricants is generally attributable to one or more of the additives.

Vegetable oils and a number of synthetic esters easily meet the ready biodegradation criterion (>60% conversion to CO_2 in 28 days) and most often have CEC test results exceeding 90% conversion after 21 days. Mineral oil based lubricant formulations typically show biodegradability results in the range of 40% (conversion to CO_2 in 28 days) but generally are unable to reach the 60% minimum threshold criteria to be considered readily biodegradable. Although this may not appear to be a significant difference from the 60% criteria, in actuality it is a major difference as seen in field conditions.

The biodegradation of PAGs is determined by the ratio of propylene oxide to ethylene oxide, with polyethylene glycols being more biodegradable. The average molecular weight of the material is also critical, with material under a molecular weight of 1000 being rapidly biodegraded. The rate and extent of biodegradation diminishes with increasing molecular weight.

A. Overview of Base Stock Options

Historically, one of the primary choices of base stocks for EALs was *vegetable oils*. This was attributed to their good natural biodegradability and very low toxicity while possessing very good lubricity characteristics. These renewable resources were also quite interesting from a cost perspective over other biodegradable and nontoxic base oils such as synthetic base stocks (esters or, more recently, PAGs). However, although some performance limitations of vegetable based lubricants have been and continue to be addressed, they are no longer the primary choice for EALs. Some of the continued performance limitations of vegetable oil are high temperature oxidation stability, low temperature performance, and a limited viscosity range. They are still less expensive relative to synthetic alternatives but can be several times the cost of conventional mineral-based lubricants, with performance debits versus these lubricants. In the future, genetic engineering will provide improved performance in areas of oxidation stability and low temperature performance by increasing the high oleic acid content as well as other genetic alterations (branching). In most applications, the vegetable-based EALs can be formulated to perform in all but the most severe equipment. It is important to note that not all vegetable-based EALs will provide the same levels of performance. Vegetable base oils derived from rapeseed plants, cotton seeds, soybean oil, sunflower seed oil, corn oil, palm oil, and peanut oils are frequently used materials with rapeseed being the most common.

Synthetic base stocks such as *polyglycols* (discussed earlier), *polyol esters, pentaerithritol esters,* and certain *polyalphaolefins* (PAOs) (see Chapter 5) are used to formulate synthetic EALs. Their advantage over vegetable-based EALs are wider temperature range application, longer oil drain capability (owing to higher oxidation stability), and excellent performance in systems with close tolerance servo valves.

The following are some of the more general performance characteristics of the various base materials:

1. *Vegetable oils.* The choice of the correct processes used to refine, bleach, and deodorize vegetable-based stocks can yield very satisfactory base materials for the formulation of finished lubricants. This renewable resource provides excellent natural lubricity, low volatility, and good environmental characteristics. Their weaknesses are in low temperature performance, hydrolytic stability, and oxidation stability in high temperature applications. They are also currently limited to low viscosity (ISO 32–68 viscosity grades) materials. Properly manufactured and formulated vegetable-based lubricants can equal conventional mineral oil based lubricants in terms of performance in all but the most severe applications.
2. *PAOs.* PAOs provide a good option for formulating low-viscosity environmental lubricants based on their ready biodegradability in the lower viscosity range. From a performance standpoint, they also provide excellent low and high temperature (oxidation stability) performance, good hydrolytic stability, low volatility, with the potential for energy efficiency benefits, which is quite interesting from an overall sustainability standpoint. Their disadvantages, however, are in initial cost and their lower rates of biodegradability as viscosities increase. To achieve the good characteristics of PAOs in finished products, they are often blended with biodegradable synthetic esters to achieve both the performance and environmental characteristics desired.
3. *Synthetic esters.* Several materials based on synthetic esters exhibit good biodegradability as well as high levels of oxidation stability, low and high temperature performance, good hydrolytic stability, and seal swell performance. The synthetic esters will also allow the formulation of higher viscosity lubricants typically used in circulating systems and some gear oils.

Table 6.5 shows a general comparison of the more common performance characteristics of fully formulated EALs using various base stocks. The actual finished product performance could vary from these ratings based on additive technology, use of blends of these base materials, manufacturing processes, and other technological advancements.

As a result of the higher costs, EALs will typically be used in areas where environmental sensitivity is an issue. In many instances of spillage or leakage that are reportable to governmental agencies, the added costs of EALs may be more than offset by the potential for lower fines, penalties, and remediation costs.

EALs are not meant for use in all applications but are recommended in cases where their use can be economically justified or where environmental sensitivity issues are of prime importance. In many cases, economic justification of EALs based solely on equipment performance is not sufficient to merit their use. The economics must be derived from the reduction in remediation costs in the event of a spill or leakage. Also, in some localities, limited legislation or regulations promote or require the use of such products. Environmental sensitivity issues generally prevail in the following areas:

- Dredging operations for waterways
- Operation of equipment for dams and locks
- Offshore drilling
- Marine equipment
- Recreation and parks
- Construction sites on or near water or ground water systems
- Agricultural operations
- Forestry and logging
- Mining
- Automotive service lifts
- Hydraulic elevators
- U.S. coastal waterways (fall under the NDPDES VGP 2013 or small vessel VGP)

V. PRODUCT AVAILABILITY AND PERFORMANCE

Because of the large volumes of hydraulic fluids used around the world and the tendency of these products to leak under conditions of relatively high pressures and the severity of some applications, the first category of EALs to be developed and widely marketed were hydraulic oils. Once readily biodegradable base oils and low toxicity additive systems had been identified, the next hurdle was to develop fully formulated oils that provide the required equipment performance as established by the builders of equipment as well as the users. Equipment builders have received many requests to approve the use of EALs by their customers and need to be assured that the EAL products will perform satisfactorily in their equipment and meet the service life requirements of the customers.

Certain EALs are formulated not only to meet environmental criteria, but also to provide performance equal to that of conventional mineral-based lubricants. Much of the performance in hydraulic systems is determined by industry standard pump tests. Three common tests are described.

- **Vickers 35VQ Pump Test (ASTM D6973)**
  - This severe industry-accepted antiwear vane pump test is based on the Vickers 35VQ25 vane pump run at 3000 psi and 203°F (95°C). Standard procedures require that the fluid be subjected to a 50-h test run, and the total ring and vane wear be less than 90 mg at the end of the run.
- **Denison Hydraulics HF-0, 1, and 2 approvals**
  - This test evaluates the multimetal compatibility of a fluid and corrosiveness to soft metals under severe operating conditions.
  - The (A-TP-30533) 600-h test uses a T6H20C Series Hybrid Piston and Vane Pump and is run in two phases of 300 h each. Phase one evaluates the fluid performance in the absence of water, and 1% of water is added at the beginning of phase two. The tests are run at a pressure of 3600 psi (250 bar) on the vane and 4000 psi (280 bar) on the

piston pump. The pump inlet temperature is controlled at 110°C during phase one and 80°C in phase two.
- The pass/fail criterion for this pump test is based on physical inspection and measurement for distress, wear, and chemical etching on working parts of the piston and vane pump. The inspection must be performed by the builder to obtain the written approval.

The approval level is HF-1 (vane pump), HF-2 (piston pump), and HF-0 for both.

- **Denison HF-6 Approval**
  - This approval is dedicated to biodegradable hydraulic oils. The fluid is evaluated according to the same procedure (A-TP-30533) as for HF-0, but only in the dry phase.

The results of these three relatively severe pump tests indicate that environmentally acceptable antiwear hydraulic fluids can be formulated not only to meet biodegradability and toxicity requirements but can also provide performance levels at least equal to conventional mineral-based antiwear hydraulic fluids in these tests. These test results have also been verified by years of field application experience. Again, some EALs may or may not be equal in performance to conventional hydraulic fluids, and suppliers should be consulted for specific test results and examples of field performance.

In addition to environmentally acceptable antiwear hydraulic fluids, several gear oils, some greases, engine oils, circulating oils, metalworking fluids, transmission fluids, and chainsaw oils are available in readily biodegradable and low toxicity formulations.

A. Vegetable Oil–Based EAL Performance Concerns

As mentioned earlier, the potential performance concerns in vegetable-based lubricants relative to comparable premium mineral oil based lubricants are oxidation stability, low temperature performance, and hydrolytic stability. These are not necessarily an issue as long as they are recognized, and proper care is taken during selection, application, and operation.

1. Oxidation Stability

The conventional laboratory oxidation tests designed for turbine oils (ASTM D2722 RPVOT and ASTM D943 TOST), but often used for other oils, yield poorer results for vegetable-based lubricants. These tests were designed for evaluating the oxidation characteristics of highly refined mineral oil based products with higher levels of oxidation inhibitors but with low levels of other additives that could interfere with achieving high RPVOT and TOST results. For example, there is a poor correlation between high RPVOT and TOST values and long-term performance of antiwear mineral oil based hydraulic oils.

In fact, some premium quality hydraulic oils with lower RPVOT and TOST values perform much better and provide longer service life than those with higher values. It is well recognized that vegetable oil base stocks do exhibit poorer oxidation stability, but with proper processing, and use of correct additives, the finished formulation can provide satisfactory performance in the majority of applications. Most manufacturers and suppliers of EALs will provide guidelines for applications such as those shown in Table 6.6.

2. Low Temperature Performance

The low temperature performance of vegetable oil based products will naturally be poorer than those based on highly refined mineral oil based products. Without the use of VI improvers and pour point depressants, paraffinic mineral oil based products also exhibit deficiencies in low temperature performance. Vegetable oils respond to VI improvers and pour point depressants to substantially lower finished product pour points and low temperature fluidity. Pour points of vegetable oil based

TABLE 6.6
Environmentally Acceptable Lubricants and Application Data

Product	Description	Application	Benefits
Vegetable-based hydraulic and circulating oil	An ISO viscosity grade 32/46 vegetable oil based antiwear hydraulic oil that is readily biodegradable and virtually nontoxic	Primary application is for industrial and mobile equipment hydraulic systems operating at temperatures of 0°F to 180°F. Meets requirements of major hydraulic pump manufacturers.	Provides good system performance while substantially reducing the negative impact on the environment when inadvertently leaked or spilled.
Synthetic-based hydraulic and circulating oil	Ester-based, readily biodegradable, and virtually nontoxic antiwear hydraulic and circulating oil. Product available in four ISO viscosity grades: 32, 46, 68, and 100, all with excellent oxidation stability and wide temperature range application capability	Primary application is for hydraulic and circulation systems operating in moderate to severe applications such as mobile equipment hydraulic systems where low and high temperatures exceed the limitations of vegetable-based EA oils. Recommended for a temperature range from −20°F to 200°F. Owing to their high degree of antiwear and high FZG (12+ stages), they can also be used in gear units not requiring EP additives.	Provides excellent performance over a wide temperature range while demonstrating excellent biodegradation and virtual nontoxicity. If inadvertently spilled or leaked, they significantly reduce the environmental impact relative to mineral oils and can help reduce or eliminate fines and the costs of remediation.
Multipurpose grease	Synthetic-based EP NLGI #1 and #2 grade greases	Designed for multipurpose outdoor applications where grease leakage or run out could contaminate soil, groundwater, or surface water systems. They can be used for indoor applications where grease leakage could enter plant water systems. They are recommended for plain and rolling element bearings and couplings that operate at temperatures from −15°F to 250°F.	Provides excellent lubrication characteristics over a wide temperature range while reducing the potential for negative impact on the environment. They are compatible with most other greases.

products are less meaningful than mineral oil based products and should not be used as an indication of the lowest application temperature.

A more meaningful indication of low temperature performance of EALs is the solidification point and low temperature pumpability. These better represent how the product performs under longer cold soak conditions. This information is important for outdoor applications such as mobile equipment where fluids may be subjected to subzero temperatures for extended periods. If the application involves low temperatures, data on low temperature performance should be obtained from the supplier.

3. Hydrolytic Stability

It is almost impossible to keep moisture out of most lubrication systems, and water can be detrimental to lubricant performance regardless of the base material used to formulate the lubricant. Vegetable oils, as well as most natural and synthetic esters, have poorer hydrolytic stability than comparable mineral oil based products. The proper formulation of EALs is the key to minimizing the potential for negative hydrolysis effects. Current studies of a specific EAL indicate that it takes severe water contamination (>0.1% allowable limit for both conventional and EA fluids in critical systems) to adversely affect the lubricants performance because of hydrolysis or additive depletion.

VI. PRODUCT SELECTION PROCESS

Of the many choices of available EALs, the selection process for a lubricant can be more difficult than that used for selecting conventional non-EALs. Although a more difficult process, many of the product performance aspects in actual equipment applications are essentially the same. The selection process needs to include the following aspects of product performance:

Environmental acceptability
Product physical specifications
Equipment builder approvals
Evidence of proven field performance
Supplier reliability
Operating and maintenance considerations

The major difference lies in the first and last items in this list: environmental acceptability and operating and maintenance considerations.

A. Environmental Acceptability

This area encompasses the greatest difference in selecting EALs versus selecting conventional mineral based lubricants. Although most conventional lubricants are inherently biodegradable and can be low in toxicity, EALs need to meet more stringent requirements as was discussed earlier in this chapter (Section II, "Definitions and Test Procedures"). Again, some industry consensus has been established for environmental acceptability, and substantial progress has been made to agree on common definitions and tests procedures over the past 15–20 years. However, at present, there are still no universal industry/regulatory standards. This should not be a deterrent to the use of EA products, particularly in areas that are sensitive to spills or leakage of conventional lubricants. Where these spills have inadvertently occurred, EALs have clearly demonstrated much less negative impact to the environment. Environmental performance requirements must also include provisions requiring that a product must maintain its environmental characteristics during its projected service life.

B. Specifications

A product that is going to be relied on in a given piece of equipment must exhibit certain physical and chemical characteristics that have been shown to be important to the performance of that equipment. The primary concern is to have adequate viscosity characteristics. Too low a viscosity could result in metal-to-metal contact and consequent wear. Too high a viscosity can result in improper flow or excessive internal shear resulting in excessive heating and energy losses. Other characteristics would include compatibility, antiwear, oxidation stability (long life), rust and corrosion protection, filterability, and demulsibility.

C. Equipment Builder Approvals

Much of the initial thrust to develop this class of lubricants came from equipment builders whose customers were requesting guidance on such products. Chances are that if the equipment builders have approved the use of these products, they have satisfactorily tested the products or have compiled test data showing that the performance requirements have been met. Also, builders have followed field applications to assure that the product not only meets laboratory test requirements but also works in the equipment under field service conditions. In some cases, builders will grant conditional approvals only if they can limit temperatures and pressures and may, in some cases, derate equipment where EALs are used. This generally means lower service pressures, speed, and/or temperatures. If the equipment is under warranty, the builders of the equipment should be consulted before its use to ensure that the requirements of the warranty are met.

In the past few years, some hydraulic equipment builders have even developed specific approval requirements for EALs. For instance, Denison HF-6 was added to Denison HF-0, HF-1, and HF-2, which have been in place for many years for conventional mineral or synthetic based hydraulic fluids.

D. Proven Field Performance

Similar to selecting conventional lubricants, testimonials of customers using the products in equipment that represents a certain application indicate that the products are likely to provide satisfactory performance. Because most major lubricant manufacturers conduct extensive laboratory testing as well (up to several years of field testing) even with the so-called "new" products, field application data should be available.

E. Supplier Reliability

A reputable supplier is important with conventional established lubricants but even more important when new technology is introduced. It is very important that suppliers support their products and recommendations. This provides an increased level of confidence that the new products will work, or at least the customer will not be handling problems alone. Reputable suppliers have generally performed adequate laboratory and field evaluations of their products before introduction so that chances of success are high. Because there are a number of tests and definitions applicable to EALs, it is quite important that suppliers be honest and forthright about the test results that they obtained on their product. Biodegradation results without any reference to the test method used are not very meaningful.

F. Operating and Maintenance Conditions

Equipment reliability and service life are strongly influenced by operating conditions and maintenance practices regardless of the lubricant. This is especially true where EALs are used. Three of the more important areas of operating conditions and maintenance practices involve control of temperatures, elimination of contamination, and good system maintenance. For example, one high-quality vegetable-based hydraulic oil could be recommended for temperatures above −10°C at startup and below 80°C for continued operation. Contamination should also be minimized to avoid loss of environmental performance capability as well as to assure long performance life. Maintenance practices should include regular inspections of the systems to correct any unsatisfactory conditions involving, for example, filtration, breathers, temperature control, pressure control, correct makeup oil, regular oil analysis, and oil/filter changes as required.

VII. CONVERTING TO EALs

In most cases, EALs will be compatible with mineral oils, elastomers, and paints used in industrial and hydraulic systems designed for mineral oils. Some products such as polyglycols may not be compatible with mineral oils or paints. A primary requirement with any conversion is to determine the compatibility of the EAL with all system components before making a conversion. Data should be obtained from the product's manufacturer or supplier. Equipment builders should also be consulted to provide input on component materials in cases where compatibility concerns may arise.

Once compatibility issues are understood, it is recommended that all systems being converted to EALs be drained, cleaned, and thoroughly flushed before adding the final charge of EAL. This will help ensure that the environmental and performance characteristics are not jeopardized by excessive contamination with previously used non-EALs or other contaminants. Even with new systems, it is important to clean and flush to remove materials such as rust preventives, assembly lubes, sealants, or other materials that may be in the system. These materials can negatively affect both the environmental and the physical performance characteristics of the EAL. For example, contamination with preservative oils and assembly lubes can affect demulsibility, air separation, oxidation stability, and antiwear performance.

The extent of cleaning and flushing is dependent on the system condition as well as the current product in service. If the system is dirty or contains deposits that have built up over time, it may be necessary to flush with a noncorrosive solvent mixed with a conventional mineral oil to remove the deposits. If this is required, care should be taken to install jumper lines across critical tolerance components such as electrohydraulic servo valves. This will reduce the potential for loosened debris to get into the close clearances and cause operational problems after startup. Where high detergent engine oils are in service, it is recommended to pay more attention to flushing particularly when vegetable-based EALs are used. This is because of the potential adverse reactions between the highly additized engine oils with the EALs.

As general guidance, no more than 3% of the previously used non-EA fluid should remain in the system before installing the final charge of the EA fluid. There are relatively simple laboratory checks that can be used to determine if this has been accomplished. One method is by determining the additive metals such as zinc, calcium, or phosphorus in the previous fluid and comparing this to the additive metals in the final charge. Another way involves analyzing the aromatic content of the fluid except where severely hydrotreated or hydrocracked base stocks are used. Again, the supplier of the EAL will be able to provide assistance in determining the success of the flushing and the degree of contamination in the final system fluid.

BIBLIOGRAPHY

European Committee for Standardization. 2011. CEN/TR 16227 Technical Report: Liquid petroleum products—Bio-lubricants—Recommendation for terminology and characterisation of bio-lubricants and bio-based lubricants.

Mudge, S.M. 2010. Comparative environmental fate of marine lubricants. Unpublished manuscript.

U.S. Environmental Protection Agency, 2011. EPA 800-R11-002. White paper on Environmentally Acceptable Lubricants.

U.S. Environmental Protection Agency. 2013. Vessel General Permit for Discharges Incidental to the Normal Operation of Vessels (VGP) http://water.epa.gov/polwaste/npdes/vessels/upload/vgp_permit2013.pdf.

Willing, A. 2001. Lubricants based on renewable resources—An environmentally compatible alternative to mineral oil products. *Chemosphere* 43:89–98.

7 Hydraulics

Hydraulics is one of the oldest branches of science and can be defined as "The transmission of force from one point to another using a fluid as a transmitter of force." Hydraulics is used to transmit force in order to, for example, change direction in a hydraulic excavator or change the magnitude of force (braking). It uses a special circulating system in which a pump drives an oil to move actuators.

Hydraulics offers major advantages over mechanical pulleys, levers, and gears:

- High power to weight ratio.
- Infinitely variable speeds.
- Eliminates complex, expensive, mechanical linkages.
- Easily rerouted and disconnected.
- Can serve many actuators/applications.
- Fluid does not break.
- Forces can be transmitted over a great distance.
- Allows large loads to be moved by small forces.
- Able to transfer large forces over distances with relatively small space requirements.
- Operations can commence from rest while the system is at full load.
- Speed can be easily controlled.
- Smooth adjustments to speed, torque, and force.
- Simple protection against system overloading.
- Suitable for both quick and slow controlled sequences of movements.
- Simple centralized drive systems available (several machines can be operated off one hydraulic system).

In service, properly selected hydraulic fluids enable effective pressure transmission and controlled flow, reduce wear and friction, improve system efficiency, provide cooling, prevent rust and corrosion, and help keep the system components free of deposits. High-quality hydraulic oils are able to maintain their initial characteristics and provide satisfactory service for long periods—often years in well-designed and well-maintained hydraulic systems.

I. BASIC PRINCIPLES

Pascal in the seventeenth century produced a fundamental law of hydraulics: "The pressure applied to a confined fluid is transmitted undiminished in all directions and acts with equal force on equal areas and at right angles to them" (Figure 7.1).

This principle allows hydraulics to act as a force multiplier because force is equal to the pressure applied to a specific surface area. For example, the force applied to the top of the flask in Figure 7.1 is 5 N (50,000 Pa × 0.0001 m^2), whereas that applied at the bottom of the flask is 50 N (50,000 Pa × 0.001 m^2). Another example is shown in Figure 7.2, where according to Pascal's law the elephant can be lifted by a small force (e.g., a car jack): when the pump moves 100 cm, it will transfer 100 cm^3 to the other side, lifting the piston and elephant 1 cm.

A. Hydromechanics

Hydraulic systems are subject to the laws of hydromechanics. This means that balances within the hydraulic system are subject to both *hydrostatic* and *hydrodynamic* forces.

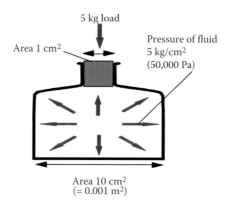

FIGURE 7.1 Pascal's law. Force applied to a stationary liquid acts in all directions within that liquid.

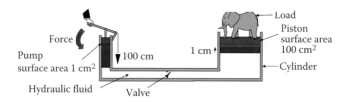

FIGURE 7.2 Basis for hydraulic actuation.

Hydrostatic force is the pressure that exists in a system. This is created by a combination of gravity, the height of the fluid in the reservoir, and the fluid density. The height of the fluid in a reservoir and its density determine the static pressure at any point within a given system, and the static pressure at the bottom of the reservoir will always be higher than the pressure at the top. Note that volume has no effect on static pressure.

Hydrostatics is important in the design and application of hydraulic systems. Hydrostatic considerations that should be recognized include: higher density fluids need more energy to pump, the location of pump suctions is critical relative to fluid levels, and the impact of how far and to what height fluids are needed to perform work. For example, pumps designed with a positive static head at their suctions are much less susceptible to cavitation than pumps with suctions located above fluid levels.

Hydrodynamics, or hydrokinetics, involves the energy of fluids in motion, and there are specific areas where they are important. Fluids in motion contain an energy component above that indicated by pressure and flow. This energy level is related to mass and acceleration. The instantaneous pressure rise generated by the sudden stoppage of motion can be very high, leading to a "shockwave." These shockwaves can be severe enough to cause failure of system components. Where these conditions exist, accumulators (hydraulic shock absorbers) should be installed at appropriate locations within the system to minimize the negative effects of these shock waves.

B. Fundamental Hydraulic Systems

Figure 7.3 shows an overall view of a hydraulic circuit. Mechanical energy is converted to fluid energy, which is converted back to mechanical energy. This principle applies to all hydraulic circuits. Although there are many different hydraulic system/circuits designs in use, almost all of them can be described by this schematic. The function and types of each component are discussed in the next section.

Hydraulics

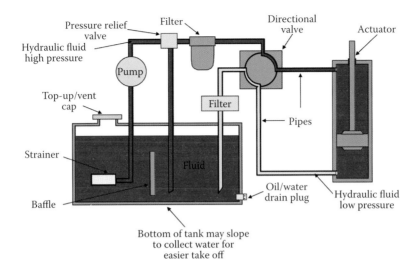

FIGURE 7.3 Hydraulic system component schematic.

II. SYSTEM COMPONENTS

A. HYDRAULIC PUMPS

All hydraulic systems use a mechanism to create system flow (motion) and resistance or stoppage of that flow creates pressure (force). The combination of flow and pressure in a hydraulic system are the basis for work. In most applications, pumps are used for this purpose, and their function is to convert mechanical energy to fluid energy. The pumps can be hand-operated or driven by electric motors, internal combustion engines, turbines, etc., that provide rotative or reciprocating motion. There are four major types of pumps:

1. Gear pumps: generally fixed displacement
2. Vane pumps: generally fixed displacement, but can be variable if necessary
3. Axial and radial piston pumps: either fixed displacement or variable
4. Centrifugal pumps: nonpositive displacement that are rarely used

1. Gear Pumps

Gear pumps are generally simple in design, lower maintenance, cheaper, and compact compared to vane and piston designs. They typically operate at low pressures (500–3000 psi or 3500–21,000 kPa), have fair to good efficiency, can tolerate high contamination, and can be noisy. Lubrication regimes are boundary lubrication for low speed and hydrodynamic lubrication for high speed. Fluid requirements include the use of antiwear additives if the original equipment manufacturer (OEM) requires it. Significant effort is being devoted to the development of "quiet pumps" of all designs to aid in overall noise reduction programs for industry.

There are four types of gear pumps: external, internal, ring gear, and screw (Figure 7.4). Gear pumps are constant discharge types, and they generate flow carrying fluid from the inlet to the outlet between the tooth spaces of the gears. The fluid pumping chambers are formed between the gears, the housing, and the side plates. The hydraulic fluid serves to help seal close clearances as well lubricating the meshing gears. Small amounts of bearing wear could result in contact of the gears with the housings or side plates, thereby reducing the efficiency of the pump. Minimizing wear is critical to optimizing pump performance.

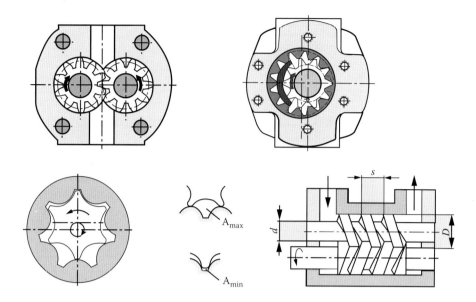

FIGURE 7.4 Four common gear pump designs: internal, external, ring, and screw.

2. Vane Pumps

Vane pumps (Figure 7.5) can be of fixed or variable displacement design. They typically operate at moderate pressures (1000–4000 psi or 7000–27,500 kPa) and are efficient. They are sensitive to contamination because of the small clearances between the vanes and the cam ring. The lubrication regime is either boundary or mixed, not hydrodynamic. Hydraulic fluid requirements for this application include the use of antiwear additives and maintaining overall good oil cleanliness.

3. Piston Pumps

Radial and axial piston pumps are positive displacement pumps used where operating pressures are high (3000–10,000 psi or 21,000–70,000 kPa), and very accurate control of discharge flow is needed. Piston pumps can be either fixed or variable displacement, where the volume and pressure can be varied by use of a tilting plate. They are very efficient, but are generally more expensive than other pumps and, in some cases, are considered to be noisy. They generally run in the hydrodynamic/mixed lubrication regime, and hydraulic fluid requirements for this pump include good oxidation stability, hydrolytic stability, and antiwear.

Although gear and vane type pumps can be designed to work with higher pressures, the life expectancy of piston pumps in severe service will be greater, making them cost-effective. The key advantages of variable displacement piston pumps include having infinitely adjustable flow along with being reversible. This means that the pump can act as a pump or a hydraulic motor if the direction of flow is changed. Circuits using these pumps can be simplified, as reversing valves are not needed. With reversing pumps, rapidly moving masses can be smoothly decelerated, reversed, and accelerated without system shocks.

4. Radial Piston Pumps

Radial piston pumps encompass two designs: in one, a rotating eccentric shaft causes the pistons to reciprocate within their cylinders; the second one uses an eccentrically mounted rotating cylinder block where the pistons and block rotate within a rigid external ring. In both designs, the reciprocating movement is perpendicular to shaft rotation (Figure 7.6). This reciprocating radial motion of the pistons within their respective cylinders creates a reduced pressure at the suction drawing fluid into

Hydraulics

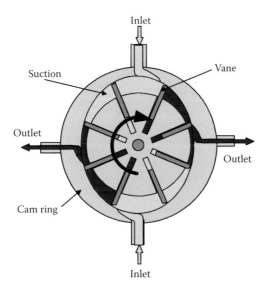

FIGURE 7.5 Vane pump. In this constant volume pump, the vanes are forced by contact with the cam ring to slide in and out of the rotor slots. The vanes are held in contact with the cam ring by centrifugal force and by oil pressure on their inner edges. Suction ports are located where the wall of the cam ring recedes from the hub because the increasing volume between the vanes at these points results in reduced pressure. Discharge ports are placed where the wall of the ring approaches the hub. At these points, decreasing volume between the vanes causes pressure to increase.

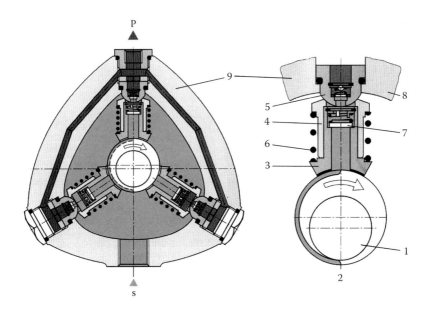

FIGURE 7.6 Radial piston pump. (1) Drive shaft, (2) eccentric shaft, (3) piston, (4) pivot, (5) cylinder sleeve, (6) compression spring, (7) suction valve, (8) pressure control valve, and (9) housing.

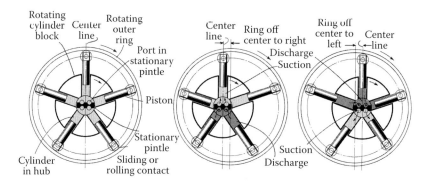

FIGURE 7.7 Principle of variable displacement radial piston pump. This type of variable stroke pump has a cylinder block that rotates about a stationary spindle, or pintle. Several closely fitted pistons are carried around the block, with their outer ends held in contact with a rotating outer ring by centrifugal force and by oil pressure. Suction and discharge ports are in the pintle and communicate with the open ends of the cylinders. When the ring and the pintle are concentric, rotation occurs without any reciprocating movement of the pistons and therefore output is "0." If the ring is moved off center, rotation causes the pistons to reciprocate, drawing oil from the suction ports on the outward stroke and delivering it through the discharge ports on the inward stroke. Moreover, by adjusting the eccentricity of the outer ring to the right or to the left (as shown), it is possible to reverse the direction of oil flow in either direction without stopping the pump or changing its direction of rotation.

the cylinder on one stroke and expelling the fluid into the discharge on the other stroke. Figure 7.7 shows a variable volume radial piston pump of the eccentric cylinder block design.

5. Axial Piston Pumps

Axial piston type pumps refer to both the swashplate (or wobble plate) and the bent axis designs. Both designs are positive displacement and are available in fixed and variable volume outputs. Axial piston pumps operate based on the same design principles as radial piston pumps except that the pump pistons reciprocate parallel to the rotor axis as shown in Figure 7.8. The angle between the cylinder block and the driving flange determines the volume output (along with speed and cylinder diameter).

Piston pumps, whether radial or axial, will be designed with an odd number of pistons. The odd number of pistons reduces the flow and pressure pulsations in the discharge when individual piston flows are added together.

B. Pump Selection Criteria

Pump selection will be based on many factors such as pressures, flows, temperatures, fluid in service, type of service, noise levels, projected service life, and initial as well as long-term costs. There are also many other factors that will influence selection. Table 7.1 shows some of the general characteristics associated with the various pump designs. Specific information on the listed characteristics as well as others should be obtained from the respective pump or system manufacturer.

III. CONTROLLING PRESSURE AND FLOW

Valves are essential components of all hydraulic systems and are used to control and direct flow as well as maintain system pressures. There are various types of valve design and function (which will be discussed shortly). They all, however, have the following in common: the need to react quickly, have narrow orifices, are susceptible to oil oxidation byproducts and contamination, and require a certain level of fluid cleanliness to operate properly. The designs range from a simple check valve, which permits flow only in one direction (Figure 7.9), to complex electrohydraulic control valves (Figure 7.10), which may be required to control the accuracy of a machining center to less than 0.0005 in (0.0127 mm).

Hydraulics

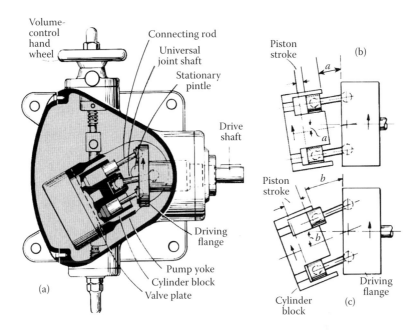

FIGURE 7.8 Variable volume axial piston pump. As in the case of the radial piston pump in Figure 7.7, variable volume (displacement) is accomplished by varying the piston strokes. The cylinder block is carried in a yoke that can be set at various angles with the drive shaft. (a) The block with its several pistons is rotated by the drive shaft through an intermediate shaft having a universal joint at each end. The drive shaft and cylinder block are supported by antifriction bearings. All working parts are submerged in oil that is continuously recirculated. (b) With the cylinder block set at angle a, rotation of the structure through 180° produces the relatively short stroke of the piston as shown. (c) With the block set at the larger angle b, the stroke (and volume of oil pumped) is seen to be increased. With the block set at "0" angle (in line with drive shaft), there is no movement of the pistons in the cylinder block and no oil is pumped. Finally, when the cylinder block is set at an angle above the drive shaft in the views shown, the direction of oil flow is reversed. In other words, flow can be adjusted in small steps from zero to maximum in either direction during operation. Suction and discharge takes place through sausage-shaped ports in the valve plate (a). Passages in the pump yoke lead from the valve plate to the stationary pintles (on which the yoke pivots) and through these to external suction and discharge connections. In addition to manual control (as shown in this illustration), volume output can be adjusted by means of a hydraulic cylinder or electric motor. Servo and automatic (pressure compensated) controls are also used. There are several makes of variable-volume axial piston pumps and the mechanism for varying piston strokes is quite different from one design to another.

TABLE 7.1
General Pump Selection Criteria

Pump Type	Speed Range	Pressure Range	Viscosity Range	Noise Level[a]	Service Life	Relative Costs
Internal gear	Good	Good	Good	Low	Good	Moderate
External gear	Very good	Good	Very good	High	Moderate	Low
Ring gear	Good	Moderate	Moderate	Moderate	Good	Moderate
Screw	Good	Moderate	Very good	Low	Good	Moderate
Vane	Moderate	Moderate	Moderate	Moderate	Good	Moderate
Piston	Good	High	Very good	Moderate	Good	High

[a] Some of the newer design pumps may contain features to reduce noise levels and therefore do not fit these general criteria.

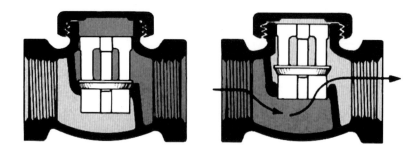

FIGURE 7.9 Check valve. A check valve permits free flow in one direction but prevents flow in the other. The one shown is one of several types used in hydraulic circuits.

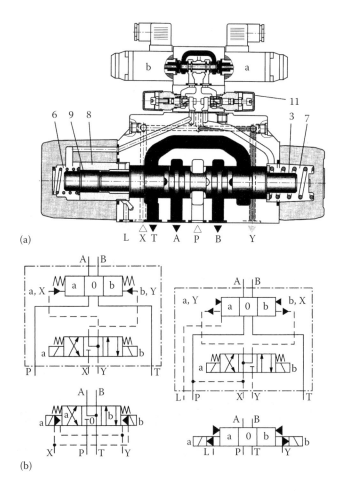

FIGURE 7.10 Electrohydraulic servo valve. The electrohydraulic servo valve is used in applications where very precise control of a machine is required. They are used to control speeds of turbines and critical machine tools where close tolerances are required in finished parts. (a) Eletrohydraulically operated directional spool valve, pressure-centered, for sandwich plate; detailed. (b) Symbol for electrohydraulically operated directional valve, pressure-centered; simplified.

Hydraulics

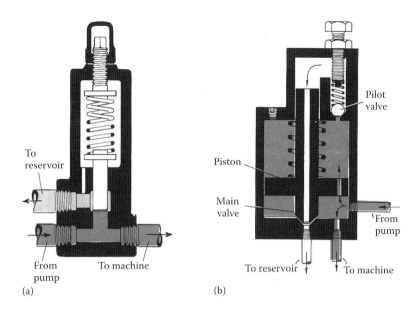

FIGURE 7.11 Relief valve. A relief valve functions when pressure in the delivery line becomes sufficient to overcome the compression of the spring (a). The valve lifts and bypasses the excess oil. As long as the pump delivers too much oil, the valve will remain open. When the pressure in the system drops below a value corresponding to the adjustment spring, the valve closes, and all of the oil discharged by the pump is delivered to the system. The *simple* relief valve at the left may permit a rather wide variation of pressure and may tend to chatter. A *compound* relief valve (b) will provide closer control of pressure and will have less tendency to chatter. Under normal conditions (as shown), the pressure on both sides of the piston of the compound relief valve is balanced, and the spring holds the main valve closed. When the pressure builds up sufficiently to raise the pilot valve, the pressure on the upper side of the piston is relieved, the main valve is lifted and the excess oil is bypassed to the reservoir.

A. Relief Valves

Where constant volume discharge pumps are used, a relief valve (Figure 7.11) is required to control the system pressure below a predetermined maximum level. This relief valve also functions as a safety device to prevent overpressure in the system by returning excessive flow back to the oil reservoir.

B. Directional Control Valves

Directional control valves control the motion of actuators. This ranges from a simple hand-operated gate valve to very complex servo valves. Directional control valves (Figure 7.12) are used to activate and deactivate hydraulic mechanisms such as cylinders and fluid motors.

C. Unloading Valves

In some circuits, two pumps may be used to meet substantially variable flow requirements (Figure 7.13). In this case, flow from the larger volume pump is used for rapid advance of the machine tool to a certain point, then only small volume flow as the work activity is performed. During the rapid advance cycle, the discharge volume of both pumps is required, but while actual work is being performed, only the small pump volume is required, causing a rise in system pressure. As the pilot pressure rises, the unloading valve opens, allowing the flow volume from the larger pump to be discharged to the reservoir at low pressure.

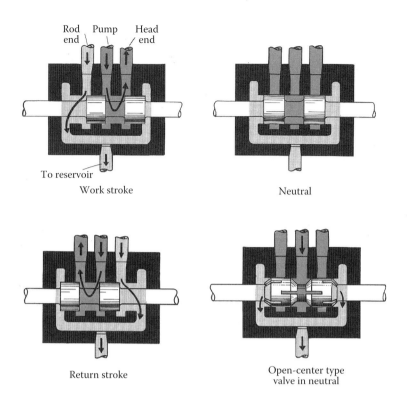

FIGURE 7.12 Four-way spool type directional control valve. The four views show a closed-center valve in work-stroke, neutral, and return-stroke positions. With the closed-center type of valve, there is very little leakage between ports that are blocked off by the valve spools. In the neutral position, there is intended to be no flow in any of the four connections. When the valve spool is shifted from one position to another, flow tends to start or stop immediately in the various connecting lines.

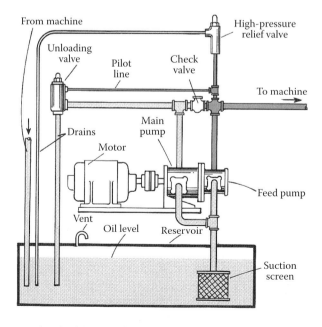

FIGURE 7.13 Two-pump circuit with unloading valve.

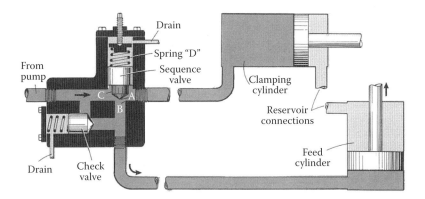

FIGURE 7.14 Sequence valve circuit.

D. Sequence Control Valve

In some machines, two or more movements may need to be hydraulically operated in sequence. When one movement must not begin until another has ended, a pressure-operated *sequence* valve may be used. Referring to the circuit shown in Figure 7.14, when oil flow stops at the end of the clamping-cylinder stroke, pressure rises sufficiently to open valve A. Full line pressure is then available to activate the feed cylinder. Sequence valves can also be activated by pressure sensing pilot valves or electronically by position or other pressure-sensing devices.

E. Flow Control Valves

Flow control valves control the speed of actuators by regulating the volume of oil to the actuators. In some applications, a slow work stroke and a rapid return stroke is desired. One simple way of accomplishing this in a nonprecise application is to use a flow restrictor in the supply line to the work stroke.

F. Accumulators

Accumulators can act as flow control mechanisms as well as store the energy of an incompressible fluid. Some presses and other machines require large volumes of oil under pressure for short duration cycles with relatively long periods between cycles. A large enough pump and motor to generate the necessary flow and pressure for such an application could be very costly. Instead, accumulators, in conjunction with small pumps and motors, can often be used as shown in Figure 7.15. Energy is stored in the accumulator by the pump during the long periods between high energy requirements. This is done by pumping hydraulic fluid into the accumulator and raising a weight, compressing a spring, or compressing a gas charge. The energy is returned to the system when required.

In addition to energy storage capacity, accumulators serve other functions. Many hydraulic systems are subject to rapid or sudden flow changes where the dynamics of fluid in motion can create high levels of system *shock*. In these situations, accumulators act as shock absorbers. They are able to reduce the severity of the system shock by allowing the instantaneous pressure rises to be taken up by the compressible mechanisms within the accumulators. This helps reduce the potential for line breakage and component failures in those high flow, high pressure systems that are subject to abrupt changes in flow. They can also be used to smooth out flow by absorbing pump pulsations, maintain constant pressures for long periods, such as in clamping operations, and make up for system internal leakage. In most hydraulic applications, gas-pressurized accumulators are used.

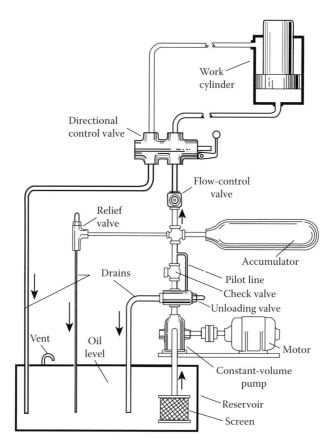

FIGURE 7.15 An accumulator circuit. In this press circuit, the pump is automatically unloaded when the accumulator pressure reaches a preset value. A flow-control valve determines the work-stroke speed, and a check valve prevents loss of high pressure hydraulic fluid.

IV. ACTUATORS

A. Hydraulic Cylinders

This is the output device that moves the load and will vary depending on the job, the conditions, and the power requirements. The device most frequently used is the *hydraulic cylinder* (Figure 7.16), which may be single acting (pressure applied in one direction) or double acting. Cylinders are usually made of steel tubing, bored and honed. Rods are often hard chrome-plated steel, ground, and polished. All surfaces in contact with seals and gaskets are finely finished. The lubrication regime for hydraulic cylinders is either hydrodynamic or boundary depending on the speed and load. In addition to single and double-acting cylinders, there are several other designs commonly used in industry:

- *Spring return* cylinders are the single-rod type where pressure activates the work stroke. The spring in the nonpressurized end returns the piston to its original position when pressure is released.
- *Telescoping* cylinders are used where long strokes are required but space constraints limit the length of the cylinder. These applications also have lower load requirements as the cylinder rod extends. These are made of multitubular rod segments that fit into each other to provide the telescoping action.

Hydraulics

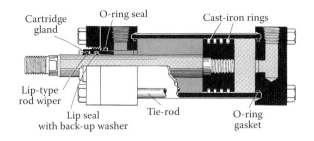

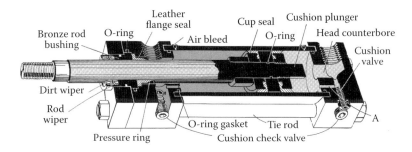

FIGURE 7.16 Hydraulic cylinders.

- *Tandem* cylinders are used when high force is required but larger diameter cylinders cannot fit in the available space although length is not limited. These are mounted in-line with a common piston rod.

B. ROTARY FLUID MOTORS

Rotary fluid motors are used instead of hydraulic cylinders to convert fluid energy to mechanical motion, especially where rotary motion is required or where continuous or long movements in one direction is needed. They compete with electric motors when: (1) a variable speed transmission having a wide range of closely controlled speeds and torque is required; (2) space limitations demand a very compact power source; (3) torque or loading could be occasionally severe enough to overload an electric motor. Rotary fluid motors can be of the gear, vane, or piston (radial or axial) design. They are similar to their pump counterparts, but differ in certain details that affect efficiency. In fact, some radial and axial piston pumps are designed to act as both a pump and a motor as was described in connection with Figure 7.8. Rotary fluid motors are supplied with oil under pressure, and rotate at a speed and torque dependent on the available volume of oil that flows through the motor. Radial and axial piston motors are usually of the constant-displacement design (Figure 7.17).

Axial piston motors exhibit high volumetric efficiencies and excellent operation over a wide range of speeds. They can be used when torque requirements are up to more than 17,000 ft lb (23,000 N m) or speeds up to 4000 rpm. Radial piston motors can attain several hundred-thousand ft lbs of torque or speeds up to 2000 rpm. Gear type motors will provide up to 6000 ft lb (8100 N m) of torque or up to 3000 rpm. It is important to recognize that the torque of hydraulic motors is inversely proportional to rotational speed, so that their highest torque will occur at low speeds.

V. HYDRAULIC DRIVES

Hydraulic drives are classified as *hydrostatic* or *hydrodynamic*. These can be designed to produce power in three ways: variable power and torque, constant power and variable torque, and variable power and constant torque.

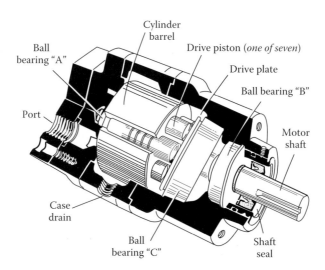

FIGURE 7.17 Axial piston fluid motor.

Hydrostatic drives use oil under pressure to transmit force, whereas hydrodynamic drives use the effects of high-velocity fluid to transmit force. Engine-driven transmissions used in main and auxiliary drives of mobile construction and farming equipment are an example of hydrostatic drives. Torque converters (sometimes referred to as *hydrokinetic* drives) are commonly found in automotive applications but have found increased use in industrial applications. Another form of a hydrokinetic drive is the *hydroviscous* drive. This form uses the viscosity characteristics of the fluid rather than the energy from fluid in motion to develop the drive torque. Hydrodynamic drives include hydrokinetic and hydroviscous drives.

A. Hydrostatic Drives

In a hydrostatic system, power from an electric motor, internal combustion engine, or other forms of prime mover is converted into static fluid pressure by the hydraulic pump. This static pressure acts on the hydraulic motor to produce mechanical power output. Although the fluid actually moves through a closed-loop circuit between the pump and motor, energy is transferred primarily by the static pressure rather than the kinetic energy of the moving fluid.

The hydraulic pump in a hydrostatic system is a positive displacement type, either fixed or predominantly variable. Axial piston pumps are the most commonly used, although radial piston pumps are used in some applications. The motor in a hydrostatic system can be any positive displacement hydraulic motor. Axial piston motors are usually used for most drives, whereas both gear motors and radial piston motors are used for specific designs. The motor is usually a fixed displacement type, but can be variable. The motor is reversible with the direction of rotation dependent on the direction of flow in the closed-loop circuit from the pump. Figure 7.18 shows a diagram of a typical hydrostatic drive.

VI. OIL RESERVOIRS

The oil reservoir (Figure 7.19) is also a very important component of the hydraulic system. It contains the oil supply, provides radiant and convection cooling, allows solid contamination and water to drop out, and helps reduce undesirable entrained air from circulating to the critical control components. In addition, in relation to oil levels and pump location, the reservoir facilitates easy flow of the oil to the pump suction, which reduces the potential for cavitation or starvation conditions. In

Hydraulics

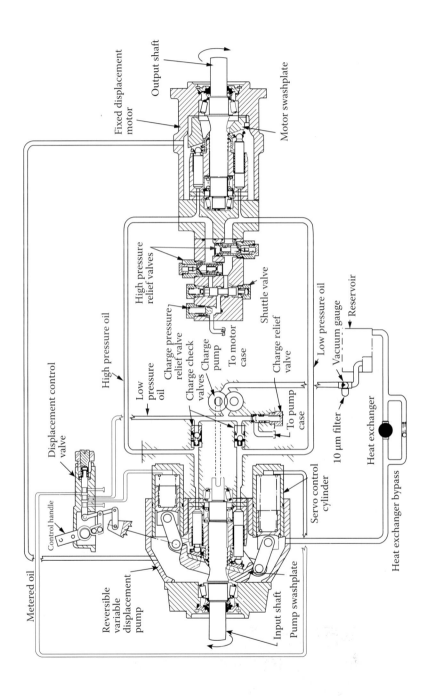

FIGURE 7.18 Diagram of typical hydrostatic drive.

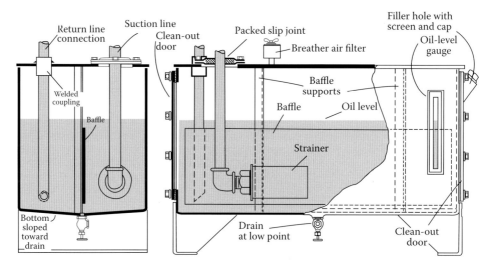

FIGURE 7.19 Oil reservoir design.

order to achieve this, the reservoir needs certain design features such as a baffle plate, drain plug, fluid level window, strainer, system/sump vent, correctly located fluid pickup, and fluid return pipes.

VII. OIL QUALITIES REQUIRED BY HYDRAULIC SYSTEMS

The primary functions of the hydraulic fluid are to transmit power, lubricate and protect metallic parts against wear, ensure good seal performance, prevent rust and corrosion, and generally ensure the efficient and smooth operation of the entire system. To perform these functions, there are certain properties that a hydraulic fluid must possess, and these are highlighted in Figure 7.20.

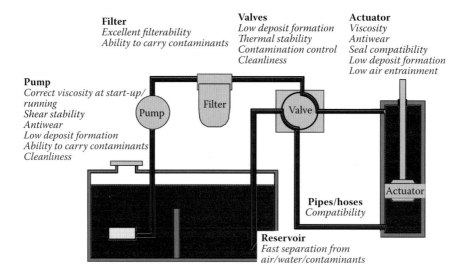

FIGURE 7.20 Important properties of a hydraulic fluid that is essential for key components in a hydraulic system.

A. Viscosity

A fluid's viscosity is defined as its resistance to flow at a given temperature. Viscosity affects the fluid's ability to be pumped, move through the system, carry load, and maintain separation between moving parts.

If the viscosity of the fluid is too high, the following problems can occur:

- The higher resistance of the fluid impacts flow
- Increases energy consumption due to high friction and input torque requirements at the pump
- Causes slow or sluggish operation
- Increases potential for pump cavitation
- Results in higher oil and system temperatures
- Increases pressure drop

If the viscosity is too low, the following problems can occur:

- Increases potential for internal leakage and loss of efficiency
- Can cause excessive wear
- Greater potential for external oil leakage

Consequently, selection of a fluid with the correct viscosity for the system is very important.

B. Viscosity Index

The viscosity of oil varies substantially with changes in temperature. In some hydraulic systems, subjected to wide variations in startup and operating temperatures, an oil that changes relatively little in viscosity for a given temperature range is desired. An oil that exhibits this property is said to have a high *viscosity index* (VI). High VIs for a fluid can be achieved by adding long-chain polymer additives (called *viscosity index improvers* [VIIs]) to the base stock or using base stocks (e.g., synthetic or mineral) with naturally high VIs.

Hydraulic fluids are subjected to high shear rate conditions, particularly as pressures and speeds increase. It is not uncommon for many long-chain polymer VIIs to shear over time. This shearing results in the loss of viscosity and, if severe enough, poorer hydraulic system performance, increased leakage, and loss of some of the friction-reducing qualities of the fluid. When selecting a hydraulic fluid with a high VI, good shear stability performance should be specified.

C. Antiwear (Wear Protection)

Component wear in a hydraulic system results in loss of mechanical efficiency as well as higher costs because of shorter component life. Antiwear fluids are generally required in gear and vane pumps operating at pressures above 1000 psi (6900 kPa) and 1200 rpm. Piston pumps may or may not require antiwear additives to be used in the fluid depending on the specific manufacturer and the metallurgy of the pump. Antiwear additives work by chemically reacting with the metal surfaces forming a strong film preventing metal-to-metal contact under boundary lubrication conditions.

D. Oxidation Stability

Severe oil oxidation and thermal degradation can cause oil blockages and component sticking (especially valves). Good oil oxidation stability is critical to the service life of the hydraulic fluid. In high-quality hydraulic oils, the natural ability of carefully selected base stocks to resist oxidation is greatly improved by the use of additives that retard the oxidation process.

E. Air Entrainment

Entrained air in the oil can disrupt the lubricating film, cause excessive component wear, and create the potential for pump cavitation. Entrained air can also cause degradation of the oil via the mechanism known as microdieseling. The compressibility of entrained air can result in erratic and inefficient operation of the system. Hydraulic oils should have good air release properties, and typically the quality and type of base stock and the viscosity and temperature of the fluid have the greatest impact on air release.

F. Antifoam

An oil that foams excessively can lead to erratic system operation along with oil losses and spills. Properly selected defoamant additives for hydraulic oils will readily dissipate the air bubbles that rise to the surface of the oil in the reservoir. Maintaining an adequate oil level in a reservoir (with a properly designed oil capacity) will help reduce the tendency for foam to form. Keeping contaminants in the oil to a minimum will also contribute to reducing the potential for foaming.

G. Demulsibility (Water Separating Ability)

Water contamination can create problems in hydraulic systems. Oil and water mixtures can form stable emulsions resulting in oil thickening, rust, and filter blockages. Some additives can react with water to form deposits and cause filter blockage. Severe water contamination results in wear and poor system function. Good hydraulic fluids should have the ability to readily separate from water so that any accumulated water can be drained off periodically.

H. Rust Protection

Water and oxygen can cause rusting of ferrous surfaces in hydraulic systems. The possibility of rusting is greatest during shutdown when surfaces that are normally covered with oil may be left unprotected when condensation forms on them as the system cools. This is particularly important in areas where operations are subject to high humidity conditions and temperature changes within the reservoirs.

Rusting results in surface destruction, and rust particles may be carried into the system where it will contribute to wear and the formation of sludge-like deposits that will interfere with the operation of pumps, actuators, and control mechanisms. Rusting of piston rods or rams can cause rapid seal or packing wear resulting in increased leakage and system contamination.

High-quality hydraulic fluids contain a rust inhibitor, which has an affinity for metal surfaces, thus forming a barrier film that resists displacement by water and protects surfaces from contact with water. The rust inhibitor must be carefully selected to provide adequate protection without reducing other desirable properties of the fluid, especially water separating ability and antiwear protection.

I. Compatibility

An often-overlooked characteristic of the hydraulic fluid during the selection process is its compatibility. Use of oils that exhibit undesirable reactions with system components such as metallurgy, elastomers, paints, or gasket materials can result in leakage or contamination within the system that can impact overall system performance. When changing from one oil to another, such as going from

TABLE 7.2
Primary Hydraulic Fluid Types and Their Major Industry Standard/OEM Specifications

Conventional (Hydrocarbon) Hydraulic Fluid Specifications/Approvals	Environmental/Biodegradable Hydraulic Fluids (More Details Can Be Found in Chapter 6)	Fire-Resistant Hydraulic Fluids	Food Grade Hydraulic Fluids
DIN 51524	ISO 15360	ISO 12922	NSF-1
ISO 11158	European Union Eco Label	Factory Mutual	
JCMAS HK	JCMAS HKB		
Eaton E-FDGN-TB002-E	U.S. Vessel General Permit		
Denison HF-0, 1, and 2	Denison HF-6		
Bosch Rexroth RD(E) 90240			

a conventional mineral oil to a fire-resistant fluid or a synthetic hydraulic fluid, attention to compatibility issues is particularly important.

VIII. HYDRAULIC FLUID TYPES

There are many different types of hydraulic fluids, with each designed to meet a specific industry, OEM, performance, or regulatory requirement. The most common types of hydraulic fluids are listed in Table 7.2. For a comprehensive list and definition of all hydraulic fluid types, users should refer to the International Organization for Standardization (ISO) 6743 classification document, Lubricants, industrial oils and related products (class L) Part 4: Family H (hydraulic systems).

A. Industry Standards and OEM Approvals

To ensure that hydraulic fluids have adequate performance capabilities to protect the equipment in use, there are several industry and OEM specifications that a high-performance hydraulic must satisfy. These specifications are listed in Table 7.2. Note that this list only provides examples of the major industry specifications and OEM approvals and is not an exhaustive list. Most of these specifications and approvals contain some form of hydraulic pump test, oxidation/thermal stability testing, rust tests, seal compatibility, etc.

IX. HYDRAULIC SYSTEM MAINTENANCE

The degree of system maintenance will be based on specific performance expectations, the fluid used, and the system operating parameters. Various hydraulic fluids ranging from mineral oil based, to synthetic, to water-containing fire resistant fluids will require various levels of maintenance to ensure performance. Water-containing fluids require higher levels of maintenance to ensure that aside from having fire protection properties, these fluids will demonstrate appropriate lubrication characteristics while in service. Matching the right fluid to specific system needs and understanding its limitations is the basic starting point for an effective maintenance program.

Once the proper fluid has been selected, the equipment and operating conditions will dictate the degree of maintenance required to keep that fluid in service for a long time while retaining its lubrication characteristics. These maintenance procedure objectives should include:

- Keeping fluid clean/controlling contamination
- Maintaining proper oil temperatures
- Maintaining proper oil levels

- Periodic oil analysis
- Routine inspections:
  - Noise levels
  - Shock loads
  - Filtration
  - Vibration
  - Leakage
  - Temperatures
    - Pressures
  - Fluid odor and color
  - Foaming

A. Filtration

The fluid in service must be clean, and the level of cleanliness depends on the system. For example, Numerically Controlled (NC) machine tools require high levels of cleanliness because of the close-tolerance servo valves, whereas the hydraulics used to operate hydraulic lifts in automotive repair shops can operate satisfactorily with minimal filtration. It should be noted that conventional filters will not remove water- or oil-soluble contaminants. Special coalescing type filters are available for removing limited amounts of water.

Full flow filtration is the most common type used on hydraulic systems to control the levels of solid contaminants. These filters are generally installed in the supply (pressure) line but can also be installed in return (low pressure) lines to the reservoir. Full flow filter housings are generally equipped with bypass valves, which will open when the pressure differential across the filter exceeds a predetermined level. This will assure that components will receive oil in the event of filter plugging or restriction of oil flow through the filter due to startup or cold oil where viscosities are high. When filters go on bypass, unfiltered oil will be supplied to components. Some filters are equipped with condition indicators or differential pressure gages to warn of restrictions or plugging.

Selection of appropriate filtration levels (fluid cleanliness) will be based on specific system components and operation. Table 7.3 shows some of the typical clearances in hydraulic system components. The vast majority of hydraulic systems will function properly on 10 μm filtration with filter

TABLE 7.3
Typical Critical Clearances for Fluid System Components

Component	Clearance (μm)	Clearance (in)
Antifriction bearings	0.5	0.000019
Vane pump: tip to vane	0.5–1.0	0.000019–0.000039
Piston pump: valve plate to cylinder	0.5–5.0	0.000019–0.000197
Gear pump: gear to side plate	0.5–5.0	0.000019–0.000197
Gear pump: gear tip to case	0.5–5.0	0.000019–0.000197
Servo valve spool (radial)	1.4	0.000055
Control valve spool (radial)	1.0–23.0	0.000039–0.000904
Hydrostatic bearings	1.0–25.0	0.000039–0.000984
Vane pump sides of vanes	5.0–13.0	0.000197–0.000511
Piston pump: piston to bore	5.0–40.0	0.000197–0.001575
Servo valves: flapper wall	18.0–63.0	0.000708–0.002363
Actuators	50.0–250.0	0.001969–0.009843

Hydraulics

efficiencies of 98.7% or greater. Filter efficiencies are often referred to as *Beta ratios* (β). A Beta ratio of 75 for a 10-μm filter would mean that 98.7% of the particles in the 10-μm and larger range will be removed. Tables 7.4 and 7.5 provide a brief explanation of Beta ratios and filter efficiencies.

Bypass filters, sometimes referred to as *polishing filters*, are usually installed in an independent system where 5–15% of the system's oil capacity is filtered to a finer degree. The oil is taken from a low point in the reservoir using an auxiliary pump, filtered, and returned to the reservoir, allowing oil purification to be continued whether the hydraulic system is in operation or shut down. An alternative to the independent system is a continuous bypass mode where a percentage of the oil flow from the pressure or return line is passed through suitable purification equipment and returned to the reservoir. Bypass purification equipment can be relatively small in size because only a portion of the total oil capacity is handled.

TABLE 7.4
Filter Efficiency/Beta Ratio

$$\beta = \frac{\text{In}}{\text{Out}}$$

Filter	Particles > 5 μm		β_5
	In	Out	
Filter A	10,000	5000	2
Filter B	10,000	100	100
Filter C	10,000	1	10,000

Note: The filtration ratio is calculated by dividing the number of particles entering the filter by the number of particles exiting the filter. β_5 represents the filtration ratio at 5 μm or the ratio of upstream to the downstream particles larger than 5 μm.

TABLE 7.5
Significance of Beta Values in Particle Counts

β Value at X μm (β_x)	Removal Efficiency (%): Particles > X μm	Downstream Count: > X μm When Filter Is Challenged Upstream with 10^6 Particles > X μm
1.0	0	1,000,000
1.5	33	670,000
2.0	50	500,000
20	95	50,000
50	98.0	20,000
75	98.7	13,000
100	99.0	10,000
200	99.5	5000
750	99.87	1333
1000	99.90	1000
10,000	99.99	100

Note: Beta values can be directly related to efficiency. To determine the relative performance of two filters with different Beta ratios, the downstream particle count from each filter can be calculated using the Beta ratio and an assumed upstream particle count (10^6 used in this example). The filter with the highest Beta value will have the lowest downstream particle count.

Portable filters (*filter carts* or *buggies*) are also used to supplement permanently installed system filters. These units consist of a motor-driven pump and filter arrangement that circulates fluid from the reservoir, through a fine filter, and back to the reservoir. The suction and return hoses should be connected to the opposite ends of the reservoir with quick-disconnect fittings. Generally, portable filters will operate for at least 24 h on each system to ensure that the full oil charge is effectively filtered. These can be used in place of bypass filters if periodic or as-needed filtration is sufficient to maintain the desired levels of fluid cleanliness. They can be a simple arrangement, or complex arrangements such as *reclamation units* consisting of motor-driven pumps, oil heating elements, vacuum chambers, and fine filters. The advantage of reclamation units is that they can remove water and some volatile contaminants (e.g., solvents) in addition to particulates.

Batch filtration may be used when the fluid volume is very large or is heavily contaminated. The large volume of oil may be removed and reclaimed as a batch through settling processes, filtration, centrifuging, and/or by use of reclamation units. The disadvantage of this process is that the machine must be shut down for removal of the fluid charge.

B. Controlling Temperatures

Excessive temperatures, in addition to reducing oil viscosity to a point where metal-to-metal contact results, will oxidize the oil, leading to varnish and sludge in the system. These deposits plug or restrict the motion of valves, plug suction screens, and cause shortened filter life.

Heat develops as the fluid is forced through the pumps, motors, tubing, and relief valves. Temperatures should be maintained between 49°C and 65°C (120–149°F) in conventional hydraulic systems. Some variable-volume pump systems, closed loop hydraulic systems, and hydraulic transmissions can operate up to 120°C (248°F), where premium fluids must be used or drain intervals shortened. Systems operating on water-based fluids should be kept below 60°C (140°F) to prevent the water from evaporating.

To allow heat to radiate from the system, the outside of the reservoir must be kept clean and the surrounding area should be clear of obstructions. Make sure the oil cooler is functioning properly and keep air-cooled radiators free of dirt and debris. Keep the reservoir filled to the proper level to allow enough fluid residence time for the heat to dissipate.

Oil degradation is even more critical in NC machine tools with electrohydraulic servo valves. Typically, these systems are designed with small reservoirs, which results in short residence times for the hydraulic fluid. Minimal rest times and high system pressures lead to entrained air bubbles causing extremely high localized temperatures because of the adiabatic compression of the air bubbles as they pass from low suction pressure to the high discharge pressure. This results in the phenomenon known as microdieseling, which—when combined with oil oxidation—can form deposits that will plug oil filters and cause servo valve sticking.

C. Maintaining Proper Reservoir Oil Levels

Systems are designed to provide a certain amount of oil residence time in the reservoir. This allows the fluid time for the separation of air, water, and solid contaminants to settle. It also provides for a degree of cooling. Operating with low oil levels reduces the effectiveness of these processes, and if levels are low enough, air could be pulled into the pump suction resulting in air entrainment leading to excessive *foaming* and *cavitation*. Foaming and air entrainment can also result from air leaks in the suction, low fluid temperatures, or fluid too viscous to release air bubbles. Oxidation and contamination also increase the fluid's tendency to foam and retain air. Entrained air is a major cause of destructive pump cavitation. The intense pressures and temperatures created by the bubbles' collapse erode metal parts of pumps and valves, resulting in excessive wear. Pump or valve cavitation may cause irregular operation or "spongy" response of the system. In addition to air leaks, a restricted suction can also cause cavitation. Causes of this restriction could be plugged strainers, oil viscosity being too high, or

Hydraulics

inadequate suction design conditions. Cavitation is noticeable through a high-pitched whine or scream in the pump with the noise sounding like there are marbles trapped in the housing.

D. PERIODIC OIL ANALYSIS

Laboratory oil analysis can be set up as a routine maintenance task or can be determined by the visual inspection of the fluid. A visual inspection may reveal the type and degree of contamination. This involves taking a sample of the fluid and allowing it to settle overnight in a clear container. The sample should then be inspected for color, appearance, and odor. If there is no evidence of water, corrosion, or excessive accumulation of debris, sediment, or sludge, and the fluid has the color and odor of a new fluid, generally, laboratory analysis is not necessary. A slight "burnt odor" is common in conventional hydraulic systems using petroleum oils. However, a burnt oil odor in an NC machine oil sample may be a cause for concern. In some cases, even conventional laboratory oil analysis cannot determine low levels of contamination, such as microdieseling (discussed earlier), under temperatures that will cause system operational problems. These systems will generally require high-quality oils as well as a specialized oil analysis program.

If the visual inspection cannot identify specific contaminants and potential sources, take a sample for laboratory analysis. It is a good practice to establish a program to periodically submit oil samples for laboratory analysis. In critical high-temperature machines, this may be as often as every 3 months. Noncritical machines will only require laboratory oil analysis every 6 to 12 months, but specific schedules should be based on operation, equipment, and oil supplier's recommendations.

E. ROUTINE INSPECTIONS

Routine inspections of operating systems can provide insight into potential problem areas that require corrective action. These corrective actions will result in longer equipment life and a lower cost of operation for the hydraulic system. The items listed below are commonly included in routine inspection programs.

Noise levels: Increases in noise levels may signal problems with cavitation. Noise levels may also increase from excessive temperatures allowing the oil's viscosity to become too low resulting in metal-to-metal contact.

Vibration: Loose mounting or misalignment of components will cause vibration resulting in accelerated wear or failure.

Shock loads: System components, such as hoses and fittings, subjected to shock loads because of abrupt changes in flows and pressures can lead to leakage and failures. Where shock loads are experienced, accumulators should be installed in the system.

Leakage: Oil leakage can result in low reservoir oil levels causing poor system performance. In addition, leakage can be costly and create safety problems.

Temperatures: As discussed earlier, controlling temperatures in the appropriate range is important from both an oil life and system performance standpoint. Excessive temperatures could be caused by plugged or dirty heat exchangers, excessive pressures, high rates of internal leakage, and low oil levels. Use of fluids with too high (excessive shearing) or too low (inadequate films) a viscosity will also result in higher temperatures.

Filtration: The condition of fill-screens, breathers, and filters (indicators or differential pressure gages) should indicate the need for cleaning or replacement.

Foaming: A little foam on top of the oil in the reservoir is normal. Excessive foam may indicate air leaks in the suction line, unsatisfactory contamination levels, or inadequate antifoam characteristics of the oil.

Fluid odor and color: Although odor and color are not characteristics used to judge the oil's ability to provide proper lubrication, changes in these physical properties may indicate contamination (solvents, wrong oils added, degradation, etc.) or the oil reaching the useful end of its service life. If in doubt as to the causes of the color and odor changes, a sample should be submitted for laboratory oil analysis. If these changes are accompanied by undesirable machine operating characteristics, an oil changeout is recommended.

BIBLIOGRAPHY

International Organization for Standardization. 1999. ISO 6743-4 Lubricants, industrial oils and related products (class L)—Classification—Part 4: Family H (Hydraulic systems). Geneva.

Mobil Oil Corporation. 1970. *Hydraulic Systems for Industrial Machines*. New York.

8 Lubricating Films and Machine Elements
Bearings, Slides, Guides, Ways, Gears, Cylinders, Couplings, Chains, Wire Ropes

The elements of machines that require lubrication are bearings—plain, rolling element, slides, guides, and ways; gears; cylinders; flexible couplings; chains, cams, and cam followers; and wire ropes. These elements have fitted or formed surfaces that move with respect to each other by sliding, rolling, advancing, and receding, or by combinations of these motions. If actual contact between surfaces occurs, high frictional forces leading to high temperatures—and possibly wear or failure—will result. Therefore, the elements are lubricated in order to prevent or reduce the actual contact between surfaces.

Without lubrication, most machines would run for only a short time. With inadequate lubrication, excessive wear is usually the most serious consequence, because a point will be reached, usually after a short period of operation, when the machine elements cannot function and the machine must be taken out of service and repaired. Repair costs—material and labor—may be high, but the cost of the lost production or lost availability of the machine may exact by far the greatest cost. With inadequate lubrication, even before failure of elements occurs, frictional forces between surfaces may be so great that drive motors will be overloaded or frictional power losses will be excessive. Finally, inadequately lubricated machines will not run smoothly, quietly, or efficiently.

Machine elements are lubricated by interposing and maintaining fluid films between moving surfaces. These films minimize actual contact between the surfaces and, by shearing easily, the frictional force opposing motion of the surfaces is low.

I. TYPES OF LUBRICATING FILMS

Lubrication films may be classified as follows:

1. *Fluid films* are thick enough that during normal operation they completely separate surfaces moving relative to each other. Fluid film lubrication includes hydrodynamic, elastohydrodynamic, hydrostatic, and squeeze types.
2. *Surface films* are not thick enough to maintain complete separation of the surfaces all the time. Surface films are often formed by components contained in the lubricant that can be physically adsorbed onto the surface or formed via a chemical reaction at the surface.
3. *Solid films* are more or less permanently bonded onto the moving surfaces.

A. Fluid Films

Fluid film lubrication is the most desirable form of lubrication because during normal operation, the films are thick enough to completely separate the load-carrying surfaces. Thus, friction is at a practical minimum, being due only to shearing of the lubricant films, and wear does not occur because there is essentially no mechanical contact. Fluid films are formed in three ways:

1. Hydrodynamic and elastohydrodynamic films are formed by motion of lubricated surfaces through a convergent zone such that sufficient pressure is developed in the film to maintain separation of the surfaces.
2. Hydrostatic film is formed by pumping fluid under pressure between surfaces that may or may not be moving with respect to each other.
3. Squeeze films are formed by movement of lubricated surfaces toward each other.

Two types of hydrodynamic film lubrication are now recognized. In plain journal bearings and tilting pad or tapered land thrust bearings operating under correct lubrication conditions, thick hydrodynamic films are formed. These films are usually more than 0.001 in (25 μm) thick. In hydrodynamic applications, the pressures are relatively low, and the areas over which the pressures are distributed are relatively large so that the amount of deformation of the load-carrying area is not large enough to significantly alter that area. Pressure in full hydrodynamic lubrication films range from a few psi to several thousand psi. Typical pressures found are in the range of 50–300 psi (344–2068 kPa). Load-carrying surfaces of this type are often referred to as "conforming," although it is obvious that in the case of tapered land thrust bearings, for example, the surfaces do not conform to the normal concept of the word. However, the term is a convenient opposite for the term "nonconforming," which quite accurately describes the types of surfaces where elastohydrodynamic films are formed.

The surfaces of the balls in a ball bearing theoretically make contact with the raceways at points, rollers of roller bearing make contact with the raceways along lines, and meshing gear teeth bearing also make contact along lines. These types of surface are nonconforming. Under the pressure (30,000–400,000 psi or 207–2758 mPa) applied to these elements by load conditions through the lubricating film, the metals deform elastically, expanding the theoretical points or lines of contact into discrete areas. The oil is able to generate films at these extremely high pressures because of the characteristic of viscosity doubling for each 5000 psi (34 mPa) increase in applied pressure. This is discussed later in this chapter.

Because a convergent zone exists immediately before these areas of contact, a lubricant will be drawn into the contact area and can form a hydrodynamic film. This type of film is referred to as an elastohydrodynamic film, where the "elasto" part of the term refers to the fact that elastic deformation of the surfaces must occur before the film can be formed. This type of lubrication is called elastohydrodynamic lubrication (EHL; the acronym EHD is also used). EHL films are very thin, in the order of 10–50 μin (0.25–1.25 μm) thick. However, even with these thin films, complete separation of the contacting surfaces can be obtained.

Any material that will flow at the shear stresses available in the system may be used for fluid film lubrication. In most applications, petroleum-derived lubricating oils or synthetic fluids are used. There are some applications for greases. Some materials, not usually considered to be lubricants, such as liquid metals, water, and gases are also used in some applications. The following discussions are concerned mainly with fluid film lubrication with oils and greases, as several rather complex special considerations are involved when other materials are used.

1. Thick Hydrodynamic Films

The formation of a thick hydrodynamic fluid film that will separate two surfaces and support a load can be described in Figure 8.1. The two surfaces are submerged in a lubricating fluid. As the upper surface moves, internal friction in the fluid causes it to be drawn into the space between the surfaces. The force drawing the fluid into space A is equal to the force tending to draw it out; but because the cross-sectional area at the outlet section is smaller than the inlet, the flow of fluid is restricted at the outlet. The moving surface tries to "compress" the fluid to force it through this

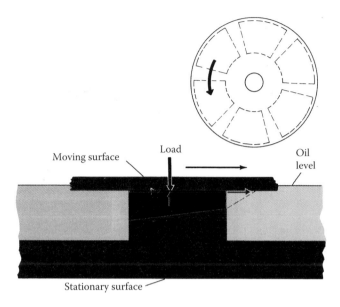

FIGURE 8.1 Converging wedge.

restricted section, and the result is an increase in pressure in the fluid. This pressure rise tends to do the following:

- Retard the flow of fluid into space A
- Increase the flow at the restricted outlet section
- Cause side leakage in the direction normal to the direction of motion

The most important effect, however, is that the pressure in this wedge-shaped film enables it to support a load without contact occurring between the two surfaces.

For the elements of Figure 8.1 to work as a bearing, the moving surface would have to be of infinite length. A more practical approach is to distribute a series of wedge-shaped sections around the circumference of a disk as shown in the inset. The moving surface then becomes a disk to which load can be applied, and the complete assembly can be used as a thrust bearing.

The pressure distribution in the wedge-shaped fluid film in Figure 8.1 is shown in Figure 8.2. In the direction of motion (left view), the pressure gradually rises more or less to a peak just ahead of the outlet section and then falls rapidly. In the direction normal to motion (right view), the pressure is at its maximum in the middle of the shoe, but because of side leakage, falls off rapidly toward each side.

In Figure 8.2, the taper of the shoe and the thickness of the fluid film are greatly exaggerated for illustrative purposes. In a practical bearing of this type, the amount of taper might be on the order of 0.002 in (50 µm) in a length of 6 in (150 mm), and the film thickness at the outlet would be on the order of 0.001 in (25 µm). With proper design and fluid viscosity, there would be no contact between the disk and shoes when the speed is above a certain minimum during normal operation. The bearing would not be frictionless, however, because a force would have to be applied to the disk to overcome the resistance to shear of the fluid films.

The same principles outlined in Figure 8.2 can be used to generate a fluid film, which will lift and support a shaft or journal in a bearing. If the moving surface and tapered shoe in Figure 8.2 are "rolled up," the journal and bearing shown in Figure 8.3 will result. A journal turning at a suitable speed will then draw fluid into the space between the journal and bearing and develop a load supporting film. This bearing is called a partial bearing and is suitable for some applications where the

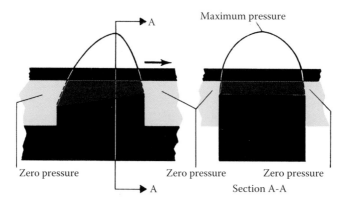

FIGURE 8.2 Pressure distribution in the wedge film. The left view shows the pressure distribution in the direction of motion whereas the right view shows the pressure distribution in the direction normal to motion.

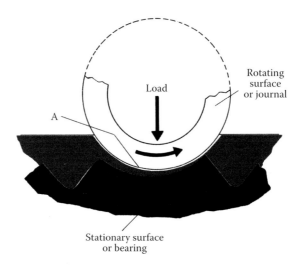

FIGURE 8.3 Partial journal bearing.

load on the journal is always in one direction. However, bearings are more commonly of the full 360° type, as shown in Figure 8.4, to provide restraint in the event of load variations and to permit easier enclosure and sealing.

The stages in the formation of a hydrodynamic film in a full journal bearing are illustrated in Figure 8.4. First, the machine is at rest with the oil shut off, and the oil has leaked from the clearance space. Metal-to-metal contact exists between the journal and bearing. When the machine has been started (left) and the oil supply is turned on filling the clearance space, the shaft begins to rotate counterclockwise and the friction is momentarily high so the shaft tends to climb the left side of the bearing. As it does this, it rolls onto a thicker oil film so that friction is reduced and the tendency to climb is balanced by the tendency to slip back. As the journal gains speed (center), drawing more oil through the wedge-shaped space between it and the bearing, pressure is developed in the fluid in the lower left portion of the bearing that lifts the journal and pushes it to the right. At full speed (right), the converging wedge has moved under the journal, which is supported on a relatively thick oil film with its minimum thickness at the lower right, on the opposite side to the starting position shown (left). Under steady conditions, the upward force developed in the oil film just equals the total

Lubricating Films and Machine Elements

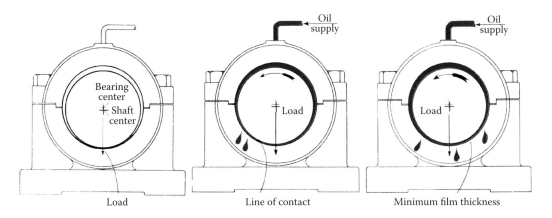

FIGURE 8.4 Development of hydrodynamic film in a full journal bearing with downward load.

downward load, and the journal is supported in the slightly eccentric position shown. Variations in speed, load, and oil viscosity will cause changes in this eccentricity, and also in the minimum film thickness. The pressures developed in the oil film are illustrated in Figure 8.5.

Thick hydrodynamic film lubrication is used in journal and thrust bearings for many applications. Characteristics and lubricant requirements of various applications are discussed more fully in later sections.

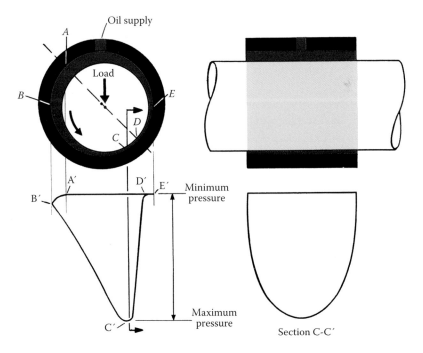

FIGURE 8.5 Pressure distribution in a full journal bearing. In the left view, from C, the point of minimum film thickness, to D the pressure drops rapidly to the minimum, which may be atmospheric, slightly above atmospheric, or slightly below atmospheric depending on such factors as bearing dimensions, speed, load, oil viscosity, and supply pressure. From A to C, the pressure increases as the clearance decreases. Pressure distribution along the line of minimum film thickness is shown at the right. From a maximum value near the center, it drops rapidly to zero at the ends because of end leakage.

2. Thin Elastohydrodynamic (EHL) Films

In the lubrication of EHL contacting surfaces, such as antifriction bearings, a necessary inherent characteristic of the oil is that its viscosity increases as the pressure on the oil increases (see Figure 8.6) in order to establish a supporting film at the very high pressures in the contact area. Thus, as the lubricant is carried into the convergent zone approaching the contact area, elastic deformation of two surfaces, such as a ball and its raceway or a roller and its raceway, occurs because of the pressure of the lubricant. As the viscosity increases, the pressure is further increased. This hydrodynamic pressure developed in the lubricant is sufficient to separate the surfaces at the leading edge of the contact area. As the lubricant is drawn into the contact area, the pressure on it rises further, thereby increasing its viscosity. Because of this high viscosity and the short time required for the lubricant to be carried through the contact area, the lubricant cannot escape, and separation of the surfaces can be achieved.

As noted, EHL films are very thin, on the order of about 10 to 50 μin (0.25–1.25 μm). The thickness of the film increases as the speed is increased, the lubricant viscosity is increased, the load is decreased, or the geometric conformity of the mating surfaces is improved. Load has comparatively little effect on the film thickness because of its increased viscosity within the contact zone. Under extreme conditions, the oil exhibits a combination of viscous (fluid like) and elastic (solid like) properties and is relatively stiff. Thus, an increase in load, for example, mainly has the effect of deforming the metal surfaces more and increasing the area of contact rather than decreasing the film thickness.

If the surfaces shown in Figure 8.7 were rougher—that is, if the asperities were higher—the opposing surface asperities would be more likely to make contact through the film. Thus, the relationship between the roughness of the surfaces and the film thickness is important in the consideration of EHL. This has led to the introduction of the quantity specific film thickness (λ) in EHL considerations. The specific film thickness is the ratio of the film thickness (h) to the composite roughness (σ) of the two surfaces:

$$\lambda = h/\sigma$$

The composite roughness of the two surfaces can be calculated as the square root of the sum of the squares of the individual surface roughness:

$$\sigma = \sqrt{\sigma_1^2 + \sigma_2^2},$$

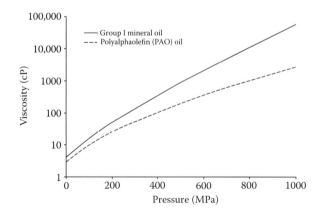

FIGURE 8.6 Effect of pressure on viscosity.

Lubricating Films and Machine Elements

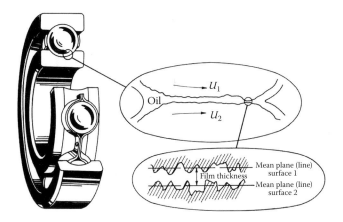

FIGURE 8.7 Schematic of contact area between ball and inner race.

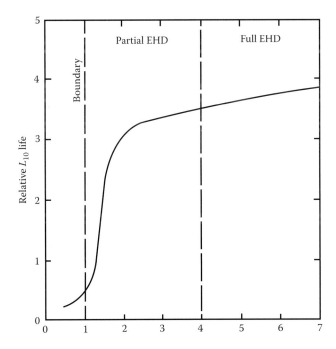

FIGURE 8.8 L_{10} fatigue life relative to book value as a function of specific film thickness. The data are for a series of cylindrical roller bearings.

where σ_1 and σ_2 are the root mean square (RMS) roughness values of the two surfaces.*

The specific film thickness is related to the ability of the lubricating film to prevent or minimize wear or scuffing. Figure 8.8 shows the effect of λ on the L_{10} fatigue life[†] of a series of cylindrical roller bearings. For λ values of approximately 3–4, the L_{10} fatigue life is relatively constant, and so it is essentially independent of the oil film thickness. This region is labeled "full EHL" and

---

* A number of calculations are given in the literature for the composite roughness of using centerline average (CLA) values. An approximate relationship between RMS and CLA values is $\sigma_{RMS} = 1.3\sigma_{CLA}$.
† The L_{10} fatigue life is the average time to failure of 10% of a group of identical bearings operating under the same conditions.

corresponds to full fluid film conditions that exist in thick hydrodynamic film lubrication. In the region labeled "partial EHL," the film becomes progressively thinner, and more surface asperities penetrate the film.

Fatigue life is decreased accordingly. This region is one of mixed film lubrication. When λ is below about 1, boundary lubrication prevails and the surfaces are in contact nearly all the time, fatigue failure is generally rapid unless lubricants with effective antiwear additives are used. For λ values above 4, full film conditions exist; however, the thicker oil films can result in higher bearing temperatures and lower overall efficiency.

These ranges for λ are considered to be applicable to rolling element bearings. As will be discussed later, somewhat lower values of the calculated safe minimum film thickness may be applicable to gears.

Increased understanding of EHL film characteristics and the use of specific film thickness have led to improved understanding of the lubricant function in many applications where unit loads are high, as well as to improved methods of selecting lubricants for such applications.

3. Hydrostatic Films

In hydrostatic film lubrication, the pressure in the fluid film that lifts and supports the load is provided from an external source. Thus, relative motion between opposing surfaces is not required to create and maintain the fluid film. The principle is used in plain and flat bearings of various types where it offers such advantages as low friction at very low speeds or when there is no relative motion, more accurate centering of a journal in its bearing, and freedom from stick-slip effects.

The simplest type of hydrostatic bearing is illustrated in Figure 8.9. Oil under pressure is supplied to the recess or pocket. If the supply pressure is sufficient, the load will be lifted and floated on a fluid film. At equilibrium conditions, the pressure across the pad will vary approximately as shown. The total force developed by the pressure in the pocket and across the lands will be such that the total upward force is equal to the applied load.

The clearance space and the oil film thickness will be such that all the oil supplied to the bearing can flow through the clearance spaces under the prevailing pressure conditions.

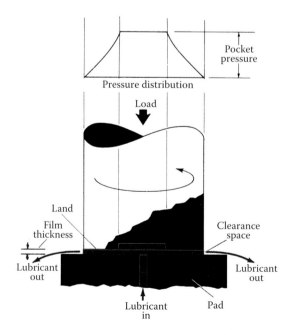

FIGURE 8.9 Simple hydrostatic thrust bearing.

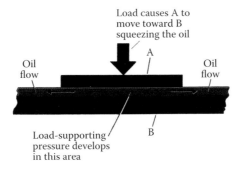

FIGURE 8.10 Squeeze film principle.

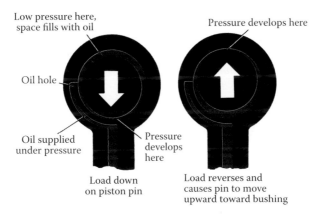

FIGURE 8.11 Squeeze film in piston pin bushings.

4. Squeeze Films

The principle of the squeeze film is shown in Figure 8.10, where application of a load causes plate *A* to move toward stationary plate *B*. As pressure develops in the oil layer, the oil starts to flow away from the area. However, the increase in pressure also causes an increase in oil viscosity, so the oil cannot escape as rapidly and a heavy load can be supported for a short period. Sooner or later, if load continues to be applied, most of the oil will flow or be forced from between the surfaces and metal-to-metal contact will occur. For short periods, however, such a lubricating film can support very heavy loads.

One application in which squeeze films are formed is in piston pin bushings used in engines. In the left-hand side of Figure 8.11, the load is downward on the pin and the squeeze film develops at the bottom. Before the film is squeezed, so thin that contact can occur, the load reverses (right view) and the squeeze film develops at the top. The bearing oscillates with respect to the pin, but this motion probably does not contribute much to film formation by hydrodynamic action. Nevertheless, bearings of this type have high load-carrying capacity.

B. THIN SURFACE FILMS

A copious, continuous supply of lubricant is necessary to maintain fluid films. In many cases, it is not practical or possible to provide such an amount of lubricant to machine elements. In other cases (e.g., during startup of a hydrodynamic film bearing), loads and speeds are such that fluid film cannot be maintained. Under these conditions, lubrication relies on very thin surface films formed by various specialized additives found in modern lubricants.

Under many conditions, enough oil is present so that part of the load is carried by fluid films and part is carried by contact between surfaces. This condition is often called *mixed film lubrication*. With less oil present, or with higher loads, a point is reached where fluid film is so low that it plays less of a role in separating the metal surfaces. This condition is often called *boundary lubrication*.

1. Nature of Surfaces

Machined surfaces are not smooth but are wavy because of the inherent characteristics of the machine tools used. Minute roughnesses—"hills and valleys"—are superimposed on these waves as depicted in Figure 8.12. The surfaces are not clean even when freshly machined as they are coated with incidental films (e.g., moisture, oxides, cutting fluids, rust preventives) and with lubricating films while in service. These films must be pierced or rubbed away before contact between clean surfaces can occur.

2. Surface Contact

When rubbing contact is made between the surface peaks, known as asperities, a number of actions take place (as shown in Figure 8.13), which represents a highly magnified contact area of a bearing and journal. These actions are as follows:

1. There is heavy rubbing action at *A*; surface films are sheared and elastic or plastic deformation occurs. Real (compared to apparent) areas of contact are extremely small and unit stresses will be very high.
2. The harder shaft material plows through the softer bearing material at *B*, breaking off wear particles, and creating new roughness.
3. Some areas such as *C* are rubbed or sheared clean and the clean surfaces weld together. The minute welds break immediately as motion continues, but depending on the strength of the welds, the break may occur at another section so that metal is transferred from one member to the other. New roughnesses are formed, some to be plowed off to form wear particles.

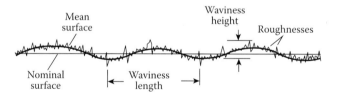

FIGURE 8.12 Magnified surface profile.

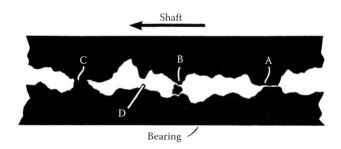

FIGURE 8.13 Actions involved in surface contacts.

These actions account for friction and wear under boundary lubrication conditions. Under mixed film conditions, the force necessary to shear the oil films at D is part of the frictional force. However, the presence of partial oil films reduces friction and wear that result from the other actions.

The actions involved in rubbing contact result in the development of very high temperatures in the minute areas of contact. These local temperatures, which are of short duration, are known as flash temperatures. With light loads and low speeds, a small amount of heat is conducted away through the surfaces without excessive rise in the general temperatures of the surfaces. However, with higher loads or speeds, or both, the number of contacts in a given area may increase to the point where the heat cannot be dissipated fast enough and the surfaces may run hot.

Under *mixed lubrication film* conditions, the moving surfaces are kept apart by very thin films of lubricant. These thin films are subjected to high shear conditions, which can lead to increased temperatures thereby reducing the effective viscosity of the oil. Quite often, because of the increase in temperature, excessive loads, shock loads, slower speeds, stop and start operation, or break-in of new components, the fluid films collapse and contact of the surfaces will occur. The collapse of the *mixed film* conditions result in *boundary lubrication*.

To help reduce the friction and wear under boundary conditions, it may be necessary to increase the viscosity of the lubricant or use a lubricant with surface active additives that provide special properties. These additives include various types of antiwear and friction-modifying additives. The two main classes of surface active additives work by either physical adsorption, often due to polar attraction to the surface, or chemical reaction forming a tenacious surface film. These surface films help reduce direct surface-to-surface contact. The polar type additives are sometimes referred to as film strength or oiliness additives, although these terms do not have exact definitions. The chemically reactive types are often also subdivided into antiwear and so-called extreme pressure (EP) additives. The latter are often used to prevent severe adhesive or scuffing wear under heavy load conditions and can be accurately be described as antiscuffing additives.

As boundary lubrication conditions become more severe, increased additive activity may be needed to prevent accelerated wear or destruction of the surfaces. Antiwear and EP additives are typically composed of chemically active ingredients such as zinc, phosphorus, sulfur, chlorine, or boron. The reacted films are typically only formed on the rubbing surfaces and act as a sacrificial layer that is removed instead of the original surface. Even under these conditions, shearing of the high spots (asperities) coming into contact may still occur. The surface films also reduce the potential for welding or fusing of these sheared surfaces. It is important to recognize that these additives often need temperatures above ambient conditions for activation and therefore do not always provide any advantages in applications where temperatures are low or the surfaces are separated by full fluid films.

C. Solid or Dry Films

In many applications, oils or greases cannot be used, either because of difficulties in applying them, sealing problems, or other factors such as environmental circumstances. A number of "permanently" bonded lubricating films have been developed to reduce friction and wear in applications of this type. These solid or dry film lubricants reduce the effective surface roughness. They attach themselves to the metal surfaces through rubbing action or by chemical reaction. This effect is dependent on load, temperature, and types of materials.

The simplest type of solid lubricating film is formed when a low-friction, solid lubricant such as molybdenum disulfide is suspended in a carrier and applied in the same manner as most lubricants. The carrier may be a volatile solvent, grease, or any of several other types of materials. After the carrier is squeezed, or evaporates from the surfaces, a solid layer of molybdenum disulfide remains to provide lubrication.

Solid lubricants can be bonded to rubbing surfaces with various types of resins, which cure to form tightly adhering coatings with good frictional properties. In the case of some plastic bearings,

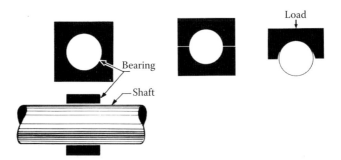

FIGURE 8.14 Elementary plain bearings.

the solid lubricant is sometimes incorporated in the plastic. This also occurs with some sintered metal bearings. During operation, some of the solid lubricant may then be transferred to form a lubricating coating on the mating surface.

In addition to molybdenum disulfide, polytetrafluoroethylene (PTFE), graphite, polyethylene, and a number of other materials are used to form solid films. In some cases, combinations of several materials, each contributing some characteristics to the film, are used.

II. PLAIN BEARINGS

The simplest types of plain bearing are shown in Figure 8.14. The bearing may be just a hole in a block (left), it may be split to facilitate assembly (center), or in some cases where the load is always carried in one direction, the bearing may consist of only a segment of a block (right). The part of the shaft within a bearing is called the journal, and plain bearings are often called journal bearings.

Plain bearings are designed for either fluid film lubrication or thin film lubrication. Most fluid film bearings are designed for hydrodynamic lubrication, but a number of bearings for special applications are being designed for hydrostatic lubrication.

A. Hydrodynamic Lubrication

The primary requirement for hydrodynamic lubrication is that oil of correct viscosity and sufficient quantity is present at all times to flood the clearance spaces.

The oil wedge formed in a hydrodynamic bearing is a function of speed, load, and oil viscosity. Under fluid film conditions, an increase in viscosity or speed increases the oil film thickness and coefficient of friction, whereas an increase in load decreases them. The separate consideration of these effects presents a complex picture that is simplified by combining viscosity Z, speed N, and unit load P, into a single dimensionless factor called the ZN/P factor.* This term is often referred to as the Stribeck number, honoring the German engineer who studied the friction response of bearings. Although no simple equation can be offered that expresses the coefficient of friction in terms of ZN/P, the relationship can be shown by a curve such as that shown in Figure 8.15. A similar type curve could be developed experimentally for any fluid film bearing.

In Figure 8.15, in the zone to the right of c, fluid film lubrication exists. To the left of a, boundary lubrication exists. In this latter zone, conditions are such that a full fluid film cannot be formed, some metallic friction and wear commonly occur, and very high coefficients of friction may be reached.

---

* The expression ZN/P is dimensionless when all quantities are in consistent units, for example, Z in poises, N in revolutions per second, and P in dynes per square centimeter; or Z in Pascal seconds, N in revolutions per second (reciprocal seconds), and P in Pascals; or Z in reyns, N in revolutions per second, and P in pounds-force per square inch.

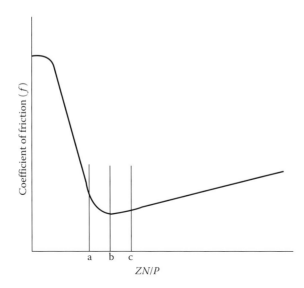

FIGURE 8.15 Effect of viscosity, speed, and load on bearing friction. For each hydrodynamic film bearing, there is a characteristic relation, such as that shown, between the coefficient of friction and the factor combining viscosity (Z), speed (N), and load (P). The curve is often called the Stribeck curve.

The portion of the curve between points a and c is a mixed film zone including the minimum value of the coefficient of friction f corresponding to the ZN/P value indicated by b. From the point of view of low friction, it would be desirable to operate with ZN/P between b and c, but in this zone any slight disturbance such as a momentary shock load or reduction in speed might result in film rupture. Consequently, it is good practice to design with a reasonable factor of safety so that the operating value of ZN/P is in the zone to the right of c.* The ratio of the operating ZN/P to the value of ZN/P for the minimum coefficient of friction (point b) is called the bearing safety factor. The common practice is to use a bearing safety factor on the order of 5.

In an operating bearing, if it becomes necessary to increase the speed, ZN/P will increase, and it may be necessary to decrease the oil viscosity to keep ZN/P and the coefficient of friction in the design range. An increase in load will result in a decrease in ZN/P, and it may be necessary to increase the oil viscosity to keep ZN/P and the coefficient of friction in the design range.

Oil film thickness can be related to ZN/P in the manner shown in Figure 8.16. The curve is typical of large, uniformly loaded medium speed bearings such as are used in steam turbines. In general, film thickness increases as ZN/P is increased, for example, in the case where the load is reduced whereas oil viscosity and journal speed remain constant. With a proper bearing safety factor, the film thickness will be such that normal variation in speed, load, and oil viscosity will not result in the film thickness being reduced to the point where metal-to-metal contact will occur.

The work done against fluid friction results in power loss, and the energy involved is converted to heat. Most of the heat is usually carried away by the lubricating oil, but some of the heat is dissipated by radiation or conduction from the bearing or journal. The normalized operating temperature is the result of a balance between the heat generated, overcoming fluid friction, and total heat removal.

The effect of increasing temperature is a decrease in oil viscosity. The reduction in viscosity results in a lower ZN/P and coefficient of friction (provided boundary or mixed-film lubrication

---

* Equations, procedures, and data for plain bearing design and performance calculations, are available in many technical papers and books. Among the latter are the following: *Bearing Design and Application* (Wilcock and Booser, McGraw-Hill Company, New York), *Theory and Practice of Lubrication for Engineers* (Fuller, John Wiley & Sons), and *Analysis and Lubrication of Bearings* (Shaw and Mack, McGraw-Hill Book Company).

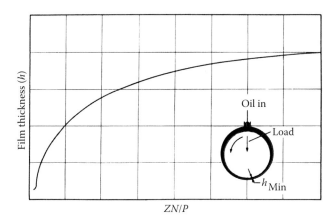

FIGURE 8.16 Effect of viscosity, speed, and load on film thickness.

conditions do not exist). Also, there is less work required to overcome fluid friction, less heat is developed, and there is a tendency for the temperature to decrease. This has a stabilizing influence on bearing temperatures.

In general, if excessive temperatures develop even though load, speed, and oil viscosity are within the correct range, it may be that there is insufficient oil flow for proper cooling. It may then be necessary to consider additional grooving of the bearing or increasing the bearing clearance in order to increase the flow of oil through the bearing.

1. Grease Lubrication

Whereas the grease in a rolling element bearing acts as a two-component system in which the soap serves as a sponge reservoir for the fluid lubricant, greases in plain bearings behave as homogeneous mixtures with unique flow properties. These flow properties are described by the apparent viscosity (see Chapter 4), that is, the observed viscosity under each particular set of shear conditions. As the rate of shear is increased, the apparent viscosity decreases, and at high shear rates, it approaches the viscosity of the fluid component used in the formulation. In many plain bearings, the shear rate in the direction of rotation is high enough to cause the apparent viscosity of a grease to be in the same general range as the viscosities of lubricating oils normally used for hydrodynamic lubrication. As a result, fluid film formation can occur with grease, and it is now considered that some grease lubricated plain bearings operate on fluid films, at least part of the time. In addition, hydrodynamic film bearings designed for grease lubrication are used in many applications.

The pressure distribution in a grease lubricated hydrodynamic film bearing is similar to that in oil-lubricated bearings (Figure 8.5). However, toward the ends of the bearing, because of reduced pressure in the film, the shear stress is lower, the apparent viscosity of the grease remains high, and end leakage is lower. This results in high pressures being maintained farther out toward the ends of the bearing; the average pressure in the film is higher, and the maximum pressure is correspondingly lower. The minimum film thickness for the same bearing load and speed will be greater. The coefficient of friction may be equal or less than that with an equivalent oil-lubricated bearing, depending on such factors as the type of grease used and the viscosity of the oil component in the grease.

Fluid film bearings lubricated with grease have several advantages compared to those lubricated with oil. As a result of the lower end leakage, the amount of lubricant required to be fed to the bearing is less, so grease-lubricated bearings can be supplied by an all-loss system with either a slow, continuous feed, or a timed, intermittent feed in conjunction with adequate reservoir capacity in the grooves of the bearing.

When a grease-lubricated bearing is shut down for a certain period with the flow of lubricant shut off, the grease usually does not drain or squeeze out completely. Some grease remains on the bearing surfaces so that a fluid film can be established almost immediately when the bearing is restarted. Starting torque and wear during starting may be considerably reduced. During shutdown periods, retained grease also acts as a seal to exclude dirt, dust, water, and other environmental contaminants, and to protect bearing surfaces from rust and corrosion. If the grease provides a lower coefficient of friction, power consumption during operation will also be lower.

When grease lubrication is used for fluid film bearings, the cooling is not as efficient as is the cooling obtained from oils. This disadvantage may be partially offset if the coefficient of friction is lower with a grease; if speeds or loads are high, however, it may be a limitation.

B. Hydrostatic Lubrication

In a hydrostatic bearing, the oil feed system used must be such that the pressure available, when distributed across the pocket and land surfaces, is sufficient to support the maximum bearing load that may be applied. The system must also be designed to provide an equilibrium condition for loads below the maximum. Three types of lubricant supply are used to accomplish this—a constant volume system, a constant pressure system with a flow restrictor, and a constant pressure system with a flow control valve.

1. Constant Volume System

In this type of system, the pump delivers a constant volume of oil at whatever pressure is necessary to force that volume through the system—that is, if the backpressure increases, the pump pressure automatically increases sufficiently to maintain the flow rate. In most cases, the volume delivered by the pump actually decreases somewhat as the pressure increases, but this has relatively little effect on the way the system operates.

A constant volume system must have adequate pressure capability to support any applied load. In Figure 8.9, when the pump is turned on, oil will flow into the pocket and the pressure will increase until the load is lifted sufficiently to establish a clearance space through which the volume of oil flowing in the system will be discharged. The clearance space and oil film thickness will be a function of the volume of flow in the system, the viscosity of the lubricant, and the applied load.

If the load is then increased, the clearance space and film thickness will decrease, and the pump pressure will have to increase in order to discharge the same volume of oil through the reduced clearance space. Only small changes in clearance space and film thickness accompany fairly large variations in load, so the bearing is said to be very "stiff."

One disadvantage of the constant volume system is that it does not compensate for variations in the point of application of the load in multiple pocket bearings. In the two pocket bearings in Figure 8.17 that are using a constant volume system, if the load is shifted to the right, the runner will tend to tilt. This will decrease the clearance at the right-hand land and increase the clearance at the left-hand land. Oil can then flow more freely out of the left pocket, the pressure in the system will decrease, and the load will sink until metallic contact might occur at the right side. This problem can be compensated for with either of the following systems.

2. Constant Pressure System with Flow Restrictor

A constant pressure system requires an accumulator or manifold to maintain the pressure at a relatively constant value. If this constant pressure is applied to the pockets of the bearing through flow restrictors (see Figure 8.17), such as capillaries or orifices, a compensating force will be developed when the runner will tend to tilt. The pressure drop across a flow restrictor increases as the flow through it increases, and decreases as the flow decreases. Thus, the pressure in the pocket, being the difference between the system pressure and the pressure drop across the restrictor, decreases as flow increases and increases as flow decreases. With this system, if an off-center load causes the runner

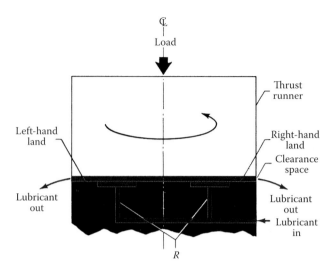

FIGURE 8.17 Two pocket hydrostatic thrust bearing. When flow restrictors are used, they are installed at the locations marked R.

to tilt toward the right, the clearance at the right-hand land will decrease, the flow to that pocket will decrease, the pressure drop across the restrictor will decrease, and the pocket pressure will increase. The opposite effect will occur in the left pocket. This will tend to lift the load over the right-hand pocket and let it sink over the left-hand pocket, restoring the runner to a more or less coaxial position.

In single-pocket bearings, a constant pressure system with a flow restrictor is not as stiff as a constant volume system. As the load is increased with the constant pressure system, the pocket pressure must increase to support it. This requires that the oil flow decreases in order to reduce the pressure drop across the flow restrictor. With less oil flow and higher pressure, a considerably smaller clearance space will be required to discharge the oil. However, the compensating feature results in wide use of constant pressure systems in both thrust and journal bearings with multiple pockets.

3. Constant Pressure with Flow Control Valve

If a flow control valve is used as the restrictor in a constant pressure system, the system becomes essentially constant pressure and constant volume. The bearing will have the stiffness of a constant volume system and also have the compensating features of the constant pressure system.

4. Hydrostatic Bearing Applications

One of the more common applications of the hydrostatic principle is "oil lifts" for starting heavy rotating machines, such as steam turbines, large motors in steel mills, and rotary ball and rod mills. Hydrostatic lifts for plain bearings are also used for turning gear operation during startup and cool-down periods of large steam and gas turbines, where the turbine rotors are rotated at speeds too slow to establish hydrodynamic films. Because metal-to-metal contact exists between the journal and the bearings when the journal is at rest, extremely high torque may be required to start rotation, and damage to the bearings may occur. By feeding oil under pressure into pockets machined in the bottoms of the bearings, the journal can be lifted and floated on fluid films (see Figure 8.18). The pockets are generally kept small in order not to seriously interfere with the hydrodynamic film capacity of the bearings. When the journal reaches a speed sufficient to create hydrodynamic films, the external pressure can be turned off and the bearings will continue to operate in a hydrodynamic manner. The reverse procedure may be used during shutdown.

The low friction characteristics of hydrostatic film bearings at low speeds are being used in a variety of ways. One application is in "frictionless" mounts or pivots for dynamometers. Another is in the

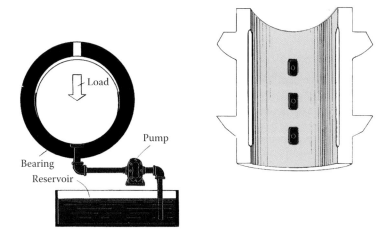

FIGURE 8.18 Hydrostatic lift. Illustration on the right side shows one type of shallow pocket through which the oil pressure can be applied.

bearings for tracking telescopes where the relative motion is extremely slow but must be completely free of stick-slip effects. The hydrostatic principle is also applied to the guides and ways of large machine tools, particularly where extremely precise movement and location of the ways is required.

The concept of controlled oil film thickness for hydrostatic film bearings is being used in high-speed applications such as machine tool spindles for high precision work. Spindles of this type are equipped with multiple pocket bearings with a constant pressure system and a flow restrictor for each pocket. With this arrangement, any change in the lateral loading on the spindle, as a result of a change in the cutting operation, is automatically compensated for by changes in pressures in the individual pockets. Lateral movement of the spindle is thus minimized, and very accurate control of the centering of the spindle in the bearings can be achieved.

C. Thin Film Lubrication

Many bearings are designed to operate on restricted lubricant feeds as the most practical and economic approach. The lubricant supplied to the bearings gradually leaks away and is not reused; thus, this type of lubrication is generally referred to as "all-loss" lubrication. Because of the restricted supply of lubricant, these bearings operate on thin lubricating films, either of the mixed film or boundary type. The simplest type of all-loss lubrication is hand oiling (Figure 8.19). Hand oiling results in flooded clearances immediately after lubrication. This condition

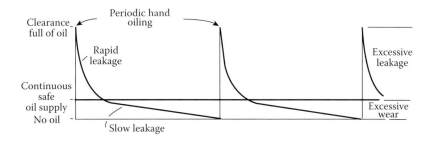

FIGURE 8.19 Hand oiling: the condition of "feast or famine," is always present with periodic hand oiling. Here, it is compared with the safe continuous supply of oil by devices that feed oil frequently in small quantities.

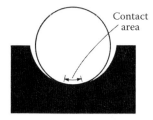 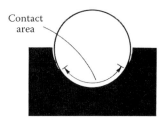

FIGURE 8.20 Contact area before and after wearing-in of plain thin film bearing.

may permit formation of fluid films for a brief period; however, the oil quickly leaks away to an amount less than what is considered acceptable for safe operation. In short, the bearing passes through the regime of mixed film lubrication and operates much of the time under boundary conditions.

A closer approach to maintaining a safe oil supply may be accomplished with application devices such as wick feed oilers, drop feed cups, waste packed cups, bottle oilers, and central dispensing systems such as force feed lubricators or oil mist systems. These devices supply oil on either a slow, constant basis, or at regular, short intervals. With greases, leakage is not as serious a concern, but the use of centralized lubrication systems will provide a more uniform lubricant supply than with a manual grease gun.

Even with regular application of small amounts of lubricant, thin film bearings require proper design and installation, and proper lubricant selection to control wear and provide satisfactory service life.

1. Wearing In of Thin Film Bearings

In a new bearing, the journal normally will make contact with the bearing over a fairly narrow area (Figure 8.20, left). Generally, lighter loads should be carried by such a bearing under thin film conditions, because unit loads beyond the ability of the oil film to prevent metallic contacts would probably exist. Under favorable conditions, wear will occur, but will have the effect of widening the contact area (Figure 8.20, right) until the load is distributed over a sufficiently large area that wear becomes practically negligible. New plain bearings generally are supplied with a thin "flashing" (approximately 0.0005 in/12.5 µm) of a softer material to help facilitate break-in. Under unfavorable conditions, this initial wear may be so rapid that bearing failure occurs.

Large bearings are often fitted before operation by hand scraping, or by counterboring the loaded area to the radius of the journal. Fitting of this type can be done only when the direction of loading is known and is constant. After fitting, the area of contact should be in the range of 90–120°.

D. Mechanical Factors

In plain bearings, there are several mechanical factors that affect lubrication. These include

1. Ratio of the length of the bearing to its diameter
2. Projected area of the bearing
3. Clearance between journal and bearing
4. Bearing materials
5. Surface finish
6. Grooving (where used)

1. Length/Diameter Ratio

The diameter of a journal is determined by mechanical requirements, which involve the torque transmitted, shaft strength, and shaft rigidity. The bearing length is governed by the load to be carried, the space available, and the operating characteristics of the particular type of machine, especially with regard to retention of bearing alignment and shaft flexing. Space limitations are often extremely important in the determination of bearing length.

As a general guide, the shorter the bearing, the higher end leakage will be, and the more difficult it will be to develop load-supporting lubricant films. On the other hand, with too long a bearing, metal-to-metal contact, high friction, high temperatures, and wear may occur at the bearing ends even with only a moderate misalignment or shaft deflection under load. Length/diameter (L/D) ratios of up to 4 have been used in past designs, but the trend for many years has been to use lower ratios. Most fluid film bearings now have L/D ratios of 1.5 or less. As an example, connecting rod bearings for internal combustion engines may have L/D ratios of 0.75 or less. Thin film bearings have L/D ratios in the same general range.

2. Projected Area

The axial length of a bearing multiplied by its diameter is called the projected area, and it is common to express unit loads in force per unit of projected area. Established practices pertaining to unit loads on this basis vary considerably for different classes of machinery and different bearing materials, ranging from as low as 15 psi (103 kPa) for lightly loaded line shafting to as high as 5000 psi (24.5 mPa) or more for internal combustion engine crankpins and wristpins. Most industrial bearings carrying constant loads—such as those found in turbines, centrifugal pumps, and electrical machinery—fall in the range of 50–300 psi (345–2700 kPa) with most being less than 200 psi (1380 kPa). Heavier loads are encountered in bearings of reciprocating machinery and in other bearings subject to varying or shock loads. Peak hydraulic pressures within the oil films (Figure 8.5) are usually 3 to 4 times these unit loads based on projected area.

To achieve optimum life in plain bearings, full film (hydrodynamic) lubrication is necessary. Other contributing factors to bearing life are speed, load, temperature, and the compressive strength of the bearing materials. If the compressive strength of the materials used for metallic plain bearings is known, a good general guide for achieving acceptable life is that bearing loads should not exceed 33% of the compressive strength of the materials. The limiting load and speed conditions can be expressed as a factor PV, where P is the pressure on the bearing (psi) multiplied by the surface speed V of the shaft [in feet per minute (ft/min)]. The PV factor varies by bearing design and materials used. Data on PV factors and compressive strengths of materials can be obtained directly from bearing manufacturers or are readily available in technical manuals if the materials used in the bearing are known.

3. Clearance

A full bearing must be slightly larger than its journal to permit assembly, to provide space for a lubricant film, and to accommodate thermal expansion, and some degree of misalignment and shaft deflection. This clearance between journal and shaft is specified at room temperature.

One of the principal factors controlling the amount of clearance that must be allowed is the coefficient of thermal expansion of the bearing material. The higher the coefficient of thermal expansion, the more clearance must be allowed to prevent binding as the bearing warms up to operating temperature. Babbitts and bearing bronzes have the lowest coefficients of thermal expansion of common bearing materials. Clearances for these materials in general machine practice range from 0.1% to 0.2% of the shaft diameter (0.001–0.002 in per inch of shaft diameter). Many precision bearings will have less clearance than this, whereas rough machine bearings may have more. Because of their higher coefficients of thermal expansion, aluminum bearings require somewhat more clearance than babbitts or bronzes, and some of the plastic bearing materials require considerably more, in some cases as much as 0.8% of the shaft diameter.

4. Bearing Materials

During normal operation of a fluid film lubricated bearing under constant load, the most important property required in the bearing material is adequate compressive strength for the hydraulic pressures developed in the fluid film. When cyclic loading is involved, as in reciprocating machines, the material should have adequate fatigue strength to operate without developing cracks or surface pits. With shock loading, the material should be of such ductility that neither extrusion nor crumbling occurs. Under boundary lubrication conditions, the material also requires:

1. Scoring resistance, requiring appreciable hardness and low shear strength
2. The ability to conform to shaft irregularities and misalignment
3. The ability to embed abrasive particles

If operating temperatures are high, resistance to corrosion and softening may be important.

Although these properties may be somewhat conflicting, numerous materials have been used to obtain satisfactory bearing life for the wide range of conditions encountered.

Plain bearing materials most often encountered in industrial machines are bronzes and babbitts. Suitable bronzes and babbitts are available for practically all conditions of speed, load, and operating temperature encountered in general practice. Steel and cast iron are used for a limited number of purposes, usually involving low speeds or shock loads. There has been considerable growth in the use of plastic and elastomeric materials such as nylon, thermoplastic polyesters, laminated phenolics, PTFE, and rubber for bearings, particularly in applications where contamination of, or leakage from, oil-lubricated bearings might result in high maintenance costs or short bearing life. Some of these materials can be lubricated with water or soluble oil emulsions in certain applications. Allowable unit loads for these bearing materials usually are lower, although in a number of cases, nylon-filled bearings have been used as direct replacements for bronze bearings.

For internal combustion engines, babbitt bearings are made with a very thin layer of babbitt over a backing of bronze or steel to increase the load-carrying capacity. Even then, the loads may be greater than babbitts can handle, so a number of stronger bearing materials have been developed. Aluminum bearings are being used in some diesel and gas engine applications because of their longer potential life and greater resistance to acid attack. Because the aluminum is harder, it will not embed particles as well as the softer bearing materials, and therefore contamination control is more critical. Engine bearings are usually fabricated in the form of precision inserts (see Figure 8.21), which are interchangeable and require no hand fitting machining when installed.

Precision insert bearings, which are usually constructed of layers of different materials, provide:

- A thin surface layer (sometimes as little as 0.0003 in or 0.0075 mm) having good surface characteristics—such as low friction, scoring resistance, conformability, and resistance to corrosion
- A thicker layer (0.008–0.025 in or 0.2–0.6 mm) of bearing material having adequate compressive strength and hardness, suitable ductility, and good resistance to fatigue
- A still thicker (usually 0.05–0.125 in or 1.25–3.2 mm) back or shell made of bronze or steel

Some of the more common combinations used with this type of construction are babbitt over leaded bronze over steel; lead alloy over copper–lead over steel; silver alloy over lead over steel; and tin over aluminum alloy over steel. These bearings all require smooth hardened journals, rigid shafts, and minimum misalignment.

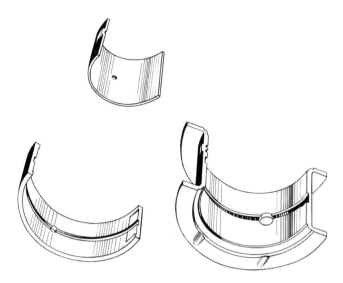

FIGURE 8.21 Precision insert bearings. These are typical main and connecting rod insert bearings used in internal combustion engines. Bearing at lower right is designed to carry thrust as well as radial loads.

5. Surface Finish

Machined surfaces are never perfectly smooth. The peak-to-valley depth of roughness in machined surfaces ranges from about 160 μin (4 μm) for carefully turned surfaces to about 60 μin (1.5 μm) for precision ground surfaces. Finer finishes, of approximately 10 μin (0.25 μm), can be obtained by other commercial methods.

Finely finished surfaces would, in general, be damaged less than rough surfaces by the metal-to-metal contact that occurs under boundary lubrication conditions. However, some degree of "wearing in" of new bearing surfaces always occurs. New surfaces, which may be relatively rough, tend to become smoother under favorable conditions and careful wearing-in. In some cases, certain degrees of initial roughness aid this wearing-in process (by holding the lubricant in place), as long as loads are minimized for break-in.

Under fluid film conditions, the minimum safe film thickness is a function of the roughness of the surfaces. Rougher surfaces require thicker films in order to prevent contact of surface asperities through the film. On the other hand, the finer the surface finish, the lower the minimum safe film thickness and the less clearance is necessary. Because the film thickness decreases with increases in unit loading, if the minimum safe film thickness is lower as a result of finer surface finishes, the allowable unit loading is higher, all other factors being equal. Conversely, it can be said that bearings designed for high unit loads and small clearances must have finely finished surfaces. Tests show that fluid films may also be formed at a lower speed when starting up a bearing with a smooth finish than when starting one with a rough finish.

6. Grooving of Bearings

In all plain bearings, some provision must be made to supply the lubricant to the bearing and distribute it over the load-carrying surfaces. Lubricant is generally admitted through an oil port or ports and then distributed by means of grooves cut in the bearing surface. The location of the supply port and the type of grooving used depend on several factors, including the type of supply system, the direction and type of load, and the requirements of the bearings. Certain basic principles apply to all cases.

a. Grooving for Oil

The distribution of oil pressure in a typical fluid film bearing with steady load is shown in Figure 8.5. Usually, oil should be fed to a bearing of this type at a point in the no-load area where the oil pressure is low. When the shaft is horizontal and the steady load is downward, it is usually convenient to place the supply port at the top of the bearing, as shown.

Generally, grooves should not be extended into the load-carrying area of a fluid film bearing. Grooves in the load-carrying area provide an easy path for oil to flow away from the area. Oil pressure will be relieved and load-carrying capacity will be reduced. This effect for an axial and a circumferential groove is shown in Figures 8.22 and 8.23. However, to provide increased oil flow for better cooling in certain force feed lubricated bearings, it is sometimes necessary to extend the grooves through the load-carrying area. With variable load direction, it may also be necessary to extend the grooves through the load-carrying area. This is done in some precision insert bearings for internal combustion engines mainly to increase cooling and oil distribution.

With constant load direction, a single oil supply hole may be sufficient. To increase oil flow or improve distribution, an axial groove cut through the oil supply port (Figure 8.24) is often all that is required. Normally, the groove should not extend to the ends of the bearing, because that would allow the oil to flow out of the ends rather than being carried into the oil wedge. An exception to this is in certain high-speed bearings, where carefully sized ports or orifices are provided at the ends of the groove to permit increased flow for cooling purposes. Also, in medium-speed equipment, end bleeder ports frequently are provided to give a continuous flushing of dirt particles.

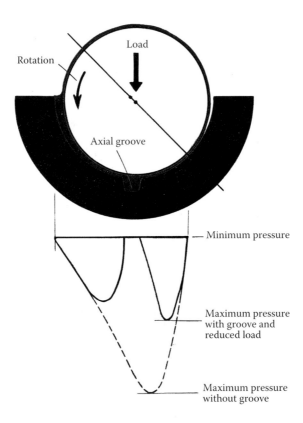

FIGURE 8.22 Axial groove reduces load-carrying capacity. An axial groove through the pressure area of a fluid film bearing provides an easy path for leakage and the relief of oil pressure. In the lower figure, the solid lines represent the approximate pressure distribution when the groove is present, and the dashed line represents approximately what it would be without the groove.

Lubricating Films and Machine Elements

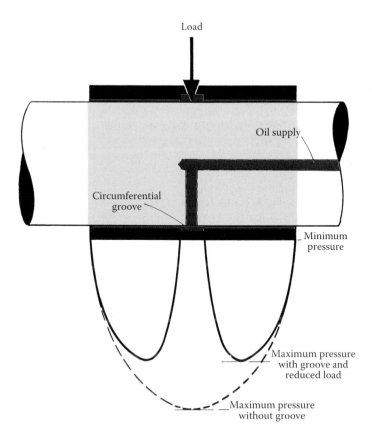

FIGURE 8.23 Circumferential groove reduces load-carrying capacity. As in Figure 8.22, the solid lines show the pressure distribution with the groove present, whereas the dashed line shows what it would be without the groove.

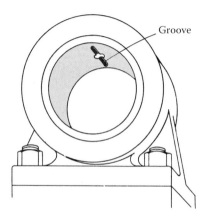

FIGURE 8.24 Axial distribution groove in a one-part bearing.

The grooving needed for distribution in a two-part bearing is usually formed by chamfering both halves at the parting line (Figure 8.25). Both sides are chamfered in cases where it is necessary to provide for rotation in both directions. These chamfers should also stop short of the ends of the bearings. Frequently, oil inlet port and distribution grooves are combined with the chamfer at the split. Where a top inlet port is used, an overshot feed groove may be machined in the upper half as shown in Figure 8.26.

If a stationary journal and a rotating bearing are used, oil may be fed through a port and axial groove in the journal. Again, the groove should be placed on the no-load side.

Where heavy thrust loads are to be carried, fluid film bearings of the tilting pad or tapered land type are often used. Tilting pad bearings require no grooving because the oil can readily flow out around the pad mountings. Tapered land bearings require radial grooves located just ahead of the point where the oil wedge is formed. If the thrust load is carried by one end face of a journal bearing, the axial groove or chamfers may be extended to the thrust end so that oil will flow directly to the thrust surfaces. The end of the bearing should be rounded or beveled to aid in the flow of oil between the end face and thrust collar or shoulder.

Circumferential grooves are sometimes cut near one or both ends of a bearing to collect end leakage and drain it back to the sump or reservoir. This oil might otherwise flow along the shaft and leak through the shaft seals. When collection grooves are used, they mark the effective ends of the bearing.

Vertical shaft bearings often require only a single oil port in the upper half of the bearing in the no-load area. In general, the lower the supply pressure, the higher the port should be. Sometimes,

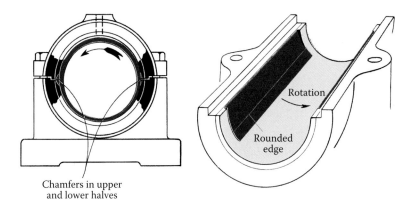

FIGURE 8.25 Distribution grooves in a two-part bearing.

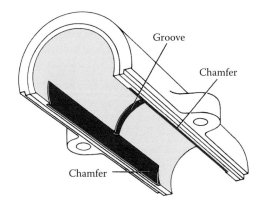

FIGURE 8.26 Overshot feed groove and chamfers.

a circumferential groove may be added near the top of the bearing to improve distribution (Figure 8.27, left). If leakage from the bottom of the bearing is excessive, a spiral groove is sometimes cut in the bearing in the proper direction relative to shaft rotation so that oil will be pumped upward (Figure 8.27, right).

Increased oil flow to cool a hot running bearing can be obtained by simple forms of grooving. An axial groove on the no-load side, for example, will increase oil flow by 3 to 4 times compared to a single port alone. Circumferential grooves also increase oil flow, but not as much as an axial groove. They also have the disadvantage of reducing the load-carrying capacity of the bearing. Increased clearance can often be used in lightly loaded, high-speed bearings to increase oil flow. When increased clearance might reduce load-carrying capacity too much, extra grooving or a clearance relief in the unloaded portion can be used to increase oil flow for cooling (Figure 8.28).

Where the direction of bearing load changes as in reciprocating machines, it is still essential that oil be fed into an unloaded or lightly loaded area. One way of doing this is with a circumferential groove. While this, in effect, divides the bearing into two shorter bearings of reduced total load-carrying capacity, it may be the most effective alternative. Also, it may be desirable to provide a path

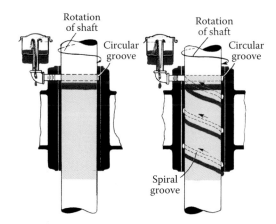

FIGURE 8.27 Grooving for a vertical bearing.

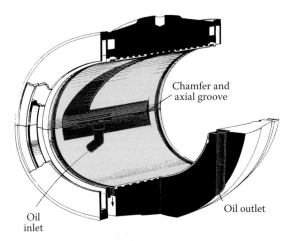

FIGURE 8.28 Grooving to increase oil flow for cooling. This cutaway view of a large turbine bearing shows a wide groove cut diagonally in the top (unloaded) half of a bearing to permit a large flow of oil for cooling purposes. A relatively small part of the oil passing through this bearing would be needed for the fluid film.

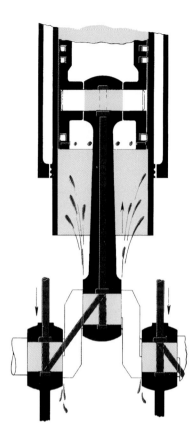

FIGURE 8.29 Circumferential grooves. In this circulation system, an oil pump, driven from the crankshaft, takes oil from the crankcase sump and delivers it under pressure to the crankpin and wristpin bearings through passages in the crankshaft and connecting rods.

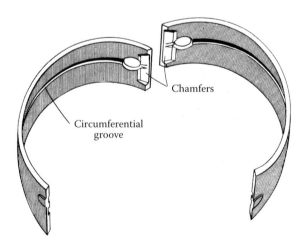

FIGURE 8.30 Circumferential groove and axial chamfer. An axial groove (also called a spreader groove) is often used with a circumferential groove in precision bearings.

Lubricating Films and Machine Elements   167

for oil flow to other bearings, for example, as in many internal combustion engines (Figure 8.29). An axial groove or chamfer may be used with a circumferential groove to improve oil distribution or increase oil flow (Figure 8.30).

Ring-oiled bearings are a special case with regard to grooving. Where the direction of load is fixed and is toward the bottom of the bearing, a simple axial groove connecting with the ring slot (Figure 8.31) is adequate. In a two-part bearing, the axial groove may be formed by chamfering the bearing halves at the parting line. However, if loads are sideways because of belt pull or gear reaction, this type of groove can be blocked as shown in Figure 8.31. To overcome this problem, grooving such as that shown in Figures 8.32 and 8.33 is often used. Grooving of this type is used for electric motors, which may be belted or geared to the load: thus, the motor manufacturer does not need to know the contemplated direction of loading and shaft rotation.

Grooving for thin film lubricated bearings generally follows the same pattern as for fluid film lubricated bearings. Because of the restricted supply of oil to this type of bearing, it may be necessary to provide additional reservoir capacity in the grooves, and to design the grooves to retain and distribute the oil (Figures 8.34 through 8.36).

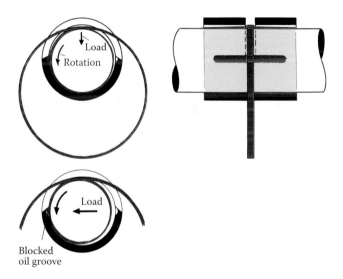

FIGURE 8.31 Ring oiled bearing, downward load. A side thrust on the shaft can cause blockage of this type of grooving (lower left).

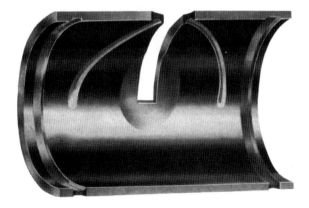

FIGURE 8.32 Section of ring oiled bearing with X grooves, which cross at the top of the bearing. The area around the end of the ring slot is relieved to aid in the distribution of oil.

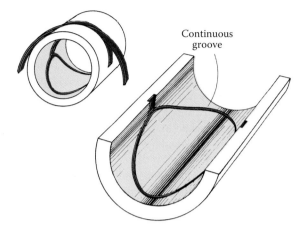

FIGURE 8.33 Ring oiled bearing, continuous groove. With this pattern of grooving, oil feed cannot be blocked and there is little interruption of load-carrying surface along any axial line.

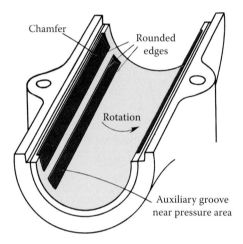

FIGURE 8.34 Auxiliary groove. Where oil application is infrequent, extra storage capacity is sometimes needed in a bearing. This can be provided via an auxiliary groove cut just ahead of the load-carrying area.

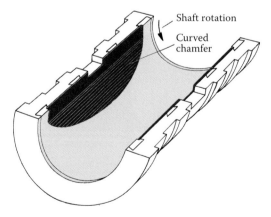

FIGURE 8.35 Curved chamfer. Where clearance at the horizontal centerline is large, end leakage of oil can be reduced by using a curved chamfer. This tends to guide a restricted supply of oil toward the center of the bearing so that it will be drawn into the load-carrying area.

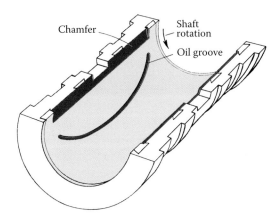

FIGURE 8.36 Straight chamfer and curved groove. The chamfer provides oil storage capacity and aids distribution. The curved groove catches oil flowing toward the bearing ends and guides it back toward the center of the bearing.

b. Grooving to Grease

Grooving principles for grease are practically the same as for oil. Because grease has high resistance to flow at the low shear rates in the supply system, grooves need to be made wider and deeper than oil grooves. Annular grooves cut near the ends of the bearing, similar to oil collection grooves but without the drain hole, are often used to reduce grease leakage. Some of the grease forced into these grooves in the load zone flows back out in an unloaded area and enters the bearing for reuse.

E. Lubricant Selection

Lubricants for plain bearings must be carefully selected if the bearings are to give long service life with low friction and minimum power losses.

1. Oil Selection

Most of the lubrication supply systems for fluid film bearings are essentially circulation systems; thus, the oil is reused many times over long periods.

Moisture and other contaminants may enter the system and should be accounted for. These conditions require that the oil must have:

- Good chemical stability to resist oxidation and deposit formation in long-term service
- Provide protection against rust and corrosion
- Separate readily from water to resist the formation of troublesome emulsions
- Resist foaming
- Provide viscosity control in applications where dilution could occur

These properties are generally found in high-quality circulation oils.

To select the proper viscosity of oil, three operating factors must be determined—speed, load, and operating temperature (including ambient temperatures at startup). Speed (rpm) is usually readily determined. Bearing loads may be calculated from the total weight supported by the bearings divided by the total projected area of the bearings. For simplicity, in industrial-type applications, bearing loads of 200 psi (1380 kPa) can often be assumed for bearings operating under light to moderate loads with no shock loads, and 500 psi (3450 kPa) for heavily loaded bearings or where shock loads are present. Operating temperatures can be approximated by using the oil exit temperature of the hottest bearing.

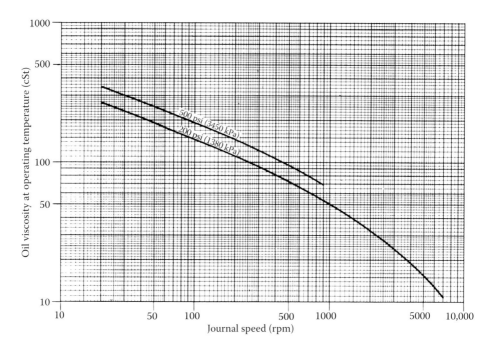

FIGURE 8.37 Oil viscosity chart for fluid film bearings. To determine the viscosity required, follow up from the appropriate journal speed to either of the bearing load lines, then to the left to read the oil viscosity at the operating temperature. The 200 psi line should be used for bearings operating under low to moderate loads without shock loads, and the 500 psi line should be used where loads are heavy or shock loads are present.

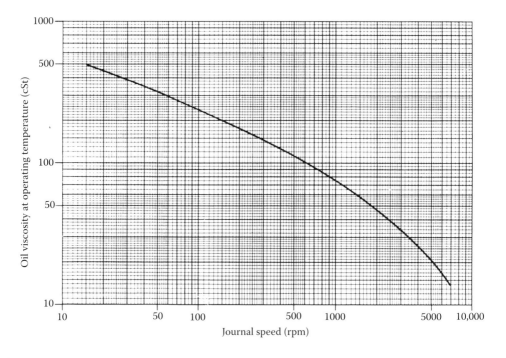

FIGURE 8.38 Oil viscosity chart for thin film bearings. As loads can be assumed to be heavy enough for the bearing to operate in a thin film regime, only journal speed and operating temperature affect oil viscosity selection.

When these three factors are known, the optimum viscosity of oil can be determined from a chart such as that shown in Figure 8.37. Because this chart gives the viscosity at the operating temperature, American Society for Testing and Materials (ASTM) viscosity–temperature charts must then be used to determine the desired viscosity at one of the standard measuring temperatures (see discussion of viscosity in Chapter 3).

Where bearings are to be started or operated at low temperatures, the pour point and low-temperature fluidity of the oil selected must be considered. Synthetic fluids are often used for low-temperature bearing lubrication because of their ability to flow.

In thin film lubricated bearings, the oil is normally not reused, so the important characteristics of the oil selected are that it distributes readily and forms tough, tenacious films on the bearing surfaces. Generally, oils with good film strength and friction-reducing properties are preferred.

With thin film lubricated bearings, it may be assumed that loads are high enough for the bearings to operate under boundary conditions at least part of the time. Therefore, only speed and operating temperature must be determined to select the correct viscosity of oil.

For thin film bearings, once the speed and operating temperature are known, the oil viscosity may be obtained from a chart such as that shown in Figure 8.38. Again, the viscosity must be converted to a value at one of the standard reporting temperatures by means of viscosity–temperature charts. Ambient temperatures must also be considered, and so must the type of application device, because some of these devices are limited as to a maximum viscosity of oil they will dispense.

2. Grease Selection

The greases used for fluid film and thin film bearings are essentially similar. Enhanced film strength is required more frequently in thin film bearings, but where heavy, shock, or vibratory loads are encountered in fluid film bearings, this property is usually required.

As the methods used to grease plain bearings are all-loss systems, the grease used is not subjected to long-term service that might cause oxidative breakdown. On the other hand, operating temperatures may be higher in grease-lubricated bearings than in comparable oil lubricated bearings because of the poorer cooling ability of grease. Thus, there may be exposure to high temperatures while the grease is in the bearings. There is also the potential for severe mechanical shearing of the grease, particularly as it passes through the load-carrying zone, which may cause softening and lead to increased end leakage.

The method of application has considerable influence on both the type of grease and the consistency selected for plain bearings. With centralized lubrication systems, the grease must be a type suitable for dispensing through such systems. Where bearings must be lubricated at low ambient temperatures, softer greases or greases with good low temperature properties are usually needed, because they pump more easily at low temperatures.

These requirements generally dictate the use of greases with good mechanical stability, adequate dispensing and pumpability characteristics, corrosion protection properties, adequate film strength, and adequate high-temperature performance for the operating temperatures. For fluid film bearings, the apparent viscosity of the grease must be such that fluid films can be formed at the shear rates prevailing, but not so high that friction losses will be excessive.

III. ROLLING ELEMENT BEARINGS

The term "rolling element bearings" is used to describe that class of bearings in which the moving surface is separated from the stationary surface by elements such as balls, rollers, or needles that can roll in a controlled manner. These bearings are often referred to as "antifriction" bearings.

The essential parts of a rolling element bearing (Figure 8.39) are a stationary ring (cup or raceway), a rotating ring (cup or raceway), and a number of rolling elements. The inner ring fits the shaft or spindle, and the outer ring fits in a suitable housing. Shaped raceways are machined in the rings to confine and guide the rolling elements. These rolling elements are usually held

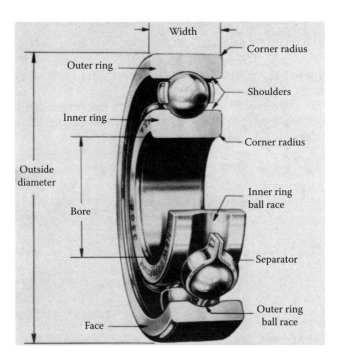

FIGURE 8.39 Ball bearing terminology.

apart from each other by a cage or a retainer, and their relative positions are maintained to keep the shaft or spindle centered by the separator. In full complement bearings, the rolling elements completely fill the space between the rings, and no separator is used. Rolling element bearings ranging in size from smaller than a pinhead to 18 ft (6 m) or more in outside diameter have been manufactured.

Manufacturers supply a wide variety of designs of rolling element bearings with the more popular types of ball and roller bearings being represented in Figures 8.40 and 8.41. The bearing in Figure 8.39 is a single-row, deep-groove ball bearing, which is usually the starting point when selecting a bearing for any application. When the fatigue life of this design is inadequate, space is limited, self-alignment is required, thrust loads must be carried, or any of a variety of other conditions must be met, one of the other types can be selected. When one of these other types is selected, the desired performance characteristics are very often obtained at the expense of higher cost, a lower speed limit, or more severe lubrication requirements.

When properly lubricated, the load capacity and life of a rolling element bearing are limited primarily by the fatigue strength of the bearing steel. Normal rolling action applies repeated compressive loading in the contact area with stresses up to about 400,000 psi (2.8 GPa). In addition to causing elastic deformation of the rolling element and raceway, this stress induces shearing stress in the steel in a zone approximately 0.002–0.003 in (0.05–0.075 mm) below the surface. This shearing stress induces fatigue cracks in the steel, which gradually grow and intersect. Small surface areas loosen and break away, forming pits. The actual time required for this to occur depends on many factors, including load, speed, and continuity of service, as well as the fatigue strength of the bearing steel.

Rolling element bearings operate under elastohydrodynamic films where inadequate lubrication can substantially reduce the fatigue life of the bearing. Research has shown that chemical effects have a considerable influence on fatigue life. The chemical composition of the lubricant additives, the base stock, and the contact surface materials are the major chemical variables. The water content of the atmosphere and the lubricant can also significantly contribute to chemical effects. At

Lubricating Films and Machine Elements

FIGURE 8.40 Popular types of ball bearings.

operating temperatures below 212°F (100°C), most industrial oils contain dissolved water. Because of the small size of the water molecules compared to oil molecules, they readily diffuse to the tips of the initial microcracks resulting from cyclic stressing. Although the precise mechanism by which this water accelerates fatigue cracking is not clear, there is evidence that the water in the microcracks breaks down and liberates atomic hydrogen, which attaches itself to the metal below the surface. This causes hydrogen embrittlement. Research data indicate that water-induced fatigue can reduce bearing life by 30–80%.

A. NEED FOR LUBRICATION

Operation of rolling element bearings always results in the development of a certain amount of friction that comes from three main sources:

1. The fluid friction from shearing of the lubricating films
2. Displacement or churning of lubricant in the path of the rolling elements
3. Sliding or contact between various elements of the bearing

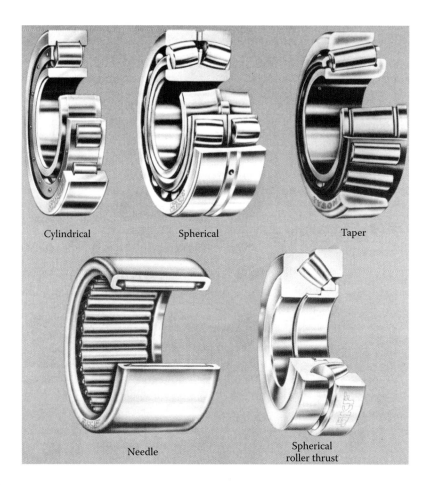

FIGURE 8.41 Popular types of roller bearings.

The main areas where sliding occurs are as follows:

- Gross sliding between rolling elements and the raceways (sometimes referred to as skidding).
- Sliding between the rolling elements and the separator (cage).
- Sliding between the end of rollers and the raceway flanges of roller bearings. This type of sliding may be particularly severe in certain types of bearings designed for thrust loads.
- Inherent microscale sliding within the contact associated with variation in surface speeds induced by bearing geometry (e.g., in deep-groove ball bearings).
- Sliding between the shaft or spindle and contact-type housing seals.
- Sliding between adjacent rolling elements in full-complement bearings.

The purpose of lubrication is to maintain suitable films between all of these sliding parts—EHL films for the contacts between the rolling elements and raceways, or thin films for other sliding parts. These films must be adequate to minimize friction and protect against wear. In addition, lubrication is expected to protect against rusting or other corrosive effects of contaminants and may provide part of the sealing against contaminants. In circulation systems, the lubricant also acts as a coolant to remove both the heat generated by friction in the bearing and any heat directed to the bearing from the nearby environment.

Lubricating Films and Machine Elements

B. Factors Affecting Lubrication

Whether oil or grease is used for rolling element bearings depends on a number of factors. Where bearings are installed in machines that require oil for other machine elements, the same oil is often supplied to the bearings. In other cases, it may be more convenient to lubricate the bearings separately, and grease is often used. Groups of similar bearings may be lubricated with oil by a circulation system, an oil mist system, or with grease in a centralized lubrication system. Many bearings are now "packed for life" with grease by the bearing manufacturer and need no further lubrication while in service. The attributes of the various methods of lubricant application have some influence on the selection of oil or grease.

The thickness of the elastohydrodynamic films formed in the contact areas between rolling elements and the raceways is a function of the speed at which the surfaces roll together, the load, the oil viscosity, and the operating temperature. The film thickness increases with increases in speed or oil viscosity and decreases with increases in load or operating temperature (because higher temperatures reduce the oil viscosity). For any given set of operating conditions, the viscosity of the oil should be selected to provide a safe minimum film thickness.

The lubrication requirements of the other sliding surfaces of a bearing must also be considered in the selection of a lubricant, but in general, the primary consideration for lubrication selection is governed by the needs of the EHL film.

1. Effect of Speed

The speed at which the surfaces of a rolling element bearing roll together is a fairly complex calculation, so an approximation called the "bearing speed factor" (nd_m) is commonly used. The bearing speed factor is determined by multiplying the rotational speed n (in revolutions per minute) by the pitch diameter d_m (in millimeters). For convenience, the pitch diameter is taken as one-half the sum of the bearing bore d and the bearing outside diameter D.* That is,

$$nd_m = \frac{n(d+D)}{2}$$

Manufacturers have established maximum bearing speed factors for both oil- and grease-lubricated bearings. These factors are shown in Table 8.1 for commercial grade bearings. Precision bearings are rated 5–50% higher. The higher speed factors allowed for oil lubrication generally reflect the better cooling available with oil, and the lower fluid friction of low viscosity oils.

For bearings lubricated with grease, the physical properties of the grease are the main concern relative to bearing speed (besides the viscosity of the oil in the grease). At low to moderate speeds, the grease must be soft enough to slump slowly toward the rolling elements, but must not be so soft that excess grease gets into the path of the rolling elements. Excess grease increases shearing friction and can cause high operating temperatures. For high speeds, a relatively stiff grease is sometimes used, but it must not be so stiff that the rolling elements, after cutting a channel through the grease, are unable to pick up and distribute sufficient amounts of grease, to provide continuous replenishment of the lubricating films. Moreover, to keep shearing friction low and prevent leakage from seal housings, the grease must have good resistance to softening as a result of mechanical shearing.

---

* The older DN factor [bearing bore D (in millimeters) multiplied by speed N (in rpm)] is still found in some references, although it has not been officially used by the bearing manufacturers' associations for many years. It is not a satisfactory factor to use because for a given bearing bore there are bearings with considerable variations in outside diameter and size of the rolling elements. Thus, for any given bore diameter, there can be considerable variation in the speeds with which the elements roll together.

TABLE 8.1
Bearing Speed Factors (nd_m)

Bearing Type	Oil Lubricated[a]	Grease Lubricated
Radial ball bearings	500,000	340,000
Cylindrical roller bearings	500,000	300,000
Spherical roller bearings	290,000	145,000[b]
Thrust–ball and roller bearings	280,000	140,000

[a] Oil lubrication is preferred where heat dissipation is required.
[b] Grease lubrication is not recommended for spherical roller thrust bearings.

2. Effect of Load

Under EHL conditions, the effect of load on film thickness is not as great as the effect of speed or oil viscosity. For example, although doubling the speed or oil viscosity might result in more than a 50% increase in film thickness, doubling the load might result in only about a 10% decrease in film thickness. As a result, under steady load conditions, the viscosity of oil can generally be selected on the basis of the bearing speed factor and operating temperature without regard to the load.

When shock or vibratory loads are present, somewhat greater film thickness is necessary to prevent metal-to-metal contacts through the film; thus, higher viscosity oils are usually required. Under severe shock loading conditions, lubricants with enhanced antiwear properties may be desirable.

Pressures between the rolling elements and separators (cages) are not high enough to form EHL films. Additionally, these pressures act more or less continuously on the same spots in the separator pockets. As a result, thin film lubrication exists and a higher viscosity oil can sometimes improve the resistance to wiping of the oil films, which will reduce wear of the separator. When grease is the lubricant used, wear can only be reduced if the lubricant can penetrate the clearances between the rolling elements and separators. This is especially important with machined separators where clearances are small.

In tapered and spherical roller thrust bearings, heavy loads sometimes force the ends of the rollers against the flanges of the raceways with considerable pressure. Severe wiping of the lubricant films and wear of the rollers and flanges may occur. In these types of bearings, a lubricant with enhanced antiwear properties may be required, even though the rollers and raceways can be satisfactorily lubricated with conventional products.

3. Effect of Temperature

Because both viscosity and grease consistency are functions of temperature, the operating temperature of bearings must always be considered when selecting lubricants. Bearing operating temperatures may be increased above what is considered normal by any heat that is directed to the bearing from a hot shaft or spindle, or by heat radiated to the housing from a hot surrounding atmosphere. Excessive churning of grease, resulting from overfilling, can also raise bearing temperatures (see Figure 8.42). Higher-than-normal temperatures reduce oil viscosity, whereupon film thickness may decrease below a safe level, and may soften grease so that excessive churning and further frictional heating occur.

High temperatures also increase the rate at which both oil and grease deteriorate because of oxidation. Oxidation may result in thickening of oil, and eventually could lead to deposits that will interfere with oil flow or bearing operation. Oxidation may also lead to deposits with greases, and in severe cases, to hardening such that the grease cannot properly lubricate.

Under low temperature conditions, the lubricant must be chosen such that the bearing can be started with the power available, and it must distribute sufficiently to prevent excessive wear before frictional heating warms the bearing and the lubricant.

Lubricating Films and Machine Elements

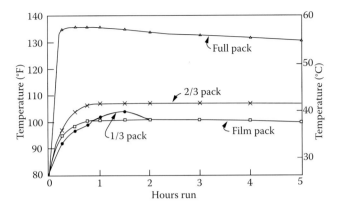

FIGURE 8.42 Temperature rise in grease-lubricated bearings. As shown by the curves, the final operating temperature of a bearing is a function of the amount of grease packed in it.

4. Contamination

Solid particles of any kind that are trapped between the rolling elements and raceways are the most frequent cause of shortened bearing life. Consequently, dirt should be kept out of bearings as much as possible. This includes maintaining bearing seals, using and maintaining filters to achieve proper oil cleanliness levels, ensuring that clean grease is applied to clean grease fittings, and keeping grease-packed bearings dirt-free during periods of storage before installation. Lubricants should be changed out before oxidation has progressed to a point where deposits begin to form. The use of premium-quality, oxidation-inhibited lubricants can greatly extend the length of time that lubricants may be left in service without excessive oxidation.

Water that gets into a bearing tends to reduce the fatigue life and cause rusting that can quickly ruin a bearing. When water is mixed with grease, it may cause the grease to soften enough to potentially leak from the bearings. Large quantities of water can wash the lubricant out. Sometimes fluids such as acids get into the bearings and cause corrosion. Any of these conditions usually involves special precautions and possibly specialty lubricants.

C. Lubricant Selection

Rolling element bearings generally require high-quality lubricants, many of which are specially formulated for the purpose.

1. Oil Selection

Oils for the lubrication of rolling element bearings should have the following characteristics:

1. Excellent resistance to oxidation at operating temperatures to provide long service without thickening or forming deposits that would interfere with bearing operation.
2. Antirust properties to protect against rust when moisture is present.
3. Good antiwear properties where required because of heavy or shock loads, or where thrust loads cause heavy wiping between the ends of rollers and raceway flanges.
4. Good demulsibility to allow for the separation of water in circulation systems.
5. Proper viscosity at operating speeds and temperatures to protect against friction and wear. Two charts (see Figure 8.43a and b) may be used to select the proper viscosity at the operating temperature from the type of bearing, the operating speed, and the outside bearing diameter (D). The speed factor (SF) is determined for the speed and type of bearing in

Figure 8.43a. The bearing size/speed factor (BS/SF = SF × D) is used in Figure 8.43b to determine the correct oil viscosity for a given specific film thickness (λ). For steady loads, the minimum viscosity line corresponding to the specific film thickness λ = 1.5 (see discussion on EHL) will provide adequate film thickness. Where shock or vibratory loads are present, higher viscosity and thicker films are desirable. Therefore, the optimum viscosity line corresponding to a specific film thickness λ = 4.0 should be used. ASTM D341, Standard Viscosity–Temperature Charts for Liquid Petroleum Products, should be used to translate this viscosity to the viscosity at the standard reporting temperatures. EHL calculations should be used for a more accurate determination of viscosity requirements.

In addition to these characteristics, there is emphasis on the interrelationship between oil attributes and fatigue. The specific lubricant properties that influence fatigue are not easily defined, but where average bearing life is lower than expected, special lubricants formulated to help minimize fatigue may be desirable. The combination of select base stocks and special additives can enhance the antifatigue properties of lubricants.

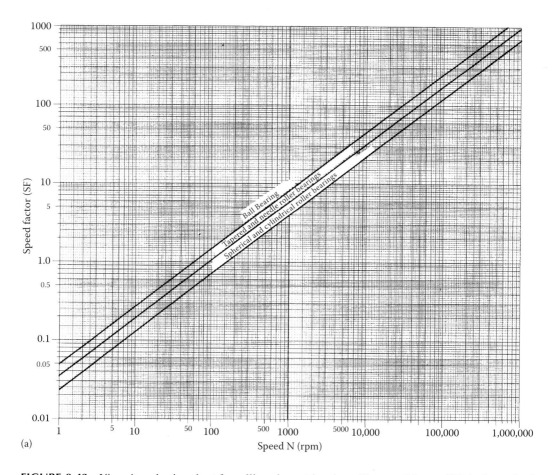

FIGURE 8.43 Viscosity selection chart for rolling element bearings. The speed factor (SF) is determined from Chart (a). The viscosity of the oil is selected from Chart (b) using the SF from Chart (a). For normal loads, the oil viscosity can be selected from the minimum viscosity line. For shock or vibratory loads, thicker oil films are required and the optimum viscosity line should be used to select the viscosity. *(Continued)*

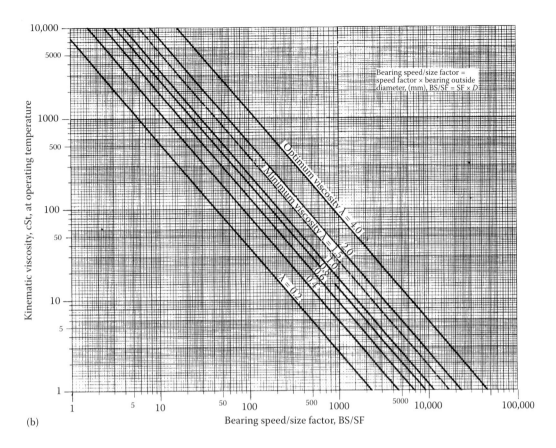

FIGURE 8.43 (CONTINUED) Viscosity selection chart for rolling element bearings. The speed factor (SF) is determined from Chart (a). The viscosity of the oil is selected from Chart (b) using the SF from Chart (a). For normal loads, the oil viscosity can be selected from the minimum viscosity line. For shock or vibratory loads, thicker oil films are required and the optimum viscosity line should be used to select the viscosity.

2. Grease Selection

Grease for the lubrication of rolling element bearings should have the following characteristics:

1. Excellent resistance to oxidation to resist the formation of deposits or hardening that might shorten bearing life. "Packed for life" greased bearings should provide trouble-free performance for the life of the bearing.
2. Mechanical stability to resist excessive softening or hardening as a result of shearing in service.
3. Proper consistency for maintaining an adequate oil film without excessive slumping in the application.
4. Controlled oil bleeding, particularly for high-speed bearings, to supply the small amount of oil needed to form EHL films.
5. Antiwear properties to resist the wiping action between roller ends and raceway flanges, in bearings with heavy radial or axial thrust loads.
6. Ability to protect surfaces against rusting, absorb small quantities of water without appreciable softening or hardening, and resist being washed out by large quantities of water.
7. Good compatibility with seals, cages, and other system components.
8. Depending on the application, the ability to resist corrosion where small quantities acids or caustic materials can get mixed into the grease.

IV. SLIDES, GUIDES, AND WAYS

This category of bearings, sometimes called flat bearings, includes all slides, guides, and ways used on forging and stamping presses; as crosshead guides on certain compressors, diesel, and steam engines; and on machine tools such as lathes, grinders, planers, shapers, and milling machines. The service conditions under which these bearings operate vary widely. Crosshead guides operate at relatively high speeds under conditions that permit the formation of fluid lubricating films most of the time. The requirements for lubrication usually are not severe, and the guides are usually designed to operate on the same oil used for the main and connecting rod bearings of the machine.

The lubrication of ways and slides of machine tools can present special challenges. At low speeds and under heavy loads, the lubricant tends to be wiped off so that boundary lubrication prevails. Although this results in higher friction, boundary films have the advantage in that they are of almost constant thickness. With low loads and high traverse speeds, the oil viscosity has to be high enough to allow the formation of fluid films that will lift and float the slide. With variations in speed or load, these fluid films can vary enough in thickness that they can affect the quality of the part being produced. This can cause a wavy surface on the parts being machined or impact the size of the parts. Thus, precision machining generally requires that the slides and ways operate under boundary conditions at all times. This frequently means that friction-reducing and antiwear additives must be included in the oil.

A phenomenon known as stick-slip can be encountered in the motion of slides and ways. If the static coefficient of friction of the lubricant is greater than the dynamic coefficient, more force will be required to start the slide from rest than will be required to maintain it in motion after it has started. There is always some amount of free play in the feed mechanism, so when force is applied to start the slide in motion, it will initially resist. When the force becomes high enough, the slide will begin to move. As soon as motion begins, the force required to maintain motion decreases, so the slide will jump ahead until the free play in the feed mechanism is taken up. At low traverse speeds, this can be a continuous process, producing chatter marks on the workpiece. With cross slides, stick-slip effects can make it extremely difficult to set feed depths accurately. Stick-slip effects can be overcome with additives that reduce the static coefficient of friction to a value equal to or less than the dynamic coefficient.

With vertical guides, there is a tendency for the lubricant to drain from the surfaces. To resist this tendency and secure adequate films, special adhesive characteristics are needed in the lubricant.

A. Film Formation

When the two parallel surfaces move with respect to each other, there is some tendency to draw oil in and form a fluid film between the surfaces. However, such a film does not have a high load-carrying capacity, and when the surfaces are long, the film is usually wiped or squeezed away rapidly. Traverse grooves in the slide divide it into a series of shorter surfaces, where lubricant supplied at the grooves can improve film formation and load-carrying ability. In some cases, with precision machine tools, hydrostatic bearings are being used. Pockets are machined in the slide, and oil is supplied from a constant pressure system with a flow restrictor for each pocket. This arrangement eliminates stick-slip effects, and if properly arranged, maintains the slide parallel to the way. Hydrostatic systems are used in industry particularly where stick-slip characteristics can contribute to reduced precision of machined parts.

1. Grease Lubrication

On a few machine tools, grease is used for the lubrication of the slides and ways. Relative to oil, grease lubrication provides several advantages and disadvantages that should be recognized. Its advantages are better sealing, improved stay-put properties, and the fact that it is generally less susceptible to wash off by water or metalworking fluids. The disadvantages of grease can be found in applications where a high degree of accuracy in machined parts is necessary, where the grease can act as a binder for debris such as metalworking chips, and can pose potential compatibility issues with metalworking fluids.

The potential disadvantage in accuracy of machined parts would be in applications where tables or ways had to operate with a specific film thickness at all times. Greases have a tendency to build thicker films when initially applied, and these are reduced as the grease is squeezed out or temperatures rise. Greases can also act as a binder for chips, grinding compounds and other materials present in machining operations. These could damage slides and ways if pulled into the contact areas. Excessive grease that could contaminate coolants or metalworking fluids may shorten batch life or reduce performance.

2. Lubricant Characteristics

The requirements that influence the characteristics of suitable lubricants for slides, guides, and ways can be summarized as follows:

- Proper oil viscosity at the operating temperature required for ready distribution to the sliding surfaces and for forming the necessary boundary films
- Antiwear and friction modifier additives to maintain the required boundary films under heavy loads and control wear
- Proper frictional characteristics to prevent stick-slip and chatter
- Adequate adhesiveness to maintain films on intermittently lubricated surfaces, especially vertical surfaces, and resist the washing by metalworking coolants

Slides and guides in some machines such as open crankcase steam engines and forging machines may be exposed to considerable amounts of water. Lubricants for these applications must be specially formulated to resist being washed off.

V. LINEAR MOTION GUIDES

Linear motion guides (LMGs) and ball screws have been supplanting the traditional box ways and screw guides in machine tools and other machines with transverse motion. In simple terms, LMGs are an antifriction bearing that is laid out flat (Figure 8.44) as opposed to a round configuration in most bearings. A ball screw translates rotational motion to linear motion with little friction using

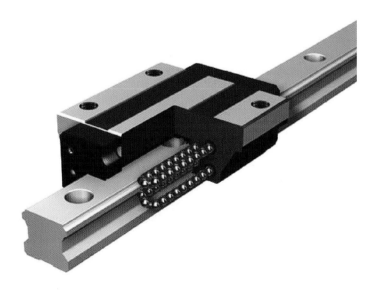

FIGURE 8.44 Cutaway view of a linear motion guide (LMG). (Courtesy of THK.)

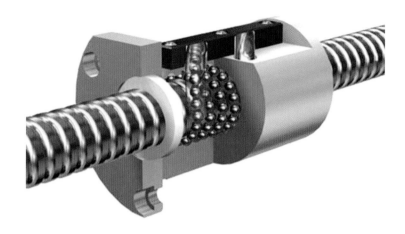

FIGURE 8.45 Cutaway view of a ball screw. (Courtesy of THK.)

a threaded shaft to provide a helical raceway for ball bearings that act as a precision screw (Figure 8.45). LMGs and ball screws can greatly enhance moving accuracy and handle the increased speeds, feeds, precision, and multidirectional load requirements of modern manufacturing.

LMGs may be mounted on the X, Y, or Z axis. When the LMG is mounted in the vertical position, special care needs to be taken to ensure that the lubricant is being transferred throughout the guide.

A. Lubricant Selection

The lubrication design for most LMGs and ball screw have been based on the concept of having delivery of the right lubricant, in just the right amount, at the right time. Without lubrication, the rolling elements or raceway would wear faster and shorten the service life. The purpose of the lubricant is to

- Minimize friction in moving elements to reduce wear and prevent seizure
- Form a film on the raceway to reduce stresses acting on the surface and extend rolling fatigue life
- Cover and protect the metal surfaces to prevent rust formation

LMGs and ball screws may be lubricated with either grease or oil. Some of the key characteristics for LMG and ball screw lubricants include:

- High film strength
- Low friction
- High wear resistance using antiwear or EP additives
- Noncorrosive
- High thermal stability
- Long life of the thickener and of the oil (for greases)

Specialized oils and greases may be required depending on the application. For LMGs and ball screws in the machine tool industry, good compatibility with metalworking coolants may be needed. In "clean room" environments, such as may be found in the electronics industry, low water and particle content of the lubricant is desirable. In food machinery applications, the lubricant may need to be National Sanitation Foundation (NSF) H1 or H2 compliant depending on whether there is a possibility of incidental contact with food.

1. Oil Lubrication

Oil-lubricated LMGs and ball screws usually use all-loss oil mist systems. The oil mist is delivered into the pathway of the rollers in the LMG or ball screw. The recommended lubricant should be an antiwear type oil with high rust and oxidation resistance. Some of the more traditional box way lubricants may not be the best lubricant for these precision LMGs and ball screws because of the tackiness, or stickiness, additives used. These special tackiness additives may interfere with the speed with which the roller moves in the LMG or ball screw. The oil mist feed rate for each application is determined by providing just the right amount of oil to wet the rollers and avoid flooding the roller tracks with oil. The viscosity of the oil should be in the range of International Standardization Organization (ISO) viscosity grade range of 32 to 68 for this application.

2. Grease Lubrication

LMGs that are grease lubricated may be supplied by a central grease system, but more commonly a smaller, individual self-contained grease system is used such as a bellows lubricator. Greasing intervals vary depending on the condition and environment. For normal use, LMGs and ball screws should be relubricated approximately every 100 km (325,000 ft) of travel distance. A grease with the correct oil viscosity for the application should be selected. The viscosity of the oil in the grease can range from an ISO VG 68 to an ISO VG 460. The oil may be a conventional mineral or a synthesized hydrocarbon (polyalphaolefin). The most common grease thickeners used are lithium/lithium complex and polyuria. It is important not to mix different thickener types in these applications. The choice of National Lubrication Grease Institute (NLGI) Grade is dependent on the type of grease dispensing unit used, with NLGI Grades of #2 and #1 being most common.

VI. GEARS

Gears are used to transmit motion and power from one rotating shaft to another or from a rotating shaft to a reciprocating machine. With respect to lubrication and the formation and maintenance of lubricating films, gears can be classified as follows:

- Spur (Figure 8.46), bevel (Figure 8.47), helical (Figure 8.48), herringbone (Figure 8.49), and spiral bevel (Figure 8.50) gears
- Worm gears (Figure 8.51)
- Hypoid gears (Figure 8.52)

The difference in the action between the teeth of these three classes of gears can have a considerable influence on both the formation of lubricating films and the properties of the lubricants required for satisfactory lubrication.

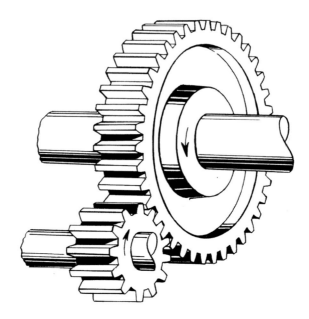

FIGURE 8.46 Spur gears.

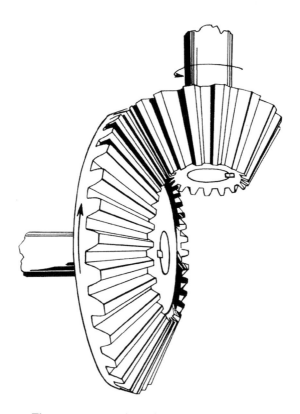

FIGURE 8.47 Bevel gears. These gears transmit motion between intersecting shafts. The shafts need not meet at a right angle.

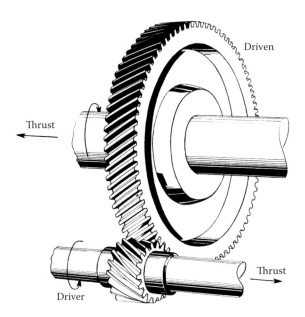

FIGURE 8.48 Helical gear and pinion.

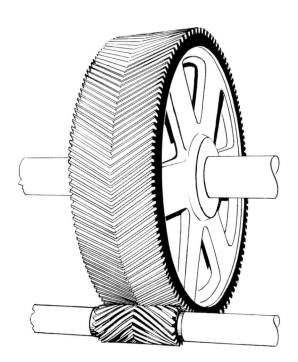

FIGURE 8.49 Herringbone gear and pinion.

FIGURE 8.50 Spiral bevel gears. These gears transmit motion between intersecting shafts.

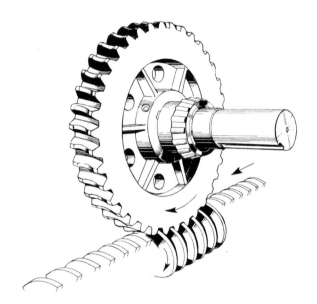

FIGURE 8.51 Worm gears. The worm here is represented by an endless rack.

A. Action between Gear Teeth

As gear teeth mesh, they roll and slide together. The progression of contact as a pair of spur gear teeth engage is shown in Figure 8.53. The first contact is between a point near the root of the driving tooth (upper gear) and a point at the tip of the driven tooth. In Figure 8.53 View A, these points are identified as 0–0 lying on the line of action. At this time, the preceding teeth are still in mesh and carrying most of the load. As contact progresses, the teeth roll and slide on each other. Rolling is from root to tip on the driver and from tip to root on the driven tooth. The direction of sliding at each stage of contact is as indicated by the small arrows.

In Figure 8.53 View B, contact has advanced to position 3–3, which is approximately the beginning of a "single tooth" contact when one pair of teeth picks up the entire load. To reach this point

Lubricating Films and Machine Elements 187

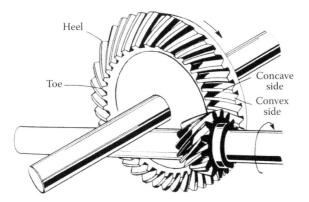

FIGURE 8.52 Hypoid gears. These gears transmit motion between nonintersecting shafts crossing at a right angle.

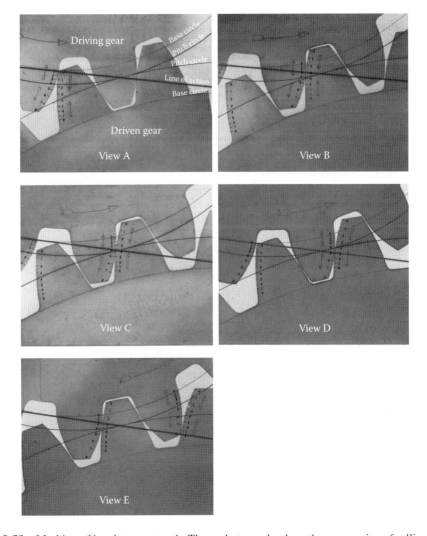

FIGURE 8.53 Meshing of involute gear teeth. These photographs show the progression of rolling and sliding as a pair of involute gear teeth (a commonly used design) pass through mesh. The amount of sliding can be seen from the relative positions of the numbered marks on the teeth.

of engagement, as the distance 0–3 on the driven gear is greater than the distance 0–3 on the driver, there must have been sliding between the two surfaces. In Figure 8.53 View C, position 4–4 shows contact at the pitch line where there is pure rolling—no sliding. It should be noted that the direction of sliding reverses at the pitch line. Moreover, sliding is always away from the pitch line on the driving teeth, and always toward it on the driven teeth. Figure 8.53 View D shows contact at position 5–5, which marks the approximate end of a single tooth contact. As shown, another pair of teeth is about to make contact. In Figure 8.53 View E, two pairs of teeth are in mesh, but as shown at position 8–8, the original pair of teeth is about to disengage.

Rolling is continuous throughout mesh. Sliding, on the other hand, varies from a maximum velocity in one direction at the start of mesh, through zero velocity at the pitch line, then again to a maximum velocity in the opposite direction at the end of mesh.

This combination of sliding and rolling occurs with all meshing gear teeth regardless of type. The two factors that vary are the amount of sliding in proportion to the amount of rolling, and the direction of sliding relative to the lines of contact between tooth surfaces.

With conventional spur and bevel gears, the theoretical lines of contact run straight across the tooth faces (Figure 8.54). The direction of sliding is then at right angles to the lines of contact. With helical, herringbone, and spiral bevel gears, because of the twisted shape of the teeth, the theoretical lines of contact slant across the tooth faces (Figure 8.55). Because of this, the direction of sliding is not at right angles to the lines of contact, and some slide sliding along the lines of contact occurs.

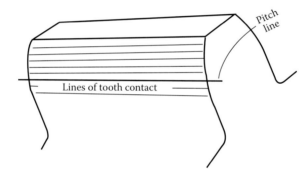

FIGURE 8.54 A spur gear tooth showing lines of tooth contact. On the driving tooth, this contact first occurs below the pitch line. As the gear turns, this contact progressively sweeps upward to the top of the tooth. The action is reversed on the driven tooth.

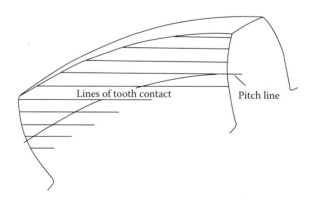

FIGURE 8.55 Tooth contact of low angle helical gear.

With worm gears, as with spur gears, the same sliding and rolling action occurs as the teeth pass through mesh. Usually, this sliding and rolling action is relatively slow because of the low rotational speed of the worm wheel. In addition, rotation of the worm introduces a high rate of side sliding. The combination of two sliding actions produces a resultant slide, which in some areas is directly along the line of contact.

In addition to the usual rolling action, hypoid gears have a combination of radial and sideways sliding that is intermediate between the motions of worm gears and spiral bevel gears. The greater the shaft offset, the more the sliding conditions approach those found in worm gears.

B. Film Formation

The requirements of EHL film formation for gears are that the pressure must be high enough to cause elastic deformation of the contacting surfaces. Thus, the theoretical lines or points of contact are expanded into areas, and a convergent zone exists ahead of the contact area. In the case of industrial gears, in all except very lightly loaded applications, tooth loading is high enough to produce elastic deformation along the line of contact. A convergent zone immediately ahead of the contact area also exists at all times (Figure 8.56), and the conditions for the formation of thick EHL films are complex. However, certain factors in these calculations are of importance in the general consideration of selection of lubricants for industrial gear drives.

The equations used do not consider the effect of tooth sliding action on the formation of the EHL films. The entraining velocity tending to carry the lubricant into the contact zone is considered to be the rolling velocity alone. The rolling velocity, for convenience, is usually calculated at the pitch line and considered representative for the entire tooth.

The critical specific film thickness λ for gears is not only considerably lower than that for rolling element bearings, but is also a function of the pitch line velocity. The curve in Figure 8.57, developed from experimental data, shows at low speeds that values of λ, as low as 0.1 or lower, can be tolerated without surface distress in the form of pitting or wear. At higher speeds, higher values of λ (up to 2.0 or higher) may be required to ensure freedom from tooth distress.

A detailed analysis has not been made to better understand the reasons for these lower specific film thicknesses being able to provide satisfactory results in gears. However, it is generally accepted that in the range where $\lambda < 1.0$, lubricants containing EP and antifatigue additives are required.

In the selection of lubricants for gears, tooth sliding is considered from two aspects:

1. It tends to increase the operating temperature because of frictional effects.
2. Sliding along the line of contact tends to wipe the lubricant away from the convergent zone; thus, it is more difficult to form lubricating films.

C. Factors Affecting Lubrication of Enclosed Gears

The lubricant in an enclosed gearset, which represents the majority of gears in service, is subjected to very severe conditions. The lubricant is thrown from the gear teeth and shafts in the form of a mist or spray. In this atomized condition, it is exposed to the oxidizing effect of air. Fluid friction and, in some cases, metallic friction generate heat, which raises the lubricant temperature. The violent churning and agitation of the lubricant by the gears of splash-lubricated sets also raises the temperature. Raising the temperature increases the rate of oxidation. Sludge or deposits, formed as a result of oil oxidation, can restrict oil flow, or interfere with heat flow in oil coolers or heat dissipation from the sides of the gear case. Restrictions in the oil flow may cause lubrication failure, whereas heat-insulating deposits decrease cooling and cause further increases in the rate of oxidation. Eventually, lubrication failure and damage to the gears may result.

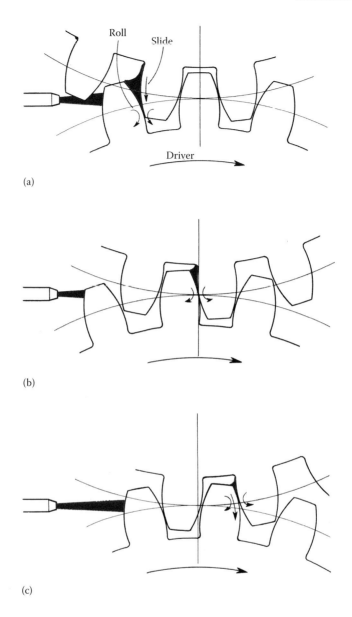

FIGURE 8.56 Convergent zone between meshing gear teeth. Clearly, if oil is present between meshing gear teeth, it will be drawn into the convergent zone between the teeth; and the point of this wedge-shaped zone is always positioned toward the roots of the driving teeth. (a) Interval of approach, (b) contact on pitch line, and (c) interval of recession.

In selecting the lubricant for enclosed gearsets, in addition to the requirement for adequate oxidation resistance, the following factors of design and operation require consideration:

1. Gear type
2. Gear speed
3. Reduction ratio
4. Operating temperature
5. Transmitted power and load

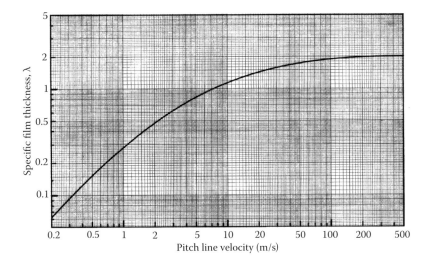

FIGURE 8.57 Critical specific film thickness for gears. This curve is based on a 5% probability of surface distress to define the target film thickness, which is adjusted to reflect the root mean square (RMS) surface roughness, $\left[\sigma = \left(\sigma_1^2 + \sigma_2^2\right)^{1/2}\right]$.

6. Surface finish
7. Drive type
8. Application method
9. Potential contamination (water, metalworking fluids, dirt, etc.)
10. Lubricant leakage
11. Ambient and startup temperatures

1. Gear Type

With spur and bevel gears, the line of contact runs straight across the tooth face, and the direction of sliding is at right angles to the line of contact. Both these conditions contribute to the formation of effective lubricating films. However, only a single tooth carries the entire load during part of the meshing cycle, resulting in high tooth loads. Additionally, if one tooth wears, there is no transfer of load to other meshing teeth to relieve the load on the worn tooth, and wear of that tooth will continue.

Helical, herringbone, and spiral bevel gears always have more than one pair of teeth in mesh. This results in better distribution of the load under normal loading. Under higher loading, the individual tooth contact pressures may be as high as in comparable straight tooth gears under normal loading. The sliding component along the line of contact, because of time and high viscosity of the lubricant in the contact area, has little or no effect on the EHL film in the contact area. In the convergent zone ahead of the contact area, the sliding component tends to wipe the lubricant sideways. Therefore, not as much lubricant is available to be drawn into the contact area, and the resultant pressure increase in the convergent zone may not be as great. These effects may contribute to a need for slightly higher viscosity lubricants, although, in general, oils for these types of gears are selected on the same basis as for straight tooth gears.

An additional factor with helical, herringbone, and spiral bevel gears is that if one tooth wears, the load is transferred simultaneously to other teeth in mesh. This relieves the load on the worn tooth and may make these types of gear somewhat less critical of lubricant characteristics. The important factor with all of these types of gears is that the lubricant should have a high enough viscosity to provide an effective oil film, but not so high that excess fluid friction will occur.

The high rate of side sliding in worm gears results in considerable frictional heating. Generally, the rolling velocity is quite low so the velocity tending to carry the lubricant into the contact area is low. Combined with the sliding action that tends to wipe the lubricant along the convergent zone, this makes it necessary to use high viscosity lubricants (typically ISO 460 or 680 viscosity grade). EP additive type gear oils are not normally recommended for worm gears, but in order to help reduce the wiping effect and reduce friction, lubricants containing friction-reducing materials are usually used. Because of their friction-reducing and long life characteristics, synthetic lubricants (such as synthesized hydrocarbon or polyalkylene glycols) are the lubricant of choice for most worm gear applications.

Hypoid gears are constructed of steel and are heat treated. They are designed to transmit high power in proportion to their size. Combined with the side sliding that occurs, these gears operate under boundary or mixed film conditions essentially all of the time, and require lubricants containing active EP additives.

2. Gear Speed

The higher the speed of meshing gears, the higher will be the sliding and rolling speeds of individual gear teeth. When an ample supply of lubricant is available, the gear speed assists in forming and maintaining fluid films. At high speed, more oil is drawn into the convergent zone; in addition, the time available for the oil to be squeezed from the contact area is less. Therefore, comparatively lower viscosity oils may be used (despite their fluidity, there is insufficient time to squeeze out the oil film). At low gear speeds, however, more time is available for oil to be squeezed from the contact area, and there is less oil drawn into the convergent zone; thus, higher viscosity oils are required.

3. Reduction Ratio

Gear reduction ratio influences the selection of the lubricating oil because high ratios require more than one step of reduction. When the reduction is more than about 3:1 or 4:1, multiple reduction gearsets are usually used, and above about 8:1 or 10:1, they are nearly always used. In a multiple reduction set, the first reduction operates at the highest speed and so requires the lowest viscosity oil. Subsequent reductions operate at lower speeds, and so they require higher viscosity oils. The low-speed gear in a gearset is usually the most critical in the formation of an EHL film. In the case of a gear reducer, this would be the output gear. In very high-speed gear reducers, both the lowest speed and highest speed gears should be checked to determine the more critical condition. In some cases, a dual-viscosity system may be effective, where a lower viscosity oil is used for high-speed gears and a higher viscosity oil is used for low-speed gears. In some gearsets, this can be automatically accomplished by circulating the cool oil first to the low-speed gears, and then, after it is heated and its viscosity decreased, to the high-speed gears.

4. Operating and Startup Temperatures

The temperature at which gears operate is an important factor in the selection of the lubricating oil, because viscosity decreases with increasing temperature and oil oxidizes more rapidly at high temperatures. Both the ambient temperature where the gearset is located and the temperature rise in the oil during operation must be considered.

When gearsets are located in exposed locations, the oil must provide lubrication at the lowest expected starting temperature. In splash-lubricated units, this means that the oil must not channel at this temperature, and in pressure-fed gearsets the oil must be fluid enough to flow to the pump suction. At the same time, the oil must have a high enough viscosity to provide appropriate lubrication when the gears are at their highest operating temperature.

During operation, the heat generated between the tooth surfaces and by fluid friction in the oil, will cause the temperature of the oil to rise. The final operating temperature is a function of both this temperature rise in the oil and the ambient temperature surrounding the gearcase.

Thus, a temperature rise of 90°F (32°C) and an ambient temperature of 60°F (16°C) will produce an operating temperature of 150°F (66°C), whereas the same temperature rise at an ambient temperature of 100°F (38°C) will produce an operating temperature of 190°F (88°C). In the latter case, an oil of higher viscosity and better oxidation stability would be required to provide satisfactory lubrication and oil life at the operating temperature. For gearsets equipped with heat exchangers in the oil system, both the ambient temperature and the temperature rise are less important because the operating temperature of the oil can be adjusted by varying the amount of heating or cooling.

5. Transmitted Power and Load

As noted in the discussion on EHL film formation, load does not have a major influence on the thickness of EHL films. However, it cannot be ignored. As load is increased, the viscosity of the lubricant may have to be increased to adjust for the small effect of load on film thickness, particularly where λ values were marginal before increasing load.

Load also has an influence on the amount of heat generated by both fluid and mechanical friction. Gears designed for higher power ratings will have wider teeth or teeth of larger cross section, or both. Nevertheless, a greater surface area will be swept as the teeth pass through mesh, so mechanical and fluid friction will be greater. At the same time, the relative area of radiating surface in proportion to the heat generated is usually less in a large gearset than it is in a small one. As a result, larger gearsets, transmitting more power, tend to run hotter unless they are equipped with oil coolers. However, if the operating temperature of a gearset is properly taken into account during the selection of lubricant viscosity, the heating effects based on the amount of power transmitted will be taken care of.

The nature of the load on any gearset has an important influence on the selection of an oil. If the load is uniform, the torque (turning effort), and the load carried by the teeth, will also be uniform. However, excessive tooth loads due to shock loading may tend to momentarily rupture the lubricating films. Therefore, where the shock factor has not been considered in the design or selection of a gearset, a higher-han-normal oil viscosity may be required to prevent film rupture.

In some operations, the conditions may be more severe owing to overloads or to a combination of heavy loads and extreme shock loads. Rolling mill stands are an application where gears are started under heavy load and/or have the capability of reversing direction. In such cases, it may be impossible to maintain an effective oil film. Hence, during a considerable part of mesh, boundary lubrication exists. This condition generally requires the use of EP gear oils.

Occasionally, owing to space constraints or other limiting and unavoidable factors, gears are loaded so heavily that it is difficult to maintain an effective lubricating film between the rubbing surfaces. Such a condition is quite usual for hypoid gears used in automotive applications. When operating under this condition of extreme loading, the potential for metal-to-metal contact can be so severe that wear cannot be completely avoided. However, it can be controlled by the use of special EP lubricants containing additives designed to prevent welding and surface destruction under severe conditions. Only slow wear of a smooth and controlled character will then take place. Synthetic hydrocarbon lubricants formulated with EP additives have proven to be ideal lubricants for hypoid gears.

6. Surface Finish

Surface roughness has a major influence on oil film thickness required for proper lubrication. Rougher surfaces require thicker oil films and higher viscosity oils to facilitate complete separation of metal surfaces. On the other hand, smoother surfaces can be successfully lubricated with lower viscosity oils. As some smoothing of the surfaces results from the running-in period, some designers recommend using an estimated "run-in" surface roughness rather than the "final" surface roughness values for determining oil film thickness calculations and for selection of oil viscosity.

7. Drive Type

Electric motors, steam turbines, hydraulic turbines, and gas turbines are generally used in applications where the requirement is for uniform torque. Therefore, when the power transmitted by gears is developed by one of these prime movers, gear tooth loading is uniform. Reciprocating engines, however, produce variable torque, so some variation in gear tooth loading results. When gears are driven by prime movers that vary in torque, higher viscosity oils may be required to assure effective oil films. Higher viscosity oils may not be necessary when the type of drive has been considered and compensated for in the design or selection of the gearset.

8. Application Method

When lubricating oil is applied to gear teeth by means of a splash system, the formation of an oil film between the teeth is less effective than when the oil is circulated and sprayed directly on the meshing surfaces. This is particularly true of low-speed, splash-lubricated units in which only a limited amount of oil may be carried to the meshing area. A higher viscosity oil is needed to offset this condition, because higher viscosity will result in more oil clinging to the teeth and being carried into the mesh.

When a gearset is lubricated by a circulation system rather than a splash lubrication system, there is better dissipation of heat. This is because the oil under pressure tends to force the oil onto all internal surfaces of the gear case, and more heat is conducted away by these radiating surfaces. With a splash system, particularly a low-speed unit, the oil may dribble over only a small part of the internal surface of the gear case, thus restricting heat dissipation. As a result, splash-lubricated units usually run hotter and require higher viscosity oils.

Another key consideration with splash-lubricated gear systems is the proper maintenance of oil levels. If oil levels are too low, insufficient oil will be available to splash on the gears resulting in increased friction and wear and higher oil temperatures. When oil levels are too high, the gears must churn through the oil, creating a channel effect that can cause excessive oil heating. This will also impact overall efficiency.

9. Water Contamination

Water sometimes finds its way into the lubrication system of enclosed gears. This water may come from heat exchanger coil leaks, condensed steam, pressure washing of equipment, or condensation of moisture in the atmosphere. In the latter case, it is often an indication of inadequate venting of the gear case and oil reservoir. Water contamination is likely to occur in gearsets operated intermittently, with warm periods of operation alternating with cool periods of idleness causing moisture to condense. Applications in high humidity conditions where temperatures can drop to levels at or below the dew point may need to be equipped with desiccant breathers. Where moisture contamination may occur, it is necessary to use an oil with good demulsibility, that is, an oil that separates readily from water.

Water and rust also act as catalysts that increase the rate of deterioration of the oil. Water separates slowly, or not at all, from oil that has been oxidized or contaminated with dirt. Water in severely oxidized or dirty oil usually forms stable emulsions. Such emulsions may cause excessive wear of gears and bearings by reducing the lubricant's ability to provide proper lubrication. The fact that oxidized oil promotes the formation of stable emulsions constitutes another reason for using an oxidation-resistant oil in enclosed gears. To protect gear tooth surfaces and bearings, the oil must not only separate rapidly from water when new, but must also have the high chemical stability necessary to maintain this rapid rate of separation even after long service in a gear case.

10. Lubricant Leakage

Although most enclosed gear cases are considered oil tight, extended operation or more severe operating conditions may result in lubricant leakage at seals or joints in the casing. Where the amount of oil leakage is high and cannot be controlled by other methods, special lubricants designed to resist leakage, such as semifluid greases, may be required. Special considerations may be required when using antileak oils or semifluid greases because these may not be consistent with the manufacturers' lubricant recommendations.

VII. LUBRICANT CHARACTERISTICS FOR ENCLOSED GEARS

The necessary characteristics of lubricants for enclosed gears may be summarized as follows:

1. Use of the correct viscosity at operating temperature to assure distribution of oil to all rubbing surfaces and formation of protective oil films at prevailing speeds and pressures
2. Adequate low-temperature fluidity to permit circulation at the lowest expected start temperature
3. Good chemical stability to minimize oxidation under conditions of high temperatures and agitation in the presence of air, and to provide long service life for the oil
4. Good demulsibility to permit rapid separation of water and protect against the formation of harmful emulsions
5. Antirust properties to protect gear and bearing surfaces from rusting in the presence of water, entrained moisture, or humid atmospheres
6. Noncorrosive to nonferrous metals to prevent gears and bearings from being subjected to chemical attack by the lubricant
7. Foam resistance to prevent the formation of excessive amounts of foam in reservoirs and gear cases
8. Exhibit good compatibility with system components such as seals, paints, and gear metallurgy

In addition to these characteristics, many modern gearsets operate under severe conditions where loads are heavy or shock loads are present, requiring lubricants with EP properties to minimize scuffing and destruction of gear tooth surfaces. Worm gears usually require lubricants with mild wear and friction-reducing properties. It is important to note that some highly additized EP gear oils may have negative effects on worm gears particularly where different metallurgy such as bronze on steel is used.

A. AGMA Standard for Industrial Gear Lubricants

AGMA (American Gear Manufacturers Association) Standard 9005-E02 (American National Standards Institute [ANSI]/AGMA 9005-E02) combines the specifications for enclosed and open gear lubricants (AGMA, 2002). Note that AGMA 9005-E02 is in the process of being updated, so readers should refer to the new AGMA 9005 standard once it is published. This AGMA Standard provides specifications for all gear lubricants used in industrial gearing. The viscosity grade ranges correspond to those in ASTM D2422 (Standard Recommended Industrial Liquid Lubricants—ISO Viscosity Classification) and B.S. 4231 (British Standards Institution). This specification uses gear pitchline velocities as the primary parameter for determining lubricant selection in other than double-enveloping worm gears. Previous specifications were based on gear center distances. The former AGMA Lubricant Number and the corresponding ISO viscosity grades are shown in Table 8.2. These former AGMA Lubricant Numbers may be referenced in older maintenance manuals or equipment nameplates and are shown here as a convenient cross reference.

General lubricant guidelines for enclosed gears, with the exception of worm gears, can be found in Table 8.3. The guideline is based on the use of a mineral oil with a viscosity index of 95 and assumes a bulk oil temperature of 45°C above ambient temperature. Guidelines like this

TABLE 8.2
Viscosity Ranges for Former AGMA Lubricant Numbers

Former AGMA Lubricant No. for Rust and Oxidation Inhibited and Extreme Pressure Gear Oils	Viscosity Range[a] mm²/s (cSt) at 40°C	Equivalent ISO Grade[a]
0	28.8 to 35.2	32
1	41.4 to 50.6	46
2	61.2 to 74.8	68
3	90 to 110	100
4	135 to 165	150
5	198 to 242	220
6	288 to 352	320
7	414 to 506	460
8	612 to 748	680
8A	900 to 1100	1000
9	1350 to 1650	1500
10	1980 to 2420	2200
11	2880 to 3520	3200
12	6120 to 7480	–
13	190 to 220 cSt at 100°C (212°F)[c]	–
Residual Compounds[d] **Former AGMA Lubricant No.**	**Viscosity Ranges**[c] **cSt at 100°C (212°F)**	
14R	428.5 to 857.0	
15R	857.0 to 1714.0	

Note: The diluent evaporates leaving a thick film of lubricant on the gear teeth. Viscosities listed are for the base compound without diluent.

Caution: These lubricants may require special handling and storage procedures. Diluent can be toxic or irritating to the skin. Do not use these lubricants without proper ventilation. Consult lubricant supplier's instructions.

[a] Per ISO 3448, *Industrial Liquid Lubricants—ISO Viscosity Classification*; also ASTM D 2422 and British Standards Institution B.S. 4231.
[b] Extreme pressure lubricants should be used only when recommended by the gear manufacturer.
[c] Viscosities of AGMA Lubricant #12 and above are specified at 100°C (210°F) as measurement of viscosities of these heavy lubricants at 40°C (100°F) would not be practical.
[d] Residual compounds-diluent types, commonly known as solvent cutbacks, are heavy oils containing a volatile, nonflammable diluent for ease of application.

should only be used when a detailed EHL analysis is not available. More detailed guidelines for this and higher viscosity index mineral and synthetic oils can be found in the AGMA Standard 9005-E02. Original equipment manufacturer lubrication recommendations should always be followed (Table 8.4).

VIII. LUBRICATION OF OPEN GEARS

In contrast to enclosed gears that are flood lubricated by splash or circulation systems, there are many gears for which it is not practical or economical to provide oil-tight housings. These so-called open gears are often designed to be sparingly lubricated and perhaps only at infrequent intervals. Gears of this type are lubricated by either a continuous or an intermittent method. Some of the more sophisticated open gearing systems may be lubricated with a full circulation system that captures and filters the oil for reuse.

TABLE 8.3
ISO Viscosity Grade Selection General Guidelines for Enclosed Helical, Herringbone, Straight Bevel, Spiral Bevel, and Spur Gear Drives

| | ISO Viscosity Grade | | | |
| | Ambient Temperature, °C (°F) | | | |
Pitch Line Velocity of Final Reduction Stage	−35 to −10 (−31 to + 14)	−10 to 15 (14 to 59)	15 to 35 (59 to 95)	35 to 55 (95 to 131)
Less than 5 m/s (1000 ft/min)[8]	32–46	100–460	460–2200	1000–3200
5–15 m/s (1000–3000 ft/min)	32–46	46–220	150–680	460–2200
15–25 m/s (3000–5000 ft/min)	Consult OEM	32–68	100–220	220–680
Above 25 m/s (5000 ft/min)[8]	Consult OEM	32–68	68–220	150–460

Note: These general guidelines are based on the use of a 95 VI mineral oil and a bulk oil temperature above 45°C. The pour point of lubricant selected should be at least 5°C (9°F) lower than the expected minimum ambient starting temperature. If the ambient starting temperature approaches lubricant pour point, oil sump heaters may be required to facilitate starting and ensure proper lubrication. The original equipment manufacturer (OEM) should be consulted in very low or high temperature cases where oils below an ISO VG 32 or above an ISO VG 3200 may be required.

TABLE 8.4
ISO Viscosity Grade Selection General Guidelines for Enclosed Cylindrical Worm Gear Drives

| | ISO Viscosity Grade | | |
| | Ambient Temperature, °C (°F) | | |
Pitch Line Velocity of Final Reduction Stage	−40 to −10 (−40 to + 14)	−10 to +10 (14 to 50)	10 to 55 (50 to 131)
Less than 2.25 m/s (405 ft/min)	220	460	680
Above 2.25 m/s (405 ft/min)	220	460	460

Source: Extracted from AGMA. ANSI/AGMA 9005-E02, Industrial Gear Lubrication, American Gear Manufacturers Association, Alexandria, VA, 2002. With permission from the publisher.

Note: These general guidelines are based on the use of a 95 VI mineral oil. Higher VI oils, such as synthetic oils, may allow wider temperature ranges. Any worm gear applications involving temperatures or speeds outside the limits shown above should be referred to the original equipment manufacturer for a recommendation. The pour point of the oil used should be at least 5°C (9°F) lower than the minimum ambient temperature expected.

The three most common continuous methods for open gear lubrication are splash, idler gear immersion, and pressure. In the first two, the lubricant is lifted from a reservoir or sump (sometimes referred to as a slush pan) by the partially submerged gear or an idler. Pressure systems require a shaft or independently driven pump to draw oil from a sump and spray it over the gear teeth. The continuous methods of lubricant application for open gears is similar to that used with enclosed gears. Because the gears are usually large and relatively slow moving, very high viscosity lubricants are required. Most gears lubricated in any of these ways are equipped with relatively oil-tight enclosures.

Many different intermittent methods of application are used. Some are arranged for automatic timing, whereas others must be controlled manually. Methods used include automatic spray, semi-automatic spray, forced feed lubricators, gravity or forced drip, and hand application by brush.

When grease-type lubricants are used, hand or power grease guns or a centralized lubrication system can be used.

With the intermittent methods of application, fluid films may exist when lubricant is first applied to the gears. However, these films quickly become thinner as the lubricant is squeezed aside until only extremely thin films may remain on the metal surfaces. During much of the time, therefore, these gears operate under conditions of boundary lubrication. Under boundary conditions, the extremely thin film must resist being rubbed or squeezed off the surfaces of the teeth. The ability of the film to resist the rubbing action depends both on its viscosity and the action that takes place between the lubricant and the metal surfaces. The lubricating film must bond so strongly to the tooth surfaces that metal-to-metal contact (and resultant wear) is minimized. With a properly selected lubricant applied at sufficiently frequent intervals, gear tooth wear may be kept to a negligible amount.

A. Factors Affecting Lubrication of Open Gears

1. Temperature

The ability to maintain adequate boundary films under the pressures existing between the meshing teeth depends on the lubricant characteristics and the operating temperature to which the gears are exposed. Heat has a tendency to cause the lubricant to thin out, to drop off, or be thrown off the gear teeth, all of which decrease the amount of available lubricant remaining on the rubbing surfaces. This results in thinner oil films and affects the resistance needed by the lubricant to stay on the contact surfaces. Thus, high temperatures require more viscous lubricants. Conversely, when gears operate at low temperatures, the lubricant must not become so viscous or hard that it will not distribute properly over the tooth surfaces. The lubricant should not harden or chip or peel from the gear teeth at the lowest temperatures encountered.

2. Dust and Dirt

Many open gears, whether operating outdoors or indoors, are exposed to dusty and dirty conditions. Dirt and dust, adhering to oil wetted surfaces, can form an abrasive lapping compound that causes excessive wear of the teeth. When viscous lubricants are used, the dirt may pack into the clearance space at the roots of the teeth, forming hard deposits. Packed deposits between gear teeth can also have a tendency to spread the gears and overload the bearings. These deposits that build up in the tooth root area, if hard enough, can also lead to tooth wear and possible breakage.

3. Water

Open gears operating outdoors may be exposed to rain and snow, whereas some open gears may be susceptible to the splash of various process fluids or solids. The lubricant must resist being washed off the gears by these fluids in order to protect the gear tooth surfaces against wear or rust and corrosion.

4. Method of Application

The method of application must be considered when selecting a lubricant for open gears. If the lubricant is to be applied by drip force feed lubricator, or spray, it must be sufficiently fluid to flow through the application equipment. For brush applications, the lubricant must be sufficiently fluid to be brushed evenly on the teeth. In any case, during operation, the lubricant should be viscous and tacky to resist squeezing from the gear teeth. Some types of very viscous lubricants can be thinned for application by heating or by using diluent-type products. These diluent-type open gear products typically contain a nonflammable diluent that reduces the viscosity sufficiently at the time of application. Shortly after application, the diluent evaporates, leaving the film of the viscous lubricant to

TABLE 8.5
Minimum Viscosity Recommendation for Open Gearing—Intermittent Method of Application

Ambient Temperature, °C (°F)	Intermittent Spray System		Gravity Feed or Forced Drip Method
	Nonresidual Lubricant	Residual-Type Lubricant	
−10 to 5 (14 to 41)	4140 cSt at 40°C[a]	428.5 cSt at 100°C[b]	4140 cSt at 40°C[a]
5 to 20 (41 to 68)	6120 cSt at 40°C[c]	857 cSt at 100°C[d]	6120 cSt at 40°C[c]
20 to 50 (68 to 122)	190 cSt at 100°C[e]	857 cSt at 100°C[d]	190 cSt at 100°C[e]

Source: Extracted from AGMA. ANSI/AGMA 9005-E02, *Industrial Gear Lubrication*, American Gear Manufacturers Association, Alexandria, VA, 2002. With permission from the publisher.
[a] Formerly AGMA 11 EP and 11S.
[b] Formerly AGMA 14R.
[c] Formerly AGMA 12 EP and 12S.
[d] Formerly AGMA 15R.
[e] Formerly AGMA 13 EP and 13S.

protect the gear teeth. When open gears are lubricated by dipping into a slush pan, the lubricant must not be so heavy that it channels as the gear teeth dip into it. When open gears are lubricated by grease, the consistency and pumpability must permit easy application under the prevailing ambient conditions. Lubricant quantity guidelines for intermittent methods of open gear application can be found in a publication (9005-E02) by ANSI/AGMA (Table 8.5).

5. Load Characteristics

Open gears are often heavily loaded, and shock loads may be present. These conditions usually require the use of lubricants with enhanced antiwear and EP properties.

B. AGMA Specifications for Lubricants for Open Gearing

The AGMA Standard 9005-E02 also contains specifications for open gear applications. The specifications include R&O gear oils (compounded gear oils are included in this specification) and EP gear oils. AGMA does publish specifications for *residual gear compounds* for open gearing included in Tables 8.2 and 8.5. Generally heavier grades of straight mineral oils and EP oils with viscosities ranging from 400 to 2000 cSt at 100°C (without diluent) are considered residual compounds. Other heavy bodied lubricants designed for open gear applications would be included in this category. These are generally mixed with diluent for ease of application. The AGMA lubricant number guidelines for open gearing using the continuous method of application are shown in Figure 8.58. (Note that AGMA 9005-E02 is in the process of being updated, so readers should refer to the new AGMA 9005 standard once it is published as this table is likely to change.) Table 8.5 shows the AGMA lubricant number guidelines for open gearing with intermittent application of lubricant.

IX. CYLINDERS

Cylinders are found in reciprocating compressors, reciprocating internal combustion engines, some older steam engines, hydraulic systems, some air tools, and linear actuators. For the purposes of this brief discussion of cylinder lubrication, reference will only be made to compressors and internal combustion engines.

Ambient Temperature, °C (°F)	Type of Operation	Splash Lubrication		Pressure Lubrication		Idler Immersion
		Pitch Line Velocity[a]		Pitch Line Velocity[a]		Pitch Line Velocity[a]
		<5 m/s	>5 m/s	<5 m/s	>5 m/s	Up to 1.5 m/s
−10 to 10 (14 to 50)	Continuous	220	150	220	150	680–1500
	Reversing or start stop	460	320	220	150	680–1500
10 to 30 (50 to 86)	Continuous	460	320	460	320	1500–2200
	Reversing or start stop	1500	680–1000	460	320	1500–2200
30 to 50 (86 to 122)	Continuous	2200	1500	460	320	4600
	Reversing or start stop	2200	1500	460	320	4600

[a] Pitch line velocity = (Pitch diameter in mm × RPM) ÷ 19,098 = m/s. 1 m/s = 200 ft/min.

FIGURE 8.58 Minimum viscosity recommendation for open gearing—continuous method of application. Note: all viscosities shown are expressed in mm²/s at 40°C. (Extracted from AGMA. ANSI/AGMA 9005-E02, Industrial Gear Lubrication, American Gear Manufacturers Association, Alexandria, VA, 2002. With permission from the publisher.)

Pistons and piston rings play an integral part in the lubrication of cylinders in reciprocating equipment such as compressors or internal combustion engines. In engines and compressors, the pistons are the main component used to compress air or other gases. Resultant pressures and temperatures can be high, and therefore some method of sealing the compressed gases is needed to avoid bypassing the piston and cylinder. This sealing is provided by the piston rings and the lubricant. The rings also serve to maintain alignment of the piston in the cylinder bore and act as the bearing surface for controlling wear and the oil films on the cylinder walls.

The lubrication of cylinders involves conditions that range from boundary films to full fluid films. For example, at the top and bottom or piston travel in the cylinder, the piston and rings actually stop and reverse direction.* These areas are called *ring reversal areas*. In these areas, the full fluid films developed by the rings in motion are lost, and boundary lubrication conditions exist until the motion of the pistons and rings reestablish the films. In the ring reversal area at the bottom of travel, the reestablishing of the films is not as critical as at the top of travel (assuming vertical travel). At the top of travel, pressures and temperatures are much higher and, as a result, most of the wear in this type of equipment occurs at the top ring reversal area.

The lubricant's primary function is to reduce wear and provide sealing. In addition, the lubricant must also minimize formation of deposits, provide protection against rust and corrosion, and handle moisture and other liquids and gases entering the cylinders as a result of compression or combustion. In the compression of some gases, the solubility of the gases in the lubricant film may result in a decrease in the oil's viscosity. This may necessitate the use of higher viscosity lubricants. Depending on application and operating conditions, the selection of a correct lubricant can be complex. This selection process is discussed in greater detail in Chapter 10 (Internal Combustion Engines) and Chapter 18 (Compressors).

X. FLEXIBLE COUPLINGS

When two rotating shafts are to be connected, some degree of misalignment is almost unavoidable. This is either because of static effects such as deflection of the shafts or thermal effects causing

---

* Engines and compressors can have vertical, horizontal, or "V" configurations. Some reciprocating engines also have opposed pistons, where there is an upper and lower piston in each cylinder. For discussion purposes, vertical travel of pistons has been selected but refers to the outward-most position in horizontal or other angles of the piston travel. Discussion also is limited to single-acting compressor cylinders.

the shafts to change relative positions in their supporting bearings. Misalignment may be angular (where the shafts meet at an angle that is not 180°), parallel (where the shafts are parallel but displaced laterally), or axial (such as results from endplay). In some cases, all three types of misalignment may be present. To accommodate misalignment, flexible couplings of various types are used to connect shafts together.

In addition to protecting the machines against stresses resulting from misalignment, flexible couplings transmit the torque from the driving shaft to the driven shaft, and help absorb shock loads.

Universal joints and constant velocity joints may be considered a type of flexible coupling. Both types of joints will accommodate relatively large amounts of misalignment and are used to some extent in industrial applications.

Several types of lubricated flexible couplings are in use. Gear type couplings (Figure 8.59) have hubs with external gear teeth (or splines) keyed to the shafts. A shell or sleeve with internal gear teeth at each end meshes with the teeth on the hubs and transmits the torque from the driving shaft to the driven shaft. Misalignment is accommodated in the meshing gears. Where large amounts of misalignment must be accommodated, special tooth forms that permit greater angular motion are used.

Chain couplings (Figure 8.60) have sprockets connected by a chain wrapped around and joined at the ends. The chain may be of either the roller or silent type. Roller chain couplings depend on the relative motion between the rollers and sprockets for their flexibility. Barrel-shaped rollers and special tooth forms may be used where large amounts of misalignment must be accommodated. Silent chain couplings depend on the clearance between the chain links and sprocket teeth for their flexibility.

Spring-laced couplings (Figure 8.61), also called steel grid or spring-type couplings, are made of two serrated flanges laced together with metal strips. Flexibility is obtained from movement of the metal strips in the serrations.

FIGURE 8.59 Typical gear coupling components.

FIGURE 8.60 Chain coupling with cover.

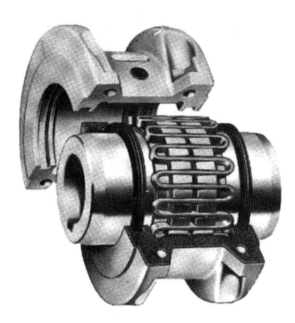

FIGURE 8.61 Spring coupling—horizontally split cover.

There are two types of sliding couplings. Sliding block couplings have two C-shaped jaw members connected by a square center member. The jaws are free to slide on the surfaces of the block to accommodate misalignment. Sliding disk couplings have flanged hubs with slots machined in the flanges. These slots engage jaw projections on a center disk. The floating center disk allows relative sliding movement between the flange slots and the disk jaws.

Slipper-type couplings, used mainly on rolling mills, are actually a type of universal joint. A C-shaped jaw attached to one shaft is fitted with pads that can turn in their recesses. A tongue on the other shaft fits between the pads. The tongue can move sideways in the pads to accommodate lateral misalignment, and the pads can rotate in their recesses to accommodate angular misalignment.

Flexing member couplings compensate for misalignment both by flexing of the components and sliding between them. The radial spoke coupling uses thin steel laminations to connect the coupling flanges. Clearance in the slots of the outer flange and flexing of the laminations allows for misalignment. Axial spoke couplings have flexible pins mounted in sintered bronze bushings connecting the flanges. The pins can flex or move in and out in their bushings to accommodate misalignment.

A. Lubrication of Flexible Couplings

All motion in flexible couplings that require lubrication is of the sliding type. Motion is continuous when a coupling is revolving, but the amount of motion may be small and can vary with the amount of misalignment between the shafts. Contact pressure may be high, particularly when there is considerable misalignment, because misalignment tends to reduce the surface area over which the load is distributed. The thickness of the lubricating film formed, and the type of lubricant required, depend to a considerable extent on the type of coupling.

AGMA defines three major operating groups for grease-lubricated flexible couplings as a function of shaft diameter, rotational speed, misalignment, torque, and coupling surface temperature. For the three operating groups, three coupling grease specifications are defined: AGMA CG-1, CG-2, and CG-3. These specifications are defined in AGMA Standard 9001-B97 (Table 8.6) as are

TABLE 8.6
Grease Lubricated Coupling Operating Classifications

Operating Conditions	Operating Groups		
	I	II	III
1. Rotational speed (rpm)			
d (in)	≤3600	≥2800^a/$d^{1/2}$	≤2800^a/$d^{1/2}$
d (mm)	≤3600	≥14,100/$d^{1/2}$	≤14,100/$d^{1/2}$
2. Misalignment (°)	≤0.75	≤0.5	≥0.75
3. Continuous torque			
T (lb in)	≤1200d^3	≥1200d^3 ^{b}	≥1200d^3 ^{b}
T (N m)	1200d^3/8.8 (25.4)^3	≥8.3 × 10^{-3}d^3	≥8.3 × 10^{-3}d^3
4. Peak torque	≤2.5 T	≤2.5 T	≥2.5 T
5. Maximum coupling surface temperature	150°F (65°C)	170°F (77°C)	212°F (100°C)
6. Normal relube interval (months)^c	6–12	12–36	1 or less

Source: AGMA, *Standard for Lubrication of Flexible Couplings* (AGMA-9001-B97), American Gear Manufacturers Association, Alexandria, VA, 1997. With permission from the publisher.

Note: d, shaft diameter; T, torque; G, ratio of actual acceleration to gravitational acceleration [$G ≈ 28.4d$ (rpm)2/106 when d is in inches, 1.12d (rpm)2/106 when d is in millimeters]; d is not an exact value because it relates approximately to its pitch radius of a coupling.

[a] Relates to centrifugal force on the lubricant of approximately 200 g.
[b] Relates to shaft torsional stress of approximately 6000 psi (0.207 MPa).
[c] The actual relube interval is dependent on experience with the specific application.

AGMA flexible coupling operating classifications and grease specifications (Table 8.7) (AGMA, 1997).

1. Gear Couplings

Gear couplings can transmit more torque than any other type of flexible coupling of equal size. Because they rotate continuously, a supply of lubricant is usually contained in the assembly. As the unit rotates, centrifugal force cause the lubricant to be thrown outward, forming an annulus of lubricant that should completely submerge the gear teeth. The relative motion between the gear teeth may be sufficient to create and maintain appreciably thick lubricating films.

Oil or grease may be used for gear couplings. Greases are often used where speeds are normal to high and temperatures are moderate. A soft grease made with a high viscosity oil, often with EP properties, is usually used for normal speeds. For higher speeds, a stiffer grease may be required (Tables 8.6 and 8.7).

2. Chain Couplings

Chain couplings (see Figure 8.59) are usually grease lubricated. Couplings without dust covers are generally lubricated by brushing with grease periodically. The grease must resist throw off and must penetrate into the chain joints to provide lubrication. In general, soft greases with good adhesive properties are usually required. Couplings with covers are packed with grease, usually of a type similar to those used in geared couplings.

TABLE 8.7
AGMA Coupling Grease Specifications

Characteristic (Test Method)[a]	Type CG-1	Type CG-2	Type CG-3
Minimum base oil viscosity:			
In centistokes	198 at 40°C (104°F)	288 at 40°C (104°F)	30 at 100°C (212°F)
In SSU (approx.)	900 at 100°F (38°C)	1300 at 100°F (38°C)	150 at 210°F (99°C)
Separation characteristics[b]	K36 ≤ 60/24, or 8% maximum fluid insoluble material	K36 ≤ 24/24	No restriction
National Lubrication Grease Institute (NLGI) grade			
(a) Metallic grid	1–3	1–3	1–3
(b) Gear or chain			
where rpm ≥ $200/\sqrt{d}$ in $1008/\sqrt{d}$ mm	0–3	0–1	1–2
where rpm ≤ $200/\sqrt{d}$ in $1008/\sqrt{d}$ mm[c]	0–1	Not applicable	1–2
Minimum dropping point	190°F (88°C)	195°F (91°C)	302°F (150°C)
Compatibility[d]	The coupling grease must be compatible with coupling seals and gaskets.		
Oxidation resistance—max pressure drop at 100 h	20 psi (13,790 Pa)	20 psi (13,790 Pa)	20 psi (13,790 Pa)
Antirust properties	Not required	ASTM rating pass	ASTM rating pass
Antiwear additives[d]	Not required	Not required	Required[e]
Extreme pressure (EP) additives[d]	Not required[f]	Not required[f]	Required
Timken OK load	Not required	Not required	40 lb minimum
Four-ball EP test	Not required	Not required	Weld point 250 kg minimum

Source: AGMA, *Standard for Lubrication of Flexible Couplings* (AGMA-9001-B97), American Gear Manufacturers Association, Alexandria, VA, 1997. With permission from the publisher.

[a] Accepted test methods: Viscosity, ASTM D 445; Grease composition, ASTM D 128; Centrifuge test, ASTM D 4425; NLGI grade, ASTM D 217; Dropping point, ASTM D 566 or D 2265; Antirust properties, ASTM D 1743; Oxidation resistance, ASTM D 942; Four-ball EP test, ASTM D 2596; Timken OK load, ASTM D 2509.
[b] ASTM centrifuge test.
[c] Relates to a centrifugal force on the lubricant of approximately 10 g.
[d] No test method.
[e] Experience has shown that a minimum of 5% (by weight) MoS_2 (molybdenum disulfide) is beneficial for couplings with hardened teeth.
[f] Extreme pressure (EP) additives recommended by some coupling manufacturers.

3. Spring-Laced Couplings

These couplings (see Figure 8.60) are grease packed. Again, greases similar to those used in geared couplings are usually required.

4. Sliding Couplings

Sliding block couplings may have the block formed from an oil-impregnated sintered metal. In other cases, the block contains a reservoir. In the latter case, high viscosity oils with good adhesiveness are required to minimize throw off. Sliding disk couplings require similar oils.

5. Slipper Joint Couplings

Slipper joint couplings are usually large and carry both heavy loads and shock loads so the lubrication requirements are severe. All-loss systems are usually used to apply the lubricant. High viscosity oils with enhanced film strength are required.

6. Flexible Member Couplings

Axial spoke couplings have prelubricated bushings for the pins. When enclosed in a dust cover, a soft adhesive grease is required for packing. Radial spoke couplings require high viscosity oil or a semifluid grease.

7. Lubrication Techniques

The preferred method of lubrication is to manually pack flexible couplings before closing. This procedure should include a coating of grease on all working surfaces including seals.

To lubricate a coupling that is assembled, remove the highest lube plug. Slowly pump the quantity of grease specified by the coupling manufacturer via the lowest lube plug. It is preferred that the coupling be rotated so that the higher lube plugs are located at the 3 and 9 o'clock position to assure adequate filling of the bottom half of the coupling. All working surfaces including seals should be coated with grease.

Key fits and keyways should be sealed with an oil-resistant sealing compound.

The proper quantity of grease is specified by the manufacturer because couplings of different series and shaft diameter require varying amounts. If in doubt, the coupling should not be filled more than 75% of capacity to allow for thermal expansion.

If the seal is damaged or distorted, replace the seal to avoid premature coupling failure from lack of lubrication.

Oil is generally preferred for gear couplings operating at normal to high speeds and high temperatures when the coupling is large enough to hold a reasonable supply of oil. Couplings may be oil filled, in which case they require only enough oil to cover the gear teeth during operation. Continuously supplied couplings may be designed to recirculate the oil, or to collect it so it can be drained and disposed of after use. Continuous supply systems are generally required where operating temperatures are high. Regardless of the method of applying the oil, high viscosity oils with good antiwear properties are required.

XI. DRIVE CHAINS

Drive chains are used to transmit power in a wide variety of applications. Chains fall into two general categories:

1. Machined surface chains suitable for high-speed, precision drives
2. Cast or forged link chains that are usually made without machined surfaces and are suitable for lower speed and power or where environmental conditions dictate the use of low-cost drives

A. SILENT AND ROLLER CHAINS

These chains are precision mechanisms that, when properly lubricated, will transmit high power at high speeds and will give long service life. Silent chains, also called inverted tooth chains, operate in such a manner that the links of the chain almost completely fill the clearance space in the sprocket throughout the arc of contact, greatly reducing backlash and minimizing noise. Roller chains are available in various designs and with up to 20 strands or more for use where high power must be transmitted.

With silent chains, the lubricant must penetrate between the links and distribute along the seat pins to provide lubricating films on the rubbing surfaces. With roller chains, the lubricant must coat the outsides of the rollers to lubricate the contact surfaces of the rollers and sprockets, and must also penetrate and distribute along the rubbing surfaces between the rollers and bushings. As a result, in both types of chains, the viscosity of the lubricating oil used is extremely important. Oils with enhanced film strength are also generally desirable, both to improve protection against wear and to permit the use of lighter oils with better penetrating properties.

Silent and roller chains are usually installed in oil-tight housings where the lubricant is applied by force feed, oil mist, slinger, or oil ring, or by the chain or sprocket dipping into a bath of oil. Occasionally, these chains are used without housings for low-speed, low-power drives. For applications of this type, the lubricant is applied by one of the all-loss methods such as drip, wick feed oiler, brush, oil mist, aerosol, or manual pour. The characteristics of the method of application and the requirement for controlling drip and throw off when the chain is not enclosed must be taken into account in the selection of the lubricant.

Normally, drive chains of these types are not grease lubricated. Under certain environmental conditions, such as corrosive atmospheres, a properly selected grease may provide better protection than oil. If maintaining proper lubrication is difficult because of chain location or if excess lubricant is frequently thrown off, a lube-free chain should be considered when replacement becomes necessary.

B. Cast or Forged Link Chains

Detachable link, pintle, fabricated, and cast roller types are among the chain forms available. Detachable link chains may be of cast malleable iron, or medium carbon steel hardened to improve strength and wear characteristics. Pintle links may be either cast or forged.

These chains usually operate at low speeds transmitting relatively low power; however, fabricated chains suitable for speeds up to 1000 ft/min (300 m/min) are also available. The chains may be open, or fitted with dust covers. Under most conditions, they require viscous, tacky oils, but under dusty or dirty conditions lower viscosity oils may reduce dirt pickup. Occasionally, diluent-type open gear lubricants that have a dry, rubbery surface after application may provide better performance.

C. Viscosity Selection

General guidelines for lubricant viscosities for silent and roller chains are shown in Table 8.8. Where drip or throwoff may be a problem, higher viscosity oils or oils with special adhesive properties may be desirable. Where water washing occurs, oils compounded to resist water washing and

TABLE 8.8
Guidelines for Oil Viscosity for Silent and Roller Chains

Ambient Temperature		Recommended Viscosity	ISO Viscosity Grade
°F	°C		
−20 to +20	−30 to −7	SAE 10	46
20 to 40	−7 to −4	SAE 20	68
40 to 100	4 to 38	SAE 30	100
100 to 120	38 to 49	SAE 40	150
120 to 140	49 to 60	SAE 50	220

Source: American Chain Association. *Identification, Installation, Lubrication and Maintenance of Power Transmission Roller Chains*, American Chain Association, Naples, FL, 1993.

protect against rusting are required. For additional information, the American Chain Association has developed a guideline for the lubrication of chains.

XII. CAMS AND CAM FOLLOWERS

There are many cases in machines where rotary motion must be converted to linear motion. Where the linear motion required is comparatively short, a common method of accomplishing this is with a cam and cam follower combination (cam follower is sometimes referred to as a tappet). Probably the best known applications of these mechanisms is for operating intake and exhaust valves in a reciprocating internal combustion engine, but many other applications also exist. As the cam will only lift the cam follower, some arrangement must be made to keep the cam follower in contact with the cam. This may be accomplished by transmitting load from the mechanism being actuated, or directly with a loading mechanism such as a return spring.

There are three general types of cam followers that can be selected according to such factors as load, speed, and the complexity of the cam shape:

1. Flat surface cam followers are used in valve systems of some internal combustion engines. To conserve fuel, numerous automotive internal combustion engines use roller tappet camshafts designs.
2. Roller cam followers may be used where there are extremely high loads or where reduced frictional characteristics are desired.
3. Spherical cam followers may be used where precise conformity to a complex cam is required.

Of these, the roller generally has the least severe lubrication requirements, because the rolling contact tends to develop and maintain an oil film.

In many cases with flat cam followers, the cam lobe is tapered and the cylindrical cam follower, which is free to rotate, is located with one axis toward one side of the cam. With this arrangement, the cam follower will be rotated as it is forced upward by the cam. This spreads the wear more uniformly over the surface of the cam follower, and may promote the formation of oil films.

In heavily loaded applications, both flat and spherical cam followers operate on EHL films at least part of the time. Even then, lubricants with antiwear additives are often necessary if rapid wear and surface distress are to be avoided. In the case of four-cycle gasoline engines, for example, one of the functions of the oil additive zinc dithiophosphate is to provide antiwear activity for the camshaft and valve lifters (cam followers). With the increased use of roller follower cams in automotive-type engines, the requirements for antiwear additives have been reduced slightly over time. The purpose is to limit the phosphorus levels in these engine oils that could adversely affect the life of emission control devices such as catalytic converters.

Most cam and cam follower machine elements are included in lubrication systems supplying other machine elements as well. For this reason, lubricants are rarely selected to meet the requirements of cams and cam followers alone. Generally, the lubricant is selected to meet the basic requirements of the system, and if special requirements exist for the lubrication of cams and followers, as in the case of a gasoline engine, these must be superimposed on the basic requirements.

XIII. WIRE ROPES

Wire rope or cable is used for a wide variety of purposes ranging from stationary service, such as guys or stays and suspension cables, to service involving drawing or hoisting heavy loads. In these various services, all degrees of exposure to environmental conditions are encountered. These range from clean, dry conditions in applications such as elevator cables in office buildings, to full exposure to the elements on outdoor equipment. This could include immersion in water that can be encountered on dredging equipment to exposure to corrosive environments, such as acid water, found in many mining applications. These and other operating factors require that wire ropes be properly lubricated to provide long rope life and maximum protection against rope failure where the safety of people is involved.

A wire rope consists of several strands laid (helically bent, not twisted) around a core. The core can be a rope made of hemp or other fiber, or may be an independent wire rope or strand. Each strand consists of several wires laid around the core, which usually consists of one or more wires but may be a small fiber rope. The number of wires per strand typically ranges from 7 to 37 or more.

A. Need for Lubrication

A number of factors contribute to the need for proper lubrication of wire ropes.

1. Wear

Each wire of a wire rope can be in contact with three or more wires over its entire length. Each contact is theoretically along a line, but this line actually widens to a narrow band because of a deformation under load. As load is applied, and as a rope bends or flexes over rollers, sheaves, or drums, stresses are set up that cause the strands and individual wires to move with respect to each other under high contact pressures. Unless lubricating films are maintained in the contact areas, considerable friction and wear result from these movements.

2. Fatigue

One of the principal causes of wire rope failures is metal fatigue. Bending and tension stresses, repeated many times, cause fatigue. Eventually, individual wires break and the rope is progressively weakened to the extent that it must be removed from service. If lubrication is inadequate, the stresses are increased by high frictional resistance to the movement of the wires over one another, fatigue failures occur more rapidly, and rope life is shortened.

3. Corrosion

Another principal cause of rope failure is corrosion. This covers both direct attack by corrosive materials, such as acid water that may be encountered in mines, to various forms of rusting. To protect against corrosion, lubricant films that resist displacement by water must be maintained on all wire surfaces.

4. Core Protection

Wear, deterioration, or drying out of the core result in reduction of the core diameter and loss of support for the strands. The strands then tend to overlap, and severe cutting or nicking of the wires may occur. The lubricant applied in service must be of a type that will penetrate through the strands to the core to minimize friction and wear at the core surface, seal the core against water, and keep it soft and flexible.

B. Lubrication during Manufacture

During manufacture, wire rope cores are saturated with lubricant. A second lubricant, designed to provide a very tenacious film, is usually applied to the wires and strands to lubricate and protect the wires and to help keep (seal) the lubricant in the core as they are laid up. These lubricants protect the rope during shipment, storage, and installation.

C. Lubrication in Service

Much of the core lubricant applied during manufacture is squeezed out when the strands are laid, and additional lubricant is lost from both the core and strands as soon as load is applied to a rope. As a result, in-service lubrication must be started almost immediately after a rope is placed in service.

Proper lubrication of wire ropes in service is not easy to accomplish. Some of the types of lubricants required for wire ropes may not be easy to apply, and often wire ropes are somewhat inaccessible. Various methods of applying lubricants are used, including brushing, spraying, pouring on a running section of the rope, drip or force feed applicators, and running the rope through a trough or bath of lubricant. Generally, the method of application is a function of the type of lubricant required to protect a rope under the conditions to which it is exposed.

D. Lubricant Characteristics

Wire rope lubricants should be able to do the following:

1. Form a durable, adhesive coating that will not be thrown or wiped off from the rope as it operates over pulleys or drums
2. Penetrate between adjacent wires in order to lubricate and protect them against wear and to keep the rope core from drying out and deteriorating
3. Provide lubrication between pulleys and sheaves and the rope
4. Resist being washed off by water
5. Protect against rusting or corrosion by acid, alkaline, or salt water
6. Form nonsticky films so that dust and dirt will not build up on ropes
7. Remain pliable and resist stripping at the lowest temperatures to which the rope will be exposed
8. Resist softening or thinning out to the extent that throwoff or drippage occurs at the highest temperatures at which the rope will operate
9. Be suitable for application to the rope under the service conditions encountered

These requirements necessitate some compromises. Wire rope lubricants may be formulated with asphaltic or petrolatum-based material and contain rust preventives and materials to promote metal wetting and penetration. Diluent products are used in some cases for ease of application. Grease products containing solid lubricants such as graphite or molybdenum disulfide are also used. The challenges with greases are the ability of the lubricant to penetrate to the inner core strands and its attraction for dust and dirt buildup. Wire ropes are often used in applications operating near or on an ocean, bay, river, lake, or other waterway, and as a result require environmentally acceptable wire rope lubricants to minimize their impact on the environment.

BIBLIOGRAPHY

AGMA. 1997. ANSI/AGMA 9001-B97 *Standard for Lubrication of Flexible Couplings.* American Gear Manufacturers Association, 1001 N. Fairfax Street, Suite 500 Alexandria, VA 22314-http://www.agma.org.

AGMA. 2002. ANSI/AGMA 9005-E02 *Industrial Gear Lubrication.* American Gear Manufacturers Association, Alexandria, VA. http://www.agma.org.

American Chain Association. 1993. *Identification, Installation, Lubrication and Maintenance of Power Transmission Roller Chains.* American Chain Association, Naples, FL.

ISO 3448. Industrial liquid lubricants—ISO viscosity classification. 1992. International Organization for Standardization. Geneva.

Mobil Oil Corporation. 1966. Mobil Technical Bulletin: Wire Rope Lubrication. New York.

Mobil Oil Corporation. 1971. Gears and Their Lubrication. New York.

Mobil Oil Corporation. 1973. Mobil Technical Bulletin: Grooving of Plain Bearings. New York.

Mobil Oil Corporation. 1973. Mobil Technical Bulletin: Rolling Element Bearings, Care and Maintenance. New York.

Mobil Oil Corporation. 1973. Plain Bearings Fluid-Film Lubrication. New York.

Mobil Oil Corporation. 1986. Mobil Lubrication Service Guides: Flexible Coupling Lubrication. New York.

Mobil Oil Corporation. 1992. Mobil EHL Guidebook. Fourth Edition. Fairfax, VA.

9 Application of Lubricants

After selecting the proper lubricant for an application, it must be delivered, in the correct quantities and under the specific application conditions, to the elements that require lubrication. Two categories of lubricant application are prevalent: *all-loss* methods, where a relatively small amount of lubricant is applied periodically and after use leaks or drains away to waste; and *reuse* methods, where the lubricant leaving the elements is collected and recirculated to lubricate again. Application methods such as some pedestal bearings and enclosed splash-lubricated gear cases are considered a form of reuse methods. Various centralized application systems, such as centralized grease systems and mist systems (not to be confused with circulation systems), are actually specialized all-loss systems, and are discussed separately. Reuse application systems are preferred because they conserve lubricant, minimize waste, and help control environmental pollution.

I. ALL-LOSS METHODS

Most open gears and wire ropes, many drive chains and rolling element bearings, and some cylinders, bearings, and enclosed gears are lubricated by all-loss methods. Nearly all grease lubrication (except sealed-for-life rolling element bearings) is an all-loss method. Only relatively small amounts of grease are applied, mainly to replenish the lubricating films, but in some cases to flush away some or all of the old lubricant and contaminants.

A. OILING DEVICES

One of the oldest known methods of applying lubricants is by using an oil can. With high viscosity lubricants such as those used on open gears and some wire ropes, a paddle, swab, brush, or caulking gun may be required in place of an oil can. These are all variations of hand oiling.

Although still widely used, there are several disadvantages to hand oiling. Immediately after application there is usually an oversupply of oil, and excessive leakage or throw-off can occur. Then follows a period when more or less of the appropriate quantity of oil is present, and finally—depending on the frequency of application—there is usually a period when too little oil is present (starvation). During this latter period, friction and wear may be high. Also, with hand application, lubrication points may be neglected, either because they are overlooked or because they are difficult or hazardous to reach. Oil leakage onto machine parts, floors, or goods being processed can be hazardous and costly in terms of safety and/or materials wasted. Hand oiling is costly both in terms of labor and because of lost production if machines must be shut down for the purpose. Many devices have been developed and are in wide use to overcome some of the disadvantages of hand oiling. The objective of these devices is to feed oil continuously or at regular, frequent intervals in small amounts with as little attention as possible.

1. Drop Feed and Wick Feed Cups

Drop feed and wick feed cups, shown in Figures 9.1 and 9.2, are often used to supply the small amount of oil required by high-speed rolling element bearings, thin film plain bearings and slides, and some open gears. The rate of oil feed from the drop feed cup can be adjusted with a needle valve, whereas the wick feed can be adjusted by changing the number of strands in the wick. Both have the disadvantage of requiring to be started and stopped by hand when the machine is started or stopped. Some drop feed cups are controlled by solenoid-operated valves, which eliminates the problems of manual actuation.

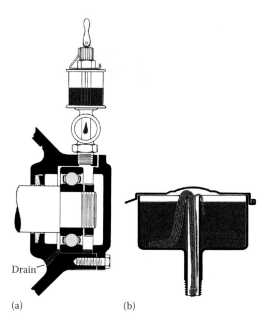

FIGURE 9.1 Drop feed (a) and wick feed (b) cups. In the drop feed application, the oil drops fall on the lock nut, which throws a spray of oil into the bearing.

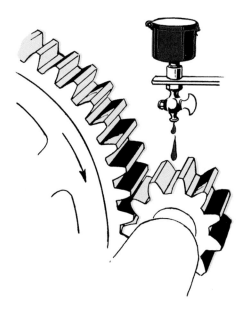

FIGURE 9.2 Drip oiling system.

2. Bottle Oilers

In a typical bottle oiler (Figure 9.3), the spindle of the oiler rests on the journal and is vibrated slightly as the journal rotates. This motion results in a pumping action that forces air into the bottle, causing minute amounts of oil to feed downward along the spindle to the bearing. The oil feed is more or less continuous but stops and starts when the machine is stopped or started.

Application of Lubricants 213

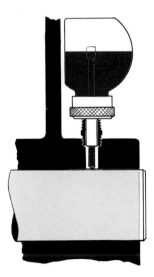

FIGURE 9.3 Bottle oiler.

3. Wick and Pad Oilers

In one type of wick oiler (Figure 9.4), a felt wick is held against the journal by a spring. The wick draws oil up from the reservoir by capillary action, and the turning journal wipes oil from the wick. No or little oil is fed when the journal is not turning. In another variation (Figure 9.5), the wick carries oil up to the slinger, which throws it into the bearing in the form of a fine spray. In both cases, oil leaking out along the shaft drains back to the reservoir so the devices have some elements of a reuse system.

Pad oilers are sometimes used for long, open bearings. One end of the pad rests in an oil reservoir, whereas the other end rests along the journal. Capillary action draws oil up the pad where it is wiped off by the turning journal.

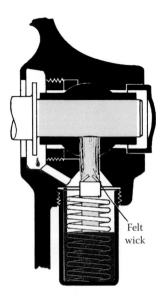

FIGURE 9.4 One type of wick oiler.

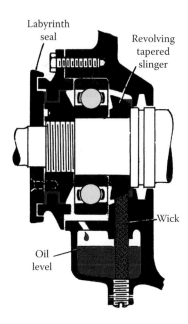

FIGURE 9.5 Wick-fed spray oiler.

4. Mechanical Force Feed Lubricators

Force feed lubricators are used in applications requiring positive feed of lubricants under pressure. A variety of force feed lubricators are in use. In the type shown in Figure 9.6, oil is drawn from the reservoir in the base on the downstroke of the single plunger pump and forced under pressure on the upstroke through the liquid-filled sight glass to the delivery line. The pump is operated by an eccentric cam and lever, which can be driven from a shaft on the machine or can be operated by a hand

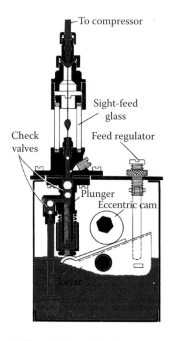

FIGURE 9.6 Force feed lubricator with liquid-filled sight glass.

Application of Lubricants

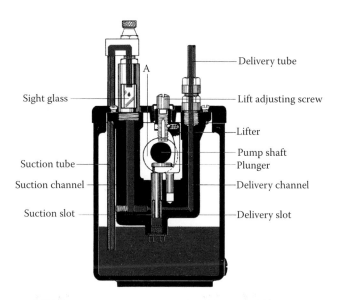

FIGURE 9.7 Force feed lubricator with air drop sight glass.

crank. The stroke of the pump can be regulated to adjust the oil feed rate, which can be estimated by counting the drops as they pass through the liquid-filled sight glass.

Some highly additized oils can cloud the liquid in the sight glasses of this type of lubricator, so lubricators with dry sight glasses are sometimes used. One of these is shown in Figure 9.7. The pump shaft is cam shaped in cross section and causes the plunger to move up and down. Specially shaped surfaces, at right angles to the pump shaft axis, also act on the eccentric head of the plunger, causing it to oscillate about its axis. On the upstroke of the plunger, the head is turned so that the suction slot registers with the suction channel port, while the delivery slot, which is 90° away, is blanked off. Oil is drawn up the suction tube and over to the sight glass where the falling drops may be observed. From the sight glass, the oil flows to the space below the plunger. On the downstroke of the plunger, the head rotates to align the delivery slot with the delivery channel port; the suction slot is blanked off. As the plunger is pushed downward, it forces oil out through the delivery tube. The plunger stroke, and amount of oil pumped, can be adjusted via the lift adjusting screw. The lubricator can be driven by the machine or turned by a hand crank.

Mechanical force feed lubricators are used to distribute lubricants to the cylinders of large stationary and marine diesel engines, reciprocating steam engines, some older gas engines, and reciprocating compressors. One pumping unit is used for each cylinder feed, and all of the pumping units are usually mounted on a single reservoir.

5. Air Line Oilers

Air-powered cylinders and tools are often lubricated by "lubricating" the compressed air supply.* Pneumatic cylinders used for actuating parts of machines may be lubricated by means of an oil fog lubricator such as that shown in Figure 9.8. The flow of air through the lubricator creates an air–oil fog that carries sufficient oil to lubricate the cylinders.

Air tools such as rock drills and jackhammers are often lubricated by means of air line oilers. The air line oiler consists of an oil reservoir that contains a device for feeding a metered amount of atomized oil into the air stream. It is installed into the air supply hose, usually just a short distance from the drill or hammer, and a fine spray of oil is carried in the air stream to lubricate the wearing surfaces. A method of varying the rate of oil feed is provided so that the oil feed can be adjusted.

---

* Air line oilers and air spray lubrication is a specialized variation of oil mist lubrication, which is discussed at the end of this chapter.

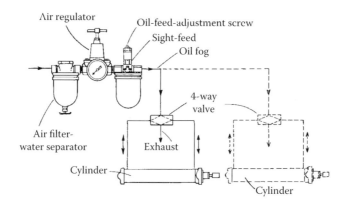

FIGURE 9.8 Fog lubricator for pneumatic cylinders.

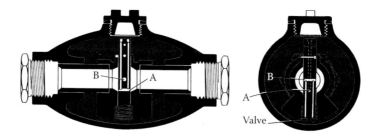

FIGURE 9.9 Air line oiler. Oil is drawn into the air stream through the port B by the Venturi effect. The enlarged section around the valve acts as a pendulum to swing the valve and keep the oil intake submerged in the oil regardless of the position of the outer housing.

In the air line oiler (Figure 9.9), air for rock drill actuation passes through the center tube, and line pressure is applied to the oil reservoir via port A and the vertical drilled passages shown in the left view. As a result of the Venturi effect, a reduced pressure prevails at port B. Because of the difference in pressure, oil feeds through the valve (right view) and port B into the air stream.

6. Air Spray Application

Some open gear and wire rope lubricants, including some grease-type materials, are applied by means of hand-operated air spray equipment using either external mixing nozzles or airless atomizing equipment.* These have some of the same disadvantages as hand oiling. A number of automatic or semiautomatic units such as that shown in Figure 9.10 have been developed to overcome some of these problems.

B. Grease Application

Greases are primarily used to lubricate rolling element bearings, flexible couplings, and thin film plain bearings and slides. Greases are also used in some open and enclosed gear applications. Because of the lower leakage tendencies of greases, periodic application with a hand- or air-powered grease gun does not have many of the disadvantages of hand oiling, but—as with any manual lubrication method—grease fittings can be overlooked and the cost of application is usually high. Therefore, central grease lubrication systems are being used more frequently, and a number of bearings are being "packed for life."

---

* See preceding footnote.

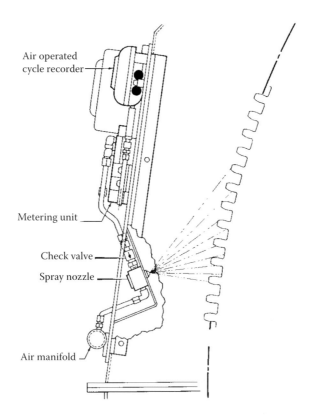

FIGURE 9.10 Automatic spray panel for open gear lubrication.

Many rolling element bearings are manually packed, at the time they are installed and periodically thereafter, as is the case with some automotive wheel bearings. Grease is forced into the spaces between the rolling elements by hand or with a bearing packer, and a moderate amount of grease placed in the hub. The housing must not be filled completely, because this will leave the grease that is trapped between the raceways with no place to go when the bearing is operated. Excessive churning of the grease and overheating can then result.

Bearings filled by a power grease gun must have a provision for pressure relief to prevent any pressure buildup and permit the purging of old grease. This is sometimes accomplished by allowing sufficient clearance in the seals, but can also be achieved with a relief valve or a drain plug such as that shown in Figure 9.11. When a bearing of this design is relubricated, the drain plug should be removed and left out until after the bearing has been operated long enough to expel all excess grease.

A wide range of self-contained grease lubrication devices exist, from spring-loaded reservoirs that feed small amounts of grease continuously to battery- or gas-operated units that can be programmed to activate at specified intervals. These electromechanical lubricators contain a motor, a piston pump, a gearbox, and a microprocessor capable of delivering precise lubricant quantities at specified intervals. These units are available in a range of sizes capable of single or multiple point distribution.

"Packed for life" bearings are used for many light to medium duty services. These bearings have seals and are packed at the factory with a special long life grease intended to last for the life of correctly applied bearings. In some cases, the service range of bearings of this type is extended by making provision for relubrication, using one of the following methods: removing one seal and applying grease through the seal by means of a gun with a hollow needle, or forcing grease through holes in the outer ring. Care is necessary with the latter method because it only takes a small amount of pressure to dislodge some seals.

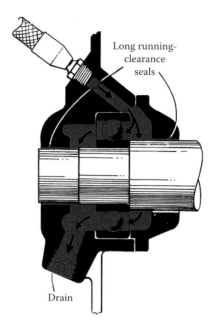

FIGURE 9.11 Bearing housing design with free purging. This design permits good purging of old grease from both sides of the bearing when new grease is applied.

FIGURE 9.12 Bellows type grease dispenser used to lubricate linear motion guides. (Courtesy of THK.)

Another grease application device that has been used to lubricate the linear motion guides on machines tools is a bellows grease dispenser. In this device, the grease is packed into a bellows container, which helps to keep the grease clean, and a pump is periodically activated to regulate the feed rate of the grease to the intended grease point (see Figure 9.12).

II. REUSE METHODS

Reuse methods of oil application include circulation systems supplying lubricant for one or more machines and self-contained systems such as bath, splash, flood, and ring oiling.

Application of Lubricants 219

A. Circulation Systems

The term "circulation system" generally refers to a system in which oil is delivered from a central reservoir to all bearings, gears, and other elements requiring lubrication. All the oil, disregarding minor leakage, drains back to a central sump and is reused. Two principal variations of this type of system are used—pressure or gravity feed. In pressure feed systems, a separate sump and reservoir may be used, or the two may be combined (Figure 9.13). The oil is pumped directly to the parts requiring lubrication. Where a separate sump is used, it may be either "wet," in which case the drain is located so that a certain amount of oil remains in the sump at all times, or "dry," in which case the drain is located and sized so that the sump remains essentially empty at all times. In gravity feed systems, oil is pumped to an overhead tank and then flows under gravity head to the elements requiring lubrication.

Although the many types of applications for circulation systems can vary in size, arrangement, and complexity, in general, all circulation systems can be classified into three groups:

1. Systems comprising a compact arrangement of pump, reservoir, and oil passages that are built into the housing of the lubricated parts such as that shown in Figure 9.13
2. Systems using a multicompartment tank combining the reservoir and oil purification equipment (Figure 9.14)
3. Systems composed of an assembly of individual units including a reservoir, an oil cooler, oil heater, oil pumps, purification equipment, etc.

The system illustrated in Figure 9.15 is fairly typical of the third type of circulation system. The returning oil drains into a settling compartment and enters the reservoir at or just above the oil level. Water and heavy contaminants settle, and the sloping bottom of the reservoir helps to concentrate these impurities at a low point from which they can be drained. This oil flows over a baffle plate to the clean oil compartment side of the reservoir. In some large reservoir systems, the baffles may be omitted as the contaminants will naturally settle because of the huge oil volume. The oil pump takes the oil through a suction strainer and then pumps it to an oil cooler, usually through oil filters,

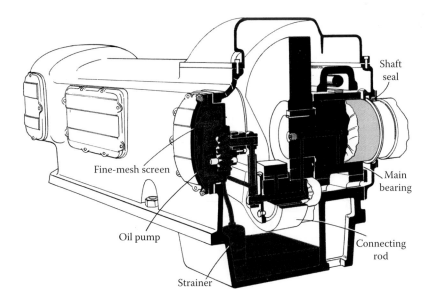

FIGURE 9.13 Pressure feed circulation system for horizontal two-stage compressor. The illustration shows a cutaway view of the main and connecting rod bearings. The oilgear pump draws oil from the reservoir through a strainer and forces it through a fine mesh screen and the hollow pump arm to the bearings. Excess oil is bypassed to the reservoir through a relief valve (not shown).

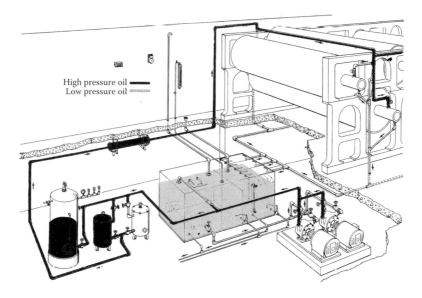

FIGURE 9.14 Paper mill dryer circulation system.

and then to bearings, gears, and other lubricated parts. The pressure desired in the oil supply piping is maintained via a relief valve that discharges it to the reservoir at a point below the oil level. A continuous bypass purification system is shown in Figure 9.15. A separate pump takes 5% to 15% of the oil volume in circulation from a point above the bottom of the reservoir where contaminants settle out and pumps it through a suitable filter back to the clean oil compartment. The following

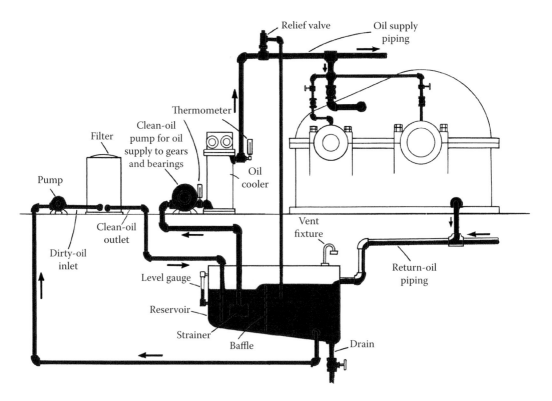

FIGURE 9.15 Oil circulation system diagram. The various components of an oil circulation are identified.

1. Oil Reservoirs

The bottom of an oil reservoir should slope 4% or more toward a drain connection, which should be located at the lowest point in the reservoir. This construction promotes the concentration of water and settled impurities and permits their removal without excessive loss of oil. In cases where oil is taken directly from the reservoir for purification, it should be removed close to the low point in the reservoir but above any water or settled impurities. An opening or openings above the oil level, adequate for inspection and cleaning, should be provided. Large reservoirs should have an opening large enough for a person to enter. These and any other openings should have well secured, dust-tight covers. All safety precautions should be taken before entering enclosed reservoirs.

A connection should be provided at the highest point in the reservoir for ventilation. Proper ventilation results in the removal of moisture-laden air and thereby reduces condensation on cooler surfaces above the oil level and subsequent rusting of these surfaces. Ventilation fixtures should be designed with air filters or desiccant breathers to prevent the entrance of airborne contaminants and excessive moisture. Instead of natural ventilation, where a source of water contamination is common—for example, large steam turbine systems or paper mill dryer systems—medium- and large-size reservoirs should be provided with a "vapor extractor," capable of maintaining a slight vacuum in the air space above the oil level. Too high a vacuum should be avoided, because it may have the undesired effect of pulling plant atmospheric contaminants into the lubrication system.

The main oil return connection should be located at or slightly above the oil level and away from the oil pump suction. Returning oil should not be permitted to drop from a considerable height directly into the oil body, because this action tends to whip air into the oil, which causes foaming and has a tendency to hold water and contaminants in suspension. Instead, the returning oil should be broken and dispersed by means of a baffle, sheet metal apron, or fine screen. A returning line that may carry any air should never be allowed to discharge below the oil level. A connection for the return to the reservoir of a "solid" stream of oil, as from a pressure relief valve, should be placed about 6 in (150 mm) below the oil level.

It is convenient to consider reservoir size in terms of the oil volume flowing in the system. The reservoir should be large enough so that oil velocity in it will be low, and the oil will have sufficient residence time to assure adequate separation of water and entrained solids, separation of entrained air, and be able to collapse any foam that may form. In practice, reservoir sizes typically range from a suggested minimum of 2 times the system flow per minute to more than 40 times. Many representative systems have reservoirs of 5 to 10 times system flow per minute. Newer design large steam turbine systems and large paper mill dryer systems will typically be in the range of 40 plus times total pumping capacity.

2. Pump Suction

The clean oil pump suction opening should be above the bottom of the reservoir to avoid picking up and recirculating settled impurities. However, it must be below the lowest oil level that may occur during operation. Where there is considerable variation in the oil level in the reservoir and it is desired to take oil at or near the surface, a floating suction may be used. Floating suctions are frequently used in reservoirs of systems exhibiting constant, extreme water contamination. The oil supplied to the circulation systems from the top of the reservoir will have the least water contamination.

3. Bearing Housings

The floors of bearing housings should have a slope of about 2% toward the drain connection. The design should be such that there are no pockets to trap oil and prevent complete oil drainage. Shaft seals should be adequate to prevent any loss of oil or the entrance of liquid or solid contaminants.

Any type of breather or vent fixture on a bearing housing should be provided with an air filter to keep out dust and dirt.

4. Return Oil Piping

All gravity return oil piping should be sized to operate about half-full under normal conditions and should have a slope of at least 1.7% toward the reservoir. Any unavoidable low spots should have a provision to allow periodic water removal. Severe piping bends, such as 90° or more out of the bearing housings, should be avoided to minimize the potential for oil backing up into the bearings causing overheating and oil leakage. Smooth flow passages of return oil is also important in avoiding any buildup of deposits in the return piping.

5. Metallurgy Composition of Circulation Systems

Exclusive of bearings, the parts of a circulation system should preferably be made of cast iron or carbon steel. Fittings of bronze are acceptable. Where tubing is used, stainless is considered ideal for most applications. Aluminum alloy tubing is chemically inert with oil but will not have sufficient structural strength for high pressure lines. No circulation system parts should be galvanized. As a general guide, no parts, exclusive of bearings, should be made of zinc, copper, lead, or other materials that may promote oil oxidation and deterioration. Copper tubing may be used for oil lines in some installations, but should not be used in systems such as those for steam turbines where extremely long oil life is desired. The use of copper should be confined to those applications where the oil is specifically formulated to inhibit the catalytic effects of copper. The use of copper tubing with active sulfur metalworking oils should be avoided.

6. Oil Filtration

Older equipment was often equipped with very coarse filtration (40 μm or larger) or in some cases, no filtration at all. As machinery became more complex over the years, the importance of oil

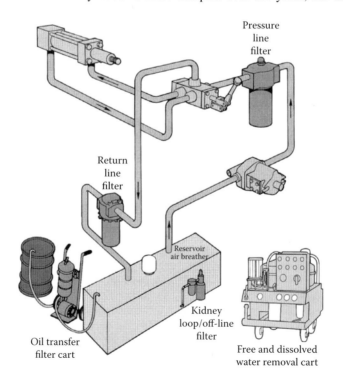

FIGURE 9.16 Oil filtration and purification methods.

filtration in helping to provide long equipment life has been well recognized. This is particularly true in close tolerance equipment such as servo valves in the machine tool industry or other precise control mechanisms such as governor controls in large turbines. Figure 9.15 shows filtration in a bypass loop, also known as a kidney loop. The bypass loop filtration system is commonly found on the pressure side of the supply line to equipment components. In addition to bypass (or kidney loop) filtration, another type of filtration can also be in low-pressure return lines. Figure 9.16 shows a primary oil filtration method and two alternative methods to keeping system oils clean: oil transfer equipment and a freestanding oil reclamation unit.

7. Oil Coolers

Oil coolers should be properly located so that all connections and flanged covers are accessible, and the tube bundles can be removed conveniently for cleaning. Cooler capacity should be adequate to prevent oil temperatures from rising above a safe maximum even during the hottest conditions. Means should be provided to control oil temperature at all times by regulating the flow of the water or cooling medium. Where practical, the oil pressure should always be higher than the water pressure to prevent water from entering the oil circulation system in the event of a cooler leak.

8. Oil Heating

Heating of the oil is sometimes desirable or required in oil circulation systems—for example, when the oil is too cold, particularly at startup, to provide adequate flow or lubrication to critical components.

The two most common methods of heating oils in industrial applications are steam and electricity. Steam is readily available in many large plants such as fossil fuel fired power plants or paper mills and, most frequently, steam is used for oil heating requirements. When steam is used, caution should be taken so that stagnant oil is not exposed to the full temperatures of the steam. Even saturated condensate steam at 15 psi will have a temperature of 335°F. Superheating of these steams can raise temperatures considerably. It is a good practice to maintain heating element surface temperatures below 200°F unless higher quality oils are used and/or oil flow across the heating elements can be maintained.

Oil degradation caused by contact with high heater skin temperatures can take various forms, including:

Additive depletion
Additive decomposition
Oil oxidation
Hydrocarbon cracking

When electrical immersion heaters are used, maximum safe heater watt densities should be determined. This information is available from oil suppliers and manufacturers of immersion heaters. As a general guide, a safe watt density to keep surface element temperatures below 200°F would be about 5 W/in^2. In many applications, it may be desirable to heat oil quickly, and this will necessitate either the use of multiple lower watt density heating elements or fewer elements with much higher watt densities (higher heating element surface temperatures). In these instances, it will be necessary to maintain sufficient oil velocity across the heating elements to minimize the time that the thin films of oil are in contact with the high temperature surfaces of the heaters.

9. Monitoring Parameters

The two most common parameters to measure on circulating oil systems are oil temperature and oil level. Oil reservoir temperatures should be measured along with the oil supply temperature and the oil discharge temperature out of main system components. In addition, the oil temperature from the oil inlet and outlet from heat exchangers should also be measured. Changes in "normal" operating temperatures could signal a malfunction in the system or be used to predict a potential equipment issue.

Maintaining oil at the proper level in the reservoir is important in several respects. Proper maintenance of the oil level will allow an adequate retention time in the reservoir to drop out contaminants such as water and solid contaminants, facilitate the dissipation of entrained air in the oil, and provide some radiant cooling of the system return oil. Maintaining proper oil levels and temperatures will go a long way to improving oil life, reducing filter costs, and protecting equipment components.

The more sophisticated oil circulation systems will monitor many more parameters such as oil pressure, oil flow, and differential pressures across filters and heat exchangers. Alarms may be used to indicate low levels, high temperatures, low flows, or low pressure. These alarms can be audible (bells) or visual (warning lights) and also tied into computer systems to monitor or alert personnel remote from the equipment location.

III. OTHER REUSE METHODS

In addition to circulation systems, a number of other methods of oil application involve more or less continuous reuse of the oil. These are differentiated from integral circulation systems primarily in that pumps are not used to lift the oil.

A. Splash Oiling

Splash oiling is mainly encountered in gearsets or in compressor or steam engine crankcases. Gear teeth, or projections on connecting rods, dip into the reservoir and splash oil to the parts to be lubricated or to the casing walls where pockets and channels are provided to catch the oil and lead it to the bearings (Figure 9.17). In some systems, oil is raised from the reservoir via a disk attached to a shaft, removed by a scraper, and led to a pocket from which it is distributed (Figure 9.18). This variation may be called a flood lubrication system. In either case, the oil returns to the reservoir for reuse after flowing through the bearings or over the gears. Accurate control of the oil level is necessary to prevent either inadequate lubrication or excessive churning and splashing of oil.

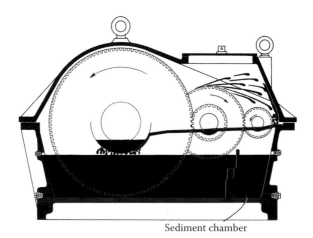

Sediment chamber

FIGURE 9.17 Splash oiling system. The gear teeth carry oil directly to some gears and splash it onto other gears and into collecting troughs, which lead it to bearings that cannot be reached by the splash.

Application of Lubricants 225

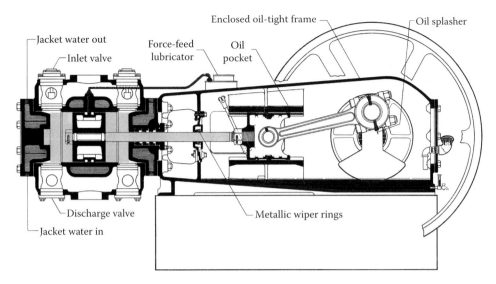

FIGURE 9.18 Flood lubrication system. In this single-cylinder compressor, the running parts are lubricated by the splash and flood method. Crankshaft counterweights and an oil splasher dip into the oil reservoir and throw oil to all bearings. Oil pockets over the crosshead and over each double row tapered roller main bearing assure an ample oil supply for these parts.

B. Bath Oiling

The bath system is used for the lubrication of vertical shaft hydrodynamic thrust bearings and for some vertical shaft journal bearings. The lubricated surfaces are submerged in a bath of oil, which is maintained at a constant level. When necessary, cooling coils are placed directly in the bath. The bath system for a thrust bearing may be a separate system or may be connected into a circulation system.

C. Ring, Chain, and Collar Oiling

In a ring oiled bearing, oil is raised from a reservoir by means of a ring that rides on and turns with the journal (Figure 9.19). Some of the oil is removed from the ring at the point of contact with the journal and is distributed by suitable grooves in the bearing. After flowing through the bearing, the oil drains back to the reservoir for reuse.

Ring oiling is applied to a wide variety of medium-speed bearings in stationary service. At high surface speeds, too much slip occurs between ring and journal, and not enough oil can be delivered. Furthermore, at high speeds, there is insufficient cooling for large heavily loaded bearings.

Oil rings are usually made about 1.5 to 2 times the journal diameter. Where bearings are more than about 8 in (200 mm) long, two or more rings are usually required. The oil level in reservoirs is usually maintained so that the rings dips into the oil at less than one-quarter of their diameter. The oil level, within a given range, is not usually critical. However, too low a level may result in inadequate oil supply, and too high a level may cause ring slip or stalling because of excessive viscous drag. As a result, too little oil may reach the bearing and "flats" may wear on the rings to the extent that they will no longer perform satisfactorily.

Chains are sometimes used instead of rings in low-speed bearings, because they have greater capacity for lifting oil at low speeds.

Where oils of very high viscosity are required for low-speed and heavily loaded bearings, a collar that is rigidly attached to the shaft may be used instead of a ring or chain. A scraper is required at the top of the collar to remove the oil and direct it to the distribution grooves in the bearing.

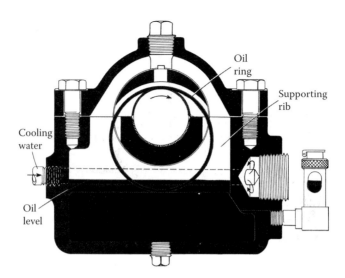

FIGURE 9.19 Ring oiled bearing for a small mechanical drive steam turbine. The plug in the cap can be removed to observe the turning of the oil ring. The running oil level is indicated. This would be raised during shutdown as a result of oil draining from the bearing and walls of the housing. The bearing is cooled by means of water passages through supporting ribs on either side of the oil ring slot. The spring cap on the fill and level gauge fitting (right) helps keep out dirt.

IV. CENTRALIZED APPLICATION SYSTEMS

A number of factors have contributed to the widespread use of centralized lubricant application systems. Among these are improved reliability, reduced cost of labor for lubricant application, reduced machine downtime required for lubrication, and a reduction in the amount of lubricant used through the reduction of waste and more efficient use of lubricants.

A. Central Lubrication Systems

A number of types of central lubrication systems have been developed. Most can apply either oil or grease, depending on the type of reservoir and pump used. Greases generally require higher pump pressures because greater pressure losses occur in the lines, metering valves, and fittings. Pump and reservoir capacities vary depending on the number of application points to be served. These range from small capacities of hand-operated lubricators (Figure 9.20), to units that install on standard drums (Figure 9.21), and systems that operate directly from bulk tanks or bins where large volumes of lubricant are required.

In *direct* systems, the pump serves to pressurize the lubricant and also to meter it to the application points. In *indirect* systems, the pump pressurizes the lubricant but valves in the distribution lines meter it to the application points.

Two basic types of indirect systems are in common use, and each type has two variations. In *parallel* systems, also called header or nonprogressive systems, the metering valves or feeders are actuated by bringing the main distribution line up to operating pressure (Figure 9.22a). All of the metering valves operate simultaneously. The disadvantage of this type of system is that, if one valve fails, no indication of failure is given at the pumping station. However, all of the other application points will continue to receive lubricant. In series, or progressive, systems the valves are "in" the main distribution line (Figure 9.22b). When the main distribution line is brought up to pressure, the first valve operates. After it has cycled, flow passes through it to the second valve, and then to each succeeding valve in turn. Thus, if one valve fails, all fail, and the pressure rise at the pump or the distribution block can be used to signal that a failure had occurred. The two variations of each of these basic types are as follows.

Application of Lubricants

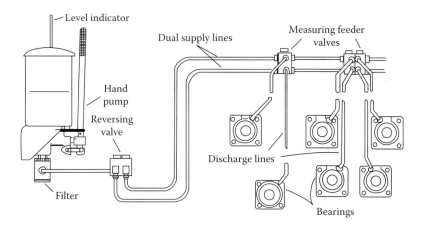

FIGURE 9.20 Typical hand operated centralized lubrication system. This type of system is for lubrication of a limited number of points. It can be either a dual-line system (shown) or a single-line system.

FIGURE 9.21 Drum mounted pump and control unit. A standard 400-lb drum serves as the reservoir. The pump can be air, electric, or hydraulic type, and develops 3500 psi (24 MPa) pressure. The unit is designed for a single-line, spring return lubrication system.

1. **Two-Line System**

In this variation of the parallel system, two supply lines are used (Figure 9.23). The four-way valve directs pressure alternately to the two lines, at the same time relieving pressure in the line that is not receiving flow from the pump. The valve can be operated manually from the machine, or cycled by a timer or by a counter that measures the volume delivered by the pump. The valves are designed to deliver a charge of lubricant to the application point each time the flow in the lines is reversed.

2. **Single-Line Spring Return**

In this variation of the parallel system, only a single distribution line is used. The layout of this type of system is shown in Figure 9.24. As with the four-way valve of the two-line system, the three-way valve may be operated manually from the machine, or by a timer or by a counter measuring pump

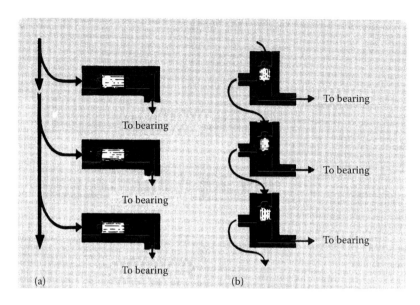

FIGURE 9.22 Parallel and series lubrication system. In the parallel system (a), the valves are "off" the supply line, whereas in the series system (b) the valves are "in" the supply line.

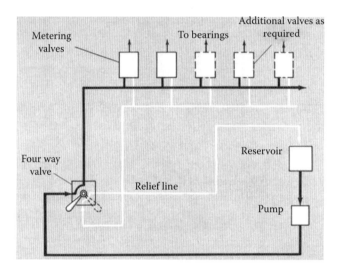

FIGURE 9.23 Two-line parallel system. The four-way valve, operated manually or automatically, alternately directs pump pressure to one line and then the other. When one line is pressurized, the other line is relieved.

output. The valves deliver a charge of lubricant when system pressure is applied to them, and reset themselves when the system pressure is relieved.

3. Series Manifolded System

In this type of system, a single supply line is used. No relief valve is required because the manifold type valves used automatically reset themselves and continue cycling as long as pressure is applied to them through the supply line. The system can be cycled by starting stopping the pump.

4. Series System, Reversing Flow

This type of series system uses a single supply line with a four-way valve to reverse the flow in it (Figure 9.25). The valves are designed to deliver a charge of lubricant, then permit lubricant flow to

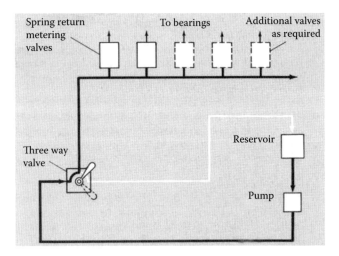

FIGURE 9.24 Single-line spring return system. The three-way valve, operated manually or automatically, either directs pump pressure into the supply line or relieves the pressure in the line to permit the spring return valves to reset.

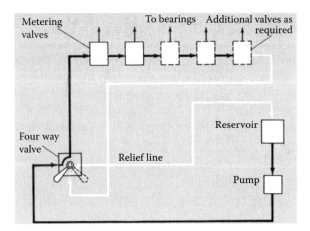

FIGURE 9.25 Series reversing flow system. The four-way valve, operated manually or automatically, directs pump pressure to one end of the closed loop supply line while relieving the pressure at the other end.

pass through to the next valve. When the flow in the supply line is reversed, the valves cycle again in sequence in the reverse order.

B. Mist Oiling Systems

In oil mist lubricators, oil is atomized by low pressure (10–50 psi or 70–350 kPa) compressed air into droplets so small that they float in the air, forming a practically dry mist, or fog, that can be transported at relatively long distances in small tubing. When the mist reaches the application point, it is condensed, or coalesced, into larger particles that wet the surfaces and provide lubrication. Condensing can be accomplished in several ways. Oil mist systems have proven their reliability in an increasing variety of applications. They are used in all types of industries—from the very light duty service of lubricating dental handpieces to the heavy-duty service of lubricating steel mill backup roll bearings. In the past, the systems were usually built onto or adapted to existing equipment. Machine tool builders are now designing them into their newer machines, primarily for spindle bearing lubrication, to provide greater reliability and productivity.

An oil mist lubrication system is simply a means to distribute oil of a required viscosity from a central reservoir to various machine elements.

A true oil mist is a dispersion of very small droplets of oil in smoothly flowing air. The size of these droplets averages from 1 to 3 µm (1 µm = 0.000039 in) in diameter. In comparison, an ordinary air line lubricator produces an atomized mixture of droplets up to 100 µm in diameter, which are suspended temporarily in turbulent air flowing at high velocity and pressure. In an air line lubricator system, the air is a working fluid that is transmitting power, whereas in an oil mist system, air is used only as a carrier to transport the oil to points where it is required for lubrication.

The droplet size is a very important consideration in the proper design of an oil mist system. The larger the droplets, the more likely they are to wet out and form an oily film at low impingement velocities. At low flow rates, the size of the droplets where they are likely to wet out is 3 µm. Droplets larger than this size will wet or spread out on surfaces quite readily whereas particles less than this diameter will not.

A dispersion of droplets less than 3 µm in diameter will form a stable mist and can be distributed for long distances through piping. At the points requiring lubrication, these drops can be made to wet metal surfaces by inducing a state of turbulence, causing small droplets to collide and form into larger diameter drops. These larger drops wet metal surfaces to provide the necessary lubricant film. The formation of larger drop sizes that wet metal surfaces is referred to as condensation, although terms such as reclassification, condensing, and coalescing, are also used.

Different degrees of condensation may be achieved by using different adapters at the points requiring lubrication. These adapters are usually classified as mist nozzles, spray or partially condensing nozzles, and completely condensing or reclassifying nozzles. When high-speed rolling element bearings create sufficient turbulence in the bearing housing to cause the droplets to join and wet out, a mist-type nozzle may be used. When gears are being lubricated, it is usually necessary to partially condense the oil mist so the limited amount of agitation within the gear housing will cause the droplets to coalesce and wet out. When slow-moving slides or ways are being lubricated, it is usually necessary to completely condense the oil mist into a liquid, which is then applied to the surface.

In a typical oil mist system (Figure 9.26), compressed air enters through a water separator, a fine filter, and an air regulator to the mist generator (a). From the generator, the mist is carried to a manifold (b) and then to the various application points (c).

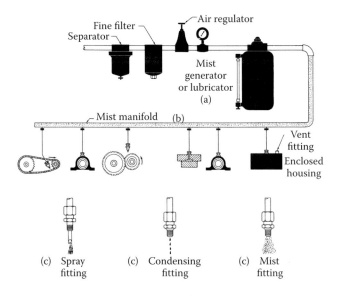

FIGURE 9.26 Oil mist lubrication system.

Application of Lubricants

To produce an oil mist, liquid oil is blasted with air to mechanically break it up into tiny particles. Droplets larger than 3 μm are screened or baffled out of the flow and returned to the sump or reservoir. The resultant dispersion, containing oil droplets averaging 1 to 3 μm in diameter, is the oil mist to be fed into the distribution system.

The size of the Venturi throat, oil feed line, and pressure differentials impose a physical limit on the viscosity of oil that can be misted. By the judicious use of oil heaters in the reservoir, and in some designs air line heaters to heat incoming air, the viscosity of normally heavy-bodied oils can be lowered to make misting possible. Systems without heaters can usually handle oils up to approximately an ISO 220 viscosity grade. If ambient temperatures are much lower than 70°F (21°C), then heat will likely be needed to reduce the oil's effective viscosity. Also, oils of over an ISO 220 viscosity grade will usually require heating to lower their effective viscosity and make the formation of a stable oil mist possible.

If the immersion elements in the oil reservoir are not properly adjusted, additional heating of the oil by the heated air will raise the bulk oil temperature to the point where it can oxidize quite readily, and varnish or sludge may form in the generator. When heated air is being used, it should be no hotter than necessary to allow misting, which is a maximum temperature of 175°F (80°C). Also, the oil reservoir immersion elements used in conjunction with heated air should be used primarily at startup and later, only if necessary, to maintain oil temperature during operation. Naturally, the immersion elements should be designed for oil heating and have a low watt density (e.g., 6 W/in^2, or 0.9 W/cm^2).

In addition to lubrication, mist systems provide other benefits: (1) the flow of air through bearing housings provides some cooling effect; and (2) the outward flow of air under positive pressure helps prevent the entrance of contaminants. Proper venting must be provided to permit this air to flow out of housings.

BIBLIOGRAPHY

Mobil Oil Corporation. 1969. Mobil Technical Bulletin: Rock Drill Lubrication. New York.
Mobil Oil Corporation. 1970. Mobil Technical Bulletin: Oil Mist Lubrication. New York.
Mobil Oil Corporation. 1971. Gears and Their Lubrication. New York.
Mobil Oil Corporation. 1973. Mobil Technical Bulletin: Rolling Element Bearings, Care and Maintenance. New York.
Mobil Oil Corporation. 1973. Plain Bearings Fluid-Film Lubrication. New York.
Mobil Oil Corporation. 1981. Mobil Lubrication Service Guides: Centralized Oil Lubrication Systems. New York.
Mobil Oil Corporation. 1981. Mobil Lubrication Service Guides: Oil Mist Systems. New York.
Mobil Oil Corporation. 1983. Compressors and Their Lubrication. Fairfax, VA.
Mobil Oil Corporation. 1990. Mobil Technical Bulletin: Handling, Storing and Dispensing Industrial Lubricants. Fairfax, VA.

10 Internal Combustion Engines

The term "internal combustion" describes engines that develop power directly from the gases of combustion. This class of engines includes reciprocating piston engines, used in a wide variety of applications, and most gas turbines. However, because closed system gas turbines are not truly internal combustion engines, gas turbines are discussed in a separate chapter. The following discussion is concerned primarily with reciprocating piston engines.

Piston engines range in size from the fractional horsepower units used to power model equipment, such as model airplanes, to engines for marine propulsion and industrial use that develop 100,000 hp (~75,000 kW) or more. Although this wide range of engine sizes and the types of applications present a variety of lubrication challenges, certain factors impacting lubrication are common to all reciprocating engines.

The primary objective of lubrication for reciprocating engines is to minimize wear as well as contribute to the power-producing ability and efficiency of the engine. These objectives require that the lubricant function effectively to lubricate, cool, seal, and maintain internal cleanliness. How well these factors can be achieved depends on the engine design, fuel, combustion process, operating conditions, the quality of maintenance, and the engine oil itself.

I. DESIGN AND CONSTRUCTION CONSIDERATIONS

Among the design and construction features that affect lubrication are the following:

1. **Combustion cycle**: whether two stroke or four stroke
2. **Mechanical construction**: whether trunk or crosshead type piston
3. **Supercharging**: whether the engine is supercharged (via supercharger, turbocharger, or blower) or naturally aspirated
4. **General characteristics**: describing the lubricant application system as a whole

A. Combustion Cycle

In a reciprocating engine, the combustion cycle in each cylinder can be completed in one revolution of the crankshaft (i.e., one upstroke* and one downstroke of the piston) or in two revolutions of the crankshaft (i.e., two upstrokes and two downstrokes). The first engine is referred to as a two-stroke cycle or a two-cycle engine, whereas the second is referred to as a four-stroke cycle or four-cycle engine. Either cycle can be used for engines operating with spark ignition (gasoline or gas) or compression ignition (diesel or residual fuel). As it is more widely used, the four-stroke cycle will be described first.

1. Four-Stroke Cycle

The sequence of events in the four-stroke cycle is illustrated in Figure 10.1. On the inlet or intake stroke at A, the intake valve is open and the piston is moving downward. Air or an air–fuel mixture is drawn in through the intake valve filling the cylinder. In diesel engines, only air is drawn or forced into the cylinder intake stroke and fuel is introduced through a high-pressure injector at the top of the compression stroke. In some natural gas and gasoline engines, an air–fuel mixture is introduced through the intake valve. Many four-cycle engine designs use injection of the fuel

---

* As not all reciprocating engines have vertical cylinders, this term is not strictly accurate; however, it does describe the stroke in which the piston approaches the head end of the cylinder.

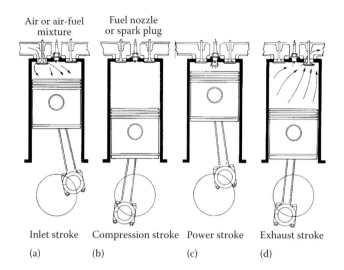

FIGURE 10.1 Four-stroke cycle. (a) The piston is moving downward, drawing in a charge of air or air–fuel, mixture. (b) The valves are closed and the piston compresses the charge as it moves upward. (c) Fuel has been injected and ignited by the high-temperature compressed air, or a spark is passed across the spark plug igniting the air–fuel charge, and the piston is pushed downward on the power stroke by the expanding hot gases. (d) The burned gases are forced out of the cylinder through the open exhaust valve.

directly into the cylinder similar to diesel engines. As the piston starts moving up at B, the intake valve closes and the air (or charge) is compressed in the cylinder. Near the top of this compression stroke, fuel is injected and/or a spark is passed across the spark plug. The fuel ignites and burns, and as it expands, it forces the piston down on the power stroke. Near the bottom of this stroke, the exhaust valve opens so that on the next upward stroke of the piston the burned gases are forced out of the cylinder. The assembly is then ready to repeat the cycle.

Most four-cycle engines are equipped with poppet valves in the cylinder head for both intake and exhaust. Various arrangements are used to operate these valves. In the conventional arrangement, a camshaft is located along the side or center of the cylinder block depending on engine configuration. It is driven from the crankshaft by gears, a silent chain, or in some engines by a toothed belt. Roller-, solid-, or hydraulic-type cam followers (often called valve lifters) ride on the cams and operate push rods, which in turn operate the rocker arms to open the valves. Valve closing is accomplished by springs surrounding the valve stems. This type of arrangement can result in some mechanical lag in valve operation at high speeds; thus, some high-speed automotive engines have the camshafts located above the cylinder head so that the valve stems bear directly on the cams or on short rocker arms. This arrangement is called overhead camshaft construction. The cam drive for many of these uses a chain or a toothed rubber belt, but a few designs incorporate gear drives for the overhead cams. Some large, medium, and low-speed diesel engines now are equipped for direct operation of the valves in a somewhat similar manner, and fully hydraulic valve actuation is also used on a few engines. The complete valve operating mechanism is often referred to as the valve train.

Loading on the rubbing surfaces in the valve train may be high, particularly in high-speed engines where stiff valve springs must be used to ensure that the valves close rapidly and positively. This high loading can result in lubrication failure unless special care is taken in the formulation of the lubricant.

2. Two-Stroke Cycle

The sequence of events in the two-stroke cycle is illustrated in Figure 10.2. Near the bottom of the stroke, the exhaust valves open and the piston uncovers the intake ports, allowing the scavenge air

Internal Combustion Engines

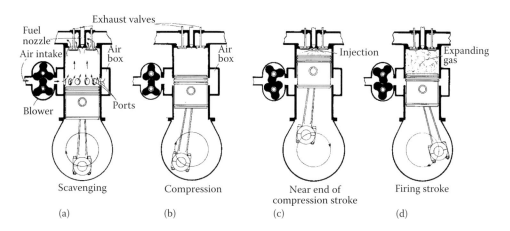

FIGURE 10.2 Two-stroke cycle. (a) Air from the blower is driving exhaust gas from the cylinder. (b) The charge of fresh air is trapped in the cylinder and compressed as the piston moves upward. (c) Fuel is injected. The fuel is ignited when the piston is close to the top of its stroke. (d) The hot expanding gases of combustion force the piston down on the power stroke.

to force the exhaust gases from the cylinder. As the piston starts on the upstroke at B, the exhaust valves* close and the piston covers the intake ports so that air (or charge) is trapped in the cylinder and compressed. Near the end of this compression stroke, the fuel is injected and begins to burn (or the charge is ignited). The expanding gases then force the piston down on the power stroke.

Two-cycle engines are usually built with either ports for intake and valves for exhaust, or with ports for both intake and exhaust. With the combination of ports and valves (Figure 10.2), the scavenge air and exhaust gases flow more or less straight through the cylinder, so this is referred to as uniflow scavenged. When the intake ports are on one side of the cylinder and the exhaust ports are on the other side, the engine is referred to as cross scavenged. The scavenge air flows more or less directly across the cylinder. Where the exhaust ports are located on the same side of the cylinder as the intake ports, the engine is referred to as loop scavenged. The scavenge air must flow in a loop into the cylinder and then back to the exhaust ports. Most small, two-cycle gasoline engines are of the cross scavenged type, whereas all three types of scavenging are used for two-cycle diesel engines. In large engines, the types of scavenging has some influence on piston temperatures and the type of piston cooling that can be used—oil or water; therefore, it has an indirect influence on lubricant selection.

Because the two-stroke cycle does not have a full positive exhaust stroke to rid the cylinder of combustion gases, the scavenging process must be assisted by pressure developed outside the cylinder. This can be accomplished by means of a separate blower or compressor, driven either by the engine or an outside source, or by what is known as crankcase scavenging. With crankcase scavenging, the section of the crankcase under each cylinder is sealed, except for a check valve to admit air (or charge) on the upstroke of the piston, and a transfer passage to permit the air (or charge) to be delivered to the intake ports as the piston approaches and goes through its bottom dead center position. A variation of this principle is used on some large, crosshead diesel engines. The lower end of the cylinder is closed and a packing gland installed where the piston rod passes through. With the addition of a check valve and transfer piping to carry the air to the intake ports, the piston then will pump air for scavenging or supercharging purposes. This may be referred to as pulse charging.

In theory, for the same bore, stroke, and rotational speed, a two-stroke cycle engine will develop twice the power of a four-stroke cycle engine. As a result, more fuel is consumed per unit of time,

---

* Some larger two-cycle engines do not use exhaust valves in the cylinder heads but instead use exhaust ports on the opposite side of intake ports. Examples of some large two-cycle gas engines are discussed later in the chapter.

the cylinder temperature and average engine temperature tend to be higher, and more contaminants may find their way into the oil. On many two-stroke cycle engines, the oil is mixed with the fuel.

B. Mechanical Construction

Most piston engines, including all automotive engines, are of the trunk piston type. Large, low-speed diesel engines used in marine propulsion and industrial applications are usually of the crosshead type.

In a trunk piston engine, the piston is connected directly to the connecting rod by a piston pin or wrist pin. Thus, the side thrust from the crankpin and connecting rod must be carried through the piston and rings to the cylinder wall. This may result in high rubbing pressures on one side of the piston (thrust side) and on the mating section of cylinder wall.

In a crosshead type of engine (illustrated in Figure 10.3), the piston is connected rigidly to a piston rod, which is connected to the crosshead containing the wrist pin. The crosshead also has a sliding guide bearing to absorb the side thrust from the crankshaft and connecting rod. This bearing is usually generously proportioned so that these thrust loads are readily carried by the lubricating oil films. No side thrust is carried by the piston. The cylinder assembly is usually completely separated from the crankcase, whether by means of a diaphragm containing a packing gland, commonly called a "stuffing box," or by having the lower end of the cylinder closed so that it can be used for pulse charging.

Crosshead-type engines are also built with double-acting pistons, or with two pistons acting in opposite directions in the same cylinder—this configuration is called an opposed piston engine. One of the chief disadvantages of crosshead construction is that it results in an engine that has a

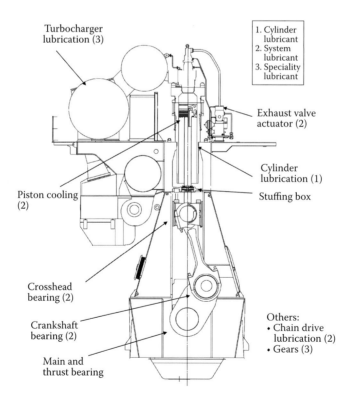

FIGURE 10.3 Schematic representation of the two-stroke crosshead diesel engine and its lubrication. (Courtesy of MAN Diesel & Turbo.)

greater overall height than a trunk piston engine of the same horsepower. However, most engines with bores larger than 600 mm (24 in) are of the crosshead type.

C. Supercharging

One of the factors that limit the power developed by an internal combustion engine is the amount of air it can "breathe." It would be easy to supply more liquid fuel, but it would be undesirable to exceed the quantity that the available air can burn more or less completely. Supercharging is a method of increasing the available combustion air by supplying air at higher pressure, thus making it possible to burn more fuel and produce more power.

The air for supercharging is provided by a supercharger, turbocharger, or blower, which may be engine driven, motor driven, electrically driven, or exhaust gas turbine driven. The latter is probably most frequently used today. With some large engines, supercharging by the blower may be supplemented by pulse charging.

Because supercharging increases the amount of air in the cylinders, and thus the amount of fuel that can be burned, it tends to raise combustion temperatures and pressures and increase the deteriorating influences on the lubricating oil. To increase supercharging efficiency, the compressed air is sometimes cooled between compression and engine intake.

D. Methods of Lubricant Application

Small- and medium-size trunk piston engines usually are lubricated by a combination of pressure and splash. Oil under pressure is fed to the main and crankpin bearings and, from the crankpin bearings, through drilled passages to the wrist pin bearings. Oil under pressure is also fed to an oil gallery from which it is distributed to the camshafts, valve lifters, and rocker arms. Oil is splashed to the lower cylinder walls to lubricate the rings and cylinders and to the undersides of the pistons for cooling. In some larger engines, oil is sprayed under pressure to cool piston undersides. Eventually all of this oil, neglecting the small amount that finds its way into the combustion chamber and is burned, drains back into the crankcase supply carrying contaminants with it.

Crosshead engines are equipped with at least two separate lubrication systems. One system supplies the lubricant for the cylinders. The relatively small portion of the used cylinder oil that is not burned in the cylinder is scraped down the cylinder liner by the piston rings and collects above the stuffing boxes from where it is drained away to a slop oil storage tank on the marine vessel. This slop oil is then properly disposed while the vessel is in port. A second lubrication system in crosshead diesel engines supplies the main and crankpin bearings and crossheads, lubricates the fuel pumps, provides piston cooling, and acts as a hydraulic fluid for engine control systems. A third system may be provided to lubricate the supercharger, or the supercharger may be lubricated from the crankcase system. There is typically only slight contamination of the crankcase oil of crosshead engines by cylinder oil, through minor leakage past the packing glands (stuffing box). A complicating factor in high-output, late model, crosshead engines with oil-cooled pistons is that piston temperatures may be high enough to cause thermal cracking of the oil while it is in the piston cooling spaces. This can lead to buildup of deposits in the piston cooling spaces, overheating of the pistons, and eventual cracking or other mechanical failure of the piston crown.

An important consideration with any internal combustion engine is the volume of oil in the system in proportion to the amount of contaminants that must be carried by the oil. Generally, large engines used in marine and stationary applications can be arranged so that the volume of oil is large, and because operation is at essentially constant speed, combustion efficiency can be kept high. Also, because of the relatively high oil consumption rates in these engines, the engine oil is typically not changed, but instead is replenished by make-up additions. The rate of contaminant buildup is then quite low, and usually can be by means of sophisticated purification equipment.

With automotive engines, the amount of oil that can be carried in the system is proportionally much lower, for a variety of reasons. With automotive gasoline engines, particularly, it is desirable to restrict the volume of oil to promote more rapid warmup. In addition, the lower oil volume is more cost-effective when frequent oil changes must be performed. The smaller volume oil sumps help to evaporate both condensed water buildup and any fuel dilution that may have resulted from start–stop type operations or cold starting. However, the variable speed conditions under which automotive engines operate, along with frequent starts and stops, tend to reduce combustion efficiency. This results in a relatively high rate of contaminant buildup in the oil. Fuel efficient engine designs incorporating port or direct fuel injection systems are helping reduce contaminant levels in the oils due to cleaner and more efficient combustion even in start–stop operation. Even more highly fuel efficient gasoline direct injection (GDI) systems can generate combustion-derived soot, and can thus contribute additional contamination to the engine oil. Smaller oil capacities also lead to higher operating temperatures for the oil, which can expose the oil to more severe oxidizing conditions.

Most small two-cycle gasoline engines, such as those used for marine outboard and utility purposes, are arranged for crankcase scavenging. To control the amount of oil carried into the cylinders with the charge, these engines are designed for dry sump operation. A small proportion of oil is premixed with the fuel, or injected, so the charge consists of a mixture of air, fuel, and oil. Some of the oil condenses on the crankcase surfaces where it provides lubrication of the main, crankpin, and wrist pin bearings. Also, some of it is carried into the combustion chambers where it is burned with the fuel. Incomplete combustion of this oil, or the use of the wrong type of oil, can cause spark plug fouling and combustion chamber and port deposit buildup.

II. FUEL AND COMBUSTION CONSIDERATIONS

When a hydrocarbon fuel is burned, certain products and residues are formed. One of the most important of these is water. In the case of liquid fuels, the volume formed is slightly greater than the volume of fuel burned. If the engine is fully warmed up, virtually all of this water passes out of the exhaust as vapor. If the engine is relatively cold, or under certain pressure and temperature conditions in the cylinders, some of the water may condense and can eventually find its way into the crankcase, where it will mix with the engine oil. The amount of condensation that occurs and the other residues that are formed are largely dependent on the type of fuel and the operating conditions of the engines in which the fuel is burned. High-temperature thermostats in the cooling system help reduce the levels of moisture and improve combustion conditions.

A. Gasoline Engines

Most four-cycle gasoline engines are used in automotive or similar applications. In these applications, the engine is generally started cold. Under some circumstances, an engine may operate for an extended time at below normal temperatures as a result of either short stop-and-go type operation, or carrying a light load. Under these conditions, a significant amount of the water formed from combustion of the gasoline may condense on the cylinder walls, and eventually reach the crankcase by way of blowby past the pistons and rings. This has a tendency to wash the lubricant films off the cylinder walls and promote rusting or corrosion. The rust formed is scuffed off almost immediately, but in the process metal is removed, and what appears to be mechanical wear occurs. In the crankcase and in other areas where temperatures are low, the water can combine with the oil and other contaminants to form sludge that usually has a soft, sticky consistency.

When a gasoline engine is cold, the air–fuel mixture must be enriched in order to provide enough vaporized fuel for starting. Some of the excess liquid fuel is blown out the exhaust, but a proportion of it drains past the pistons and into the crankcase, usually carrying with it some partially burned components and fuel soot. As the engine warms up, much of the gasoline is evaporated off, but some of the heavier ends, some of the partially combusted materials, and any solid residues remain in the oil. Even when the engine is fully warmed up, some fuel decomposition products may blow by into the oil.

Fuel soot and partially burned fuel and water, when present in the oil, can lead to the formation of varnish, sludge, and deposits. The potential for rusting of ferrous surfaces and corrosion of bearings can be promoted. The solid residues, such as fuel soot, when combined with small amounts of burned or partially burned oil in the combustion chambers, can form deposits that adhere to piston tops and the combustion chamber surfaces. Deposits of this type can increase the knocking tendency of an engine, creating combustion chamber hot spots that may induce detrimental preignition of the fuel.

The use of unleaded gasoline, required for those engines equipped with catalytic converters, may have the effect of reducing the amount of rust and corrosion that occurs. In some older engines (before 1980), the absence of the lead antiknock additives in the gasoline may result in accelerated valve wear, called valve recession. Generally, this relates more to metallurgy than to lubrication. Somewhat higher operating temperatures result from the emission controls, but these have not presented a major problem from a lubrication point of view. The use of exhaust gas recirculation (EGR) to control nitrogen oxide (NO_x) emissions, along with other emission control changes, may result in a greater tendency to form valve stem and engine deposits.

Two-stroke cycle gasoline engines are used in applications such as outboard motors, snowmobiles, chainsaws, lawn and garden equipment, and utility purposes such as pumps and lighting plants. These engines are typically lubricated by oil mixed with the fuel with a fuel/oil ratio ranging from 16:1 to 50:1, so some oil is always present in the charge. Under low-speed, low-load conditions (such as trolling a boat), poor combustion can result in a heavy buildup of port and piston crown deposits. Ashless and low ash oils tend to help control the level of deposit formation that is attributable to the ash-containing additives in the formulation.

B. DIESEL ENGINES

Diesel engines in trucks generally operate at more nearly constant speed and higher load factors compared to gasoline engines. Thus, combustion conditions are closer to optimum, and water condensation and fuel dilution associated with normal operating conditions are not as serious a problem. Where excessive fuel dilution occurs, it is usually the result of a mechanical problem, such as a faulty fuel injector or defective turbocharger. Diesel fuel is not as readily evaporated from the engine oil as is gasoline; thus, if a problem exists, the concentration of diesel fuel will tend to increase steadily. This can lead to deposit formation and a reduction in oil viscosity, which can cause mechanical wear if excessive.

One of the major problems for many diesel engine applications is the level of sulfur in the fuel. Sulfur in diesel fuel used to be at a considerably higher concentration than sulfur in gasoline. Today, many jurisdictions, particularly in developed countries, have regulations requiring sulfur levels in diesel fuel to be below 15 ppm, which may be equal or less than the regulated sulfur levels in some gasoline (e.g., ≤15 ppm in much of Europe and 30 ppm in many parts of the United States). The global trend for lower diesel fuel sulfur levels continues. There are a few areas of the world where diesel fuel sulfur levels are considerably higher (e.g., >5000 ppm in parts of Africa and the Middle East). When diesel burns, it forms sulfur dioxide, part of which may be further oxidized to sulfur trioxide. Depending on the fuel sulfur level, these sulfur oxides can combine with water to form strong acids (such as sulfuric acid) that are not only corrosive in themselves but also have a strong catalytic effect on oil degradation. Because piston temperatures are also high, this may result in heavy deposits of carbon and varnish on the pistons and in the ring grooves. Under severe conditions where higher sulfur fuels are used, deposits may build up in the piston ring grooves to the point where the rings cannot function properly, and may even stick, causing higher oil consumption, high wear, blowby, and a loss of power.

Many large diesel engines in marine and stationary industrial service are operated on residual-type (bunker) fuels with sulfur content in the 0.2–3.5% range (2000–35,000 ppm). The strong acids formed from the combustion of these high sulfur content fuels can be extremely corrosive to piston rings and cylinder liners, with the result that metal removal may be rapid and wear rates excessive. Selection of the correct oil characteristics for the type(s) of fuel burned is critical.

Modern cars, trucks, and buses using diesel engines, in regions with strict emission regulations, commonly use a range of exhaust aftertreatment technologies. These technologies include selective catalytic reduction (SCR), diesel particulate filters (DPF), diesel oxidation catalysts (DOC), and continuously regenerating traps, which combine the concepts deployed in both DPF and DOC. A level of EGR is also common, typically used to lower fuel combustion temperatures and so control potential exhaust gas species. Strategies using one or more of these technologies are combined by original engine manufacturers (OEMs) to achieve the required level of emission control.

A new generation of engine oils, called low SAPS oils (low in sulfated ash, phosphorus, and sulfur), have been developed to be more compatible with aftertreatment systems. These oils are typically specified by OEMs for use when certain types of exhaust aftertreatment device, such as DPF and DOC, are in place.

Some regions have requirements for a minimum bioderived fuel content in diesel fuel. These biofuels are frequently derived from crops such as rapeseed or soybean. The resultant fuel is typically referred to as "biodiesel." A fuel containing 10% bioderived material is termed "B10." The use of biodiesel fuel is impacting lubricant requirements and specifications. For example, in Europe, the Association des Constructeurs Européens D'Automobiles (ACEA) is progressively introducing bench and engine tests into its Sequences, reflecting concerns regarding the impact of biodiesel fuel on lubricant performance areas such as oxidation and piston deposits.

C. Gaseous Fueled Engines

Engines burning clean gaseous fuels, such as liquefied petroleum gas (LPG; propane, butane, or mixtures) or compressed natural gas (CNG), are comparatively free of the contaminating influences encountered in liquid fueled engines. Although water is formed from combustion, most of it passes out of the exhaust as vapor. Products of partial combustion that blow by into the crankcase have a tendency to polymerize with the engine oil in extended service, which may cause an increase in viscosity and, under severe conditions, may result in varnish and lacquer formation.

III. OPERATING CONSIDERATIONS

The operating conditions in an engine have a major influence on the severity of the service the lubricating oil is exposed to in the process of performing its functions of minimizing wear, reducing corrosion, reducing friction, assisting in cooling and sealing, and controlling deposits.

A. Wear

Normally, most engine wear occurs in cylinders, piston rings, and pistons. Less frequently, wear may be experienced in bearings and on camshafts. The three principal wear mechanisms include:

1. Abrasion
2. Metal-to-metal contact
3. Corrosion

Any dust and dirt mixed with intake air, which does get past the air filter into the engine cylinders, is hard and abrasive. Some fuels (particularly the residual fuels used in many large engines) may also contain abrasive materials. Abrasive particles carried by the oil onto load-supporting surfaces of the cylinder walls and other areas can cause piston ring and cylinder wear no matter how persistent the lubricating oil films are. Fortunately, wear due to abrasives can be held to almost negligible values by the use of effective air, fuel, and oil filters and/or purifiers. Proper maintenance of these filters and/or purifiers is important to prevent free passage of abrasives.

As a result of the conditions of boundary lubrication that exist on the upper cylinder walls, metal-to-metal contact cannot be avoided entirely. The rate of wear from this depends, to a large degree, on the suitability of the lubricating oil. When the oil is of the correct viscosity and has adequate antiwear characteristics, wear due to metal-to-metal contact can be kept to a low rate. However, reduced rates of oil flow, poor oil distribution, and/or excessive cylinder temperatures can contribute to increased wear rates in these areas.

In large, highly turbocharged, trunk piston engines, the side thrust on the cylinder walls is high. It has been found that some of these engines require oils with high load-carrying ability in order to keep wear rates to an acceptable level.

Corrosion and corrosive wear typically results from water contamination or a combination of water, severe oil degradation, and corrosive products of combustion. When operating temperatures are low, either during the warmup period or as a result of low load or stop-and-go operation or engine design, condensation can increase and corrosive wear may be rapid. Where the use of high sulfur fuels is unavoidable, the acidic compounds formed from their combustion promote corrosive wear. Generally, it has been found that the use of alkaline additives in the lubricating oil acts to retard this type of wear. These materials neutralize the acidic materials so that their corrosive and catalytic effects are reduced.

Some hard alloy-type bearings are susceptible to corrosive attack by certain oxidation products resulting from oil degradation. Oils for engines, where corrosion-susceptible bearings are used, are formulated with additives that provide protection against these oil oxidation products.

Metallic wear may be encountered on highly loaded valve train parts and fuel pump cams of some engines. Antiwear agents are incorporated in oils for these engines in order to minimize wear. Metallic wear is not generally a problem with engine bearings as long as sufficient oil of the proper viscosity is available, oil flow is not interrupted, and contaminants are kept to a minimum. Current engine technology using rollerized valve train components and electric fuel pumps also helps reduce the wear.

B. Cooling

Engine cooling is necessary to avoid engine damage and failure through overheating and thermal distortion. This is primarily a function of the cooling system, but the engine oil also has a critical role to play in cooling. This is especially important in large diesel engines where forced oil cooling of the pistons is used. Heat picked up by the oil is dissipated by natural radiation from the walls of the crankcase, or by means of an oil cooler where the oil is relied on for substantial engine cooling.

The specific heat of all petroleum oils is essentially the same and has no relevance in the choice of a lubricant for cooling purposes. Of importance are the chemical stability and the ability of the oil to resist the formation of deposits that might interfere with heat transfer, either from the engine parts to the oil or from the oil to its cooling medium. Under the severe conditions encountered in large diesel engines with oil-cooled pistons, thermal stability is also important.

C. Sealing

Effective cylinder sealing is necessary to minimize blowby and thereby maintain power and economy. Blowby, which cannot be entirely prevented, is a function of engine design (speed, piston rings, cylinder conformity, and provision for cylinder lubrication) and oil characteristics. The greater responsibility lies with the piston rings and their ability to adjust themselves to the varying cylinder contours throughout the length of ring travel, but the oil has an important complementary role. Generally, as far as the oil is concerned, there will be maximum sealing if the ring grooves are clean and unobstructed (so that the rings are free to move as required), and if there is no excessive removal of oil from the cylinder walls by the oil control rings. Too little oil on the cylinder walls will not only result in poor sealing but also may result in rapid wear. This, in turn, leads to even poorer sealing. On the other hand, for trunk piston engines (with recirculating lubrication systems), too much oil on the cylinder walls results in more of the oil being exposed to combustion conditions so that oil consumption and rate of contaminant buildup in the oil may be high.

D. Deposits

Control of deposits in an engine is a fundamental need in order to realize long life and efficient engine performance. Deposit formation is affected by engine design, operating conditions, maintenance, fuel type and combustion, and the performance of the oil. Deposits affect engine power output, wear, noise, smoothness, economy, life, and maintenance cost.

Two important sources of engine deposits have been discussed briefly, namely, dirt entering with the fuel and combustion air, and the fuel combustion process. Dirt in air or fuel causes abrasive wear. Deposits on piston crowns, in ring grooves, and on valves may often be referred to as "carbon," but this is a loose term. Usually, analysis will show that such deposits consist of dirt, solid combustion residue such as fuel soot, lubricating oil in various stages of decomposition, and residues from lubricating oil additives. It is important to note that any analysis of an engine deposit will show lubricant additive metals. This does not necessarily mean that the deposits are caused by an "oil" problem.

Temperatures in the combustion zone are high; thus, continued exposure of oil reaching this area causes it to oxidize, crack, and polymerize to heavier hydrocarbons. Large amounts of deposits in the combustion zone may be caused by excess amounts of oil reaching the upper cylinder area. Other possible causes include malfunctions of the fuel system, improper fuels that cause less complete combustion, poor ignition, faulty or worn valve guides, or worn cylinders and rings.

Diesel engines are not as sensitive to deposits in the combustion zone as gasoline engines are, especially those with high compression ratios. Depending on the character of the deposits, continued exposure to the combustion process may cause them to glow, which may provide an unwanted source of ignition. Undesirable combustion phenomena such as knock, preignition, low-speed preignition, rumble, and engine run-on can result. In gasoline engines, use of a higher octane fuel is usually beneficial in alleviating these problems. Similarly, use of an oil that has less tendency to form combustion chamber deposits can be beneficial.

Deposits in ring grooves are similar in origin to those in combustion chambers, but because they are not exposed to the direct flame of combustion, they may be more carbonaceous. Toward the bottom of the pistons (cooler areas), deposits tend to have a higher oil content. In severe cases, ring groove deposits can be packed so tightly in the clearance spaces behind the compression rings that the rings cannot operate freely. Compression pressure cannot then be maintained, and blowby becomes excessive. Excessive piston ring and cylinder liner wear may also result. Furthermore, oil control ring slots may become plugged, resulting in loss of oil control.

Piston varnish can also be formed from fuel and oil decomposition products. It may vary from a smooth, shiny, almost transparent coating often called lacquer, to a dark, opaque coating that becomes progressively more carbonaceous with continued operation.

Valve deposits are sometimes observed (Figure 10.4). These are generally the decomposition products of both fuel and oil, similarly formed in the same manner as combustion chamber deposits. Intake valves may show more deposits than exhaust valves, particularly when they run cool because of low load operation. Under such condition, there is a greater opportunity for droplets of gasoline and EGR substances to form gums on the valve stems. Dirt and solid residues will then adhere to these gummy deposits. In some designs of overhead valve engines, oil that flows down the valve stems can be excessive and this speeds up deposit formation, as well as being wasteful of oil.

Internal Combustion Engines

FIGURE 10.4 Engine valve deposits.

FIGURE 10.5 Top deck (cylinder head) sludging. This photograph shows heavy sludge buildup resulting from severe engine operation combined with an inadequate quality oil.

The other major type of deposit is the emulsion or sludge formed by water, fuel decomposition residues, and solid residues. Sludge generally deposits on cooler engine surfaces, such as the bottom of the crankcase pan, the valve chambers, and the top decks (Figure 10.5). The main problem with this type of deposit is that it can be picked up by the oil and carried to areas such as the oil pump inlet screen or oil passages where it can obstruct oil flow and cause lubrication failure.

Although the physical appearance of deposits varies greatly throughout an engine, basically these deposits result from the combustion process and lubrication oil deterioration. The exact chemical and physical nature of deposits depends on the area where they are formed along with the duration of exposure and any relative motion present. In the case of sludge, water is an essential factor, and any condition that encourages the entrance and retention of water in an engine oil promotes sludge formation.

The contribution of the engine oil to the control of deposits is discussed more fully in the following section.

IV. MAINTENANCE CONSIDERATIONS

The quality of maintenance of engine components that affect the lubrication process is of considerable importance. It is necessary to have clean combustion and crankcase ventilation air, which means that the air filters and positive crankcase ventilation system must be serviced regularly. A clogged air cleaner, although it may be effective in cleaning the air, can restrict the volume of air reaching the engine to such an extent that power output is reduced significantly. In order to maintain power, the natural tendency is to open the throttle, which only adds to the difficulty because the extra fuel is not burned. Diesel engines are quite sensitive to such a condition, which can result in more smoke, a rapid buildup of engine deposits, and increased soot in the oil. In addition to keeping the filters serviced, it is important that the piping connecting the filter to the engine be unobstructed and not have any leaks. This is particularly important in installations where the air filter is located a considerable distance from the engine such as found in some off-highway equipment.

The carburetor and ignition systems of spark ignition engines and the fuel injection systems of both gasoline and diesel engines should be in good working order and properly adjusted to assure as clean and complete combustion as possible. Malfunction or incorrect adjustment of these systems can result in increased amounts of unburned fuel in the cylinders, fuel dilution or more rapid buildup of contaminants in the oil, deposits in the engine, and an increase in exhaust emissions.

Proper engine oil drain schedules require that the oil be drained before the contaminant load becomes so great that the oil's lubricating function is impaired, or heavy deposition of suspended contaminants occurs. The correct drain interval for a given engine in a given service is best established by means of a series of used oil analyses and inspections of engine condition. As a practical matter, this is an economical approach for very large engines or fleets of similar engines in similar service. For smaller engines, such as those used in passenger cars, the oil drain intervals established by the manufacturer should be followed.

Oil drain intervals can vary considerably depending on the engine and type of service. For example, a large diesel engine in central station service, with a relatively large crankcase oil supply, may operate for thousands of hours between oil changes or may renew the oil through regular make-up additions. Such engines are usually in good adjustment, temperatures are moderate, and the contamination rate is low in comparison to the volume of oil in the system. On the other hand, a passenger car engine may require an oil change every few thousand miles. Such engines are physically small with relatively small crankcase capacity, and operate under conditions conducive to rapid oil contamination. Load factors are often low, and they may be engaged mostly in short runs and start–stop service, and operate over a wide range of ambient temperatures. These factors all favor the accumulation of oil contaminants and the risk of deposit formation.

The presence of an oil filter does not necessarily permit an extension of the oil drain interval. Filters do not remove oil-soluble contaminants and water, which are important factors in deposit formation. Regular filter changes are, however, important in keeping the filter operable so that it can perform its function of removing insoluble contaminants from the oil.

V. ENGINE OIL CHARACTERISTICS

Some of the more important characteristics of engine oils are discussed in the following sections. Additional information on physical and chemical characteristics, additives, and evaluation and performance tests can be found in Chapter 3.

A. Viscosity, Viscosity Index

The viscosity of the engine oil is extremely important for reciprocating engines. It can impact the degree of wear, sealing, oil consumption, frictional power losses (fuel economy), and deposit formation. For many engines, it is also a factor that affects cranking speed and the ease of cold startups. Too high a viscosity may cause excessive viscous drag, reduce cranking speed, and increase fuel consumption.

The viscosities of engine oils are usually reported according to the Society of Automotive Engineers (SAE) J300—Engine Oil Viscosity Classification standard. Although this system was originally intended only for automotive engine oils, its use has now been extended to include most oils for internal combustion engines. Engine manufacturers normally specify the viscosities of oils for their engines, according to ambient temperature and operating conditions, by SAE grade.

The viscosity index (VI) of the engine oil is important for engines that must be started and operated over a wide temperature range. Oils with higher VIs will give less viscous drag during startup, will generate thicker oil films at higher temperatures for better sealing and wear prevention, and may result in lower oil consumption.

In the past years, the VI of an oil was of limited importance for an oil that was used for engines not subject to frequent cold starts. This is still partially true today, except that many engine manufacturers are recommending the use of multiviscosity oils with high VI for year-round service. This is primarily attributable to the viscous drag of an oil being proportional to its viscosity impacting fuel economy. Other materials such as friction modifiers can be added to the oil to help reduce friction but at present, the biggest single oil characteristic that affects fuel economy is viscosity.

B. Low Temperature Fluidity

When an oil is used in engines operating at low ambient temperatures, the oil must have adequate low temperature fluidity to permit immediate flow to the oil pump suction when the engine is started. The pour point of an oil is an indicator of whether the oil will flow to the pump suction. Most conventional oils will flow to the pump suction at temperatures below their pour points because the pump suction creates a considerably greater pressure head than is present in the pour point test. However, many multigrade oils will not circulate adequately at temperatures considerably above their pour points. The correlation between pour point and flow in instrumented engines is poor, so that at best, pour point can only be considered to be a rough guide to the minimum temperature at which an oil may be used safely in engines.

The American Society for Testing and Materials (ASTM) has developed two tests that measure an oil's low temperature performance. The test that simulates cold cranking of an engine is called the ASTM D5293, Standard Test Method for Apparent Viscosity of Engine Oils and Base Stocks Between −5°C and −35°C Using Cold-Cranking Simulator, known as the Cold-Cranking Simulator (CCS) test. The test for measuring low temperature pumping is the ASTM D4684, Standard Test Method for Determination of Yield Stress and Apparent Viscosity of Engine Oils at Low Temperature, also known as the Mini-Rotary Viscometer (MRV) test. Both the CCS and MRV show good correlation to low temperature performance. The maximum values for these tests are listed for the "W" grades in the SAE J300 Specification for Engine Oil Viscosity Classification. The CCS is designed to reproduce the elements of viscous drag that affects cranking speed. The MRV is designed to predict low temperature oil pumpability and therefore the ability of the oil to reach critical components under low-temperature starting conditions.

C. Oxidation Stability (Chemical Stability)

High resistance to oxidation is an important requirement of a good engine oil considering the high temperatures the oil is exposed to and the agitation of the oil in the presence of air. Deterioration of engine oil by oxidation tends to increase an oil's viscosity, create deposit-forming materials, and promote corrosive attack on some hard alloy bearings. Where the engine sump capacity and corresponding oil volume is relatively small, the rate of oil deterioration tends to be higher with all other factors being equal.

An oil's natural oxidation stability is determined in part by the quality of the base stocks used in the finished oil product, whether they are produced by refining and hydroprocessing processes from crude oils, or are produced by synthesis processes and chemical transformations of simple hydrocarbons such as ethylene or methane. Where engine design or operating conditions require a high degree of oxidation stability, oxidation inhibitor additives are used in the oil. As a general rule, the need for greater oxidation stability increases as oil service temperatures and oil drain intervals increase. Among the principal factors that make enhanced oxidation stability necessary are high engine specific power output (high horsepower per unit of displacement), high stroke/bore ratios, cylinder oil exposed to high temperatures for longer durations, small crankcase oil volumes, long oil drain intervals, and modifications or devices to control emissions that result in high operating temperatures. For example, a heavily loaded truck engine requires an oil with excellent oxidation stability because of the high operating temperatures involved. Similarly, a large, low-speed diesel engine in stationary service generating electricity, with its large crankcase oil supply at a moderate temperature, requires an oil with good stability because the oil is expected to remain in service for thousands of hours. Regardless of the type of operating condition, good oxidation stability is always desirable in view of its helpful influence on engine cleanliness.

Oxidation stability plays only a minor role in the process in which combustion chamber deposits are formed. Under the severe thermal conditions of combustion, deposit tendency of entrained lubricating oil may be partially controlled by the selection and use of higher quality base stocks in the finished oil.

D. Thermal Stability

Thermal stability, or resistance to cracking and decomposition under high temperature conditions, is a fundamental characteristic of the lubricating base oil that cannot be substantially improved by means of additives. However, careful selection of additives is important in formulating thermally stable oils because decomposition of the additives, and their diluent oils, can contribute to the formation of deposits under operating conditions where thermal cracking of the oil occurs.

As noted earlier, thermal stability of the engine oil is of special concern in certain large, highly turbocharged, two-cycle diesel engines used for marine propulsion. Thermal cracking of the system oil can be experienced in the piston cooling spaces, resulting in deposits that interfere with heat transfer. Oils for these applications require carefully chosen base oils and additives to maximize the thermal stability of the finished oil.

E. Detergency and Dispersancy

The natural detergent and dispersant properties of most base stocks is minimal, and where these characteristics are important, they are obtained through the use of additives.

In nearly all current internal combustion engine applications, oils with enhanced detergency and dispersancy are necessary to control engine deposits and maintain engine performance. The levels of detergency and dispersancy required depend on a number of factors such as engine design, operating temperatures, type of fuel, continuity of operation, and exposure to low ambient temperatures. In general, conditions that tend to promote oil oxidation, such as turbocharging or the use of high sulfur fuels, dictate the use of oils with higher levels of detergency. Conditions that promote condensation of water and unburned or partially burned fuel in the engine require the use of oils with higher dispersancy.

Detergency and dispersancy are not clearly differentiated properties. There is a trend in additive development to improve the dispersancy of the so-called "detergents" and the detergency of the "dispersants." Thus, some oils formulated only with dispersants may give entirely adequate control of high-temperature deposits in some applications, whereas other oils formulated only with detergents may give good control of low-temperature emulsions and sludge in other applications. In general, most engine oils contain a mixture of the two types of additives with the concentrations and relative proportions depending on the type of engine service for which an oil is designed.

One of the main functions of both detergents and dispersants is to suspend potential deposit-forming materials in the oil. In this form, these materials are relatively harmless and may be removed from the system by draining the oil. Regular oil drains for this purpose are important, particularly for engines with a relatively small oil capacity. If the oil drain intervals are too long, the ability of the additives to suspend the deposit-forming materials may be exceeded, deposits will begin to form on engine surfaces, and engine performance will deteriorate. This is the reason manufacturers specify more frequent oil drains for "severe service" such as short trips, sustained high-speed driving, operation in dusty environments, and trailer towing.

There are no absolute measures of detergency and dispersancy. It is now customary to describe engine oils, on the basis of performance testing as was discussed in Chapter 3 (Section IV, "Engine Tests for Oil Performance"). Although this description includes an evaluation of detergency and dispersancy, the intent is to provide a more comprehensive description that includes all of the factors that make a particular oil suitable for a particular type of engine service.

F. Alkalinity

Most detergents, and to a lesser extent many dispersants, have some ability to neutralize the acidic products of fuel combustion and oil oxidation. However, where a considerable ability to neutralize acids is required, such as in oils for diesel engines burning high sulfur fuels, highly alkaline (over-based) detergent-type additives are used. The concentration of these materials in an oil, and an indication of the oil's ability to neutralize acids, is given by the total base number (TBN), or alkalinity value as it is also called. There is only a general relationship between TBN and the ability of an oil to control wear and corrosion caused by strong acids. This is because some newer additive systems have been found to be more effective in this respect than would be predicted by considering the TBN value alone.

Many of the highly alkaline oils are used as cylinder oils in large, low-speed crosshead diesel engines, which is a once-through or "all-loss" use. In these engines, as well as in engines where the same oil serves as both the cylinder and crankcase oil, it is desired to monitor the alkalinity of the oil as a method of determining if it is still capable of performing its acid neutralization function. This is particularly important where high sulfur or other acid-producing constituents are present in the fuel. In marine and industrial engines with once-through cylinder lubrication, samples of the cylinder scrape-down oil are taken for monitoring purposes.

G. Antiwear

In addition to the corrosive wear caused by acidic products of combustion, metallic wear may occur in areas where, because of loads or operating conditions, effective lubricating films cannot be maintained. The main areas where this occurs are on cylinder walls and rings of large, high-output trunk piston engines, and the valve train mechanisms of small, high-speed engines. In these cases, it is usually necessary to use oils that are formulated with additives that provide enhanced protection against wear and scuffing under boundary lubrication conditions.

H. Rust and Corrosion Protection

All petroleum oils have some ability to prevent rusting and corrosion of engine parts. However, in most cases, this natural ability is not sufficient to fulfill the following functions:

1. Protect hard alloy bearings from corrosion caused by oil oxyacids
2. Prevent rusting and corrosion due to condensation of water and combustion products in low temperature or stop-and-go service
3. Control the corrosive wear caused by acidic end products of combustion

Because one or more of these conditions, which can lead to troublesome corrosion, are encountered to some extent in nearly all internal combustion engine service, most oils for internal combustion engines are formulated to provide additional protection against corrosion.

Automotive engine oils usually are formulated to provide protection against corrosion and, particularly, corrosion of hard alloy bearings. In the case of oils intended for gasoline engine service, protection against corrosion and rusting caused by condensation of water and unburned or partially burned fuel components is emphasized. Diesel engine oils are usually formulated to provide protection against corrosion caused by acidic end products of combustion. As stated previously, in these latter oils, the function of corrosion protection is closely related to detergency and alkalinity.

Special preservative oils are available for the protection of engines that are to be idled seasonally or "mothballed" for extended periods. These oils are formulated to provide protection against rust and corrosion caused by atmospheric conditions and are usually formulated to work in engines for short-term use under moderate operating conditions. As a result, an engine can be run safely on the preservative oil before being laid up to distribute the oil over the internal engine surfaces. Also, an engine can be run for a short period after being taken out of storage before the preservative oil is drained and replaced by the normally recommended type of oil. This assumes that excessive contamination has not entered the engine or oil during storage.

I. Foam Resistance

All oils will foam to some extent when agitated. If excessive foaming occurs in an internal combustion engine, several problems may result. Overflow and spillage of oil is one of the most obvious consequences, but foaming can also cause oil starvation at the oil pump inlet. Foam in the oil can cause failure of lubricating films, which may cause, for example, noisy, erratic operation of hydraulic valve lifters, or under certain circumstances bearing failures. Foaming can be a concern with oils particularly intended for small, high-speed engines where agitation is severe or the crankcase oil volume is small. To reduce foaming, a defoamant additive is often included in the oil formulation. Overfilling the crankcase can cause foaming even with sufficient defoamant. Defoamant additives may also be used to minimize foaming due to product transfer, pumping, and handling.

J. Effect on Gasoline Engine Octane Number Requirement

The characteristics of lubricating oil can have a considerable influence on the octane number requirement (ONR) of gasoline used in engines. Some oil always finds its way past the piston rings into the combustion chamber, where it is partially burned to form deposits that adhere to piston tops and combustion chamber surfaces. As these deposits build up, the ONR of the engine gradually increases. The increase in ONR is a function of both the quantity and nature of combustion chamber deposits. Heavy, ragged deposits (Figure 10.6) cause the ONR of the gasoline to increase because sharp projections on the ragged deposits can be more easily heated to incandescence. Such a condition may

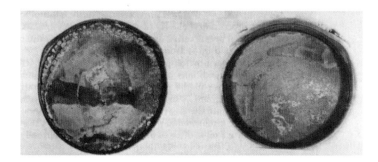

FIGURE 10.6 Combustion chamber deposits. Heavy, ragged deposits, seen on the piston crown on the left, generally contribute significantly to increasing the engine octane number requirement (ONR). Light, smooth deposits, as seen on the right, have less impact on increasing the engine ONR.

cause a deposit-related hot spot to ignite the fuel charge, either before the spark plug fires or before the normal flame front reaches portions of the charge. In either case, knock or power loss can result.

The contribution of lubricating oil to increased ONR may be reduced considerably by careful attention to oil cleanliness performance and to engine oil formulation factors such as the quality of the base stock(s) used. Heavy hydrocarbon base stocks, such as bright stock, are rarely used anymore in engine oil formulations. The increasing quality and performance specifications for modern light-duty automotive engine oils (as demonstrated by American Petroleum institute [API], International Lubricant Standardization and Approval Committee [ILSAC], and ACEA specifications) have generated increasing demand for higher quality base stocks, such as highly refined and hydroprocessed Group II and Group III base stocks, as well as synthetic base stocks (Groups IV and V). The overall cleanliness and wear-protection performances of modern multigrade automotive engine oils have helped to diminish the impact of ONR increase in engines.

Although oil formulation can help control combustion chamber deposits, the quality of fuel, type of service, operating conditions, engine conditions, and maintenance practices generally are the major contributors to combustion chamber deposits. These conditions will also affect engine emissions and fuel economy.

K. Engine Oil Identification and Classification Systems

The use of the SAE viscosity classification system (SAE J300) has already been discussed in Chapter 3. It is important to remember that this system is concerned only with viscosity. Thus, it may be used to describe the viscosity characteristics of oils intended for widely different services. Regarding engine performance, however, several other systems in general use that describe the performance qualities of automotive engine oils are API's Engine Oil Service Categories, ILSAC's Engine Oil Specification, and ACEA's Performance Specification.

1. API Engine Oil Service Categories

As originally set up by the API, this system included three gasoline engine service categories—ML, MM, and MS—and two diesel engine categories—DG and DS. A third diesel category, DM, was added later, and in 1969 the ML category for gasoline engines was dropped. However, certain inherent difficulties with the system resulted in a joint effort by API, ASTM, and SAE to develop a new system that would be more effective in communicating engine oil performance and engine service category information between the petroleum and automotive industries. This current system, which was finalized in 1970 and consists of two categories, S (Service Station) and C (Commercial), is described in API Bulletin 1509 and SAE Recommended Practice SAE J183. Table 10.1 shows the various API service categories. The API service categories primarily apply to engine oils manufactured for vehicles designed in North America and Asia. Most European-designed vehicles follow a different set of minimum standards for oils as designated in the European Automobile Manufacturers' Association ACEA European Oil Sequences.

2. API Classification System

The API engine oil system is open-ended, so that additional categories and classifications can be added when needed. When the letter designations are used by oil marketers or engine manufacturers to indicate the service classification for which oils are suitable or required, it is intended that they be preceded by the words, "API Service." To illustrate, an oil suitable for servicing new 2015 model cars under warranty would be referred to as "for API Service SN." If oils are suitable for more than one type of service, it is appropriate that these oils be designated, for example, "for API Service CJ-4/SM."

There is a relationship between API engine oil service classifications and some of the oil quality standards previously outlined. These relationships result from the fact that, in addition to the definitions for oil service classifications, there are also minimum performance requirements for

TABLE 10.1
API Service Category Chart

Gasoline Engines[a]	Diesel Engines
SA (obsolete) For older engines, no performance requirement. Use only when specifically recommended by manufacturer.	CA (obsolete) For light-duty engines (1940s and 1950s).
SB (obsolete) For older engines. Use only when specifically recommended by manufacturer.	CB (obsolete) For moderate-duty engines for 1949–1960.
SC (obsolete) For 1967 and older engines.	CC (obsolete) For engines introduced in 1961.
SD (obsolete) For 1971 and older engines.	CD (obsolete) Introduced in 1955. Certain naturally aspirated and turbocharged engines.
SE (obsolete) For 1979 and older engines.	CD-II (obsolete) Introduced in 1987. Two-stroke engines.
SF (obsolete) For 1988 and older engines.	CE (obsolete) Introduced in 1987. High-speed, four-stroke, NA, and turbocharged engines. Can be used in place of CC and CD oils.
SG (obsolete) For 1993 and older engines.	CF-4 (obsolete) Introduced in 1990. High-speed, four-stroke, NA, and turbocharged engines. Can be used in place of CE oils.
SH (obsolete) For 1996 and older engines.	CF (obsolete) Introduced in 1994. Off-road, indirect-injected, engines using fuel with >0.5 wt.% sulfur. Can be used in place of CD oils.
SJ (current) For 2001 and older engines.	CF-2 (obsolete) Introduced in 1994. Severe duty, two-stroke engines. Can be used in place of CD-II oils.
SL (current) For 2004 and older engines.	CG-4[b] (obsolete) Introduced in 1995. Severe duty, four-stroke engines. Can be used in place of CD, CE, and CF-4 oils.
SM (current) For 2010 and older engines.	CH-4[b] (current) Introduced in 1998. High-speed, four-stroke engines, meeting 1998 emission standards. Can be used in place of CD, CE, CF-4, and CG-4 oils.
SN (current) Introduced in October 2010, designed to provide improved high temperature deposit protection, sludge control, and seal compatibility.	CI-4[b,c] (current) Introduced in 2002. High-speed, four-stroke engines, meeting EGR and impending 2004 emission standards. Can be used in place of CD, CE, CF-4, CG-4, and CH-4 oils.
	CJ-4[d] (current) Introduced in 2006. High-speed four-stroke engines, meeting emissions (2007–2010), after treatment, and low-sulfur fuel standards. Can be used in place of CI-4, CH-4, CG-4 and CF-4 oils.

Source: Reproduced courtesy of the American Petroleum Institute: Motor Oil Matters, 2013; API 1509, 17th Edition, Addendum 1 (October 2014).

[a] Each gasoline engine category exceeds the performance properties of all the previous categories and can be used in place of a lower one. For example, an SN oil can be used for any previous "S" category.
[b] For severe duty, four-stroke cycle engines. Intended for use with diesel fuels with sulfur content ≤0.5 wt.%.
[c] Exhaust gas recirculation (EGR).
[d] For high-speed four-stroke cycle diesel engines designed to meet exhaust emission standards (spanning 2007–2010), especially with aftertreatment devices. For use with diesel fuels with sulfur content <500 ppm. Standard is directed to providing catalyst and particulate filter compatibility, engine durability and cleanliness, plus lubricant stability and robust service life.

TABLE 10.2
API Engine Oil Classification System for Automotive Gasoline Engine Service—Service "S" Oils

API Automotive Gasoline Engine Service Categories	Industry Definitions	Engine Test Requirements
SA/Obsolete	Straight mineral oil	None
SB/Obsolete	Inhibited oil only	CRC L-4[a] or L-38
		Seq. IV[a]
SC/Obsolete	1964 Models	CRC L-38
		Seq. IIA[a]
		Seq. IIIA[a]
		Seq. IV[a]
		Seq. V[a]
		Caterpillar L-1[a]
		(1% sulfur fuel)
SD/Obsolete	1968 Models	CRC L-38
		Seq. IIB[a]
		Seq. IIIC[a]
		Seq. IV[a]
		Seq. VB[a]
		Falcon Rust[a]
		Caterpillar L-1[a] or 1H[a]
SE/Obsolete	1972 Models	CRC L-38
		Seq. IIB[a]
		Seq. IIIB[a] or IIID[a]
		Seq. VC[a] or VD[a]
SF/Obsolete	1980 Models	CRC L-38
		Seq. IID
		Seq. IIID[a]
		Seq. VD[a]
SG/Obsolete	1989 Models	CRC L-38
		Seq. IID
		Seq. IIIE
		Seq. VE
		Caterpillar 1H2
SH/Obsolete	1994 Models	CRC L-38
		Seq. IID
		Seq. IIIE
		Seq. VE
SJ	1997 Models	CRC L-38
		Seq. IID
		Seq. IIIE
		Seq. VE
SL	2002 Models	Seq. IIIF
		Seq. IVA
		Seq. VE or VG
		Seq. VIII

(*Continued*)

TABLE 10.2 (CONTINUED)
API Engine Oil Classification System for Automotive Gasoline Engine Service—Service "S" Oils

API Automotive Gasoline Engine Service Categories	Industry Definitions	Engine Test Requirements
SM	2006 Models	Seq. IIIG, IIIGA
		Seq. IVA
		Seq. VG
		Seq. VIII
SN	2011 Models	Seq. IIIG, IIIGA
		Seq. IVA
		Seq. VG
		Seq. VIII

Source: Reproduced courtesy of the American Petroleum Institute: Motor Oil Matters, 2013; API 1509, 17th Edition, Addendum 1 (October 2014).

Note: (a) Energy Conserving/Resource Conserving options are available for service oils. These options are based on Sequence VI (fuel economy) engine testing and align with ILSAC claims. (b) Classifications "SL-SN" obtained and summarized from ASTM D4485-11c.

^a Obsolete test.

each classification that are defined in terms of performance in prescribed engine tests (Tables 10.2 and 10.3).

This approach of providing a complementary definition of minimum performance requirements for the service classifications will provide the user with better assurance that oils marked as being suitable for a particular service classification will provide an acceptable minimum performance level in service. However, performance differences can still exist among oils designated for the same service classification, because many oils are, and will be, formulated to exceed the minimum performance levels. These oils may provide additional benefits in service.

3. ILSAC Performance Specifications

In addition to API Service Categories, the ILSAC, in conjunction with automobile manufacturers, created the GF (gasoline fueled) service specification for passenger car engine oils in 1992. The "GF" categories (see Table 10.4) are principally designed to measure fuel economy benefits of engine oils, but also include critical areas of performance that are essentially equivalent to corresponding API categories (Tables 10.1 and 10.2). ILSAC "GF" categories only apply to engine oil viscosity grades SAE 0W-xx, 5W-xx, and 10W-xx.

Fuel economy performance is measured using the Sequence VI engine test (VI, VIA, VIB, VID; planned VIE) that has evolved in stepwise progression to increasingly higher percentages of fuel economy improvement (% FEI) targets. The Sequence VI fuel economy test is a multistage, high-temperature engine test, where % FEI performance is measured against a standard reference oil. Meeting the % FEI target for a particular "GF" category allows the use of a claim of "Energy Conserving" in the API donut, the API Service Symbol, and meeting all performance requirements of a "GF" category can be indicated by the licensed use of the API starburst, the API Certification Mark. Figures 10.7 and 10.8 are examples of the API Service Symbol and API Certification Mark that are typically displayed on the back and front labels of an engine oil bottle.

As of this writing, the ILSAC GF-6 specification is in proposal stages and targets the introduction of the next higher level of ILSAC GF engine performance standards. ILSAC GF-6 plans to

TABLE 10.3
API Engine Oil Classification System for Commercial Diesel Engine Service—Commercial Oils "C"

API Commercial Engine Service Categories	Related Military or Industry Designations	Engine Test Requirements
CA/Obsolete	MIL-L-2104A	CRC L-38
		Caterpillar L-1[a]
		(0.4% sulfur)
CB/Obsolete	MIL-L-2104A	CRC L-38
		Caterpillar L-1[a]
		(0.4% sulfur)
CC/Obsolete	MIL-L-2104B	CRC L-38
	MIL-L-45152B	Sequence IID
		Caterpillar 1H2[a]
CD/Obsolete	MIL-L-45199B, Series 3	CRC L-38
	MIL-L-2104C/D/E	Caterpillar 1G2
CD-II/Obsolete	MIL-L-2104D/E	CRC L-38
		Caterpillar 1G2
		Detroit Diesel 6V53T
CE/Obsolete	None	CRC L-38
		Caterpillar 1G2
		Cummins NTC-400
CF-4/Obsolete	None	CRC L-38
		Caterpillar 1K
		Cummins NTC-400
		Mack T-6; Mack T-7
CF-2/Obsolete	None	CRC L-38
		Caterpillar 1M-PC
		Detroit Diesel 6V92TA
CF/Obsolete	None	CRC L-38
		Caterpillar 1M-PC
CG-4/Obsolete	None	CRC L-38
		Caterpillar 1N
		GM 6.2L
		Mack T-8
		Sequence IIIE
CH-4	None	CRC L-38
		Caterpillar 1P
		Cummins M11
		GM 6.5L
		Mack T-9
CI-4	None	Caterpillar 1R
		Caterpillar 1K or 1N
		Cummins M11-EGR
		GM 6.5L
		Mack T-10
		Mack T-8E
		Sequence IIIF
		Engine Oil Aeration

(Continued)

TABLE 10.3 (CONTINUED)
API Engine Oil Classification System for Commercial Diesel Engine Service—Commercial Oils "C"

API Commercial Engine Service Categories	Related Military or Industry Designations	Engine Test Requirements
CJ-4	None	Caterpillar 1N
		Caterpillar C13
		Cummins ISB
		Cummins ISM
		GM 6.5L
		Mack T11
		Mack T12
		Sequence IIIF or IIIG
		Engine Oil Aeration

Source: Reproduced courtesy of the American Petroleum Institute: Motor Oil Matters, 2013; API 1509, 17th Edition, Addendum 1 (October 2014).

a Caterpillar L-1 and 1H2 Tests are obsolete.

include upgrades to engine oil performance previously required by ILSAC GF-5 (fuel economy, engine sludge protection, cam wear protection, piston cleanliness, oxidation control). ILSAC GF-6 is expected to add new engine oil tests in turbocharged, GDI engines, with oil performances required for low-speed preignition protection, and timing chain wear protection against GDI-type soot.

4. ACEA European Oil Sequences

The ACEA, also known as the European Automobile Manufacturers Association, introduced its first performance sequences for engine oils in 1996. These new ACEA sequences replaced the CCMC (Comite des Constructeurs du Marche Commun) specifications previously used by European engine manufacturers. The year number for ACEA Sequence indicates the year of implementation, for example, ACEA 2012. The ACEA sequences originally covered three ranges of engines and applications: "A" class for service fill oils for gasoline engines; "B" class for service fill oils for light duty diesel engines; and "E" class for service fill oils for heavy-duty diesel engines. With the 2004 release of the ACEA European Engine Oil Sequences, "A" and "B" classes were combined to form the "A/B" class. The "C" class was introduced to support gasoline and diesel engines with aftertreatment devices. Categories within each general class indicate oils having different performance levels, and are designated with numbers (e.g., A1/B1, C2, E4). Table 10.5 shows the engine tests used for each of the current sequences. New ACEA sequence specifications are under development. These new sequences will increase focus on lubricant performance in the presence of biodiesel. A new fuel economy category, ACEA C5, is also anticipated.

5. U.S. Military Specifications

Since 1941, the U.S. Army developed and maintained specifications for lubricants to be used in military equipment. The lubricant specifications are commonly referred to as MIL SPECS. Currently, some of the MIL SPECS are being converted to Commercial Item Description (CID) and to performance specifications. These are still entirely military specifications. Changes in military procurement regulations now allow the purchase of lubricants meeting applicable industry specifications for use in most military equipment.

TABLE 10.4
ILSAC Engine Oil Classification System for Gasoline Engine Service

ILSAC Gasoline Engine Service Category	Nearest Equivalent API Gasoline Engine Service Category	Engine Test Requirements
GF-1/Obsolete	SG	CRC L-38
		Seq. IID
		Seq. IIIE
		Seq. VE
		Seq. VI
GF-2/Obsolete	SJ	CRC L-38
		Seq. IID
		Seq. IIIE
		Seq. VE
		Seq. VIA
GF-3/Obsolete	SL	Seq. IIIF
		Seq. IVA
		Seq. VE + VG
		Seq. VIB
		Seq. VIII
GF-4	SM	Seq. IIIG, IIIGA
		Seq. IVA
		Seq. VG
		Seq. VIB
		Seq. VIII
GF-5	SN	Seq. IIIG, IIIGA, IIIGB
		Seq. IVA
		Seq. VG
		Seq. VID
		Seq. VIII
GF-6/Proposed	–	Seq. III (Update)
		Seq. IV (Update)
		Seq. V (Update)
		Seq. VI (Update)
		Seq. VIII
		Low-Speed Pre-Ignition
		Timing Chain Wear

Source: Reproduced courtesy of the American Petroleum Institute: Motor Oil Matters, 2013; API 1509, 17th Edition, Addendum 1 (October 2014); ILSAC draft document of February 2014, for proposed ILSAC GF-6A and GF-6B.
Note: GF, gasoline fueled engine oils.

Brief descriptions of MIL engine oil specifications, many being obsolete specifications, are given in the following list. These may still be used to some extent as performance references.

a. *U.S. Military Specification MIL-PRF-2104J.* This active specification describes lubricating oils for internal combustion engines used in combat tactical service. This specification superseded MIL-PRF-2104H. There is no comparable API or ILSAC specification that matches this specification. The API equivalent to this specification is approximately CI-4, CJ-4.
b. *CID A-A-52306A2.* This obsolete specification describes lubricating oils for heavy-duty diesel engines used in nontracked vehicles. The API equivalent to this specification is approximately CI-4, CJ-4.

FIGURE 10.7 Example of the API service symbol. (Reproduced courtesy of the American Petroleum Institute.)

FIGURE 10.8 Example the API Certification Mark (starburst). (Reproduced courtesy of the American Petroleum Institute.)

c. *CID A-A-52039.* This obsolete specification describes lubricating oils for use in military automotive engines. It is equivalent to a minimum performance level of API service category SJ.
d. *U.S. Military Specification MIL-PRF-46167.* This active specification describes lubricating oils for use in engines exposed to arctic conditions.
e. *U.S. Military Specification MIL-PRF-21260.* This active specification describes preservative break-in oils to be used in equipment that will be subject to long-term storage.
f. *U.S. Military Specifications MIL-L-2104 D thru F.* These obsolete specifications for lubricating oils for internal combustion engines used in combat tactical service has been superseded by MIL-PRF-2104G, where API CG-4 lubricants are considered the approximate equivalent.
g. *U.S. Military Specification MIL-L-46152.* This obsolete specification described oils for both gasoline and diesel engines in commercial vehicles used in U.S. federal and military fleets. In contrast to earlier U.S. Military specifications, it places strong emphasis on gasoline performance in addition to diesel performance. The gasoline engine performance requirements are the same as API Service SE, and the diesel engine performance requirements are the same as the former MIL-L-2104B (API Service CC). It was superseded by CID A-A-52039.
h. *U.S. Military Specification MIL-L-2014C.* This obsolete specification described oils for gasoline and diesel engines in tactical vehicles in U.S. military fleets. The diesel engine performance requirements are the same as the former U.S. Military Specification

TABLE 10.5
ACEA Engine Oil Sequences

Sequence	Partial List of Engine Tests	Performance Parameter
A1/B1$_{-12}$	CEC L-088-02 (TU5JP-L4)	High temperature deposits, ring sticking, and oil thickening
A3/B3$_{-12}$	ASTM D6593-00 (Seq. VG)	Low temperature sludge
A3/B4$_{-12}$	CEC L-038-94 (TU3M)	Valve train scuffing wear
A5/B5$_{-12}$	CEC L-054-96 (M111)[a]	Fuel economy
	CEC L-093-04 (DV4TD)	Medium temperature dispersivity
	CEC L-099-08 (OM646LA)	Wear
	CEC L-078-99 (VW TDI)	DI diesel piston cleanliness, ring sticking
	CEC L-104[b]	Effects of biodiesel
C1$_{-12}$	CEC L-088-T02 (TU5JP-L4)	High temperature deposits, ring sticking, and oil thickening
C2$_{-12}$	ASTM D6593-00 (Seq. VG)	Low temperature sludge
C3$_{-12}$	CEC L-038-94 (TU3M)	Valve train scuffing wear
C4$_{-12}$	CEC L-054-96 (M111)[a]	Fuel economy
	CEC L-093-04 (DV4TD)	Medium temperature dispersivity
	CEC L-099-08 (OM646LA)	Wear
	CEC L-078-99 (VW TDI)	DI diesel piston cleanliness, ring sticking
	CEC L-104[b]	Effects of biodiesel
E4$_{-12}$	CEC L-099-08 (OM646LA)	Wear
E6$_{-12}$	ASTM D5967 (Mack T8-E)[c]	Soot in oil
E7$_{-12}$	Mack T11[d]	Soot in oil
E9$_{-12}$	CEC L-101-08 (OM501LA)	Bore polishing, piston cleanliness
	Cummins ISM[e]	Soot induced wear
	Mack T12[f]	Wear (liner-ring-bearings)

Source: European Automobile Manufacturers Association. ACEA European Oil Sequences, 2012. Brussels.

Note: With the 2004 release of the ACEA European Oil Sequences, A and B categories were combined to A/B sequences. The C sequences were introduced to support gasoline and diesel engines with aftertreatment devices. The E sequences are intended for heavy duty diesel engines.

All of the engine tests listed for each sequence apply to all categories under each engine classification.

[a] Not required for A3/B3$_{-12}$ or A3/B4$_{-12}$.
[b] Not required for A1/B1$_{-12}$ or A3/B3$_{-12}$.
[c] Not required for E9$_{-12}$.
[d] Only required for E9$_{-12}$.
[e] Not required for E4$_{-12}$ or E6$_{-12}$.
[f] Not required for E$_{4-12}$.

MIL-L-45199B (API Service CD). Gasoline engine performance in Sequences IID and VD is required at a level approximately intermediate between API Service SC and SD.

i. *MIL-L-2104A*. This specification became obsolete in 1964. The performance requirements of this specification were required for API Service CA.

j. *MIL-L-45199B*. This specification became obsolete in 1970. The performance requirements of this specification were essentially equivalent to those required for API Service CD.

6. Manufacturer Specifications

Many engine manufacturers issue specifications for oils for their engines. Some of these specifications apply only during the break-in or warranty period, whereas most others represent the type of oil that the manufacturer believes should be used for service fill during the life of the engine.

These specifications are generally compatible with standard oil qualities available in the market. More often, however, manufacturers may require performance standards exceeding the minimum requirements of industry specifications (API, ILSAC, or ACEA). For example, in 2011, General Motors introduced specifications for GM passenger vehicles named "dexos® 1" (for gasoline) and "dexos 2" (for diesel), both of which exceeded the requirements of comparable API quality standards. In other cases, selected manufacturers may specify special oil formulations designed to meet performance conditions that are specific to certain engines in certain applications.

VI. OIL RECOMMENDATIONS BY APPLICATION

The remainder of this chapter constitutes a guide to the types and viscosities of oils usually recommended for internal combustion engines used in the various major fields of application. It must be remembered that a particular engine may have different requirements when used in different fields of application.

A. Passenger Car

Global regulatory and environmental trends to greatly reduce CO_2 emissions and the corresponding vehicle fuel consumption are dramatically changing the technology landscape of both vehicle and engine designs and performance. Increasing standards for improving fuel economy performance has fostered the broad introduction of energy efficient engine strategies such as engine downsizing, engine downspeeding, turbocharging or supercharging, GDI, lean-burn fuel combustion, engine thermal management, oil viscosity reduction, alternate fuels, and hybrid powertrain systems. Such changes have increased the severity of lubricant service conditions in many modern engines, with increasing power density, higher temperatures, higher pressures, and greater loads. Oil performance specifications, and corresponding oil quality, have increased to ever-higher levels. Most passenger cars have four-stroke gasoline engines. A growing trend is the use of diesel engines (already popular in Europe), whereas a few cars have rotary engines or are powered by electric motors. Two-stroke gasoline engines were phased out many years ago but may still be seen in some parts of the world. The recommendations for latest production four-stroke gasoline engines, for North American and comparable global vehicles, are oils with API Service SN and ILSAC GF-5 qualification. These oils provide good protection against low-temperature deposits and corrosion, protect against wear, and provide protection against oxidation, thickening, and high-temperature deposits under the most severe conditions. They also provide improved fuel economy and help assure the long-term performance of emission control systems. More emphasis is also being placed on oil quality aspects to provide longer drain intervals.

Many older, non-European, model four-stroke gasoline engines can be satisfactorily operated on oils for API service SJ or SL, although oils for API Service SM or SN may provide better overall performance and are more readily available.

The viscosities usually recommended by the automotive manufacturers for passenger car engines encompass a range of grades including SAE 10W-30, 5W-30, 5W-20, and 0W-20. The main reasons for the choice of lower viscosity products is to help achieve fuel economy requirements, such as CAFE (Corporate Average Fuel Economy), and to help assure quick oil supply to critical areas such as rocker arms during cold starts. For extremely low temperature operations, and for improved energy efficiency, SAE 0W-XX (XX = 20, 30) oils are often recommended and are currently available. In support of the ongoing trends to higher fuel efficiency performance, extremely low viscosity engine oil classifications (e.g., SAE 8, 12, 16), according to SAE J300 specification, have been defined and industry approved as of 2015 (see Table 3.1). The trend toward use of lower viscosity oils such as 0W-16, 0W-12, or SAE 0W-8 is expected to continue.

Internal Combustion Engines

The recommendations for non-European diesel car engines are generally similar to those for four-stroke gasoline engines, that is, oils for API Service SM/CJ-4 or CJ-4 being recommended. Most manufacturers recommend the use of multiviscosity oils, often SAE 15W-40, 10W-30, with future trends toward lower viscosities. There are some select cases where manufacturers may prefer single-viscosity grades.

The rotary engines used in passenger cars generally require SAE 10W-30 oils, API Service SJ quality or higher, which have shown evidence of satisfactory performance in service. The performance of oil additives in these engines is not necessarily the same as in reciprocating engines, hence the desire of the manufacturers for assurance that the particular oil formulation recommended has given evidence of satisfactory performance.

Two-cycle gasoline engines are used to a minor extent in various areas of the world. These engines are lubricated either by premixing the oil with the fuel, or by injecting oil into the fuel at the carburetor. The oil used is usually of either SAE 30 or 40 viscosity and is specifically formulated for this service. Generally, four-stroke engine oils are not satisfactory in these applications because they have ash-containing additives that will form undesirable deposits in the combustion chambers of two-cycle engines.

B. Truck and Bus

A significantly larger proportion of diesel engines are used in trucks and buses than in passenger cars. Both two-cycle and four-cycle diesel engines are used with the gasoline engines being all four-cycle.

The recommendations for gasoline engines in trucks and buses are similar to those of passenger cars, that is, oils for API Service SJ or higher. The viscosities used are somewhat heavier, typically SAE 30, 40 or multiviscosity SAE 10W-30, 15W-40, 20W-40, and 20W-50 oils. Recommendations for lower viscosity oils are increasing in order to take advantage of improved starting and fuel economy performances.

There is considerable variation in the oils recommended for diesel truck and bus engines. As shown in Table 10.3, there are several current API Service Classifications for commercial diesel engines. These service classifications are differentiated by a suffix (-2 or -4) to signify two- or four-cycle diesel engine requirements. Those "C" categories without a suffix can be used in either engine type. The higher performance and quality levels for diesel engines are API CF, CF-2, CH-4, CI-4, and CJ-4. Two-cycle engines are somewhat more sensitive to the ash content in the oil, and the typical recommendation is an oil for API Service CF-2 with certain restrictions on the ash content regardless of whether they are supercharged or naturally aspirated. Some of the four-cycle engines have also shown sensitivity to the additive system, so some select manufacturers may have special requirements over and above the basic API Service Classifications.

The recent wide availability and low cost of unconventional gaseous fuels—that is, methane or natural gas (such as CNG, or liquefied natural gas [LNG] and LPG)—have spurred the development of dedicated gaseous-fueled engines as well as dual-fuel engine capable of using both gaseous and conventional liquid fuels. In particular, CNG (i.e., methane) offers the advantage of a lower carbon footprint, with lower CO_2 emissions, than conventional liquid gasoline/diesel fuels, thus offering additional incentive for its use. Dedicated fleets of trucks or buses are showing interest in taking advantage of these alternate gaseous fuels. Engine oils useful for CNG/LNG service are typically high-quality API oils, from a minimum SJ up to SN. In many cases, oils specially formulated for CNG/LNG use may be required, which could include low-ash or zero-ash oils (obtained by using ashless additive systems).

C. Farm Machinery

The engines used in farm machinery include gasoline, diesel, and LPG. Some two-cycle gasoline engines are used for utility purposes such as water pumps and lighting plants. In general, the oil recommendations parallel those of cars and trucks. Typical oil recommendations for naturally aspirated and turbocharged diesel engines are for API Service CF, CJ-4 or ACEA E-7. For gasoline engines, API SN is typically the recommendation. Some manufacturers express a preference for special oils without metallo-organic detergents (low ash formulations) for LPG engines. Recommended viscosities closely parallel those used for truck and bus engines.

The small two-cycle gasoline engines used for utility purposes are lubricated by oil mixed into the fuel. Special oils designed for use in two-cycle gasoline engines should be used. These oils are formulated for the conditions encountered in the two-cycle engines and will generally give lower port, combustion chamber, and spark plug deposits. Viscosities are usually in the SAE 30 or 40 viscosity range, and in some cases there is a recommendation to predilute the oils with approximately 10% of a petroleum solvent so they will mix more readily with the fuel. Usually, fuel oil mixtures on the order of 16:1 to 50:1 are used, but some newer engines are designed for the use of higher ratios.

D. Aviation

Primarily, reciprocating piston engines for aviation use are four-stroke gasoline types. By far, the majority of these engines are relatively small and are used in personal and civil aircraft.

In the past years, aircraft engines were operated on high-quality, straight mineral oils. Over time, dispersant type oils have been developed that offer benefits in engine cleanliness with most aircraft engine manufacturers now accepting or recommending the use of these oils. Straight mineral oils may still be recommended for the break-in period. Multigrade oils such as 20W-50 or 15W-50 are more common, and they are often semisynthetic oils using a mixed base stock system of mineral and synthetic base oils.

Piston aircraft engine oils are usually formulated from special base oils made from selected crudes by carefully controlled refining processes. Quality is usually controlled to meet U.S. and U.K. government specifications, although some of the engine manufacturers have their own closely related specifications. Dispersant-type oils are usually manufactured by combining an approved additive system with proven straight mineral aircraft engine oils.

The viscosities of aircraft engine oils may be designated by SAE viscosity grade or by a grade number that is the approximate viscosity in SUS at 99°C (210°F). The specification numbers and viscosity grades covered are shown in Table 10.6. The common grade viscosity designations were part of the naming convention of the obsolete U.S. Military Specification system, and now the approximate SAE viscosity grade is covered by SAE J1966 and J1899.

TABLE 10.6
Aircraft Piston Engine Oil U.S. Specifications

Commercial Grade Designation	Straight Mineral Oil (Formerly MIL-L-6082E)	Dispersant (Formerly MIL-L-22851D)	Approximate SAE Grade
65	SAE J1966	SAE J1899	30
80	SAE J1966	SAE J1899	40
100	SAE J1966	SAE J1899	50
120	SAE J1966	SAE J1899	60

Internal Combustion Engines

E. Diesel Engines Used in Industrial and Marine Applications

Diesel engines range in size from just a few horsepower up to engines rated 100,000 hp or more. Generally, they can be divided into three classes: high-speed engines, medium-speed engines, and low-speed engines.

The lubrication requirements differ considerably for these classes, to some extent, because of engine designs and the different types of fuels that the various classes of engines can burn.

1. High-Speed Engines

Generally, these engines are considered to be high speed if they are designed to operate at speeds of 1000 rpm or higher. Engine sizes range up to about 300 mm (11.8 in) bore, with power outputs up to about 400 hp per cylinder. Multicylinder engines with outputs up to about 7000 hp are available. These engines are all of the trunk piston type, may be either turbocharged or naturally aspirated, and may be either two-stroke or four-stroke. The rings and cylinder walls are lubricated by oil splashed from the crankcase sprayed on piston undercrowns or from oil supplied to wrist pins through passages in connecting rods.

High-speed engines require high-quality fuels, and therefore are usually operated on distillate fuels similar to those used in automotive diesel engines. As a result, the sulfur content of the fuel rarely exceeds 1%, and is often considerably lower. With fuels of this quality, corrosive wear of cylinders and rings can be satisfactorily controlled by the types of oils developed for automotive diesel service, such as oils for API Service CF, CG, or CH (followed by the appropriate suffix). If the engine is one of the types developed for railroad service, one of the special railroad engine oils may be used. Viscosities recommended are usually SAE 30, SAE 40, or SAE 20W-40.

2. Medium-Speed Engines

These engines are designed to operate at speeds ranging between 375 and 1000 rpm. Engine sizes and outputs in this classification range from about 225 mm (8.85 in) bore with an output of 125 to 135 hp per cylinder, to 600 mm (23.6 in) or larger bore developing 1500 hp or more per cylinder. Large medium-speed engines with power outputs to 30,000 hp or more from V type configurations are available with up to 20 cylinders. All engines in this class are of the trunk piston type, typically four-cycle. Newer engines are usually turbocharged.

For smaller engines in this class, the rings and cylinder walls are lubricated by oil splashed or thrown from the crankcase to the lower parts of the cylinder walls. Larger engines have separate cylinder lubricators to supply supplemental oil to the cylinder walls.

Medium-speed engines are operated on a wide range of fuels, from LNG to residual fuels, including high-viscosity bunker fuels. Many of the smaller engines are operated on high-quality distillate fuels. Somewhat heavier fuels, such as marine diesel oils, may be used in some engines. The disadvantages of this heavier fuel include higher boiling ranges and higher sulfur contents. Also, some residual fuel components may be included in the blend. Many of the large engines are designed to operate on residual fuels although the residual fuels used are sometimes lower in viscosity than the heavy, bunker-type fuels that are often used in low-speed engines. Sulfur contents of the residual fuels may range upward to about 3.5%.

Although some medium-speed engines may be operated with automotive diesel engine oils of API Service CG or CH type quality, in general, the oils used in these engines were developed specifically for them and for similar engines used in marine propulsion service along with the fuels that they burn.

For convenience, these oils are often described in terms of their TBN. Smaller and larger medium-speed engines that are burning high-quality fuels are usually lubricated with oils of 10 to 20 TBN. Where residual fuels are used or the operation is severe, oils of 30 to 40 TBN are usually used. In some cases where poor quality or high sulfur fuels are used, 50 TBN oils may be used.

Viscosity grades used are usually SAE 30, 40, or 15W-40. In some cases, where the engines are used in intermittent service in exposed locations, subject to low temperatures, SAE 20 or multiviscosity oils may be used in the crankcase.

Oils for medium-speed engines burning residual fuel must be formulated to keep residual fuel contamination solubilized in the oil so that it does not separate from the oil and form deposits in high-temperature areas of the engine such as the piston undercrown. It must also provide good oil film thickness and integrity (low volatility), boundary lubrication, and demulsibility so that contaminants can be removed by the lube oil purifiers. Careful selection of the base oils and additives and formulation expertise are required to provide the required performance.

3. Low-Speed Engines

The low-speed engine class consists of the large, crosshead-type engines, most of which operate at speeds below 100 rpm. Engines in this class usually range from about 500 mm (19.7 in) to 980 mm (38.6 in) bore, with outputs for the largest engines as high as 9100 hp per cylinder, or 110,000 hp from a 12-cylinder engine, the largest built. Almost all of these engines operate on the two-stroke cycle. Separate cylinder lubricators are always used.

Low-speed diesel engines can be operated on a wide range of fuels, ranging from LNG to the highest viscosity residual fuels. Marine vessels using low-speed engines are typically operated on residual fuels except when operating in the regulated Emissions Containment Areas (ECAs). In these ECAs, they must operate on 0.1% maximum sulfur content fuel, which is typically distillate fuel, or be equipped with exhaust gas scrubbers to yield equivalent SO_x emission levels. Residual fuels can range in sulfur content from 0.1% to 3.5% and are typically in the range of 2.5–3.5% sulfur. Because the combustion of fuels of this quality results in the formation of large amounts of strong acids, highly alkaline cylinder oils in the range of 60–100 TBN are needed to control corrosive wear of piston rings and cylinder liners. Where the fuel sulfur content is relatively low (e.g., 0.1–1.5%), cylinder oils in the 15–40 TBN range may be used. The viscosity grade is typically SAE 50. In addition to providing the correct level of acid neutralization for the fuel burned, cylinder oils must be formulated to provide good oil film thickness and integrity (low volatility), good boundary lubrication, and control of deposits on the piston rings and on the high temperature areas of the piston. Careful selection of the base oils and additives and formulation expertise are required to provide the required cylinder oil performance.

The crankcase oil, known as system oil, in these engines lubricates the bearings and crossheads and typically lubricates the turbocharger bearings, and the camshaft and hydraulic pump drive gears, and serves as a hydraulic fluid for actuation of the exhaust valves. It also provides the very important function of piston cooling by circulating the system oil through the piston rod to the internal passages of the piston. As a large volume of oil is involved, extremely stable oils designed for long service life are desirable. Because the lubrication of the cylinders is physically separated from the lubrication of the crankcase, little or no contamination of the system oil by combustion products should occur. Therefore, high levels of detergency and alkalinity are not required and system oils are typically only 5–10 TBN. However, because system oils must protect against deposits in the extreme high temperature conditions found inside the pistons of modern low-speed engines, they must be formulated to provide excellent thermal stability and deposit control under these conditions. Deposits in these cooling spaces can impede the heat transfer required to prevent excessive piston crown temperatures and subsequent piston failure. They must also provide good antiscuffing performance (e.g., FZG fail stage 11) for adequate gear protection as well as good water separation performance so that contaminants can be removed by the oil purifiers. Careful selection of the base oils and additives and formulation expertise are required to provide the required performance.

Usually, system oils are of the SAE 30 viscosity grade, although SAE 40 oils may occasionally be used.

F. Natural Gas Fired Engines

Reciprocating, spark-ignited internal combustion engines burning natural gas (methane) are composed of a wide variety of configurations, designs, and applications. These engines can be two-stroke cycle, four-stroke cycle, stoichiometric, lean-burn, naturally aspirated, or turbocharged, and they operate over a wide range of loads and speeds. Speeds range from as little as 200 rpm to more than 3000 rpm, and range in size from 100 hp to well over 20,000 hp. The lubrication needs of these engines vary considerably based on the designs, applications, operating conditions, and fuel quality used to fire these engines. The application range includes mainline gas compression, field gathering gas compression, combined heating and power, and driving pumps. It is increasingly common to use gas engines at landfill or biogas generation sites to convert methane into power. There is also increased use of natural gas fuel in the transportation sector, for example, light and heavy-duty trucks, buses, railroad, and marine.

Low-speed gas engines are designed to operate at speeds below 500 rpm, and high-speed gas engines are designed to operate above 900 rpm. All engines designed to operate at speeds between 500 and 900 rpm will be defined as medium-speed engines. As an example, most high-speed gas engines are designed to operate at speeds of 1000 rpm or higher but can satisfactorily operate at speeds below 900 rpm. These engines are still defined as high-speed engines. The definitions of speed used here will vary slightly depending on industry sources but will not affect the materials covered in this section.

As far as selecting lubricants for gas engines is concerned, there are no industry standards or widely accepted specifications to define performance requirements. Although there are specifications for internal combustion engines using gasoline and diesel fuel, these specifications do not apply to natural gas fueled engines. Engines operating on gaseous fuels, other than converted automotive engines using LPG or propane, require lubricating oils designed and formulated to meet the unique requirements of the gas engine. Most gas engine OEMs have their own lubrication specifications. Both base stocks and additive combinations are critical in balancing the performance needs of these engines.

Most lubricating oils used in gas engines today were developed specifically for this type of service. Most contain dispersants to control varnish type deposits resulting from oxidation and nitration. Other additives include detergents, antiwear, oxidation inhibitors, corrosion inhibitors, and metal deactivators. Some engine manufacturers recommend only oils containing ashless additives (no metallo-organic detergents), whereas others recommend the use of oils containing metallo-organic oxidation and corrosion inhibitors in combination with ashless dispersants. Still, others prefer metallo-organic detergents to be included in the formulations. The amount of metallo-organic detergent required, as measured by sulfated ash content, is an indication of the detergent level as is TBN. The requirements can vary considerably by engine manufacturer, engine design, fuel quality, and operating conditions. Where clean, dry natural gas is burned as the fuel, the purpose for including ash-containing additives in the formulations is to control valve and valve seat wear in four-cycle engines and to neutralize the acids formed during the combustion process. The residue, from burning the ash-containing additives (mainly detergents) during combustion, produces a solid lubricant to help protect the valve and seat surfaces. Depending on such factors as metallurgy and operating conditions, varying the amount of ash in the oil, and the residue it produces during combustion has been found to be effective at controlling wear in different engines. However, using oils with too high an ash level can have negative consequences on engine performance that includes the formation of excessive ash deposits in the combustion chamber, which can increase the potential for detonation. Selecting the optimum oil for an application requires balancing many factors such as engine makes and models with operating conditions and the fuel qualities. As the result of the various needs of the different engines, premium gas engine oils can be classified as follows:

> **Ashless oils**: These oils have sulfated ash levels of 0.00% and contain ashless inhibitors (oils below 0.11% sulfated ash are considered to be in the ashless oils category).

Low ash: These oils have sulfated ash levels between 0.4% and 0.6%. They contain metallo-organic detergents and ashless dispersants in combination with other inhibitors.

High ash oils: These are oils with sulfated ash levels exceeding 0.6%. These oils contain higher levels of metallic detergents.

Landfill gas oils: These are a specially formulated class of gas engine oils to handle the unique requirements, and often severe engine conditions, resulting from burning landfill gas in the engines. These oils should have good corrosion resistance and the ability to deal with deposits stemming from siloxanes. Ash levels typically range from 0.50% to >1.00%.

The oil viscosity typically recommended is an SAE 40 grade. However, some engine manufacturers recommend SAE 30 grades for their engines. An SAE 30 grade or a multiviscosity oil (e.g., 15W-40) can be used where low temperatures are experienced. An SAE 30 grade oil can also provide energy efficiency benefits when compared to an SAE 40 grade oil. Where extreme temperatures exist, fully synthetic gas engine oils or gas engine oils blended with combinations of Group II/Group IV base oils will provide the best protection and most reliable service. These oils may also provide benefits in terms of improved fuel efficiency relative to SAE 40 grade gas engine oil products. Applications such as remotely started and stopped engines that are subjected to low oil temperatures at startup will benefit from the use of synthetic gas engine oils. Oil and filter change intervals in many applications can be extended with synthetic gas engine oils because of their higher level of oxidation and thermal stability.

1. **Two-Stroke Gas Engines**

Large slow-speed two-stroke gas engines, such as those used for gas compression in mainline transmission stations, generally operate at about 300 rpm. These engines depend on the use of high-quality low-ash oils for maximum performance. The selection of proper base stocks for two-stroke cycle gas engines is at least as critical to performance as is the selection of additives. The oil from the crankcase is generally used to lubricate the power cylinders in these engines. Hence, lower quality lubricants, oils with too high an ash level, or excessive power cylinder oil feed rates can cause engine operating problems. These problems consist of spark plug fouling, combustion chamber deposits, ring sticking, and the plugging of exhaust and intake ports, which in turn, cause losses in power and efficiency. In some instances, cylinder and piston scuffing can occur because of improper selection of the lubricant.

The oils recommended for these large two-stroke gas engines are predominately SAE 40, ashless (0.00% to 0.11% ash) oils formulated with high-quality base stocks that provide very low carbon formation characteristics. Multiviscosity oils are not generally recommended for the large two-stroke engines but have been used in some instances. Some smaller two-stroke engines, used in field gas gathering operations subject to low temperatures, may benefit from the use of multiviscosity gas engine oils.

2. **Four-Stroke Low- to Medium-Speed Gas Engines**

The low-speed four-stroke gas engines are used in mainline transmission stations for gas compression. The medium-speed engines are used for mainline transmission as well as to drive electric generators for stationary power. Occasionally, these medium-speed engines are also used in packaged compressor units in field gas gathering. The fuel used is clean natural gas with the exception of gas gathering applications where wellhead gas is used. The wellhead gas is a generally clean dry fuel, but fuel quality needs to be determined for these applications before selecting a lubricant. The oils recommended for the engines burning clean natural gas are low ash SAE 40 or SAE 30 grade oils, depending on the particular engine manufacturer. Engines with higher BMEP (brake mean effective pressure) ratings require higher levels of antiscuff protection. Where valve life is a concern, higher ash oils should be considered.

Internal Combustion Engines

3. Four-Stroke High-Speed Gas Engines

The four-stroke high-speed gas engines are generally used to drive compressors in field gas gathering operations. These engines are also used for power generation in applications such as cogeneration. Sometimes, the fuel is a clean natural gas, whereas at other times the fuel can contain liquids and sulfur compounds. Landfill gas can contain high levels of chloride compounds. In this application, conventional low ash oils will not provide adequate protection against the corrosive materials that are formed during the combustion process where chlorides are present. The quality and chloride content of the fuel is a real key to making the proper lubricant selection. Once the fuel considerations are addressed, the next concern is in protecting valve life in the high-speed engines because low-emissions engines have significantly lower levels of oil consumption that can impact valves. Because valves depend on a certain amount of oil ash residue to provide a solid film between the valve faces and seats, higher ash level oils may be required to offset the reduced combustion ash levels caused by the lower oil consumption. The oils generally recommended for modern design (primarily lean-burn engine type) four-stroke cycle high-speed gas engines are SAE 40 grade oils with low (0.4–0.6%) ash levels. However, in some cases, the level of ash required can vary by engine manufacturer and even by models by the same manufacturer. SAE 30 grade oils can be used for lower temperature applications and although monograde oils are preferred, multiviscosity oils can be used in some engines.

4. Gas Engine Oil Selection

The following is a summary of some of the items that should be considered when selecting an oil for natural gas engine applications:

- Engine manufacturers recommendations
- Oil suppliers recommendations
- Makes and models of engines
- Fuel used
- Operating conditions
  - Engine loads
  - Ambient conditions
  - Oil and water temperatures
  - Engine speeds
- Engine condition
- Past experience
- Type of service (continuous or frequent start–stops)
- Add-on equipment
  - Catalytic converters
  - Ebullient cooling
  - Lean burn conversions
  - Stratified charge systems
- Oil consumption rates
- Desired oil drain interval

5. Dual Fuel Engines

To some extent, the lubricating oil in a dual fuel engine is subjected to the deteriorating influences found in both diesel and gas engines. The diesel fuel burned as a pilot fuel tends to produce soot and varnish deposits, whereas the gas fuel, combined with high operating temperatures, tends to cause oxidation and nitration of the oil. Although these conditions are not necessarily conflicting, there is some indication that oils specially developed for the service may provide superior engine cleanliness with both diesel and gas engine oils. In some applications, where 5% pilot diesel fuel is used with the balance of fuel being clean natural gas, these engines can use gas engine oils with good results. However, some require that their converted medium-speed marine diesel engines use higher ash marine engine oils for their dual fuel engines. SAE 30 or 40 oils are used as crankcase oils, with SAE 50 or even higher viscosity oils preferred for cylinder lubrication of engines with separate cylinder lubricators. It is important to refer to specific OEM lubrication specifications/guidelines when selecting the engine oil for dual fuel engines.

G. Outboard Marine Engines

The majority of outboard motors that were used for recreational boats were historically two-stroke gasoline engines. In recent years, this has changed with approximately 70% of new outboard motor sales being four-stroke engines. The four-stroke outboards are claimed to produce lower emissions, have better fuel economy, and are considered smoother and quieter than two-stroke outboards. Two-stroke outboard claims include having better acceleration, being lighter in weight, are easier to repair, and are often less expensive to purchase.

Because of the nature of two-stroke marine outboard engines, two-cycle oil lubricates the engine parts as it passes through the engine and is then burned along with the fuel. Two-cycle oil is either physically mixed with the fuel or, in the case of direct fuel injection, is combined with the fuel in the combustion chamber. This is in contrast to four-stroke engines, which have oil sumps that circulate the oil by pumping it throughout the engines. In two-stroke outboards, some oil is always carried into the combustion chambers where, when burned, it can form deposits. Excessive deposit buildup can affect combustion and obstruct the ports causing loss of power. Metallo-organic additives in the oil may aggravate this deposit problem and as a result, most outboard engine oils are formulated with ashless dispersants. A corrosion inhibitor to provide additional protection against rusting during shutdown periods is usually included. The base stocks are selected to minimize carbonaceous deposits in the combustion chambers.

The National Marine Manufacturers Association (NMMA) is the governing body that develops the standard and certifies outboard engine oils. The TC-W3® standard is a performance-based qualification program for two-cycle outboard oils. The requirements include various bench tests for fluidity, miscibility, rust, compatibility, as well as engine tests to evaluate the prevention of ring sticking and carbon buildup on pistons and other engine parts. The NMMA has developed the FC-W® standard for four-stroke outboard oils. The performance criteria for the FC-W standard include bench tests for viscosity, corrosion, filter plugging, foaming, and aeration. In addition, the oil must successfully pass a 100-h general performance engine test.

H. Railroad Engines

Diesel engines used in railroad locomotives, because of space limitations, often have high specific ratings in order to obtain high power output from a relatively small engine. Combined with restrictions on cooling water capacity, this results in both crankcase and cylinder temperatures being high. Long idling periods and rapid speed and load changes also contribute to severe operating conditions for the lubricating oil.

Both two and four-stroke cycle engines are used in this service, and cylinder lubrication is supplied by oil throw from the crankcase.

U.S. railroad engine manufacturers recommend oils developed specifically for their engines. These oils are usually SAE 40 grades although SAE 20W-40 multigrade oil are receiving greater acceptance. Some builders have special requirements because of specific design issues. The engine manufacturer EMD, for example, recommends zinc-free oils because of the adverse effects of the phosphorus portion of the zinc compounds on silver bearings that have been used in their engines.

I. Motorcycles

Both two- and four-stroke gasoline engines are used in motorcycles. The worldwide population of two-stroke motorcycles is much greater than that of four-stroke motorcycles, but global environmental emission regulation has seen a shift toward four-stroke motorcycles. The engines used in motorcycles have significant differences than those used in passenger vehicles. Motorcycle engines are air cooled versus liquid cooled, run at higher temperatures and higher speeds, have greater power density, and typically require engine oils to lubricate clutches and gears in addition to the engine. These and other differences highlight the need for specially formulated oils for motorcycles.

1. Two-Stroke Motorcycle Oils

The most widely recognized industry standards for two-cycle motorcycle oils are JASO M345, ISO 13738, API TC, and SAE J2116.

   a. The JASO M345 standard recognizes three levels of low ash two-cycle engine oil (FB, FC, and FD performance levels). Lubricant performance is assessed against performance criteria such as lubricity, detergency, initial torque, exhaust system blocking, and smoke. The FC and FD grades are the more common as they have stringent smoke requirements that address concerns from major cities particularly in Asia.
   b. The ISO 13738 standard has established performance criteria, similar to JASO M345, around lubricity, detergency, initial torque, exhaust system blocking, and smoke but has an additional piston skirt deposit criteria. The ISO 13738 standard includes three grades (EGB, EGC, and EGD) that correspond to the JASO M345 grades. The JASO FD and ISO 13738 EGD grades both have greater detergency criteria.
   c. The API TC two-cycle engine oil classification was established for air-cooled, high-performance engines typically between 200 and 500 cc. Two-cycle engine oils designed for API Classification TC address ring-sticking, preignition, and cylinder scuffing.
   d. SAE J2116 has established performance criteria around ring sticking, varnish, preignition, exhaust system blocking, and scuffing.

Most two-cycle oils have typical viscosity grades of SAE 20, 30, or 40.

2. Four-Stroke Motorcycle Oils

For four-cycle motorcycle oils, JASO T403 and ISO 24254 are the recognized industry standards.

   a. The JASO T403 standard introduced the MA and MB specifications to distinguish between friction-modified and nonfriction-modified engine oils. Most four-stroke motorcycles with wet clutches use a JASO MA oil. The JASO MB oil has the lowest friction specification. The JASO MA and MB oils also typically meet one of the following API Service Categories: SG, SH, SL, SM, or SN.
   b. The ISO 24254 standard has established EMA and EMB categories that are similar to JASO MA and MB categories. This standard also states that these oils must be of a quality level that is equivalent to one or more of the following specifications: API SJ, SL, SM; ILSAC GF-5; or ACEA A1/B1, A3/B3, A5/B5, C2, C3.

For four-cycle motorcycle oils, multigrade oils such as SAE 20W-50 or 10W-40 are the most commonly used, but other grades may also be used. OEM owner's manuals are the best source for oil recommendations.

BIBLIOGRAPHY

API 1509. 2014. Engine Oil Licensing and Certification System, 17th Edition, Addendum 1. American Petroleum Institute. Washington, DC.

ASTM D4684-14. 2014. Standard Test Method for Determination of Yield Stress and Apparent Viscosity of Engine Oils at Low Temperature, ASTM International, West Conshohocken, PA. http://www.astm.org.

ASTM D5293-14. 2014. Standard Test Method for Apparent Viscosity of Engine Oils and Base Stocks Between −5°C and −35°C Using Cold-Cranking Simulator, ASTM International, West Conshohocken, PA. http://www.astm.org/.

European Automobile Manufacturers Association. 2012. ACEA European Oil Sequences 2012. Brussels. http://www.acea.be/publications/article/acea-oil-sequences-2012.

ISO 13738:2011. Lubricants, industrial oils and related products (class L)—Family E (internal combustion engine oils)—Specifications for two-stroke-cycle gasoline engine oils (categories EGB, EGC and EGD). International Organization for Standardization. Geneva.

ISO 24254:2007. Lubricants, industrial oils and related products (class L)—Family E (internal combustion engine oils)—Specifications for oils for use in four-stroke cycle motorcycle gasoline engines and associated drivetrains (categories EMA and EMB). International Organization for Standardization. Geneva.

JASO M345:2004. Two Cycle Gasoline Engine Oil Performance Classification (JASO M345) Implementation Manual. Japanese Automobile Standards Organization. Japan.

JASO T 903:2011. Motorcycle Four Cycle Gasoline Engine Oil (JASO T 903:2011) Application Manual. Japanese Automobile Standards Organization. Japan.

Mobil Oil Corporation. 1974. Mobil Technical Bulletin: Engine Oil Specifications and Tests—Significance and Limitations. New York.

Mobil Oil Corporation. 1969. Diesel Engine Lubrication in Stationary Service. New York.

Mobil Oil Corporation. 1969. Mobil Technical Bulletin: Diesel Engine Operation. New York.

NMMA. 2009. TC-W3 Two Stroke Oil Product Approval System. National Marine Manufacturers Association. Chicago.

NMMA. 2009. FC-W Four-Stroke Cycle, Water-Cooled Gasoline Engine Lubricant Product Approval System. National Marine Manufacturers Association. Chicago.

SAE J183. 2013. Engine Oil Performance and Engine Service Classification (Other than "Energy Conserving"). SAE International. Warrendale, PA.

SAE J300. 2015. Engine Oil Viscosity Classification. SAE International. Warrendale, PA.

SAE J1899. 2011. Lubricating Oil, Aircraft Piston Engine (Ashless Dispersant). SAE International. Warrendale, PA.

SAE J1966. 2011. Lubricating Oils, Aircraft Piston Engine (Non-Dispersant Mineral Oil). SAE International. Warrendale, PA.

SAE J2116. 2003. Two-Stroke-Cycle Gasoline Engine Lubricants Performance and Service Classification. SAE International. Warrendale, PA.

11 Automotive Transmissions and Drive Trains

In automotive equipment, the power developed by the engine must be transmitted to the drive wheels to propel the vehicle. This is accomplished by the power train, the elements of which vary from application to application. In probably its simplest form, a bicycle equipped with an auxiliary engine, the power train may consist only of a belt drive with an idler pulley that can be actuated to engage or disengage the drive. At the other extreme, the power train may consist of some combination of a clutch or coupling, a transmission, transfer case, interaxle differential, constant velocity (CV) joints, drive shafts, front and rear differentials—possibly in tandem at the rear, driving axles—and planetary reducers at the wheel end of the axles. Almost any arrangement between these two extremes will be encountered in most types of automotive equipment.

In addition to transmitting the power from the engine to the drive wheels, the power train performs several other functions. For example, the power train provides:

1. Mechanism for engaging and disengaging the drive so the vehicle can be started and stopped with the engine running
2. Torque multiplication (speed reduction) so that sufficient torque is available to start from rest, accelerate, climb hills, and pull through soft ground
3. Mechanism for reversal of direction
4. Mechanism that allows one wheel to be driven at a higher or lower speed than the other when the vehicle is negotiating curves and turns
5. Change in direction of torque and power flow to couple a longitudinally mounted engine to the traverse drive axle (not needed in vehicles with traversely mounted engines)

The components of power trains can be conveniently considered starting from the engine and progressing toward the drive axle.

I. CLUTCHES

In order to engage and disengage the various gear ratios in a mechanical transmission, provision must be made to disconnect the engine from the power train. This is accomplished with a clutch.

Most road vehicles with mechanical transmissions are now equipped with a single plate, dry disk clutch. A machined surface on the flywheel serves as the driving member. The driven member is usually a disk with a splined hub that is free to slide lengthwise along the splines of the transmission input shaft (clutch shaft), but which drives the input shaft through these same splines. About 65% of the clutch plate area is faced on both sides with friction material. A typical passenger car clutch plate, for example, with a 15-cm (10 in) diameter, would have a band of friction material about 6 cm (4 in) wide along its outer portion. When the clutch is engaged, the pressure plate (driven member) is clamped against the driving member by a diaphragm-type spring or an arrangement of coil springs (Figure 11.1). The entire mechanism is usually called the clutch plate and cover assembly. To disengage the clutch, pressure is applied through mechanical linkage or a hydraulic system on the end of the throw-out fork that pivots on a support in the clutch housing. When release pressure is applied, the fork transmits the force to the release levers in the cover assembly, which compresses the springs and retracts the pressure plate from the driven member.

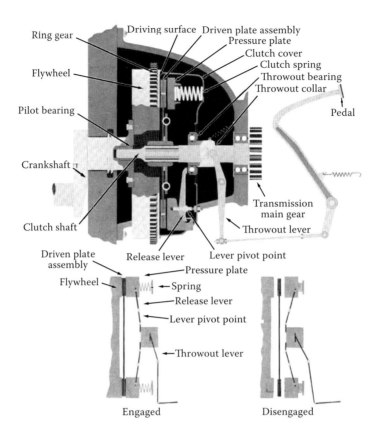

FIGURE 11.1 Single-plate, dry disk clutch.

On models with hydraulic clutch activator, the system reservoir should be kept filled to the correct level. Conventional brake fluid is often specified; however, the manufacturer's recommendations should always be followed. The only lubricated part of this type of clutch is the throw-out bearing. In most cases, these bearings are "packed for life" on assembly and do not require periodic relubrication. In a few instances, these bearings are equipped with a fitting and require periodic lubrication, usually with multipurpose automotive grease. If greased, care must be exercised to avoid getting grease on clutch faces that will result in slippage and excessive heating. Also, various pivot points in the actuating linkage may require periodic lubrication, with either a small amount of engine oil or multipurpose automotive grease.

Multiple plate clutches, usually oil immersed or "wet" type, are used in some tractors and off-highway machines. With these clutches, the frictional properties of the fluid surrounding the clutch are extremely important if the clutch is to engage smoothly, resist slipping, and provide extended service life. These fluids are discussed in this chapter.

Single-plate, double-acting, dry disk clutches are used with some torque converter transmissions for buses and similar applications. Two driven members are used, one on each side of the driving member. When engaged in one direction, the clutch connects the drive to the torque converter, which in turn drives the input shaft of the transmission. When engaged in the other direction, it connects the drive to a through shaft that provides a mechanical drive to the transmission, either to the input or output shaft depending on the arrangement. Again, the only lubrication required is for the throw-out bearing, which is in such a location that it is lubricated by the torque converter fluid.

II. TRANSMISSIONS

The transmission has three primary functions:

1. Provide a method of disconnecting the power train from the engine so that the vehicle can be started and stopped with the engine running.
2. Provide torque multiplication for conditions where greater driving torque is required at the wheels than is available from the engine.
3. Provide a method of reversing the drive.

Since torque multiplication is accompanied by speed reduction at the output end, the transmission also permits the operator to select different travel speeds for any given engine speed.

There may be more variation in transmission design and application than in any other automotive component. For ease of discussion, transmissions can be considered as mechanical, automatic, semiautomatic, or hydrostatic.

Today's transmissions are lighter in weight, have more speeds, have improved controls, and have higher torque capacities. In many cases, particularly for passenger cars, they are designed to work with smaller engines and provide higher fuel economy. Transmission manufacturers are challenged with balancing the need for better fuel economy with comfort and cost.

A. Mechanical Transmissions

A mechanical transmission is an arrangement of gears, shafts, and bearings in a closed housing such that the operator can select and engage sets of gears that give different speed ratios between the input and output shafts. In most cases, a set of gears that can be engaged to drive the output shaft in the opposite direction is also included. For a constant power, torque increases as speed is decreased; the transmission provides a series of steps of torque multiplication.

In an elementary sliding element transmission (Figure 11.2), one of each pair of gears is splined onto its shaft in such a way that it can be moved along it by a shift fork into and out of mesh with its mating gear. Drive is from the input shaft (also called the clutch shaft) through the main gear to the countershaft. The output shaft ends in a pilot bearing in the main gear, which is free to revolve at a different speed or in a different direction than the input shaft. Thus, the output shaft is driven from the countershaft by whichever pair of gears is engaged or directly from the main gear if the direct drive gear is engaged with the internal gear in the main gear.

Sliding element transmissions are typically used in only low speed applications, such as tractors. In this application, the clutch must be disengaged and the vehicle must be at a complete stop before the gears can be engaged or the gear ratio changed. For other applications, syncromesh or synchronized transmissions are used. In this type of transmission, all gears are always in mesh, except the reverse gear. One of each pair of gears is free to revolve on its shaft unless locked to it by a clutching mechanism called a synchronizer as shown in Figure 11.2. The synchronizer, which is keyed or splined to the shaft, consists of a friction clutch and a dog clutch. As the shift fork moves the synchronizer toward the gear, the friction cones make contact first to bring the shaft to the same rotational speed as the gear. The outer rim of the clutch gear then slides over its hub, causing a set of internal teeth to engage with a set of teeth (dogs) on the side of the gear. This then provides a positive mechanical connection between the gear and shaft.

Usually, synchronizers are equipped with a blocking (also called baulking) system to prevent engagement of the dog clutches until the gear and shaft speeds are fully synchronized. Generally, this is a spring-loaded mechanism, which keeps the teeth on the synchronizer from lining up with the teeth on the gear as long as there is any slip in the friction clutch.

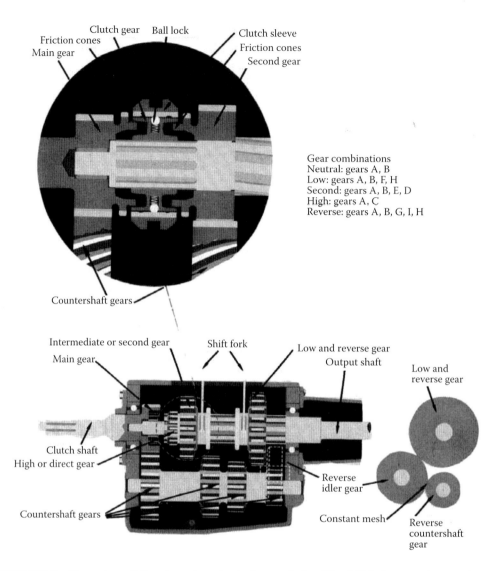

FIGURE 11.2 Elementary sliding element mechanical transmission. The shift forks move the gears into and out of mesh along with splined main shaft to change output gear ratio.

Mechanical transmissions are built with up to about six gear ratios. Where more ratios are required, as in the case of heavy trucks equipped with diesel engines, they are usually obtained by using a two- or three-speed auxiliary transmission mounted behind the main transmission. With this arrangement, for each ratio in the main transmission there are two or three ratios in the auxiliary; so, for example, a four-speed main transmission with three-speed auxiliary becomes a 12-speed transmission. Heavy-duty transmissions are often built with twin countershafts to decrease gear tooth loading.

Sliding element transmissions are built with straight spur gears. Synchromesh transmissions for over-the-road vehicles are usually built with helical gears, both because they provide greater load-carrying capacity and because they operate more quietly. Transmission for off-highway equipment may be built with either type of gearing.

B. Automatic Transmissions

Early passenger car automatic transmissions were built with a fluid coupling and a hydraulically operated power shift gearbox. The fluid coupling permitted enough slip with the engine idling so the vehicle could be stopped with the gears engaged. Power transfer efficiency was also good. However, since a fluid coupling does not multiply torque, the gearbox required four forward speeds to provide a smooth progression of gear ratios and this resulted in considerable complexity. As efficient hydraulic torque converters were developed, they replaced the fluid couplings. In the late 1960s through the early 1980s, the gearboxes were standardized on an arrangement with three forward speeds and a reverse speed. Current production conventional hydraulic automatic transmissions today have four to nine dedicated speeds including an integrated overdrive gear. Overdrive allows a vehicle to cruise at a sustained highway speed with reduced engine speed leading to improved fuel economy with less engine noise. Hydraulic automatic transmissions are among the most complex mechanical components found on a car. A fast-growing alternative to conventional hydraulic automatic transmissions for passenger vehicles is the continuously variable automatic transmission (CVT). This type of transmission has gained significant market share primarily because of the benefit of improved fuel economy.

Truck automatic transmissions may have more forward speeds, and may also have an arrangement to lock out or bypass the torque converter when the transmission is in any gear except first or reverse. Transit coach transmissions may have only a drive through the torque converter or direct drive. All of these transmissions are sometimes referred to as "hydrokinetic" transmissions, because engine power is transmitted by the kinetic energy of the fluid flowing in the torque converter.

1. Torque Converters

The simplest single-stage torque converter consists of three elements: a centrifugal pump, a set of reaction blades called a stator, and a hydraulic turbine (Figure 11.3). These three elements

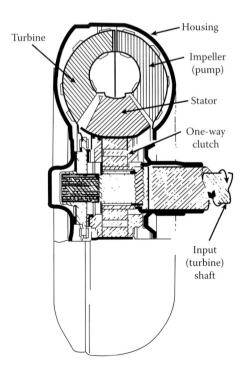

FIGURE 11.3 Three-element torque converter.

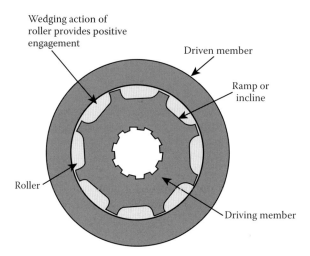

FIGURE 11.4 Overrunning clutch. When a clockwise force is applied to the movable member, the rollers wedge on the ramps to prevent rotation. When the force is released, the rollers move back down the ramps permitting the movable member to rotate freely in the clockwise direction.

are installed inside a case filled with a hydraulic fluid. The pump is driven by the engine, and the turbine drives the input shaft of the gear box. The pump blades are shaped so that they discharge the fluid at high speed and in the correct direction to drive the turbine. As the fluid flows out of the turbine, it strikes the fixed stator blades and is redirected into the inlet side of the pump, where any velocity it still retains is added to the velocity imparted to the fluid by the pump. With this arrangement, most of the power delivered to the pump is available to drive the turbine (some power is lost because of fluid friction), and as long as the turbine is running at a lower speed than the pump, torque multiplication will occur. Most single-stage torque converters are designed for maximum torque multiplication ratio of slightly more than 2:1, which—because of maximum load conditions—occurs in "stall" conditions when the turbine is stationary.

Torque converters can be built with more than one stage—that is, additional pumps and stators in pairs to give greater torque multiplication. However, this is done less frequently, with the majority of units built being single stage.

A torque converter does not transmit power very efficiently when the speed of the turbine reaches approximately the speed of the pump. To improve this power transfer efficiency, the stator is mounted on a one-way or overrunning clutch (Figure 11.4). With this addition, when the coupling stage is reached, the stator revolves (free wheels) with the turbine and the whole assembly performs as a fluid coupling.

The efficiency of power transfer through the converter when it is operating normally in the coupling phase is generally considered satisfactory. However, to gain an advantage in fuel economy, torque converter "lockup" devices may be used to eliminate all slippage in the coupling phase. The lockup mechanisms are designed to be effective when the transmissions are in direct drive and converter torque multiplication is not required.

At low engine speeds, the torque transmitted by a torque converter is low enough that it is insufficient to move the vehicle or will cause only a small amount of creep. This feature permits the vehicle to be stopped without disconnecting the engine from the power train.

2. Planetary Gears

Passenger car hydraulic automatic transmissions are built with planetary and neutral gearsets to provide the additional torque multiplication, reversal of direction, and neutral. This type of gearing

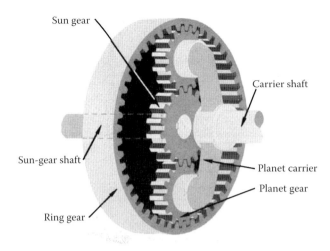

FIGURE 11.5 Planetary gearset. In this arrangement, only two planetary gearsets are used. Automatic transmissions commonly have either three or four planetary gearsets.

is also used in truck and heavy equipment automatic transmissions. Planetary gearsets have several advantages for these applications:

1. Ratio changes and reversal of direction can be accomplished through these constant mesh gears by locking or unlocking various elements of the gearset.
2. The gears are coaxial; thus, they provide a compact arrangement.
3. In addition, good load-carrying ability can be obtained from a relatively small gearset. The coaxial construction of planetary gears carries most of the operating loads. This allows the use of thin, lightweight aluminum die cast housings because they do not have to withstand extreme mechanical loads. A simple planetary gearset is shown in Figure 11.5.

A single planetary gearset can provide direct drive, two stages of forward speed reduction, reverse, and can also be operated in the overdrive phase. By locking the sun gear and using the planet carrier as input, the annular gear output rotation speed is increased. A simple three-speed automatic transmission is shown in Figure 11.6 for illustration purposes of the basic transmission components even though higher speed transmissions are used today.

3. Transmission

The gearbox of an automatic transmission is a "power shift" gearbox. That is, it can be engaged, or the gear ratio changed, while engine power transmitted by the torque converter is being continuously applied to the input shaft. These operations are performed by engaging and disengaging clutches in the drive lines to various planetary elements and by applying and releasing brakes (sometimes called "bands") to lock or unlock elements. The clutches and brakes are operated by hydraulic servomechanisms, which are controlled by a complex valve arrangement. Hydraulic pressure to operate the servos is provided by an auxiliary pump, usually of the internal–external gear type, mounted in the front end of the gearbox.

In operation, the operator selects a driving range, and the gear ratio changes are made automatically within that range. Basically, speed-sensing and load-sensing devices on the output shaft of the gearbox determine these shifts. Throttle position and, in some designs, engine vacuum is used to modulate the shift speeds. The farther the throttle is depressed, the higher the speed that shifts will occur. Normally, a forced downshift (sometimes referred to as a passing gear) is provided so the operator can downshift the transmission for additional acceleration by "flooring" the accelerator pedal.

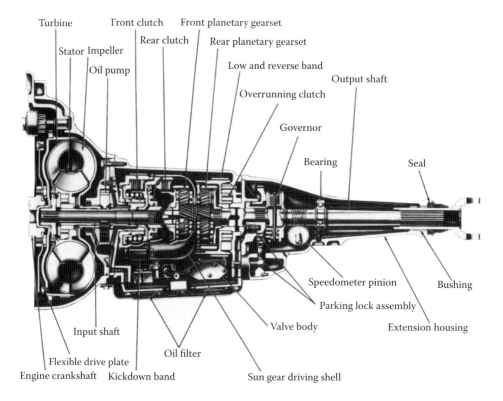

FIGURE 11.6 Cutaway view of a simple three-speed automatic transmission.

C. CVT

CVT is much more simple in design in terms of having fewer moving parts compared to the conventional hydraulic automatic transmission. The CVT uses two variable-diameter cone-shaped pulleys with a metal belt between them instead of the myriad of gears used in a hydraulic transmission. In a CVT, the transmission ratio can be varied by changing the diameter of the pulley, whereas the conventional transmission changes ratios by shifting gears in stages. The CVT continuously varies the ratio, allowing the engine to operate at peak power. This variability results in improved fuel economy compared to hydraulic transmissions and hence their gaining popularity. One disadvantage is that CVTs do not have the torque capacity of conventional transmissions, which limits their application in heavy-duty service.

D. Semiautomatic Transmissions

A number of arrangements are used to reduce the operator effort required to select and engage gear ratios. Clutch operation may be made automatic, or the gearbox arranged so that a clutch is not needed. Gear ratio selection and engagement are still performed by the operator; thus, these arrangements are considered semiautomatic.

Electronically shifted manual transmissions are one form of semiautomatic transmission where the clutch is controlled by an onboard computer that takes signals from a paddle shifter. A number of cars have used this arrangement to enhance the driving experience.

Dual clutch transmission (DCT) is another form of electronically shifted manual transmission that is an alternative to an automatic transmission. DCT can be considered as two manual transmissions in one housing. This arrangement uses two clutches for even and odd gearsets. Computers, solenoids, and hydraulics do the shifting instead of a clutch pedal, making it an automated manual transmission. Improved fuel economy and smooth shifting are some of the claims coming from proponents of DCT designs. The use of these DCTs in some passenger cars is growing globally from its initial European application.

In another type of older semiautomatic transmission, a torque converter is installed ahead of an electropneumatic clutch and three-speed, manually shifted gearbox. Electrical contacts in the shift lever knob are closed when a slight downward pressure is applied to the knob. This completes the circuit to the vacuum valve, allowing vacuum from a vacuum storage tank to act on the servo and disengage the clutch. Gearshifts can then be made in the normal manner and when the knob is released, the clutch engages again. The torque converter provides enough multiplication so that a three-speed gearbox is adequate with a small engine. It also eliminates the need for some shifting and helps to cushion shocks that may occur when the clutch engages at the end of a shift.

Some farm and construction machines are equipped with "power shift" transmissions. The gearboxes of these transmissions are somewhat similar in principle to synchromesh transmissions, except that the synchronizers are replaced by hydraulically operated, oil-immersed clutches. The hydraulic circuit is, in turn, controlled by a shift lever. In some cases, two levers are used, one for gear ratio selection and one for direct shifting from forward to reverse. No clutch is required as the gears can be engaged and disengaged under power.

There is another type of arrangement, in which there is a power shift gearbox in series with a conventional clutch and a conventional gearbox. The clutch and conventional gearbox are used to select a range. Shifts within that range may then be made with the power shift gearbox without using the clutch.

E. Hydrostatic Transmissions

Hydrostatic transmissions are now used on many self-propelled harvesting machines and garden tractors, as well as significant numbers of large tractors and construction machines. These types of drives are also used in many small lawn and garden tractors, and on-highway truck applications have also been developed. In the sense that no clutch is used and no gear shifting is involved, this type of transmission could be called an "automatic," but in all other respects the hydrostatic transmission has no similarity to the hydrokinetic automatic transmission.

Hydrokinetic transmission transfers power from the engine to the gearbox by first converting it into kinetic energy of a fluid in the pump. The kinetic energy in the fluid is then converted back to mechanical energy in the turbine. In the hydrostatic system, engine power is converted into static pressure of a fluid in the pump. This static pressure then acts on a hydraulic motor to produce the output. Although the fluid actually moves through the closed circuit between the pump and motor, energy is transferred primarily by the static pressure rather than the kinetic energy of the moving fluid. The relatively incompressible fluid acts much like a solid link between the pump and motor.

The pump in a hydrostatic system is of the positive displacement type. It may be either constant or variable displacement, but for mobile equipment applications, it is usually a variable displacement type. Axial piston pumps are the most common, although some radial piston pumps are used for small transmissions. In the variable displacement axial piston pump (Figure 11.7), the cylinder block and pistons are driven from the input shaft. Piston stroke, pump displacement, and direction of fluid flow are controlled by the reversible swash plate, which in this case is moved by a pair of balanced and opposed servo pistons. The servo pistons are, in turn, controlled by a speed control lever. On smaller units, where the forces acting on the swash plate are not as great, its position is controlled directly by the speed control lever. With radial piston pumps, a moveable guide ring is used to control piston stroke instead of a swash plate.

As shown in Figure 11.7, when the speed control lever is in neutral, the swash plate in the pump is perpendicular to the pistons and no pumping occurs. As the speed control lever is moved in one direction, the swash plate is tilted, the piston stroke is gradually increased, and fluid is pumped from

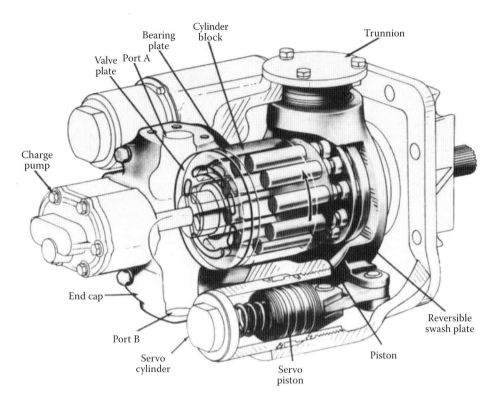

FIGURE 11.7 Variable displacement pump. The servo pistons tilt the reversible swash plate on the trunnion to vary the displacement from maximum in one direction through zero to maximum in the other direction. In many cases, motion in the reverse direction is limited so that the maximum reverse speed is one-half or less of the maximum forward speed.

one of the outlet ports. If the speed control lever is moved in the opposite direction, the swash plate is tilted in the opposite direction, and the piston stroke is moved 180° around the case. Fluid is then pumped from the other outlet port. In combination with a reversing motor, this permits a continuously variable range of speeds from full forward to full reverse with only the single lever for control.

The motor in a hydrostatic system can be any type of positive displacement hydraulic motor. Axial piston motors are usually used for larger drives, but are also used for some smaller drives. Both gear motors and radial piston motors are used for low-power drives. The motor is usually of the fixed displacement type, shown in Figure 11.8, but may be variable displacement. As noted, the motor is reversible with the direction of rotation dependent on the direction of flow in the closed loop circuit to the pump.

In addition to the pump and motor, connecting lines, relief valves, and a charge pump are required. The connecting lines may be passages where the pump and motor are in the same housing, or may be hoses where the motor is mounted away from the pump. The charge pump provides initial pressurization of the motor and replaces any fluid lost because of internal leakage. On small tractors it may also be used to supply fluid for remote hydraulic cylinders.

Various arrangements of the pump and motors are used. A variable displacement pump with a variable displacement motor may be used. Thus, the swash plates must be linked and synchronized so that the motor displacement decreases as the pump displacement is increased. Because motor displacement is maximum when pump displacement is low, motor speed will be low and torque will be high. Conversely, motor displacement will be at a minimum when pump displacement is at a maximum, so the maximum speed will be high. The arrangement gives high starting torque and the widest range of speeds for any given size of pump and motor.

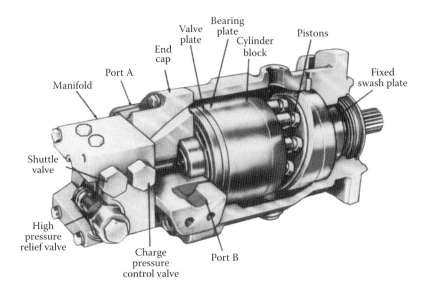

FIGURE 11.8 Fixed displacement pump. This axial piston motor has a fixed swash plate. Similar design motors are available with a movable swash plate.

Another variation has a variable displacement pump and a two-piston swash plate or guide ring on the motor. The latter is controlled by a range lever. In the low range, motor displacement is greater, thus the starting torque is higher and the maximum speed is lower.

Most drives in mobile-type equipment have a variable displacement pump in combination with a fixed displacement motor. This type of circuit gives a constant torque output, with the power output increasing as the pump displacement is increased.

The fact that the pump and motor do not need to be connected directly together permits considerable flexibility. The pump may be connected directly to the engine output shaft and motors located at the driving wheels. In another arrangement, two pumps and two motors may be used with each pump and motor driving one wheel. One wheel can then be driven forward with the other in neutral or reverse for spin turns.

One of the main disadvantages of hydrostatic drives is that they permit the operator to select any travel speed up to the maximum without varying the engine speed. The engine can be operated at governed speed to provide proper operating speed for elements such as the threshing section of a combine, but a full range of travel speeds is available to adjust to terrain or crop conditions. Operation is also considerably simplified.

F. Factors Affecting Lubrication

The differences in lubrication requirements of the various types of transmissions require separate consideration of the factors affecting lubrication. Transaxle units are discussed separately.

1. Mechanical Transmissions

The elements in mechanical transmissions requiring lubrication are the bearings, gears, and sliding elements in the synchronizers. Bearings may be either plain or rolling element. As noted, gears are usually either straight spur or helical, and gear loads are moderate to heavy. Normally, lubrication is by bath and splash, but some large transmissions may have integral pumps to circulate the lubricant.

Most mechanical transmissions are designed to be lubricated by fluid products. Soft or semifluid greases may be used in small units, such as the transmissions of some lawn and garden equipment and scooters.

Generally, the lubricant in a mechanical transmission is expected to remain in service for an extended period. Thus, the lubricant must have the chemical stability to resist oxidation and thickening under conditions of agitation and mixing with air. Operating temperatures may also be quite high. Plain bearings, thrust bearings, and synchronizer components are often composed of bronze or other copper alloys. Thermal degradation of the lubricant can result in the formation of materials that are corrosive to these components. Severe agitation also occurs; therefore, the lubricant must have good resistance to foaming.

A lubricant selected for mechanical transmissions must have adequate fluidity to permit immediate circulation and easy shifting when a vehicle is started in cold weather. At the same time, the lubricant must have high enough viscosity at operating temperature to maintain lubricating films and to cushion the gears so that acceptably quiet operation results. Synthetic lubricants address these needs and additionally provide fuel economy benefits that have led to their widespread acceptance.

A variety of different types of lubricants are recommended for manual transmissions, and original equipment manufacturer (OEM) recommendations should be followed. Straight mineral gear lubricants suitable for American Petroleum Institute (API) Service Gear Lubricant (GL)-1 had been used in the past for some manual transmissions but are no longer satisfactory for most passenger cars with API GL-1 being an inactive service category. This type of lubricant has also been used in some truck and tractor manual transmissions, but API Manual Transmission (MT)-1 is considered to be the preferred upgrade for these applications. Some manual transmission manufacturers will specify either API GL-4 or GL-5 quality gear lubricants, whereas other passenger car manufacturers will recommend the use of a specific automatic transmission fluid (ATF) or engine oil. With the wide variety of designs comes different lubrication requirements, and hence it is important to follow OEM recommendations. Manufacturers of farm and construction machines frequently install the transmission in a common sump with the final drive; the sump may also serve as the reservoir for the central hydraulic system on the machine. Special fluids designed for service as a combination heavy-duty gear lubricant and hydraulic fluid are usually required for these applications. It is important to check manufacturer's recommendations to ensure adherence to specific requirements.

2. Automatic Transmissions

In some installations, the torque converter is located in a separate housing with its own supply of hydraulic fluid. However, in most passenger car automatic transmissions, the torque converter and gearbox operate from a common fluid reservoir.

In a torque converter, the fluid serves mainly as a power transfer fluid. It also lubricates the bearings and transfers heat resulting from fluid friction and power losses to a cooler or to the transmission case for dissipation in the atmosphere. Power transfer efficiency increases with decreasing viscosity. Heat transfer efficiency also generally increases with decreasing viscosity. These factors suggest that a torque converter fluid should be designed with as low a viscosity as is practical. On the other hand, high operating temperatures and the need for long service life of the fluid require that it resist oxidation.

Where the torque converter operates from the same fluid supply as the gearbox, the physical characteristics of the fluid must be compromised in order to meet the lubrication requirements of the gearbox.

The fluid in the gearbox portion of an automatic transmission performs several functions:

1. It lubricates the gears and bearings of the planetary gearsets.
2. It serves as a hydraulic fluid in the control systems.
3. It controls the frictional characteristics of the oil-immersed clutches and brakes.
4. It provides a degree of cooling.

These functions must be performed under a variety of operating conditions that tend to make the service severe.

Automatic transmissions are expected to engage and shift properly at low temperatures when a vehicle is started in cold weather. In operation, oil temperatures of the order of 250–300°F (121–149°C) may be reached. Gear loads are relatively heavy, and the fluid is exposed to severe mechanical shearing both in the gears and in the hydraulic circuit. Some breathing of air occurs because of changes in temperature inside the unit, and this tends to promote oxidation of the fluid, particularly when operating temperatures are high. Where the gearbox operates on the same fluid as the torque converter, the severe churning in the torque converter tends to cause foaming. Seal compatibility of the fluid is also an important consideration.

To satisfy these requirements for ATFs, various highly specialized products have been developed. They are discussed in detail in this chapter.

3. Semiautomatic Transmissions

Since the type of semiautomatic transmissions discussed is a combination of torque converter and a mechanical transmission, the earlier discussions of the lubrication requirements of these units apply to it.

Power shift transmissions used in heavy equipment have lubrication requirements not unlike the gearbox section of automatic transmissions. Because of the higher torques transmitted, somewhat higher pressure may be required in the hydraulic system in order to obtain proper engagement of the clutches. This in turn may apply somewhat higher mechanical shear stresses to the fluid. Again, frictional characteristics of the fluid, and its compatibility with the clutch materials, are critical if the clutches are to engage smoothly and firmly.

4. Hydrostatic Transmissions

A hydrostatic drive is a high-pressure hydraulic system, so the basic fluid requirements correspond closely with those of industrial hydraulic systems. Good oxidation stability is required, as well as good resistance to foaming and good entrained air release. Antiwear properties are also required because operating pressures are usually in excess of 2500 psi (17.2 MPa).

In addition, the fact that hydrostatic drives must operate over a wide range of temperatures generally dictates that very high viscosity index (VI) fluids with good low-temperature fluidity must be used. Because the fluid is a major factor in proper sealing of the pump pistons, the high-temperature viscosity of the fluid is important, and excellent shear stability is required to maintain this viscosity in spite of the severe shearing that occurs in the pump and motor.

Many hydrostatic drives are operated from a common reservoir with the differential or final drive. In these cases, the fluid used must also provide satisfactory lubrication of the gears and bearings.

The hydrostatic drives used on garden tractors usually are designed to operate on ATFs. These fluids are readily available and generally provide the combination of performance characteristics required. ATFs may also be recommended for hydrostatic drives on larger machines. Engine oils, such as an Society of Automotive Engineers (SAE) 10W-30 viscosity, may also be recommended. Where a hydrostatic drive on a larger machine is operated from a common reservoir with other drive elements, the fluid recommended is usually one of the special fluids discussed later under the section "Multipurpose Tractor Fluids."

III. DRIVE SHAFTS AND UNIVERSAL JOINTS

Road vehicles have their wheels connected to the body and chassis through springs, but the engine and transmission are mounted directly on the chassis. Where a rigid axle is used and the springs flex, the position of the axle with respect to the engine and transmission changes. Thus, there must be provision in the power connection between the transmission and drive axle to accommodate these changes in angular contact. This is accomplished in the drive (propeller) shaft and universal joints.

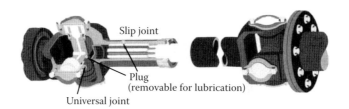

FIGURE 11.9 Cross-type universal joint. This type of joint, which is often called a cardan joint, may have needle roller bearings or plain bearings at the ends of the cross. Fittings may be used for relubrication, but the weight of the fitting can cause unbalance so a more common arrangement is to use special flush fittings, or plugs that must be removed and replaced with fittings during lubrication.

Typically, a drive shaft consists of a tubular shaft with a universal joint at each end. The universal joints allow for angular changes, and a slip joint at one end allows for changes in length. Some long drive shafts are made in two parts with a center support bearing to minimize whip and vibration. Three universal joints are then used.

Universal joints are usually of the cross or cardan type (Figure 11.9). Ball and trunnion type universal joints were used to some extent in the past. If there is any angular misalignment between the driving shaft and driven shaft, both of these types of joints will transmit rotation with fluctuating angular velocity. The amount of fluctuation increases with increasing misalignment, rising from about 7% at 15° misalignment to more than 50% at 40°. Because this fluctuation in velocity may be accompanied by vibration, the drive shafts, in which these joints are used, are designed for minimum misalignment. Another approach is to use what are called constant velocity (CV) joints.

One type of CV joint consists of two cross-type joints in a tandem assembly. Several other designs are available. CV joints are used to some extent in propeller shafts and are used as the drive axles of front engine, front wheel drive or rear engine, rear wheel drive vehicles, and conventional arrangement vehicles with independent rear suspension. In some cases, rather than CV joints, cross-type universal joints may be used at each end of the axle shaft, positioned specifically so that the changes in angular velocity cancel out.

A. Lubrication

In some designs, the transmission lubricant lubricates the universal joint and slip joint at the transmission end of the drive shaft. In other designs, the joints are lubricated with grease. Many joints are now packed for life on assembly and require servicing only if other repairs are required. Some joints require periodic disassembly and repacking, whereas some are equipped with a fitting or plug for periodic relubrication. The plug can be replaced with a special fitting while relubrication is being performed. In most cases, the grease used is a multipurpose automotive grease, sometimes with the addition of molybdenum disulfide.

B. Drive Axles

The drive axle usually contains one or more stages of gear reduction such as the gears in the differential that enable the wheels to be driven at different speeds. Also, in vehicles with a longitudinally mounted engine, the drive axle provides the gears with the capability to produce a 90° change in direction of power flow to couple the transverse axle shafts to the longitudinal transmission output shaft. In passenger cars and most trucks, the gear reduction in the drive axle is the final stage of gear reduction in the power train. In heavy trucks, and farm and construction equipment, additional stages of speed reduction, usually called final drives, may be used at the wheel ends of the drive axle.

In the most common rear wheel drive passenger car and light truck arrangement (Figure 11.10), the drive shaft couples through a universal joint to the front end of a pinion shaft of the differential.

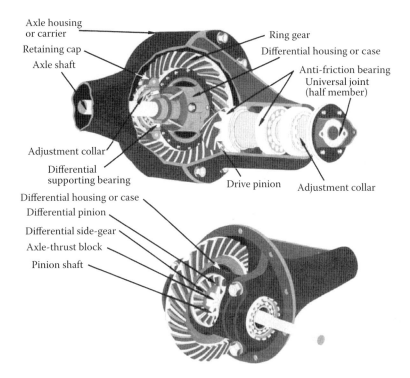

FIGURE 11.10 Hypoid-type drive axle. In the lower view, the differential case is cut away to show one of the side gears.

The pinion gear at the rear of this shaft meshes with the ring gear, which is bolted or riveted solidly to the differential case. The differential, in turn, drives the half axle shafts.

In most drive axles of this type, hypoid gears are used for the reduction stage. This type of gear design has a high load-carrying capacity in proportion to the size of the gears and operates quietly. In addition, the offset position of the centerline of the pinion, with respect to the centerline of the gear, permits the drive shaft to be located lower. This helps to lower the center of gravity of the vehicle and reduces the size of the tunnel through the floor of the passenger compartment that covers the drive shaft.

With front engine, front wheel drive or rear engine, rear wheel drive cars, spiral bevel gears are usually used for this reduction state. If the engine is mounted transversely with either of these arrangements, the 90° change in direction of power flow is not required, and straight spur or helical gears are used. In many heavy trucks, two stages of reduction are used in the drive axle. The first stage of reduction is usually a set of spiral bevel gears and the second stage either straight spur or helical gears. Some trucks are equipped with worm gears where a large total reduction is required.

C. Differential Action

As a vehicle turns, the wheels on the outside of the turn follow a longer path than those on the inside of the turn. To compensate for this and other differences in rolling distances between the driving wheels, a differential is used.

The principle of operation of a differential is shown in Figure 11.11. In this illustration, the arm represents the differential case, which is bolted to the ring gear. The differential pinion is free to rotate on its shaft, and meshes with the side gears, which drive the axle shafts. When the driving resistance is equal at both wheels, the pinion acts as a simple lever to drive the axle shafts at the same speed as the ring gear. If greater rolling resistance is encountered at one wheel than at the

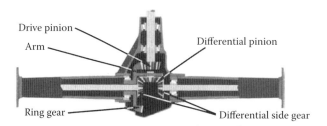

FIGURE 11.11 Elementary differential. In an actual differential at least two pinions are used and the arm is replaced by a case that more or less completely encloses the pinions and side gears.

other, the unbalanced reaction forces acting on the pinion will cause it to rotate on its shaft. The wheel encountering the least resistance will then be driven faster, and the wheel encountering the most resistance will be driven slower. The increase in speed of one wheel will be exactly equal to the decrease in speed of the other.

Differentials are built with either two or four pinions. Two pinions are mounted on a shaft, which runs across the case and is mounted in a bearing at each end. Four pinions are mounted on a cross-shaped member called a spider, and the case is split to permit assembly. Normally, straight bevel gears are used for the pinions and side gears.

D. Limited-Slip Differential

Conventional differentials have a major drawback in that exactly the same torque is delivered to both wheels regardless of traction conditions. Thus, if one wheel is on a surface with low enough traction for the applied torque to exceed the traction, that wheel will break loose and increase in speed until it is revolving at twice the speed of the ring gear, and the other wheel would stop revolving. All the power will then be delivered to the spinning wheel and no power will be delivered to the wheel with traction. Limited slip, or torque biasing, and locking-type differentials have been developed to overcome this problem.

The limited-slip differentials used in passenger cars are all similar in principle. Clutches are inserted between the side gears and the case. When these clutches are engaged, they lock the side gears to the case and prevent differential action. Either plate- or cone-type clutches may be used. A typical unit using cone type clutches is shown in Figure 11.12. Initial engagement pressure for the clutches is provided by the springs. As torque is applied to the unit, normal gear reaction forces tend to separate the side gears, which apply more pressure to the clutches. The more torque is applied, the more closely the unit approaches a solid axle. When differential action is required, the changes in torque reaction at the wheels tend to reduce the pressure on the clutches, permitting them to slip. Coil springs, dished springs, and Belleville springs are all used to provide the initial engagement pressure.

A variation of the unit shown in Figure 11.12 has the cones reversed so that increasing torque input reduces the engagement pressure on the clutches. This is referred to as an "unloading cone," spin-resistant differential. It has been found useful for the interaxle differential of four-wheel drive vehicles and some high-performance cars. Both torque biasing and locking differentials are used for trucks and off-highway equipment. Some locking differentials lock and unlock automatically, whereas others are arranged so the operator can lock them when full traction at both driving wheels is needed. Because of the higher torque inputs involved with these machines, more positive locking arrangements than the clutches used in passenger cars are required for the torque biasing differentials. One type uses cam rings and a set of blunt-nosed wedges that operate much in the manner of an overrunning clutch. Other types use special tooth profiles on the pinions such that a torque bias in favor of the wheel with the best traction is always provided.

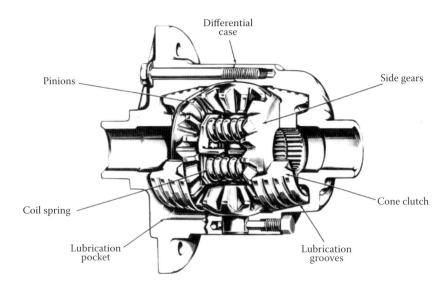

FIGURE 11.12 Limited-slip differential. A typical limited-slip differential using cone type clutches.

E. Factors Affecting Lubrication

The hypoid gears used in drive axles are among the most difficult lubricant applications in automotive equipment. The high rate of side sliding between the gear teeth tends to wipe lubricant films from the tooth surfaces, and the amount of sliding in proportion to rolling increases as the offset between the shaft center lines is increased. The gears must transmit high torques, and shock loads are often present. The drive axle is normally not sprung; thus, to keep the unsprang weight low, it is desirable to make the gears as compact and lightweight as possible. This has resulted in the use of high-strength, hardened steel for these gears, and tooth loading is usually high.

Where spiral bevel gears are used in the drive axle, conditions are more favorable to the formation of lubricant films and tooth loading is generally lower. Worm gear axles present special problems because of the high rate of sliding between the teeth and the metallurgy that must be employed. Where spur gears are used, lubrication conditions are generally not considered severe.

There are some manufacturers that no longer recommend periodic lubricant changes for passenger car drive axles. The lubricant for these axles must be suitable for extended service, often at high temperatures.

The presence of moisture in drive axles is a frequent occurrence owing to the formation of condensation and inhalation of moist air. Combined with the severe churning action of the gears, this can contribute to foaming of the gear lubricant and can also promote rust and corrosion. So a lubricant must be able to operate satisfactorily under this condition.

An important consideration for the drive axles of all vehicles operated in cold weather is the ability of the lubricant to flow to the teeth to provide lubrication of the gears, and to be carried up by the gears in sufficient quantities to lubricate the pinion shaft bearings.

In farm and construction equipment, the drive axle and differential are often installed in a common sump with the transmission, which may also serve as the reservoir for the hydraulic system. Oil immersed clutches and brakes may also be involved. In these cases, the lubricant requirements of these other elements must be considered in the selection of the lubricant for the drive axle.

Limited-slip axles present special lubrication problems because of the clutches. The clutches must engage firmly so that proper torque biasing is obtained, but the clutches must slip smoothly when differential action is required. If the clutches do not release properly, or stick and slip, chatter will result and in extreme cases, one wheel may be forced to break traction in high-speed turns. This can be an unsafe condition. To minimize these problems, lubricants for limited-slip axles must have special frictional properties, which may conflict with their ability to protect the gears against wear, scuffing, and scoring. In most cases, a small amount of chatter or noise while turning corners is considered normal for these units.

Synthetic lubricants do an excellent job of addressing many of the lubrication needs of drive axles and additionally provide fuel economy benefits that often make them a preferred recommendation.

IV. TRANSAXLES

When a transmission and drive axle are combined in a single housing, the unit may be referred to as a transaxle (Figure 11.13). The arrangement is common for front engine, front wheel drive or rear engine, rear wheel drive cars.

A. Factors Affecting Lubrication

The drive axle reduction gears used in a transaxle are either spiral bevel or, with a transverse engine, spur or helical gears. As a result, the lubrication requirements are not as severe as with hypoid gears. At the same time, the lubricant must meet the requirements of the transmission portion of the transaxle, which often has synchronizer elements that are sensitive to active extreme pressure (EP) agents designed for use in hypoid axle gears. Most builders of transaxles now recommend either specially formulated lubricants, lubricants for API Service GL-4 or GL-5, or multiviscosity engine oils.

Some engine/transaxle combinations with automatic transmission, whether at the front or rear of the vehicles, are usually equipped with separate compartments for the transmission and final drive. The transmission compartment contains ATF and the final drive compartment contains suitable gear lubricant for the spiral bevel gears. The current trend is to use front wheel drive transaxles with transverse engines. This design is usually provided with a common compartment that contains ATF for both the transmission and final drive.

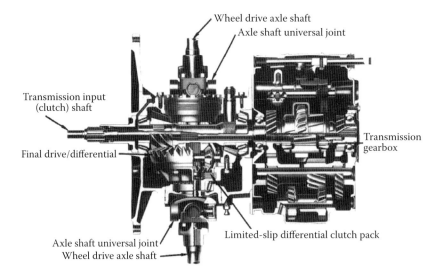

FIGURE 11.13 Typical transaxle cutaway with manual transmission.

V. OTHER GEAR CASES

A number of other gear cases may be used in various types of automotive equipment. Some of the more important are discussed in the following sections.

A. Auxiliary Transmissions

Auxiliary transmissions are used with mechanical transmissions to provide a larger choice of reduction ratios. These are usually two- or three-speed gearboxes. Shifting is by means of a range control lever, which, on trucks, usually controls a hydraulic circuit to perform the actual shifting. On low-speed equipment, shifting may be by means of a mechanical linkage. In some cases, the auxiliary may be built into the same housing as the main transmission; in other cases it may be a separate unit. Gears in auxiliary transmissions may be either spur or helical type. Bearings may be either the plain or rolling element type.

B. Transfer Cases

A transfer case is required with most four-wheel drive vehicles in order to provide a second output shaft to drive the second axle. A transfer case may also provide a power takeoff to drive accessory equipment such as a hoist. In some heavy equipment, a transfer case is not required because the main transmission is provided with both front and rear outputs.

With the conventional four-wheel drive arrangement, operation at highway speeds can cause stresses to build up in the drive line owing to the different distances traveled by the front and rear wheels. If these stresses become excessive, they may cause one wheel to break traction and "hop." To prevent this, the drive to the front wheels is normally disconnected for highway travel. A recent approach is to use a third differential in place of the transfer case. Differentiation in this unit prevents the buildup of stresses so the front wheel drive can be left engaged all the time.

C. Overdrives

An overdrive is an arrangement that drives the transmission output shaft at a higher speed than the input shaft. At cruising speeds, this reduces engine rpm and improves fuel economy. Two general approaches are used.

Many of the five-speed mechanical transmissions used in small cars have an overdrive fifth gear. The fourth gear is made with a ratio only slightly greater than 1:1, and the fifth gear has a ratio that is slightly lower. The normal progression of ratios in the gearbox is maintained, but in the fifth gear the engine rpm will be somewhat below the rpm of the output shaft of the transmission.

With three-speed transmissions, an auxiliary unit is added to the rear of the main transmission. This unit has an arrangement of planetary gears that provides a step-up ratio. The unit may be controlled electrically or hydraulically. When not engaged, it acts as a solid coupling. When the operator moves the control lever to the engaged position, the overdrive is activated but does not engage until a predetermined cut-in speed is reached, and the operator momentarily releases the accelerator pedal. The overdrive will then remain engaged until the car speed drops below the cut-in speed or until the operator forces disengagement by fully depressing the accelerator pedal. Separate overdrive units are usually bolted to the rear of the main transmission, and provision is made for the lubricant to flow from one case to the other. Separate drain plugs are usually provided, and special care in filling may be necessary to ensure that both units are properly filled.

D. Final Drives

Several types of drive units are used at the wheel ends of the drive axles to obtain additional reduction or rotate the wheels. Planetary reducers are used on many large, off-highway vehicles such as haul trucks used in mining. They are also used on many tractors and some self-propelled harvesting machines. Planetary speed increasers are used at the front wheels of conventional tractors equipped with power front wheel drive to match the travel speed of the smaller front wheels to that of the larger rear wheels. Various types of chain drives are used on self-propelled harvesting machines. Chains or gears are also used to couple a single drive axle to tandem driving wheels on certain types of heavy construction machines. Drop housings, which may or may not involve speed changes, are used on farm tractors to increase the clearance under the tractor for row crop work.

In many cases, these final drives have a separate lubricant supply. In other cases, they are lubricated from the drive axle or a common sump that supplies other units.

E. Factors Affecting Lubrication

Auxiliary transmissions, transfer cases, and overdrives are generally similar to mechanical transmissions in their lubricant requirements. As noted, auxiliary transmissions and overdrives are frequently coupled to the main transmission so that the same lubricant supply serves both units. Transfer cases are usually independent, but do not present any special lubrication challenges. If an interaxle differential of the limited-slip type is used instead of a transfer case, the frictional characteristics of the lubricant are extremely important for the proper operation of the clutches.

Final drives present a range of lubrication problems because of the diversity in design of these units. Some are lubricated from the drive axle—thus, they are designed to operate on the type of lubricant that is suitable for the axle. To simplify lubrication, many final drives with independent lubricant reservoirs are also designed to operate on one of the lubricants required for other parts of the machine. Chain drives may present special problems. Some are fully enclosed and run in a bath of oil, but others are enclosed in relatively loose fitting dust shields. In the latter case, hand oiling or a drop feed oiler may be used. In extremely dusty conditions, it may be necessary to let the chains run dry. It may also be desirable to remove the chains periodically and soak them in a bath of oil so that some lubricant can reach inside the rollers on the pins.

VI. AUTOMOTIVE GEAR LUBRICANTS

In automotive gear units, gears and bearings of different designs and materials are employed under a variety of service conditions. The selection of the lubricant involves careful consideration of these factors and their relationship to the performance characteristics of the lubricant. Some of the more important performance characteristics of automotive gear lubricants are discussed in the following sections. Additional information can be found in Chapter 3, under sections "Physical and Chemical Characteristics," "Additives," and "Evaluation and Performance Tests."

A. Load-Carrying Capacity

One of the most important performance characteristics of a gear lubricant is its load-carrying capacity—that is, its ability to prevent or minimize wear, scuffing, or scoring of gear tooth surfaces. The load-carrying capacity of straight mineral oil is adequate for the conditions under which some gears operate; however, most gears require lubricants with higher load-carrying capacity. This higher capacity is provided through the use of additives. Lubricants of this type are generally referred to as EP lubricants.

In order to provide differentiation between automotive gear lubricants with different levels of EP properties, the API prepared a series of six lubricant service designations for automotive manual transmissions and axles. These service designations describe the service in which various types of lubricants are expected to perform satisfactorily. In addition, for the two service designations intended for use in hypoid axles, antiscore protection must be equal to or better than that of certain reference gear oils. And the lubricants must have been subjected to the test procedures and provide the performance levels described in the American Society for Testing and Materials (ASTM) STP-512, Laboratory Performance Tests for Automotive Gear Lubricants Intended for API GL-4 and GL-5. The latter publication describes tests for other performance characteristics in addition to those for load-carrying capacity.

B. API Lubricant Service Designations

API publication 1560 (Lubricant Service Designations for Automotive Manual Transmissions, Manual Transaxles, and Axles) and ASTM method D7450 (Standard Specification for Performance of Rear Axle Gear Lubricants Intended for API Category GL-5 Service) provide further insight into the various API gear lubricant service designations. A summary of the gear lubricant service designations are as follows:

> **API GL-1** (not in current use): designates the type of service characteristics of automotive spiral bevel and some truck manually operated transmissions that have components sensitive to additive materials such as EP additives. They are designed for operation under such mild conditions of low unit pressures and sliding velocities, that straight mineral oil can be used satisfactorily. Oxidation and corrosion inhibitors, defoamants, and pour depressants may be utilized to improve the characteristics of lubricants for this service. Frictional modifiers and EP agents shall not be utilized.
>
> **API GL-2** (not in current use): designates the type of lubricant for automotive type worm gear axles operating under such conditions of load, temperature, and sliding velocities where API GL-1 service lubricants will not perform satisfactorily. These gear oils generally contain fatty-type additives, making them satisfactory for worm gear and other types of industrial gearing.
>
> **API GL-3** (not in current use): designates the type of lubricants for manual transmissions operating under moderate to severe conditions of speed and load and spiral bevel axles operating in mild to moderate conditions. These service conditions require a lubricant having greater load-carrying capabilities than those which will satisfy API GL-1 service, but below the requirements of lubricants satisfying API GL-4 service.
>
> **API GL-4**: designates the type of lubricants typically used for differentials and transmissions operating under moderate to severe conditions of speed and load where spiral bevel gears are used or moderate conditions where hypoid gears are used. These oils may be used in manual transmissions and transaxles where EP oils are acceptable and API MT-1 oils are unsuitable. Limited-slip differentials generally have special lubrication requirements, and the manufacturer or lubricant supplier should be consulted regarding the suitability of his lubricant for such differentials.
>
> **API GL-5**: designates the type of lubricants for gears, particularly hypoid gears in automotive equipment operated under combinations of high-speed, shock load conditions and low-speed, high torque conditions.
>
> **API MT-1**: designates lubricants for manual transmissions that do not contain synchronizers that are typically found in heavy-duty trucks and buses. These oils are formulated to provide higher levels of oxidation and thermal stability, component wear, and oil–seal compatibility that may not be provided by API GL-1, GL-4, and GL-5 category oils.

The military specification for automotive gear lubricants is the SAE Surface Vehicle Standard J2360, which has replaced the now obsolete MIL-PRF-2105E specification. Lubricants that satisfy

SAE J2360 will satisfy the requirements for API GL-5; however, the SAE J2360 standard has additional performance requirements that exceed those set by the API GL-5.

C. Viscosity

The viscosity of a gear lubricant has some effect on load-carrying capacity, leakage, and gear noise. At low temperatures, it may also determine the ease of gear shifting and will have considerable influence on flow to gear tooth surfaces and bearings.

The viscosities of automotive gear lubricants are usually reported according to the SAE Axle and Manual Transmission Lubricant Viscosity Classification (Chapter 3). This classification system provides for measurement of low-temperature viscosities according to ASTM D 2983, Standard Method of Test for Apparent Viscosity of Gear Oils at Low Temperatures using the Brookfield Viscometer. The 150,000-cP (150 Pa·S) value selected for the definition of low-temperature viscosity in this revision was based on a series of tests in a specific axle design that showed that pinion bearing failures could occur if the lubricant viscosity exceeded this value. However, other axle designs may operate safely at higher viscosities or fail at lower viscosities, so it is the responsibility of the axle manufacturer to determine the viscosity required by any particular axle design under low temperature conditions.

Other gear applications may have different limiting viscosities. For example, for satisfactory ease of shifting, many manual transmissions require a lubricant viscosity not exceeding 20,000 cP at the shifting temperature. However, this does not necessarily mean that a lubricant with a viscosity exceeding 20,000 cP at the lowest expected starting temperature cannot be used satisfactorily. At low temperatures, it may be necessary to idle the engine for a short period before driving away. During this period, the main gear and countershaft in the transmission will be revolving if the clutch is engaged. As long as the lubricant does not channel, some of it will be picked up and circulated by the gears. The resulting fluid friction may warm the lubricant, and the shear and churning conditions may increase the lubricant's fluidity sufficiently for satisfactory shifting. Gear or bearing failures usually do not occur during this period as the loads are relatively low, being only those resulting from fluid friction in the lubricant. In a drive axle, on the other hand, as soon as the drive is engaged, loads are relatively high and the need for adequate flow of lubricant is critical. Synthetic lubricants can provide superior performance at low temperatures.

Multigrade automotive gear lubricants such as 80W-90 and 85W-140 can be formulated fairly readily under this system using conventional gear lubricant base stocks. Lubricants in these viscosity grades are generally suitable for a wide range of operating temperatures in automotive gears. Multigrade lubricants covering an ever wider range such as 75W-85, 75W-90, 80W-140, or 75W-140 are formulated using either special high VI base oils such as some of the synthetics, or the use of VI improvers. VI improvers must be selected with extreme care because under the severe mechanical shearing that exists in gears, the contribution of the VI improver to high temperature viscosity can be lost rapidly if an unsuitable material is used. With an increased focus on fuel efficiency, there is a trend toward the development and use of lower viscosity gear oils.

D. Channeling Characteristics

Under low temperature conditions, a lubricant may "channel"—that is, it may become solid enough that when gear teeth cut a channel through it, the lubricant does not flow back rapidly enough to provide fresh material for the gear teeth to pick up.

The channeling temperature of a lubricant is somewhat related to its pour point, and also to its low-temperature viscosity. However, the pressure head tending to cause flow into a channel cut in a lubricant may be greater than in the pour point test, and the conditions in the Brookfield viscometer are sufficiently different that it can provide a viscosity measurement for a lubricant that might not flow under other conditions. In effect, the Brookfield viscometer will predict if a lubricant can be carried up by the gears and distributed, but it does not necessarily predict if the lubricant will flow to the gear teeth to be picked up.

E. Storage Stability

In extended storage, some of the additive materials may separate from some highly additized gear lubricants or reactions may occur that can change the properties of these materials. This is particularly true if the storage temperature is excessively high or low, or where moisture is present. Proper storage and control of inventory to prevent excessively long storage are still important.

F. Oxidation Resistance

Operating temperatures of drive axles and manual transmissions in normal passenger car service are usually moderate so that the oxidation resistance of the lubricant may not be a major concern. In heavy duty service, such as trailer towing and in many commercial vehicles, high operating temperatures may be encountered. In these applications, the lubricant must have adequate oxidation resistance to prevent excessive thickening or the development of sludge that could restrict oil flow. Oxidation may also result in the development of materials that are corrosive to some of the metals used.

G. Foaming

The churning of oil in gearsets, combined with contaminants such as moisture, tends to promote foaming of the gear lubricant. Foaming can be severe enough to cause overflow with loss of lubricant, and may interfere with lubricant circulation and the load-carrying ability of lubricant films.

Defoamants are used in gear lubricants to reduce the foaming tendency. Some of the commonly used defoamants are not soluble in oil. As a result, defoamants are carried in suspension. If the density of the defoamer is significantly greater than that of the oil, the defoamer can settle out during extended storage. As a check on this, gear lubricants submitted for U.S. Military approval must pass a foam test both when freshly prepared and after 6 months of storage. To meet this requirement, organic defoamants—rather than silicone defoamants—are now generally used in automotive gear lubricants.

H. Chemical Activity or Corrosion

EP agents function by reacting chemically with metal surfaces. This reaction is normally initiated when local overheating occurs because of rubbing of surface asperities through the oil film. If the reactivity of the EP agents is too high, some chemical reaction may occur at normal temperatures, resulting in corrosion and metal loss. Copper and its alloys are particularly susceptible to this type of corrosion and, although copper alloys are not normally used in drive axles, they are frequently used in manual transmission components.

I. Rust Protection

Some moisture enters gearsets through normal breathing, and heavy water contamination may occur in certain types of off-highway equipment. The gear lubricant must provide adequate rust protection to prevent rusting that might result in gear or bearing damage, particularly during shutdown periods.

J. Seal Compatibility

In order to maintain control of leakage, gear lubricants must be formulated to be compatible with the elastomeric materials used as seals. This means that the lubricant must not cause excessive swelling, shrinkage, or hardening. A slight amount of swelling is usually considered acceptable in that it helps to keep the seals tight.

K. Frictional Properties

As pointed out earlier, the frictional properties of the lubricant are critical in limited-slip drive axles, particularly the types that use clutches to accomplish lockup. For slip in the clutches to be initiated smoothly, the lubricant must have a lower coefficient of friction at lower sliding speeds than it has at high sliding speeds. This requires the addition of special friction modifiers to the lubricant, or the formulation with an additive system with one or more components that have these special frictional properties. However, because of the strong polar attraction of the friction modifiers to metal surfaces, they may reduce access to these surfaces by the EP agents. This may reduce the overall load-carrying ability of the lubricant. Also, the frictional properties of the friction modifiers may change relatively rapidly, requiring limited-slip additive replenishment or a complete fluid change-out.

To minimize these difficulties with limited-slip axles, some manufacturers apply special coatings to their axle gears to reduce the severity of the service on the gear lubricant. Regular lubricant changes are also recommended by some manufacturers.

Friction modification of the lubricant may also offer benefits in conventional axles, particularly during break-in or during severe service. The reduction in sliding friction between the gear tooth surfaces reduces the amount of heat generated at these surfaces. As less heat is generated, less heat will be rejected to the lubricant, and the operating temperature of the lubricant will be lower. With lower operating temperatures, oxidation and thermal degradation of the lubricant may not be as serious. During break-in, high surface temperatures have been experienced in some axle designs, which may affect the metallurgy of the gear tooth surfaces. The reduction in friction from the use of friction modifiers can also provide some relief from this problem.

L. Identification

In addition to the API lubricant service designations and the SAE viscosity classification, automotive gear lubricants have been identified by U.S. Military specifications. Some manufacturer specifications also exist, but products meeting them are primarily used only for factory fill of new machines.

M. U.S. Military Specifications

SAE J2360 is the only active automotive gear lubricant specification for U.S. Military applications. This specification contains performance requirements beyond API GL-5. The obsolete U.S. Military specification that SAE J2360 replaced was MIL-PRF-2105E. There are a number of other obsolete military specifications that are occasionally referenced, and they include MIL-L-2105D, MIL-L-2105C, MIL-L-2105B, and MIL-L-10324A.

VII. TORQUE CONVERTER AND ATFs

The specialized fluids used in passenger car, truck, bus, and off-highway automatic and semiautomatic transmissions are all manufactured to meet OEM specifications. As most OEMs use proprietary friction materials in their automatic transmissions, virtually every ATF is OEM-specific and, in some cases, transmission-specific.

A. Torque Converter Fluids

Separately housed torque converters are used with some limited passenger car automatic transmission arrangements, but the main application is in transit coach drives and some construction equipment drives. Separately housed torque converters for passenger cars are operated on ATFs.

To provide better oxidation stability, low-viscosity mineral oils were adopted for torque converters. The most commonly used products now are oxidation-inhibited fluids of about 20–21 cSt at 40°C. Two older manufacturer specifications covering these fluids are the General Motors Corporation Truck & Coach Division Specification No. C-67-I-23 for Type 2 V-Drive Fluids, and the Dana M-2003 specification for torque converter fluid. These specifications have the same physical requirements, and the performance requirements are similar. It is suggested that the OEM recommendation be followed.

B. ATFs

Commercially available passenger car ATFs are developed to meet either an individual OEM specification or group of OEM specifications from various manufacturers.

For heavy truck and off-highway automatic and semiautomatic transmissions, Allison TES-439 or Caterpillar TO-4 type fluids may be recommended depending on the transmission in service. It is suggested that the OEM recommendation be followed.

Passenger vehicle ATFs have evolved over the years as vehicle weights have changed and operating conditions have made the service on the fluids more severe. Other factors such as the oil drain interval extensions or the elimination of oil drains have dictated a need for products that are more resistant to deterioration while in service. Several automobile manufacturers have advertised 100,000 miles (160,000 km) before the first major service interval and recommended transmission fluid changes at this interval when the vehicle is driven under "normal service."

The following is an outline of some of the more common ATF specifications past and present.

1. General Motors

a. General Motors Type A Fluid

This specification and qualification procedure was introduced in 1949. These products came to be known as "Type A" fluids. They provided satisfactory performance in the transmissions of that time, and were recommended by a number of manufacturers in addition to General Motors. In 1956, the Type A specification was revised to require improved quality. They were suitable for use in transmissions where Type A fluids had been originally recommended.

b. General Motors DEXRON® Fluid

In 1967, General Motors issued the DEXRON specification representing a further improvement in quality, mainly in the areas of high-temperature stability, low-temperature fluidity, and retention of frictional properties in service. Fluids were intended to be suitable where Type A or Type A, Suffix A fluids had been recommended originally, but—as noted—some European manufacturers continued to show preference for the older Type A, Suffix A fluids. In addition to the General Motors applications, DEXRON fluids were at that time acceptable to Chrysler, American Motors, and some Borg-Warner affiliates.

c. General Motors DEXRON II Fluid

The DEXRON II specification in 1972 was originally intended as a multipurpose fluid suitable for passenger car automatic transmissions, the lubrication of rotary engines, and as a replacement for the Allison C-2 fluids recommended by the Detroit Diesel Allison Division of General Motors for automatic and semiautomatic transmissions used in trucks and off-highway equipment. These fluids proved to be deficient in corrosion protection for the brass alloys used in certain transmission fluid coolers. The specification was revised to include a brass alloy corrosion test, and the rotary engine lubrication requirements were dropped. Fluids qualified under this version of the specification are generally suitable for applications where any of the earlier General Motors approved fluids were originally recommended.

d. General Motors DEXRON IId, IIe, and III

DEXRON IId and IIe were interim specification requirement changes to enhance the specific properties of the DEXRON II. The DEXRON IIe, introduced in late 1990, was replaced with the DEXRON III specification in 1993. Both the DEXRON IIe and III stressed improved low temperature performance, improved shear stability, frictional characteristic durability, and improved oxidation, and thermal stability. The DEXRON III fluids were back serviceable to all previous vehicle transmissions that specified DEXRON IIe, IId, DEXRON, Type A-Suffix A, and Type A fluids.

e. General Motors DEXRON VI

General Motors released DEXRON VI as a fill-for-life fluid for 2006 model year cars. This DEXRON VI fluid had even greater antiwear performance compared to past DEXRON type fluids and has improved oxidative, thermal, and shear stability to provide extra protection under severe service conditions. The DEXRON VI is back serviceable to all DEXRON III and DEXRON IIe applications. Many current GM vehicles use DEXRON VI as its power steering fluid. Please refer to the vehicles owner's manual to select the proper power steering fluid.

f. General Motors DEXRON HP

General Motors introduced DEXRON HP in 2014 as a fill-for-life fluid for next-generation eight-Speed Automatic Transmissions. It met the DEXRON VI performance requirements with respect to antiwear and corrosion protection but utilized API Group IV PAO base stocks. DEXRON HP has outstanding low-temperature properties and friction properties that improve vehicle fuel economy. It is also approved for older transmissions specifying DEXRON VI fluid.

2. Ford ATFs

a. Ford Type F ATF

In 1959, Ford ATFs were issued under the designation ESW-M2C33 with a suffix letter indicating different revisions. They are usually referred to as "Type F" fluids. M2C33-E/F was issued and is suitable for use in all Ford automatic transmissions (except the C-6) manufactured before 1980. A number of other manufacturers, including some of the Borg-Warner affiliates, also recommended this type of fluid. The major difference between the early General Motors fluids and the M2C33 Ford fluids is in their frictional properties. The Ford fluids have a higher coefficient of friction at low sliding speeds than they do at high sliding speeds, whereas the General Motors fluids are the opposite. This results in faster and more positive lockup of the clutches and bands at low sliding speeds with the Ford Type F fluids. The Ford Type F fluids are physically compatible with the General Motors fluids, but they should not be mixed.

b. Ford Type CJ and Type H ATF

Early in 1980, Ford specifications M2C138-CJ and M2C166-H were issued. These are referred to as Type "CJ" and "H". Only Type CJ fluid was recommended for the C-6 and "JATCO" automatic transmissions. They differ from the earlier fluids in that they contain friction modifiers and are not recommended where Type F fluids are recommended.

c. Ford MERCON ATF

In 1988, Ford introduced its MERCON® specification designed to meet the requirements of the Ford transmissions manufactured since that date and through the mid 1990s.

d. Ford MERCON V

The MERCON V® specification was introduced in 1997. These fluids have more rigorous requirements for oxidation stability and wear protection. Although compatible, the MERCON fluids should not be used for change-outs (if required) where MERCON V is specified. The MERCON V fluids have limited back-serviceability so the manufacturers' recommendations should be followed on vehicles produced before 1997.

e. Ford MERCON SP

Ford later introduced MERCON SP as a separate specification with some enhancements to MERCON V for some specific vehicles. MERCON SP is not compatible with any of the other MERCON fluids, nor should it be used in transmissions specifying a different MERCON fluid.

f. Ford MERCON LV

Ford introduced the MERCON LV specification as a low-viscosity fill-for-life fluid for 2008 model year and newer vehicles. The recommended oil drain interval for this fluid is 150,000 miles (241,000 km).

3. Chrysler ATFs

In 1998, Chrysler introduced a specification—MS 9602 ATF+4®. This replaced Chrysler MS 7176E ATF+3® and is intended to be used for complete change-outs. The ATF+3 had recommended service intervals of 30,000 miles, whereas the ATF+4 never required changing even under severe service conditions. ATF+4 is generally considered compatible with all transmission applications where the older ATF+®, ATF+2®, and ATF+3 products were specified.

4. ATFs for Japanese Vehicles

The Japanese Automotive Standards Organization (JASO) developed a performance specification for ATF service fill referred to as JASO 1-A to address the needs of Japanese passenger vehicles. There are many individual Japanese OEM transmission fluid specifications, such as Toyota Type WS, Type T, Type T-II, and T-IV; Honda ATFDW-1 and ATF-Z1; and Nissan Matic S, J, D, K, W.

5. ATFs for European Vehicles

Some European vehicles specify some of the ATFs described earlier although the majority have their own fluid specification. It is estimated there are more than 70 ATF specifications.

6. Multipurpose ATFs

There are some products on the market that are formulated to meet the requirements or quality level of some of the major transmission fluid specifications and are often referred to as multipurpose or multivehicle ATF's. Generally, these fluids are not approved against most of the ATF specifications and as such should not be used in transmissions under warranty.

Regarding vehicle warranties: To satisfy the vehicle warranty and insure the desired performance over the lifetime of the vehicle, the OEM's recommendation for ATF must be followed. Please refer to the owner's manual of the vehicle to identify the correct approved ATF. All manufacturers recommend against adding performance enhancing aftermarket additives to fully formulated transmission fluids, and doing so will void the transmission's warranty.

7. Transmission Fluids for CVTs and DCTs

Both CVTs and DCTs require specialized transmission fluids. OEM recommendations should be followed.

8. Allison Transmission

Allison Transmission manufactures both fully automatic and semiautomatic transmissions for on-highway and off-highway equipment. For off-highway transmissions, Allison Transmissions has an oil specification for standard oil drain intervals of 1200 hours called TES-439 and an oil specification for longer oil drain intervals of 4000 hours called TES-353. The TES-439 specification replaced the now obsolete Allison C-4 (TES-228) specification in 2011. For on-highway transmissions, they have an oil specification called TES-389 for standard oil drain intervals of 50,000 miles

(80,000 km) and an oil specification called TES-295 for longer oil drain intervals of 150,000 miles (240,000 km). Both the TES-353 and TES-295 fluids are considered synthetic oils designed for the longer oil drains. Several multipurpose ATFs meeting DEXRON and MERCON V are qualified as TES-389 fluids. Several API CJ-4 engine oils are qualified as TES-439 fluids.

9. Caterpillar

The current TO-4 specification of Caterpillar Inc. is designed to meet the requirements of their oil-cooled friction compartments such as transmissions, final drives, and planetary gears including wet braking systems. The TO-4 specification does not include any requirements to meet any engine oil performance criteria. The TO-4 specification includes three monoviscosity grade products: SAE 10W, 30, and 50. The TO-4 tests include tests for high-temperature/high-shear, lower temperature requirements, and testing in both gear (FZG) and hydraulic pumps (Vickers Vane) to assure performance in hydraulic and gear systems. For Caterpillar transmissions operating where a wide range of temperatures are encountered (particularly cold weather), a multigrade oil specification called TO-4M was developed.

VIII. MULTIPURPOSE TRACTOR FLUIDS

One area of lubrication where there has been extensive development and use of manufacturer specifications is in fluids for the drive (transmissions, differentials, final drives) and hydraulic systems of farm and industrial tractors including self-propelled farm equipment. At least 25 different fluid specifications are used by the equipment manufacturers, although several are now considered obsolete. Several equipment manufacturers market many of these fluids through their parts departments and, in many cases, these fluids—or an equivalent product that meets the same specifications and performance standards—must be used during the warranty period. This can cause considerable complexity in oil inventories and application for the operator with several different makes of equipment. The risk of misapplication may also be high under these conditions.

Many reasons can be given for the large number of these fluids and their variations in characteristics. Some of the differences among the products result from differences in design philosophies and construction material among the manufacturers. The number and types of machine elements served by the fluid may be extensive, and even minor differences in materials of construction, applied loads, or operating characteristics can have a marked effect on the lubricant properties required for satisfactory performance. Most manufacturers tried to develop fluids or fluid specifications that were specific to their needs. Table 11.1 lists some of the more common, past and present, OEM tractor fluid specifications.

TABLE 11.1
OEM Tractor Fluid Specifications

OEM	Specification
AGCO (Deutz Allis)	Power Fluid 821XL
Case Corp. (Case I.H., J.I. Case)	MS-1204, MS-1205, MS-1206, MS-1207, MS-1209, MS-1210, B-6
Case New Holland	MAT 3505, MAT 3525, ESN-M2C48-B, ESN-M2C48-C
John Deere	J20C, J20D
Kubota	UDT
Massey Ferguson	M1110, M1127A/B/C, M1129A, M1135, M1139, M1141, M1143, M1145
New Holland Group (Ford New Holland)	FNHA-2-C-201.00 (134D), FNHA-2-C-200.00 (hydraulic oil 134), ESN-M2C134-C, ESN-M2C134-B, ESN-M2C134-A, ESN-M2C86-B, ESN-M2C53-A
Steiger	SEMS 17001
Volvo	WB 101, VCE 1273.03
White Farm (Oliver)	Q-1826, Q-1766B, Q-1722, Q-1705

There are continuing attempts to classify these fluids so that some simplification can be achieved, and this has been done to some extent with universal tractor transmission oils (UTTOs) and super tractor oil universal (STOUs). The STOU oils are formulated for use in engines, transmissions, wet brakes, and hydraulics. The UTTOs are more popular in North America and the STOUs are more popular in Europe. The success of these fluids depends on their ability to satisfy all of the needs of the various manufacturers and their availability to the markets where needed.

Multipurpose fluids must satisfy the lubrication requirements of hydraulic systems, gears, and bearings. They must also work with wet braking systems and provide the proper frictional characteristics for clutches in transmissions and power takeoffs. In addition, they must be compatible with all system components and capable of performing under a wide range of operating conditions.

A. Tractor Fluid Characteristics

In addition to the usual physical and chemical requirements, most multipurpose tractor fluids have requirements for seal and other component compatibility. The tests for this characteristic vary widely both in the types of elastomer and friction materials used for testing. Various other bench-type tests for such properties as wear sensitivity or tolerance, filterability, compatibility with other similar fluids, and low temperature performance are also used in many of the specifications. Many of the specifications also include performance tests, including such tests as automatic transmission oxidation tests, transmission durability tests, cold engagement tests, various hydraulic pump tests, gear wear tests, wet brake, clutch chatter and durability tests, friction tests, frictional property retention tests, and dynamometer or full-scale field tests.

The following are some of the more important properties of the multipurpose tractor fluid.

1. Viscosity and VI

The type of equipment in which these fluids are used is often operated on a year-round basis. This means that the fluids must have adequate low-temperature fluidity to flow to parts requiring lubrication and to the inlet of hydraulic pumps at typical winter temperatures. Many of the specifications prescribe a maximum allowable viscosity at 0°F (−18°C), and some also limit the viscosity at even lower temperatures. Generally, there is a high probability that the fluids will give satisfactory low-temperature performance in the equipment they were tested in or tested for. However, there may be some compromises in the maximum allowable low-temperature viscosity in order to meet a desired viscosity at high temperatures.

In high-temperature operations, the viscosity of the fluid should be sufficient to permit adequate maintenance of hydraulic pressure and to protect gears and bearings from excessive wear. Although many of the fluids are similar in viscosity at 212°F (100°C), there could be some differences owing to how the oil is formulated to have a high VI. The degradation products of some VI improver additives in a formulation are reported to cause glazing of certain types of facing materials used on oil immersed ("wet") clutches and brakes. This condition can cause slippage and other problems. Therefore, some equipment manufacturers require that VI improvers not be used in their fluids, so the viscosity characteristics of those fluids are limited to what can be obtained from high VI base stocks.

In general, VI improved fluids are desirable to maintain adequate high-temperature viscosity while maintaining the low-temperature viscosity in a range that will permit satisfactory operation at the lowest ambient temperatures encountered in service. Where used, careful selection of shear-stable VI improvers is necessary to be able to resist the mechanical shearing in hydraulic pumps and gears. Many of the specifications include shear stability tests such as high-temperature, high-shear requirements.

2. Foam and Air Entrainment Control

Churning and pumping through the system can promote foaming, which can seriously degrade the performance of hydraulic components and may interfere with the formation of proper lubricating films. Since the residence time of the fluid in the reservoir is short, the fluid must not only resist foaming, but also permit the rapid collapse of any foam that does form. Excessive entrainment of air in the fluid can also cause issues in hydraulic components, so the fluid must release entrained air as rapidly as possible. Careful selection of the base stock and use of an effective defoamer are necessary to meet these requirements.

Some ingress of moisture into these systems always occurs, often through condensation buildup or humid air contact. Since moisture can promote foaming, some of the manufacturers require that their fluids have good foam resistance when they are contaminated with a small amount of water.

3. Rust and Corrosion Protection

Construction and farm equipment will operate, at least part of the time, under wet or humid conditions. Breathing of this moisture into the lubrication systems can cause rust and corrosion of ferrous parts, particularly above the oil level during shutdown periods. Nearly all of the specifications have some requirement for protection against rust and corrosion, with some type of humidity cabinet test being the most common. The selection of very effective rust inhibitors is needed to meet this type of requirement.

Similar to gear lubricants, the EP agents used in multipurpose tractor fluids function by reacting chemically with metal surfaces. If the reactivity is too high, then corrosion may occur. Nearly all of the specifications contain a copper strip corrosion test, although the time and temperature vary considerably. Oxidation of a fluid in service can increase its acidity and corrosivity so both an oxidation and corrosion requirement must be met.

4. Oxidation and Thermal Stability

Operating temperatures often are high in these systems as a result of heavy loads encountered and continuous operation. Thermal and oxidative degradation of the fluid can result in thickening, which can reduce hydraulic pump capacity and cause the formation of varnish and sludge. These latter materials plug filters and leave deposits on clutch facings where they can cause slippage, or they can cause erratic control in hydraulic control valves.

Some type of oxidation test is used in nearly all of the specifications. These tests can be interpreted in various ways. Excessive evaporation loss usually results from the use of volatile, low-viscosity components in the base oil blend (NOACK volatility tests are incorporated to evaluate volatility). The amount of viscosity increase is somewhat indicative of the amount of oxidation of the fluid. Sludge or sediment may indicate thermal decomposition of additive components or some form of chemical reaction between additive components at high temperatures.

Various performance tests are also used in many of the specifications to evaluate the stability of fluids in long-term service or under simulated service conditions. These tests may be field or dynamometer tests or specialized tests such as transmission oxidation tests.

5. Frictional Characteristics

The use of oil immersed clutches and wet brakes require special focus on the frictional characteristics of the fluid and the retention of these characteristics in service. Generally, friction modifiers must be included in the fluid if the clutches and brakes are to engage smoothly and resist chatter and squawk. Detergent–dispersants may be desirable to help keep the facing surfaces clean. Good oxidation stability is important to reduce the amount of varnish-type materials that might deposit on the friction surfaces. Oil additives must be carefully selected as some are known to cause rapid deterioration of some types of friction materials. The overall problem is complicated by the varying levels of friction performance required in different systems, and the many types of friction materials involved. It is also very important that the fluids retain their frictional properties after extended service.

6. EP and Antiwear Properties

Multipurpose tractor fluids must provide adequate protection against wear and gear tooth surface distress for a variety of gears operating under mild to severe conditions of loading. Shock loads are often encountered. In hydraulic systems, pump wear protection must be provided when operating at hydraulic pressures of 4000 psi (28 MPa) or higher. Some of the specifications require or limit the amount of zinc dithiophosphate antiwear additive, which can cause challenges for lubricant formulators as they balance the need for antiwear protection with friction control.

In addition to a considerable number of bench tests for EP properties, there are several specialized gear tests used by equipment manufacturers. Differences in gear types, oil tests, and operating conditions make it difficult to compare the load-carrying capability of various oils. Because the fluids must satisfy the lubrication requirements of hydraulic systems, gears, and bearings, they are subjected to a range of tests such as the Vickers vane pump test (ASTM D2882) and the FZG Test for gearing. Additional pump tests such as the Denison HF-0 (piston and vane pumps) or the Sundstrand piston pump test may be also required.

BIBLIOGRAPHY

Allison Transmission. Approved Fluids. http://www.allisontransmission.com/parts-service/approved-fluids.

American Petroleum Institute. 2013. API 1560: Lubricant Service Designations for Automotive Manual Transmissions, Manual Transaxles, and Axles. Washington, DC.

ASTM D2983-09, Standard Test Method for Low-Temperature Viscosity of Lubricants Measured by Brookfield Viscometer, ASTM International, West Conshohocken, PA, 2009, www.astm.org.

ASTM STP512, Laboratory Performance Tests for Automotive Gear Lubricants intended for API GL-4, GL-5, and GL-6 Services, ASTM International, West Conshohocken, PA, 2009, www.astm.org.

ASTM D7450-13, Standard Specification for Performance of Rear Axle Gear Lubricants Intended for API Category GL-5 Service, ASTM International, West Conshohocken, PA, 2013, www.astm.org.

ASTM D2882-00, Standard Test Method for Indicating the Wear Characteristics of Petroleum and Non-Petroleum Hydraulic Fluids in Constant Volume Vane Pump (Withdrawn 2003), ASTM International, West Conshohocken, PA, 2000, www.astm.org.

Caterpillar. 2013. Special Publication SEBU6250-19: Caterpillar Machine Fluids Specifications. Peoria, IL.

SAE J2360. 2012. Automotive Gear Lubricants for Commercial and Military Use. SAE International. Warrendale, PA.

12 Automotive Chassis Components

Although the engine and power train are generally considered to be part of the automotive chassis, they are presented in detail in another section of this book. The discussion in this section is concerned only with suspension and steering linkages, steering systems, wheel bearings, and brake systems.

I. SUSPENSION AND STEERING LINKAGES

A suspension is an arrangement of linkages, resilient members—such as coil and leaf springs, struts, or torsion bars—and shock absorbers. These components attach the wheels to the frame or body of a vehicle, allowing road imperfections to be absorbed rather than being transmitted directly to the passenger compartment. The front suspension of a vehicle is designed to allow the wheels to pivot to steer the vehicle. Some off-highway and farm vehicles do not have steerable front wheels. In this type of equipment, steering is accomplished either by an articulated joint located between the front and rear sections of the machine or by disconnecting the power and applying braking to the wheels or tracks on the inside of the turn. Many self-propelled combines and some trucks and forklifts use steerable rear wheels.

Many suspension designs are used for passenger cars and trucks. Today, all light-duty passenger vehicles are equipped with "independent" front suspension, which permits the front wheels to move independently of each other with respect to the frame and body. Some medium and heavy trucks may still be equipped with rigid front axles. Most passenger vehicles are equipped with independent rear suspension systems with the exception of pickup trucks, which still use a solid rear axle. The front wheel drive passenger cars are the predominant type of vehicles being marketed today.

A. Front Wheel Suspension Systems

Front wheel suspension systems in light-duty vehicles are predominated by two designs: A-Arm and MacPherson strut. The A-Arm design is shown in Figure 12.1. In this design, the wheels are pivoted on a pair of ball joints (Figure 12.2) seated in the outer ends of the upper and lower A-shaped control arms. These joints permit the wheels to move up and down with respect to the frame and to be turned for steering. Additional pivot points at the inner ends of the control arms permit them to swing up and down against the restraint of the coil springs and shock absorbers. Front suspension arrangements for front wheel drive vehicles typically use the MacPherson strut design (Figure 12.3). In this design, a lower A-Arm is still used, but the upper is replaced with a sturdy strut containing the spring and shock absorber fixed rigidly to the wheel hub. As the wheel is steered, the spring and shock-body turn as well.

Rigid front axles are currently used on heavy vehicles such as over-the-road trucks and buses. Where rigid front axles are used, usually the wheels are mounted on a yoke and kingpin (Figure 12.4) to permit turning the wheels for steering. Although some medium-sized trucks may have the "rigid" front axle separate for each wheel, they are still referred to as rigid.

302 Lubrication Fundamentals

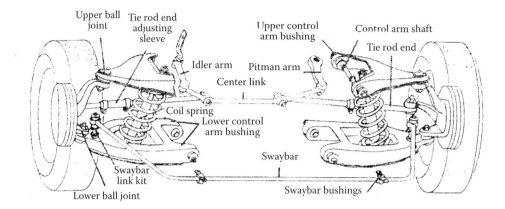

FIGURE 12.1 Passenger car front suspension (rear wheel drive). This illustration shows one typical front suspension for rear wheel drive vehicles. The lubrication points are lower and upper ball joints and steering linkages, although many of today's suspension and steering components may not require relubrication.

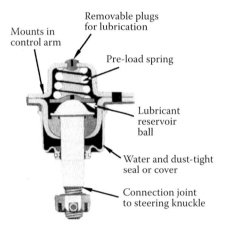

FIGURE 12.2 Suspension ball joint. Although today most ball joints in passenger vehicles are lubricated for life, this type of joint is designed for periodic relubrication, at low pressure, through the removable plug (some contain a Zerk fitting). Some similar joints have a vent on the seal so that they can be lubricated through a fitting.

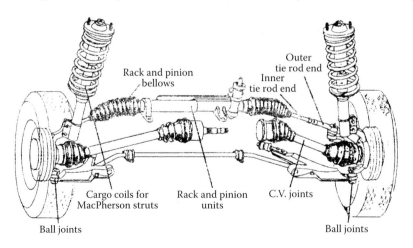

FIGURE 12.3 Front suspension for front wheel drive vehicles (MacPherson strut design).

Automotive Chassis Components

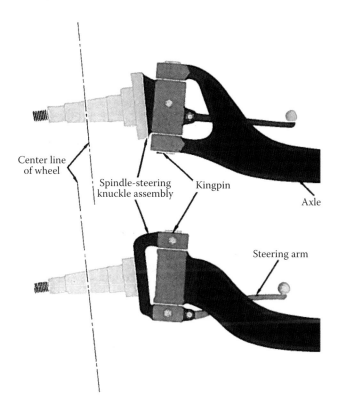

FIGURE 12.4 Rigid axle for front wheel mounting. In this design, the yoke is on the wheel spindle, steering knuckle assembly. In other designs, the yoke is on the axle. Relubrication is provided by fittings in the center of the upper illustration and fittings in the upper and lower portions of the spindle in the lower illustration.

B. Rear Suspension Systems

The usual rear suspension, because there is no requirements for steering, is simpler in design than the front suspension. Either leaf or coil springs may be used. Some additional complexity may be introduced with independent rear suspension to permit the wheels to move independently of each other while still transmitting the driving force to the wheels.

The most common types of independent rear suspension systems include the trailing arm and H-Arm designs. In some exotic cars, the rear wheels can contribute to the steering of the vehicle.

C. Active Suspension Systems

The suspension systems mentioned up to this point are simple mechanical systems that respond to vehicle dynamics and road loads with fixed set of springs and levers. Many high-performance cars can now respond actively to these driving loads. Suspension position and acceleration sensors can vary spring rates and shock absorber damping rates to improve vehicle performance. Shock absorber damping rates can be altered using electrically controlled variable valves, or by using a magnetorheological damper. The fluid in these types of shock absorbers contain ferromagnetic particles that can change the fluid properties in reaction to an induced magnetic field. A significant safety feature in many cars today is the use of active vehicle dynamics control. The vehicle can identify driving situations such as where the car begins to skid when cornering. The vehicle computers will then adjust throttle, suspension, and braking as required to retain control of the vehicle.

D. Steering Systems

The most common type of steering linkage used on passenger vehicles is the parallelogram type (see Figure 12.5). This term is derived from the parallelism maintained between the pitman and idler arm at all linkage positions. The pitman and idler arm are connected to an intermediate rod through pivot bearings, such as those shown at the right-hand side of Figure 12.6. All other connections are made through ball joints (sometimes referred to as tie-rod ends) such as those shown at the left-hand side of Figure 12.6. The pitman is operated by the steering gear that is discussed in the following section.

Power-assisted rack-and-pinion steering is now used on most light cars. In this design, no pitman or idler arm is used. The intermediate rod is replaced by a rack that meshes with a pinion on the end of the steering shaft. This arrangement results in a simple, direct-acting linkage.

Other linkage arrangements may be found on farm and off-highway equipment. For example, with the row crop tractors, no linkage is used. The wheels are mounted on a single vertical shaft

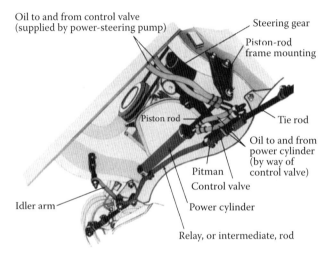

FIGURE 12.5 Linkage-type power assisted steering. The control valve may be incorporated in the cylinder, rather than being a separate unit as shown here. The valve opening is proportional to the force applied to the pitman, so the amount of hydraulic assist provided is proportional to the turning force applied to the steering wheel.

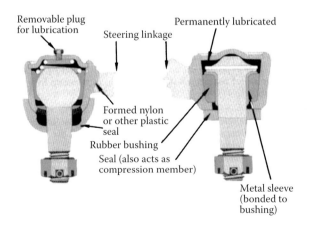

FIGURE 12.6 Typical steering linkage ball joints. The joint on the right is not truly a ball joint and is designed for rotation only in a single plane. It is suitable for connecting the intermediate link to the pitman and idler arm.

that is rotated directly by the steering gear. Generally, where steerable front wheels are used, some type of linkage similar in principle to the parallelogram type is used. However, hydrostatic steering without linkages or gears, as such, is also used.

E. Factors Affecting Lubrication

A number of factors contribute to the lubrication challenges associated with suspension and steering linkage components. Motion of the pivot points is of an oscillating nature through relatively short arcs. This does not permit the formation of fluid lubricating films and also causes wiping off of lubricant films. Because these components are located below the body, they are exposed to water (sometimes containing salt and sand used for melting ice) and dirt. Although effective sealing reduces the possibility of entry of these contaminants, seal failure or leakage is always possible. Loads are usually high, and severe shock loads are often present especially in suspension members.

Over the years, procedures for lubrication of suspension and steering linkage components have changed markedly, particularly for passenger cars. Lubrication intervals have been extended with improved designs, particularly with respect to sealing, and by the use of better quality greases. The majority of vehicle manufacturers have eliminated the requirement for periodic lubrication entirely, either by adopting designs that use flexible elastomeric bushings, or by coating the rubbing surfaces with a wear-resistant, low-friction coating. In larger commercial vehicles, there is still the provision for lubricating ball joints.

Two types of seals are used on ball joints designed for relubrication: umbrella seals and balloon seals. Umbrella seals permit excess lubricant to escape readily while providing a degree of protection against contaminant entrance. Joints with umbrella seals can be equipped with fittings and lubricated with usual high-pressure grease dispensing equipment. Balloon seals without a pressure release feature can be damaged if grease is pumped into the joint under high pressure. Thus, many joints are equipped with plugs. To relubricate the joint, the plugs must be removed and a fitting installed. Grease is then applied with a low-pressure adapter to prevent damage to seal. Balloon seals with a pressure relief feature have superseded this type of seal. With this design, lubrication fittings usually are factory installed.

F. Lubricant Characteristics

Most suspension and steering linkage pivot points designed for lubrication are lubricated with grease. Fluid lubricants are used in some types of off-highway equipment. These applications may involve the use of central lubrication systems to replenish the lubricant automatically. Lubricants used include automotive gear lubricants of American Petroleum Institute (API) Gear Lubricant GL-4 or GL-5 quality and semifluid greases. Rheopectic greases have also been used. These greases, which are semifluid as manufactured and dispensed, stiffen when subjected to mechanical shearing in bearings. If operating and ambient temperatures permit, greases may be preferred because of their better stay-put and sealing capabilities.

The grease used in suspension and steering linkage components, particularly where extended lubrication intervals are recommended, should provide:

1. Good oxidation resistance and mechanical stability
2. Protection against corrosion by both fresh water and salt water
3. Resistance to water washout
4. Good wear protection under the conditions of loading and motion involved
5. Provide some sealing against the entrance of dirt, water, and other contaminants
6. Resistance to pounding, leaking, and squeezing out
7. Provide friction reducing capabilities to decrease steering effort and provide smoother riding characteristics

Several grease formulations are used and provide acceptable performance in these applications. Most of these formulations are similar to NLGI (National Lubricating Grease Institute) No. 2 in consistency and are made with an oil component approximating ISO (International Standardization Organization) viscosity grade 150 or 220. They are intended for multipurpose use. In addition, special applications, such as wheel bearings on disc brakes, require greases that are resistant to high temperatures. Lithium soap base greases, which are used to a considerable extent, are modified by the addition of extreme pressure and antiwear additives to provide better load-carrying capabilities and lower wear rates. Colloidal or fine particle size solid lubricant materials such as molybdenum disulfide are often added, particularly when the grease is intended for use in passenger car suspensions and off-highway equipment pivot and hinge pins. These materials generally reduce both friction and wear.

II. STEERING GEAR

The steering gear that transmits motion from the steering wheel and steering shaft to the pitman arm of the steering linkage generally uses some type of worm gear mechanism acting against short levers or a gear segment connected to the pitman shaft.

In cam and lever type steering gears, the worm is referred to as the cam. It meshes with pins on a lever attached to the pitman shaft. The pins may be either solid or mounted on roller bearings to reduce friction. In other variations, a segment of, or a complete worm wheel meshing with the worm, or a throated worm meshing with a roller in the shape of a short worm, may be used. The most common manual steering gear design is the recirculating ball type, in which the worm drives a ball nut through a series of steel balls (Figure 12.7). The path of the balls includes an external guide tube that permits them to recirculate freely in either direction. This arrangement provides low friction between the steering shaft and ball nut, helping to keep steering effort low.

Most cars today are equipped with power-assisted rack-and-pinion steering (Figure 12.8). In this design, a small pinion on the end of the steering shaft meshes with a rack mounted in a guide tube. The tie rods are connected directly to the rack through a mounting bracket. This simple linkage produces better "road feel" for the driver and may reduce steering effort. Power steering, or power-assisted steering, is now widely applied to all types of automotive equipment. In addition to the rack-and-pinion steering systems, there are two other types generally used: the linkage type (Figure 12.5) and the integral type (Figure 12.9). Although these illustrations show passenger car applications, the principles are applicable to systems used on other equipment.

In linkage-type power steering, a double-acting hydraulic cylinder is attached between a point on the frame and the intermediate rod. A control valve is mounted so that it is operated by the pitman

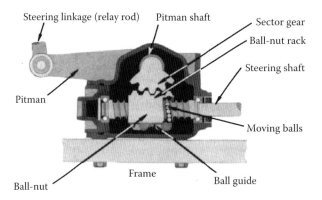

FIGURE 12.7 Recirculating ball-type steering gear. As indicated, the recirculating balls act as a low friction mechanism for transmitting the motion of the steering shaft to the ball nut.

Automotive Chassis Components

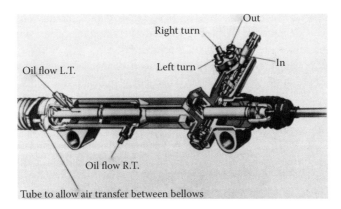

FIGURE 12.8 Rack-and-pinion steering gear.

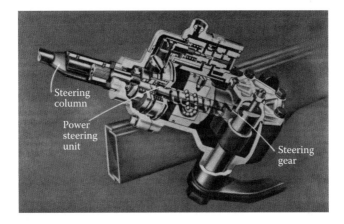

FIGURE 12.9 Integral-type power steering. In this design, longitudinal forces from the worm act on reaction members to control the opening of the control valve and the amount of assist provided. In the other common design, the control valve is actuated by a torsion bar in which the amount of twist is proportional to the turning force applied to the steering wheel. With both designs, the vehicle can still be steered if no power is available because there is a mechanical connection between the steering shaft and the ball nut.

arm through a ball joint. Steering load deflects the spool valve, allowing power steering fluid to flow into the appropriate reaction chamber in the power cylinder. The amount of power assistance (pressure) is proportional to the spool valve deflection, which increases with the steering effort. High-pressure hydraulic fluid is supplied from a small pump mounted on and driven by the engine. When the pitman moves in one direction, as a result of turning the steering wheel, the control valve allows fluid under pressure to flow into the correct end of the cylinder. This provides the hydraulic assist that forces the intermediate rod in that direction. When the steering wheel is centered, the valve is also centered and the fluid can flow freely in and out of either end of the cylinder. With this arrangement, the steering gear itself is a conventional type, although the ratio may be somewhat lower than when the hydraulic assist is not installed. The steering gear is lubricated separately with a special lubricant. The power steering fluid only serves as the hydraulic medium.

Integral-type power steering systems are the most common type in use today. In these systems, when developed from a recirculating ball-type steering gear, the ball nut is also a double-acting piston. Hydraulic fluid under pressure is admitted to one end, or the other, by means of a control valve. The control valve can be actuated by reaction forces along the steering shaft, or by torsional forces acting on a torsion bar that rotates a spool valve. In either case, the amount the valve opens, and the power assist provided is designed to be proportional to the amount of force applied to the steering wheel. In this way, the driver's feel of the road is retained.

With both types of power steering, the steering shaft is connected mechanically through a steering gear. Thus, if fluid pressure is not available, either because the engine is not running or a system failure has occurred, the vehicle can still be steered (with some difficulty).

In the usual automotive practice, hydraulic fluid is supplied from an integrated pump and reservoir mounted on the engine. Some use, such as in off-highway equipment, has been made of a central hydraulic system to supply a number of mechanisms such as power steering, brakes, transmissions, and equipment hydraulics. Power steering systems on farm and construction equipment may be supplied from a separate system, but generally, they are supplied from the main hydraulic system of the machine.

A. Factors Affecting Lubrication

The lubricant in the steering gear is required to lubricate the gears and bearings. It must withstand shock loads, which are transmitted to the steering gear from the wheels hitting bumps and obstructions, and must resist the wiping action of gear teeth. At the lowest expected operating temperatures, it must not exhibit excessive resistance to motion, yet it must have enough viscosity to lubricate properly at the highest expected operating temperatures reached.

The fluid in integral power steering systems serves both as the hydraulic fluid and as the lubricant for the gears and bearings. Thus, it must perform all the functions of the lubricant in conventional steering gears with the additional requirement of being a highly stable, shear-resistant hydraulic fluid.

The conditions under which the power steering pump operates make the hydraulic fluid service relatively severe, particularly in passenger cars. The operating speed of the pump at maximum engine speed may be more than 8 times what it is at idle speed, yet at idle speed the output must be sufficient to provide whatever power assist is required. This means that at high speeds, the pump output far exceeds the requirements of the system, and the excess must be recirculated internally through a flow control (or relief valve) and the reservoir. Because the quantity of fluid must be kept low to keep the system compact, fluid temperatures can be quite high and severe shearing of the fluid may occur. Under these conditions, wear of the pump vanes or rollers may be a concern.

1. Electric Power Steering

The increased focus on fuel economy has led to the rapid developments in the use of electric-assist power steering. By using an integral electric motor, providing the mechanical assistance to the steering effort, the hydraulic system and its associated power draw can be eliminated. The mechanical components of the electric power steering are lubricated with grease.

B. Lubricant Characteristics

In manual or linkage-type steering gear units, most U.S. manufacturers fill their units with semifluid grease to minimize leakage during installation and shipment. Some automotive units, however, are filled with a multipurpose gear lubricant of API GL-4 or GL-5 quality. For tractors and similar equipment, the lubricant may be a multipurpose fluid designed for use in tractor hydraulic systems, transmissions, and final drives. Except in the latter case, the lubricant used for make-up in the field is usually a multipurpose gear lubricant of about SAE (Society of Automotive Engineers) 80W-90 or 85W-140.

Hydraulic power steering systems require an oil with exceptional low temperature properties and a robust antiwear system to ensure reliable operation. Automotive original equipment manufacturers will often use an automatic transmission fluid for use in their power steering systems, whereas some will use a specifically formulated power steering fluid. It is critical for the safe operation of the steering system that the car manufacturer's recommended fluid always be used.

III. WHEEL BEARINGS

Wheel bearings for automotive equipment are of the rolling element type. Because the bearings of steerable wheels must carry considerable thrust loads in addition to the radial loads, they must be angular contact ball, taper, or spherical roller bearings. Usually, the bearings are installed in pairs so that each bearing is subjected to thrust loads only in one direction. Bearings of nonsteerable wheels are not subjected to high thrust loads—thus, deep groove ball bearings or cylindrical roller bearings may be used. In many cases, only a single bearing is required.

Most wheel bearings in passenger car applications are designed for grease lubrication. Most current production vehicles with front wheel drive are equipped with wheel bearings that are "packed for life" on assembly and generally do not require repacking during the life of the vehicle. Some wheel bearings on auxiliary automotive equipment have fittings for relubrication. The bearings of nonsteerable driving wheels that are connected with axles through differentials or final drives are generally lubricated with the gear lubricant from the drives and do not require relubrication from an external source.

Front wheel bearings of rear wheel drive cars and light trucks are designed to be removed periodically for cleaning, inspection, and relubrication. Repacking can be done by hand or with a bearing packer. Generally, the latter is preferred because it is quicker, somewhat less wasteful of grease, and less skill is required to accomplish a satisfactory packing job. On assembly, the shaft and housing should be coated lightly with the same grease as a corrosion prevention measure; however, the housing should never be filled with grease as this may cause overheating due to churning. Only specially trained personnel should perform the entire wheel bearing repacking. The final adjustment of the wheel bearing, running clearance, and preload is critical, and specifications vary from one vehicle to another. Further removal of the disc brake caliper, when necessary, requires mechanical expertise.

A. Lubricant Characteristics

Greases for wheel bearings are expected to provide acceptable performance over long intervals. They must resist softening and leakage, and hardening, which will cause increased rolling resistance and heating. They should also provide good rust and corrosion protection as well as reduced friction and protection against wear.

In stop-and-go city driving, the use of disc brakes will result in higher operating temperatures for the wheel bearings as compared to drum brakes. As a result, greases with higher dropping points and better high temperature stability are generally required for those vehicles equipped with disc brakes.

In the past, wheel bearing greases were specialty short fiber, sodium soap products. These single-purpose greases have all but disappeared from use and have been replaced with multipurpose automotive greases. Most of the types of multipurpose greases mentioned as being suitable for suspension and steering linkage components are being used for wheel bearing lubrication. These are usually of the NLGI No. 2 or No. 3 consistency. Where higher temperature capability is required for disc brake equipped vehicles, lithium complex or comparable products using other thickeners with higher temperature capability may be used. For consolidation purposes and to eliminate possible cross mixing of products, it might be advisable to standardize on one product, which meets all of the user's requirements.

IV. BRAKE SYSTEMS

Most service brake systems are hydraulically operated. There are two types generally in current service: (1) conventional systems that use disc brakes, drum brakes, or a combination of these (Figure 12.10) and (2) antilock braking systems (ABSs), which also use these components but are computer controlled to avoid wheel lockup upon hard braking (Figure 12.11).

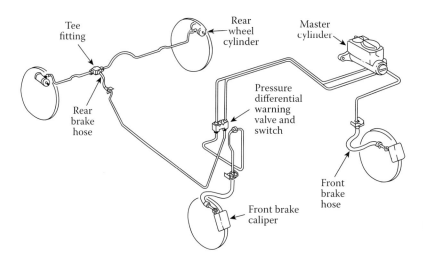

FIGURE 12.10 Conventional braking system.

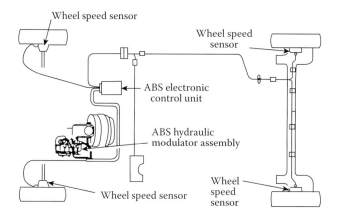

FIGURE 12.11 Antilock braking system.

U.S. National Highway Traffic Safety Administration regulations require a dual or two-sided system so that some braking capability is retained if a failure occurs on either side of the system. The common practice is to operate the front wheel brakes from one side of the system and the rear wheel brakes from the other side. Also used are "dual diagonal" systems, in which a front wheel and the diagonally opposite rear wheel are paired.

Operation of the brake pedal forces fluid from the master cylinder under considerable pressure to the wheel cylinders or disc brake calipers. In turn, the wheel cylinders act through connecting links to force the brake shoes out against the surfaces of the brake drums, or to force brake pads against the sides of the brake rotor (disc). A mechanical linkage to the rear wheel brakes is usually provided to serve as a parking brake, although some current production vehicles use a hydraulic interlock to actuate the parking brake system.

Disc brakes are widely used on front wheels of passenger cars and are installed on all four wheels on many current production vehicles. Compared to drum brakes, disc brakes are most resistant to brake fade or loss of effectiveness due to heat buildup with repeated application, and they also usually provide smoother stopping. As higher application pressure is required, disc brakes are nearly always equipped with a power assist system to reduce the pedal pressure required.

Generally, power assist systems involve a double-acting piston, or diaphragm, coupled to the master cylinder piston. Normally, both sides of the piston are under vacuum supplied by engine manifold vacuum through a reservoir. When the brake pedal is depressed, a valve is opened to allow atmospheric pressure to act on one side of the piston assisting the pedal action in moving the master cylinder piston.

A. ABSs

There are several disadvantages to the conventional braking systems. During hard stops, lockup of the wheels can occur, which makes it more difficult to steer the vehicle and also increases the stopping distances required. ABS brakes address both of these areas by regulating hydraulic pressures to each wheel to prevent lockup under hard braking. The ABS is designed to activate when the vehicle is moving above a predetermined speed (generally above 3 miles/h) and hard braking occurs. Under other conditions of low speed or normal braking, the ABS will respond similarly to conventional brake systems.

ABS components consist of a master cylinder, hydraulic lines, brakes, and booster similar to conventional brake systems. Also included are individual wheel speed sensors, hydraulic pressure modulators, solenoid valves, and the ABS control unit (Figure 12.11), which is the heart of the system. The speed sensor for each wheel generates a voltage signal that is sent to the control module. The control module converts this signal to wheel rotational speed. When the control unit senses wheel lockup under hard braking, it activates the affected wheel's braking solenoid operated valve, which then regulates the hydraulic pressure to control braking without lockup occurring.

It is noteworthy to add that when the ABS is activated, there will be pulsations felt in the pedal because of regulation of the fluid needs and pressure within the system. A rapid clicking type of noise will also be heard owing to the operation of the solenoid valves, and some vehicle body vibration may be felt because of the cycling of the individual wheel brakes (front to rear), causing suspension movements as brake pressures are modulated. The brake pedal on ABS should never be "pumped," as the control unit essentially does this function.

B. Other Braking Systems

Hydraulic retarders for dynamic braking are used on some heavy vehicles equipped with automatic transmissions. A torque converter does not transmit power well in the reverse direction, so with an automatic transmission, the engine cannot be used as effectively for dynamic braking. With a hydraulic retarder, which operates as a torque converter coupled in the opposite direction, energy from the wheel is converted into heat energy in the fluid in the retarder, and then dissipated to the atmosphere.

Another type of assistance is called "power boost" or "hydroboost," which is a hydraulically operated power brake. The hydraulic booster consists of an open center spool valve and a hydraulic cylinder combined in a single housing. A dual master brake cylinder bolted to the booster is actuated by a push rod projecting from the booster cylinder. The power steering pump provides the hydraulic fluid under pressure to the booster cylinder. The master brake cylinder and the braking system use conventional brake fluid.

Oil immersed or "wet" brakes are used on some crawler and wheeled tractors as well as on many other types of tractors and heavy construction and mining-type equipment. These brakes operate submerged in the lubricant for the final drive. Actuation is hydraulic, using fluid from the main hydraulic system on the machine. These systems are discussed in more detail at the end of Chapter 11.

C. Fluid Characteristics

Primarily, the fluid in brake systems is a hydraulic fluid, but it must also lubricate the elastomer seals in the wheel cylinders and calipers as well as protect system metal parts against rust and corrosion. In practice, it has been found almost impossible to exclude moisture completely from brake systems. To prevent this moisture from collecting and freezing in cold weather or causing rust and corrosion, which can cause brake failure, the usual approach is to use brake fluids that are miscible with water. Petroleum-based fluids do not meet this requirement, so most brake fluids are based on glycols (alkylene glycol and alkylene glycol ethers).

In operation, considerable heat is transmitted from the friction surfaces of the brakes to the fluid. If this raises the temperature of the fluid sufficiently to cause vaporization of some of the fluid, braking effectiveness will be reduced because of the compressibility of the vapor. The problem may be more severe with disc brakes, because the smaller areas and higher pressures of the friction pads may result in higher operating temperatures. With higher brake pad temperatures, more heat is transferred through the wheel calipers to the brake fluid.

The absorption of moisture by a brake fluid lowers the boiling point, because water boils at a lower temperature than the glycols used as a base for brake fluids. For this reason, brake fluid specifications include both an equilibrium reflux boiling point, which is the boiling point under reflux conditions of the pure fluid, and a wet equilibrium reflux boiling point, which is the boiling point of the fluid when it is contaminated with a specified amount of water. To some extent, the amount that these boiling points can be raised is restricted because higher boiling point glycols also have higher viscosities at low temperature and may not provide satisfactory brake performance in cold weather.

The most widely accepted brake fluid specification is the U.S. Federal Motor Vehicle Safety Standard (FMVSS) No. 116, which outlines the requirements of the Department of Transportation (DOT) brake fluid grades. Fluids meeting this standard generally are suitable for all normal brake systems designed for nonpetroleum fluids over a wide range of vehicle applications. FMVSS 116 defines the minimum dry and wet boiling points and the viscosity parameters of the DOT brake fluid grades among other things, such as elastomer compatibility and corrosion performance.

DOT 3 fluid is glycol based and is used in most cars and light trucks under normal driving conditions. It also meets the requirements of the SAE Standard J1703, Motor Vehicle Brake Fluid, which is generally similar to the DOT 3 Standard, but somewhat less restrictive in its requirements. DOT 3 fluid has the lowest minimum dry and wet boiling points as defined by FMVSS 116.

Some manufacturers specify a higher boiling point fluid for vehicles with disc brakes; and higher boiling fluids may also be required in certain types of severe service, such as mountain operations, particularly in hot climates, and road racing. DOT 4 fluid is used in these applications. It is also glycol based, but it contains borate esters to increase the dry and wet boiling points. DOT 4 fluid will also meet the requirements of SAE Standard J1704.

DOT 5 fluid is silicone based and does not absorb water the way that the glycol-based fluids do. It is sometimes referred to as a "low water tolerance fluid" or LWTF. DOT 5 fluid is generally only used in vehicles with intermittent usage that sit for long periods as well as in some military applications. DOT 5 fluid should not be mixed with any other DOT fluid as their compatibility cannot be guaranteed.

DOT 5.1 fluid is a glycol-based fluid that meets the high minimum dry and wet boiling points of the DOT 5 silicone-based fluid. It is reserved for the most severe of applications such as heavy tow vehicles or race vehicles. DOT 5.1 fluid is compatible with and can be safely mixed with DOT 3 and DOT 4 fluids.

Most brake fluids contain dye to aid in leak detection, but there is no industry standard for brake fluid colors. As a result, the color cannot be used to distinguish between the various DOT brake fluid grades. Commercial brake fluid colors include clear, pale yellow, crimson red, blue, purple, and violet. Color does not affect the quality or performance of the brake fluid.

In service, it is important that every precaution be taken to keep brake fluids as clean and free of moisture as possible. Containers should be kept sealed when not in use and should never be reused. Master cylinder reservoirs should be kept filled to the proper level because this minimizes the amount of breathing that can occur. Reservoir caps should be replaced properly after the fluid level is checked or fluid is added. A number of manufacturers use translucent plastic reservoirs so fluid levels may be observed without removing the cap. This method reduces the possibility of contaminants being introduced into the system.

V. MEDIUM- AND HEAVY-DUTY TRUCK CHASSIS

As mentioned earlier in this chapter, most medium and heavy-duty trucks are equipped with rigid front axles. The front wheel spindles are mounted on a yoke and kingpin, similar to that shown in Figure 12.4, to allow for steering. The steering system uses spindle arms at each wheel position connected to a cross tube running parallel to the rigid axle via tie rod ends. Each of the tie rod ends has a ball joint. This cross tube, or tie rod assembly, links both steering knuckles for uniform movement and steering control.

Although this arrangement mirrors that of some rear wheel drive passenger cars, because of the weight of the vehicle, the high average annual mileage, and the long life expectancy of the vehicle, each of these components is larger, heavier, and more durable. Furthermore, they will typically each have permanent grease fittings attached to allow for regular regreasing intervals.

Medium and heavy-duty trucks most often will have their front axles attached to the vehicle frame via leaf springs. The attachment points contain a spring pin and shackle assembly, which allows for lengthening of the leaf springs as the vehicle encounters road imperfections and jolts. This spring pin and shackle is actually a long pin in a longitudinal bushing. There are typically two pin connections at the front of each leaf spring and one at the rear for a total of six per vehicle. Recently some of these assemblies have been designed with rubber isolated mounts or with nylon or similar material pins that require no lubrication, but there continues to be many trucks on the road with spring pins and shackles that are regreased via grease fittings at the outboard end of each pin. Figure 12.12 shows a typical rigid front axle system for a heavy-duty truck.

The Pitman arm that transmits steering input is on the upper left, and the cross tube and tie rod ends sit behind the axle. Note the grease fittings showing at the bottom of each tie rod end. The grease fittings for the kingpins, which connect the spindle-steering knuckle assembly to the axle, are shown at the top and bottom of the left and right side joints. The flat areas with the predrilled holes on the top of the axle are the mounting positions for the leaf springs.

Another unique feature of medium and heavy-duty trucks is their pneumatic braking systems. Drum brakes with S-cams at all wheel positions are the most common. Compressed air transmits pressure from the brake pedal to the brake pads, which are extremely thick as compared to passenger car brake pads. As these thick pads wear, the distance between the face of the pad and the inside of the drum—known as slack—will increase such that the brakes become out of adjustment. This added slack can make it difficult for the truck to stop. Automatic slack adjusters are used at each wheel position between the brake air chamber and the S-cam that actuates the brake pads against the drum so that a consistent distance between pad and drum is maintained. The internal working parts of slack adjusters typically include a worm gear, clutch spring, and a spline that are grease lubricated through one permanent grease fitting installed on the housing. There will also be a grease fitting on the tube that houses the S-cam located between the slack adjuster and the back of the brake drum.

FIGURE 12.12 Rigid front axle system for a heavy-duty truck. (Courtesy of Detroit Diesel Corporation.)

FIGURE 12.13 Single rear axle for a medium or heavy-duty truck. (Courtesy of Detroit Diesel Corporation.)

Figure 12.13 shows a typical single rear axle from a medium or heavy-duty truck. The short shafts at both ends in front of the axle that terminate into the back of the brake drums contain the S-cam shafts. The slack adjusters (not shown in this figure) complete the connection between the inboard end of the S-cam shaft and the air chamber at each wheel position.

In these heavier vehicle classes, wheel bearings are usually not grease-packed. The wheel bearings on driven axles are lubricated with the gear lubricant from the differentials and do not require relubrication from an external source. Oil lubrication of nondriven wheel bearings (steer axles and trailer axles) is predominant on medium and heavy-duty trucks. Lubrication with oil requires careful attention to seal condition to prevent oil leakage onto the brakes causing loss of braking capacity or failure. For this reason, use of semifluid grease has become an option for the nondriven axle wheel bearings.

Likely the most recognizable component of a medium or heavy-duty tractor chassis is the fifth wheel. The fifth wheel maintains the tractor and trailer connection while allowing the combination freedom of movement to negotiate turns. The top plate of the fifth wheel mates to the bolster plate on the underside of the trailer, and the trailer's pin is coupled to the lock and jaw mechanism in the center of the top plate. Without effective lubrication, the top plate and bolster plate can weld to one another leading to a loss of vehicle control or even jackknifing. Although a coating of grease on the fifth wheel's top plate is a familiar sight, there are several other fifth wheel components that need lubrication. These typically include pivot points, slider mechanisms (allowing fore and aft adjustment of the fifth wheel on the tractor chassis), lock and jaw mechanisms, and the release handle. The actual lubrication points will vary based on the fifth wheel manufacturer, model, and application. There are some fifth wheels with wear pads on the top plate that do not require greasing, but the tradeoff is a very precise tractor–trailer coupling and uncoupling procedure that, if not followed properly, can lead to wear pad damage.

A. Factors Affecting Lubrication

Chassis components of medium and heavy-duty trucks suffer from the same lubrication challenges associated with light duty vehicle suspensions, but they are magnified by the heavier loading of each component because of the higher vehicle weights. The wiping off of lubricant films due to the pivot points moving through relatively short arcs is one, and the displacement of grease due to contamination with water, snow/ice, dirt, sand, and deicing chemicals is another. Other factors that are unique to these heavier vehicles include infrequent and improper lubrication.

Typically, medium and heavy-duty trucks will have their chassis regreased during a "dry" preventative maintenance (PM) interval that occurs at one-half the mileage of an engine oil change and at the "wet" PM—one that includes an engine oil change. (These are also referred to as an "A" PM and a "B" PM, respectively.) Based on the size and weight of the vehicle as well as its application, this can range from 5000 to 20,000 miles (from 8000 to 32,000 km) or even longer. Proper grease selection is the key to ensuring that the components are protected between PMs. Excessive suspension noise, poor vehicle control, and frequent chassis component replacements are symptoms of a chassis regrease interval that is beyond the capability of the chosen grease. Conversely, some brake slack adjusters suffer from too frequent regreasing, so it is a must to check the slack adjuster manufacturer's recommendation for proper regreasing intervals.

Improper lubrication techniques also account for chassis component problems. For example, it is not possible for spring pins and shackles or the upper halves of kingpins to fully accept the proper amount of grease or to have all of their internal surface contact areas completely covered with grease unless the front axle is in an unloaded position with the wheels off of the ground. Fifth wheels suffer from their grease fittings being overlooked as some are only accessible by tilting the top plate up and out of the way with a pry bar. The top plate is usually subjected to overgreasing, which will cause an accumulation of grease over time that attracts dirt and can interfere with the lock and jaws.

B. Lubricant Characteristics

The grease used in the suspension and steering linkage components of medium and heavy-duty trucks must provide the same seven performance characteristics outlined earlier in this chapter. These vehicles also will typically require grease of an NLGI No. 2 consistency, but they benefit from greases made with a higher oil viscosity (ISO viscosity 220 or 320) than those used for passenger cars and light trucks, again because of the heavier loading of components. Lithium complex based greases with extreme pressure and antiwear additives are widely used. Recently, greases manufactured using a calcium sulfonate thickener technology have been gaining favor because of their outstanding resistance to rust, corrosion, and water washout. Again, colloidal or fine particle size solid lubricant materials such as molybdenum disulfide are often added, especially when extended regreasing intervals are desired. Kingpins, spring pin and shackles, and fifth wheels can be better protected when solid lubricant additives are used in the grease, and most other components can also use the same grease, allowing the use of a single product for the entire chassis. The one possible exception is brake slack adjusters. Some slack adjuster manufacturers limit the amount of solid lubricant additives present in a grease or prohibit their use entirely. They may also have special requirements for the viscosity of the oil in the grease to ensure proper functioning of the internal mechanisms in low ambient temperatures.

Some medium and heavy-duty trucks may be equipped with a centralized automatic greasing system. These systems will vary by manufacturer, but the typical arrangement consists of a grease reservoir, a pump, and a manifold system with grease supply lines to each of the chassis grease points discussed. The reservoir may be mounted on the rear of the cab or inside the engine compartment. These systems are designed to deliver a specific amount of grease to each component at a set time interval. Because of the variety of models and settings, the system manufacturer will make the proper grease recommendation based on the type of vehicle and its duty cycle. The recommendation will likely specify the thickener type, the NLGI grade, the oil viscosity, and the presence or absence of solid lubricant additives.

Oil-lubricated wheel bearings on nondriven axles are usually lubricated with the same multipurpose, heavy-duty automotive gear lubricant being used in the truck's differential for convenience. These are typically API GL-5 multiviscosity gear lubricants such as 80W-90 or 85W-140 mineral oils or 75W-90 or 80W-140 synthetic oils. Nondriven wheel bearings do not necessarily require the extreme pressure characteristics that these gear oils provide, so the same API GL-4 SAE 50 synthetic oil frequently used in heavy manual or automated manual truck transmissions is a common alternative.

BIBLIOGRAPHY

National Highway Traffic Safety Administration, U.S. Department of Transportation. 2011. 49 CFR 571.116 — Standard No. 116; Motor vehicle brake fluids. Washington, DC.

13 Stationary Gas Turbines

There are three commercially important types of prime movers that provide mechanical power for industry: the steam turbine, the reciprocating piston engine, and the gas turbine. All types achieve their result by converting heat into mechanical energy; any machine that does this (uses chemical reactions to supply thermal energy) is properly called a heat engine.

Of the basic prime movers, the gas turbine is the most direct in converting heat into usable mechanical energy. The steam turbine introduces an intermediate fluid (steam), so the products of combustion (from the chemical reaction) do not act directly on the mechanism creating motion. Piston engines initially convert heat into linear motion that must be transmitted through a crankshaft to produce usable shaft power. But the gas turbine, in its simplest form, converts thermal energy into shaft power with no intermediate heat or mechanical redirection.

The theory of gas turbines was well understood by about 1900, and although development work started shortly after that, it was not until 1935 that engineers overcame low compressor efficiency and temperature limitations imposed by available materials and succeeded in building a practical gas turbine.

Just before World War II, the Swiss successfully built and operated gas turbine plants for industrial use. In the United States, the first large-scale gas turbine application was in a process for cracking petroleum products.

Near the end of World War II, both Germany and England succeeded in developing turbojet-propelled aircraft. Turbojet and related designs for aircraft propulsion are an extension of the industrial gas turbine. These aircraft turbines have been adapted for stationary power applications and marine propulsion, and are referred to as aeroderivative engines.

Gas turbines have been used in power generation since the 1940s. The early units had the capacity to produce 3500 kW h of power output with thermal efficiencies in the 20–25% range. Today's gas turbines can produce power outputs of more than 340 MW and attain thermal efficiencies of approximately 40% for a simple cycle open system, making them competitive with other forms of prime movers. They have gained a place in stationary applications because of several advantages: lower emissions, less operating personnel required, comparative light weight, small size, ease of installation, quick starting, multifuel capabilities (e.g., natural gas, propane, distillate), minimal cooling water requirements (for some designs), and availability of large amounts of exhaust heat. Recovery of this exhaust heat in a steam turbine or other in-plant heating requirements is raising the efficiency of the combined cycle application to more than 60%.

Gas turbines in power generation will operate at speeds that range from 3000 to 12,000 rpm and drive generators that rotate at 1800 or 3600 rpm for 60 Hz power or 1500 or 3000 rpm for 50 Hz power. Turbines operating above 3600 rpm (60 Hz) or 3000 rpm (50 Hz) are connected to generators through reduction gears. Europe and much of Asia and Africa typically operate at 50 Hz, whereas America, some Middle East, and some Southeast Asian countries have units that operate at 60 Hz.

I. PRINCIPLES OF GAS TURBINES

Gas turbines basically consist of an axial compressor, to compress the intake air, a combustor section, and a power turbine. The compressed air is mixed with fuel (liquid or gas) and burned in combustion chambers (combustors). The hot gases expand through a turbine or turbines to drive the load. There may be a single shaft with a single turbine to drive both the compressor and the load, or two shafts with a high-pressure turbine to drive the compressor and a low-pressure turbine to drive the load.

A. The Simple Cycle, Open System

The simple cycle, open system gas turbine consists of a compressor, combustor, and turbine (Figure 13.1). The compressor draws in atmospheric air, raises its pressure and temperature, and forces it into the combustor. In the combustor, fuel is added, which burns on contact with the hot compressed air, boosting its temperature and heat energy level. The hot, compressed mixture travels to the turbine where it expands and develops mechanical energy (i.e., torque applied to a shaft). The largest portion of this energy is needed to drive the compressor; the rest is available to drive a useful load such as a generator and a compressor. Figure 13.2 illustrates how temperature and pressure rise and fall within this type of gas turbine system. It is the gaseous working fluid that gives the gas turbine its name—not the fuel, which may be liquid or gaseous.

The flow of energy through the cycle is shown in Figure 13.3. The entering air contributes zero energy. (Actually, shaft work into the compressor is transferred to the compressed air to add some energy to it as it enters the combustor.) Fuel introduced into the combustor represents net energy input, balanced by work output (useful shaft work) plus exhaust energy vented to the atmosphere. The turbine work required to drive the compressor simply circulates within the cycle—from shaft to compressed air to combustor energy and back to turbine work. Surprisingly, of the total shaft work developed by the turbine, almost two-thirds must go to drive the compressor, leaving only a little more than one-third for useful shaft work. Improvements in compressor efficiencies and component metallurgy are increasing the amount of available shaft power.

A common variation of the simple cycle, open system design is the two-shaft type (Figure 13.4). The turbine is divided into two stages—a high-pressure stage that drives the compressor and a low-pressure stage that drives the load. The low-pressure (free or power) turbine is not mechanically connected to the high-pressure turbine. Its speed can be controlled independently to suit the speed of the driven unit, whereas the speed of the high-pressure turbine can vary as needed to develop the required power.

The thermal efficiency of the simple, open cycle gas turbine is improving, currently reaching a maximum of about 40%. Efforts to improve thermal efficiency have included designing for as high a turbine inlet temperature as practicable and as high a compressor pressure ratio as practicable up to an optimum value. Improvement can also be made through recovery of exhaust heat, which is discussed later.

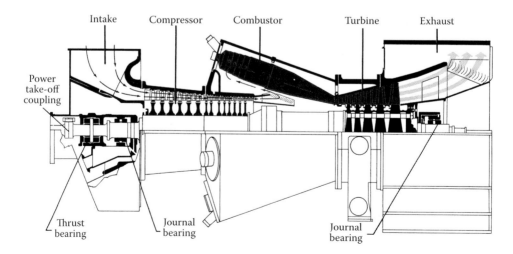

FIGURE 13.1 Gas turbine of the simple cycle, open system single shaft design.

Stationary Gas Turbines

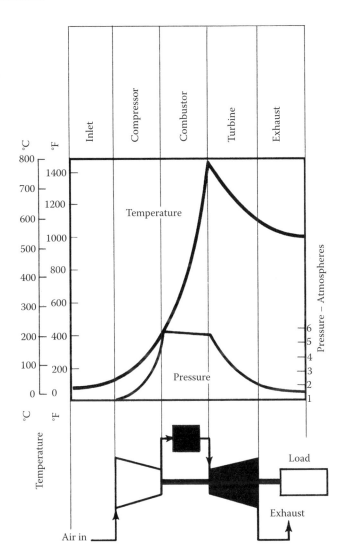

FIGURE 13.2 Graph showing approximately how gas temperatures and pressures change in flowing through a gas turbine of the type shown in Figure 13.1.

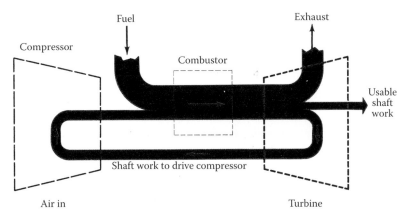

FIGURE 13.3 Diagram showing flow of energy in simple cycle, open gas turbine.

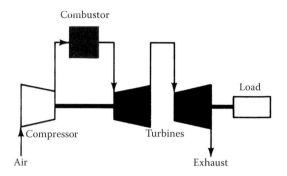

FIGURE 13.4 Simple cycle, open system two shaft gas turbine.

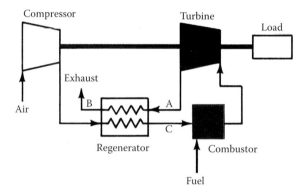

FIGURE 13.5 Regenerative cycle, open system, showing use of exhaust to heat entering air to improve thermal efficiency.

B. Regenerative Cycle, Open System

The regenerative cycle was designed to make use of exhaust heat (Figure 13.5). The regenerator receives hot, turbine exhaust gas at A and rejects it at B. Compressed air passes through the unit in a counterflow direction and picks up heat from the turbine exhaust; then the heated air leaves the regenerator at C and passes into the combustor. Hotter air in the combustor needs less fuel to reach the maximum temperature before flowing into the turbine; thus, thermal efficiency is improved. This conforms to a basic rule for heat engines, which states that the effect of raising the mean effective temperature at which heat is added while lowering the mean effective temperature at which heat is rejected increases a cycle's thermal efficiency.

C. Intercooling, Reheating

The compressor tends to be an inefficient machine; the ratio of energy at its outlet to that at its inlet is low. A detailed discussion of compressors appears later. The elementary theory, however, shows that compression in which the temperature of the gas does not rise during compression requires less work than compression with increasing temperature. To partially achieve the former, the compression system is sometimes divided into two stages and the working fluid (air) is withdrawn and cooled between stages, as shown in Figure 13.6.

Intercooling gives more net work per pound of working fluid allowing a smaller turbine plant for the same output or a greater output from the same size turbine. Also, where combustor temperature

Stationary Gas Turbines

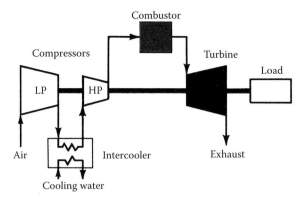

FIGURE 13.6 Intercooling between compressor stages to reduce work of compression and size of machine.

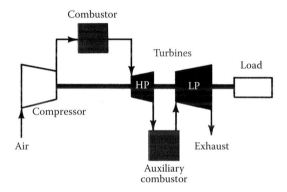

FIGURE 13.7 Reheating between turbine stages to increase power output.

is maintained, the portion of usable shaft work developed increases because less shaft work is required at the compressor.

Just as the compression process may be divided, so may the expansion process, with reheating between stages instead of intercooling (Figure 13.7). Reheating, ideally to the same top temperature in each stage, offers a straightforward way of boosting output with little increase in plant size, although a combustor is needed for each stage and the turbine is correspondingly more complex. With higher temperatures at the exhaust, the regenerative cycle is almost always used. Without a regenerator, there is little gain in thermal efficiency.

D. Essential Gas Turbine Components

1. Compressor

The key to successful gas turbine operation is an efficient compressor. Early turbine plants were troubled with units that could not handle large-volume air flows efficiently. Today, almost all compressors in industrial gas turbine applications are axial flow or centrifugal, or a combination of the two.

In the axial flow compressor, blades on the rotor have air foil shapes to provide optimum air flow transmission. Moving blades draw in entering air, speed it up, and force it into following stationary vanes. These are shaped to form diffusers that convert the kinetic energy of moving air to static pressure (Figure 13.8). A row of moving blades and the following row of fixed blades are considered

FIGURE 13.8 Axial flow compressor blading.

a stage; air flows axially through these stages. The number of axial flow compressor stages in a modern industrial gas turbine varies from 12 to 22. Pressure ratios range from 5.5:1 to 30:1. With intake air at 14.7 psia (1.03 kP/cm^2) and 60°F (15.6°C), and pressure ratio of 6.45:1, air will be compressed, in an uncooled compressor, to about 94.7 psia (6.66 kP/cm^2) and 426°F (219°C). Blade heights become smaller as air travels through the axial flow compressor because the air's specific volume diminishes at the increasingly higher pressures.

Air displacement of axial flow compressors varies from 8 to 20 cfm (13.6–34 m^3/h) of working fluid per horsepower. Large-volume, multistage machines may displace more than 300,000 cfm (510,000 m^3/h), with temperature rises of 400°F (204°C) and more. Compressor efficiencies range from 85% to 92%, although performance hinges on minimum fouling of blades and diffusers.

Although the centrifugal compressor can be multistaged, its best performance is as a single-stage machine operating at a pressure ratio not exceeding 4 or 5:1. In comparison to the axial flow compressor, the centrifugal machine operates more efficiently over a wider range of mass flow rates, is more rugged in construction, and is less susceptible to contaminant deposits left by air on flow passages. However, the machine's application in industry is limited to small gas turbines because it cannot handle large volumes of air efficiently at higher pressure ratios.

2. Combustor

For efficient combustion, the combustor, or combustion chamber, must assure low pressure losses, low heat losses, minimum carbon formation, positive ignition under all atmospheric conditions, flame stability with uniform outlet temperatures, and high combustion efficiencies.

The combustion section may comprise one or two large cylindrical combustors; several smaller tubular combustors; and annular combustor with several fuel nozzles compactly surrounding the turbine; or an annular combustor in which several tubular liners, or cans, are arranged in an annular space (Figure 13.1). For all these, the design is such that less than a third of the total volume of air entering the combustor is permitted to mix with the fuel; the remainder is used for cooling. The ratio of total air to fuel varies among engine types from 40 to 80 parts of air, by weight, to one of fuel. For a 60:1 ratio, only about 15 parts of air are used for burning; the rest bypasses the fuel nozzles and is used downstream to cool combustor surfaces and to mix with and cool the hot gases before they enter the turbine. Dry low NO_x burners that reduce emissions require more precise air/fuel ratio control that is often actuated by lubricating oil.

Natural gas is the primary fuel type for gas turbines but other commonly used alternate fuels are classified as:

1. Low heating value (LHV) gaseous fuel (e.g., blast furnace gas)
2. Medium LHV gaseous fuels (e.g., landfill gas)
3. High LHV gaseous fuels (e.g., propane)
4. High hydrogen gaseous fuels (e.g., refinery gas)
5. Liquid fuels (e.g., diesel oil or jet fuel)

3. Turbine

This component is an energy converter, transforming high-temperature and high-pressure gases into shaft output. Gas turbines follow impulse and reaction designs—terms that describe the blading arrangement. The distinction depends on how the pressure drop at each stage is divided. If the entire pressure drop takes place across the fixed blades and none across the moving blades, it is an impulse stage (Figure 13.9a). If the drop takes place in the moving blades, as well as the fixed blades, it is a reaction stage (Figure 13.9b).

A row of fixed blades ahead of a row of moving blades constitutes a stage. Many turbines are multistaged, because the amount of power that can be developed by a single stage of reasonable size, shape, and speed is limited.

For best performance, the gas turbine must work with high inlet temperatures. Manufacturers continually strive to build turbines that can accommodate higher temperatures, either through improved materials or through blade cooling designs. For the most part, design improvements focus on cooling the parts subjected to highest temperatures and stresses, particularly the first stage or two of the turbine. One cooling technique is to take air bled from the compressor outlet and channel it over parts holding the blades. Another method is to extend the shank linking blade bucket to the rotor, thus removing rotor parts from the hottest concentration of gases.

More than 30 years ago, inlet temperatures of 1200°F (649°C) were considered extreme; today, as a result of continuing research, the ability of turbine materials to withstand extremely

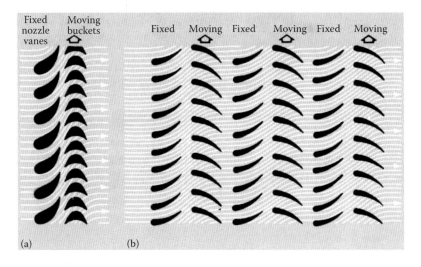

FIGURE 13.9 (a) Impulse and (b) reaction turbine blading.

high temperatures permits more efficient operation. Temperatures in the range 1600–1700°F (871–927°C) are seen in fully loaded units with some designs operating with firing temperatures at 2300°F (1260°C). By the time these gases reach the turbine exhaust, their temperature has dropped at least 600°F (333°C), and possibly by as much as 1200°F (649°C)—depending on inlet temperature and the number of stages. Today's high-output gas turbines can have exhaust temperatures of about 1100°F (594°C).

Compressor efficiencies have substantially improved over the past years, which have led to these increased efficiencies of the turbines. Compressors can have 22 stages and compression ratios that exceed 30:1. This increase in compressor efficiency has helped raise firing temperatures, which, in turn, has increased overall turbine efficiencies. Firing temperatures range from 1700°F (927°C) for sequential combustion units (e.g., ABB GT 24, where there are two burn stages) to more than 2300°F (1260°C) for single combustion units (e.g., GE MS 7001 F) that are fully loaded.

II. JET ENGINES FOR INDUSTRIAL AND MARINE PROPULSION USE

In several applications, aircraft jet engines are being used for industrial service and are often referred to as aeroderivative engines. Initially, instead of providing propulsion power directly, the hot compressed gases from the engine were fed to a power or free turbine, which converted the heat energy into rotary power (Figure 13.10). Used in this manner, the jet engine acts as a gas generator, providing the working fluid (gases) to power a turbine. As many as 10 jet engines have been used as gas generators for one large turbine. This provided a compact, lightweight design and the ability for rapid replacement of components.

The compactness and light weight of the jet engine is partially explained by its relatively high speed—usually in the range 8000–20,000 rpm (Figure 13.11). Large industrial gas turbines, on the other hand, usually run at speeds of 3000–12,000 rpm.

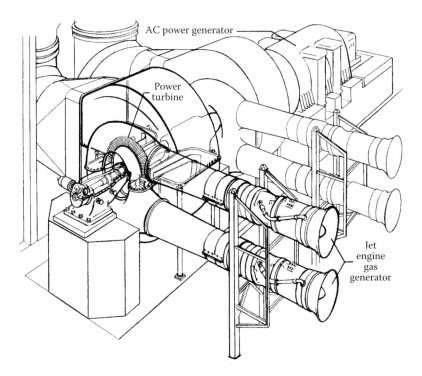

FIGURE 13.10 Two jet engines provide hot gas to each of the two power turbines coupled to a single generator.

Stationary Gas Turbines

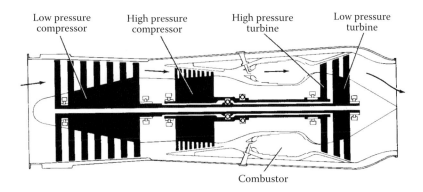

FIGURE 13.11 Aeroderivative-type generator of dual-axial or twin-spool design.

The use of direct connected aircraft jet engine designs has been adapted for industrial service, power generation, and marine propulsion. Because jet engines are mass produced, they offer several advantages. Replacement parts are readily available, and even entire engines can easily be replaced. This enables interchangeability and factory overhauls instead of maintenance in the field. This light weight reduces foundation needs and makes installation easier. Other advantages include cold startups to full-load operation in less than 2 min in most applications and operation without on-site personnel. Also, simple cycle thermal efficiencies of 40% or more are possible with some of the newer-design jet engine derivatives.

However, bearings of jet engines (practically always the rolling-element type) run hotter than those of typical industrial gas turbines. Most of the bearings are "buried" within the engines and surrounded by hot gases. Bearing housings are sealed and pressurized with hot air bled from appropriate compressor stages. The resulting problem is twofold: (1) removing the heat rejected at the bearing, and (2) finding lubricants that can stand up to bearing temperatures of 400°F (204°C) with metal temperatures up to 600°F (316°C). Sump lube oil temperatures can range from 160°F to 250°F (71°C to 121°C). To address this challenge, higher lubricant flow rates and the use of high-temperature synthetic ester lubricants, which are discussed in Chapter 5, are recommended.

Average lube oil makeup rates of 0.014 gal/h (0.053 L/h) will help rejuvenate the turbo oil under these difficult conditions. Lube oil consumption rates that approach 0.4 gal/h (1.5 L/h) are considered excessive. Current turbine oils for aeroderivative, land-based power generation turbines are often described as 5 cSt turbo oils. Aeroderivative turbines operate with much smaller lube oil sumps, typically 150 gallons (570 L) or less. A key performance characteristic for in-service turbo oil is deposit control.

A. SMALL GAS TURBINE FEATURES

Not all industrial gas turbines are large machines delivering thousands of horsepower. A great many are machines of less than 1000 hp rating, and a surprisingly large number of units are less than 100 hp in size.

Many different makes of small gas turbines are available. Most follow the design practices of larger industrial turbines, using plain (journal) bearings located in relatively cool areas. Others, however, including those designed for frequent start–stops, especially at low temperatures, use rolling element bearings, which are also located in relatively cool areas. Some makes are available with either plain or rolling element bearings, depending on the type of service anticipated.

Some relatively small gas turbines use axial flow compressors and turbines. Others, in order to achieve compactness, use centrifugal compressors (usually one but sometimes two stages) or, as in

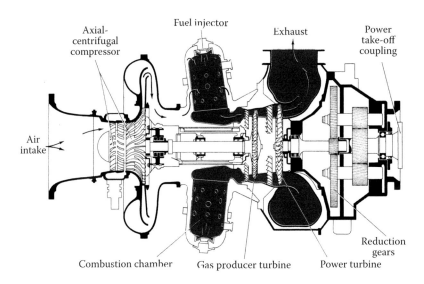

FIGURE 13.12 Small (maximum 360 hp) two-shaft gas turbine for mechanical or generator drive. Shaft bearings of the gas producer and the power section are of the plain, or sleeve, type. Both plain and rolling element bearings are used in the reduction gear box.

Figure 13.12, an axial stage followed by a centrifugal stage. Some use a radial in-flow turbine rather than an axial one. Small turbines usually run at speeds ranging from about 18,500 to 60,000 rpm.

Several small gas turbines are being developed for vehicle use, using regenerators, high turbine inlet temperatures, and other means to increase thermal efficiency and reduce fuel consumption. Most of these machines are of two shaft type, with gas generator speeds ranging from 22,000 to 65,000 rpm and power turbine speeds ranging from 12,000 to 46,000 rpm.

III. GAS TURBINE APPLICATIONS

Gas turbines, in sizes from 50 to more than 200,000 hp, have wide application in industry today. Four broad areas of usage stand out: electric power generation, pipeline transmission, process operations, and total energy applications. These four areas indicate the gas turbine's versatility and suggest its wide-ranging potential as an efficient power source. The gas turbine's success in these applications, along with increased thermal efficiencies, will doubtless make them more desirable in other areas as well.

A. Electric Power Generation

Utilities use gas turbines for peaking generation and for emergency service. Most units exceed 10,000 hp, and both conventional turbines and jet engines adapted for industrial service are used. System loads in a utility vary daily and seasonally. For efficient operation, base loads are usually supplied by steam turbines, with peak loads handled by gas turbines. In emergency shutdowns of base load equipment, gas turbines can take over when necessary. In peaking applications, a gas turbine may operate only a few hundred hours a year; this arrangement is the most efficient for year-round generation. As gas turbine operation has become more economical and convenient, the units are being used for longer periods because they are capable of giving reliable service. More and more gas turbines are being used as base loaded units because of the improved efficiencies and reliability as well as considerations for fuel costs and environmental emissions.

Outside the utility, gas turbines are used to drive generators in industrial plants and buildings on an emergency standby basis. Sometimes they are used as peaking units to supplement on-site, steam plant power. Along with diesel and gas engines, gas turbines are finding application in a new concept called continuous duty/standby, in which they are normally coupled to a nonessential load such as the compressors in an air conditioning system. In emergency blackout situations, the nonessential load is dropped and a generator supplying essential electrical loads such as elevators, exit lighting, and critical equipment is picked up. Hospitals, universities, and senior care facilities are finding this form of emergency power extremely reliable.

B. Pipeline Transmission

In the transmission of natural gas, centrifugal compressors are often driven by gas turbines that are fueled by natural gas. This is an ideal arrangement because this preferred turbine fuel is readily available, and the large loads are well suited for gas turbine drives. Units range in size from 1000 to 20,000+ hp, with capacities in the 5000–10,000 hp range being the most prevalent.

C. Process Operations

In this area, more than in any other, the full potential of the gas turbine has yet to be realized. In fact, if properly designed, the gas turbine can become an integral part of the process, as essential as any other key element in the system. In process applications, the turbine is designed not so much to satisfy unit efficiency as to satisfy system efficiency. For example, in addition to generating shaft power, a turbine with an oversized compressor can produce pressurized air for use in steel mill applications. Instead of attempting to utilize exhaust gas to raise turbine efficiency, the gas can be channeled to a process requiring heat, for example, a dryer. Such uses may be accomplished with supplementary firing using the hot exhaust gas (which contains up to 17% of oxygen) as preheated combustion air for kilns or boilers. Hot, pressurized gases form certain processes can be ducted to the turbine's combustor, reducing the amount of fuel needed; the turbine itself may drive process equipment.

D. Combined-Cycle Operation

Combined cycles can achieve thermal efficiency of up to 65% by combining a gas turbine and a steam turbine as shown in Figure 13.13. The high temperature energy in the gas turbine's exhaust gas is extracted in a heat recovery steam generator (HRSG). The steam is fed to a steam turbine, and

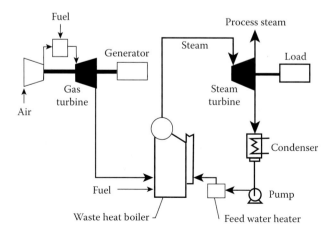

FIGURE 13.13 Gas turbine, steam turbine combined-cycle plant.

the exhaust steam is used either for process heating or returned to a condenser as boiler feed water. Combined cycle plants are offered ranging from a net plant base load output of 12 to 1200 MW.

Another method of combined cycle use passes the exhaust gases through two HRSGs in series. The first boiler produces superheated steam at 400 psia and the second, saturated steam at 104 psia. The high-pressure steam is fed to the inlet of a mixed-pressure steam turbine, and the low-pressure steam is fed at an intermediate stage of the same steam turbine. The combined steam flow exits the turbine to the main condenser. The gas turbine, steam turbine, and compressor are mounted on a single shaft.

E. Total Energy

In installations of this type, all building energy is provided at the site (Figure 13.14). Typically, natural gas supplies a gas turbine, which drives a generator to carry the electrical load. Waste heat from the turbine exhaust supplies an HRSG, which generates steam for absorption air conditioning, space heating, and domestic hot water. Some installations use diesels or reciprocating gas engines instead of gas turbines. Plant capacity depends on the size and type of building (e.g., shopping center, school, offices, apartments). Power requirements today range from 300 hp to about 5000 hp. Compact turbines, either single shaft for constant speed applications or two shafts for variable speed applications, are selected. The units are small and light (frequently less than 1500 lb) and can accept full load from startup in a minute or less. Well-designed total energy systems are capable of thermal efficiencies exceeding 65%.

F. Marine Propulsion

Gas turbines are used for marine propulsion when faster ship speed is required and where the ship operates primarily at full throttle. Reduced thermal efficiency with turbines and higher fuel costs often preclude its use in favor of diesel engines.

G. Microturbines

Microturbines are small gas turbines, ranging from 30 kW to 50 MW, used for local distributed power generation for commercial buildings such as restaurants, small shopping malls, and hospitals. Most microturbines are composed of a compressor, combustor, turbine, alternator, recuperator (a device that captures waste heat to improve the efficiency of the compressor stage), and generator.

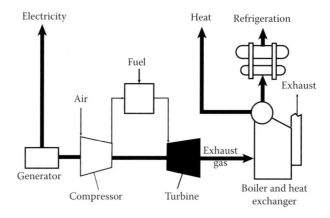

FIGURE 13.14 Use of exhaust heat from gas turbine to provide for absorption refrigeration in a "total energy" plant.

Single shaft units commonly operate at 90,000–120,000 rpm with recuperated efficiencies of just 20–30%. Note that microturbines use air for bearing support and do not require lubrication.

IV. LUBRICATION OF GAS TURBINES

The principal purposes of lubrication are to reduce wear, reduce friction, keep systems clean, remove heat, and prevent rust. The elements of gas turbines that need lubrication include plain and rolling element bearings, thrust bearings, gears, couplings, and contact-type seals.

As noted previously, aircraft-type gas turbines are equipped with rolling element bearings, where most larger frame size industrial gas turbines will use plain (sliding) bearings. Plain bearings used for radial loads may be of the following type:

1. Conventional heavy steel-backed babbitt-lined cylindrical or elliptical type
2. Precision-insert type (as in reciprocating engine practice)
3. Tilt-pad or "antiwhip" type in order to suppress self-induced vibration and improve shaft alignment, which can sometimes be a problem in high-speed machinery

Two of the antiwhip bearings are shown schematically in Figure 13.15. Gears used for speed reduction and for accessory drive are usually of spur, helical, herringbone, or bevel types.

A. LARGE INDUSTRIAL GAS TURBINES

Lubrication systems for the large, heavy-duty, industrial gas turbines are generally similar to those for steam turbines or other high-speed rotating machines.

A typical lubrication system may be described as a pressure circulation system, complete with reservoir, pumps, cooler, filters, protective devices, and connecting tubing, fittings, and valves. Its purpose is to provide an ample supply of cool, clean lubricating oil to the gas turbine, accessory and reduction gearing, driven equipment, and hydraulic control system.

An example of the arrangement of these parts is shown in Figure 13.16. A gear-type positive displacement main pump, driven by the power turbine, supplies pressurized oil to each of the main compressor and turbine bearings and to accessory gearboxes. Return oil from the bearings and gearboxes drains by means of gravity to the reservoir. Foot valves in the tank prevent oil in the supply system from draining back to the sump during shutdown periods.

The pressurized oil passes through an oil cooler and filter (specified as 5 mm by some manufacturers) before reaching the bearings. A pressure relief valve installed in the supply line maintains pressure at a constant level.

A motor-driven standby pump automatically supplies pressurized oil to the bearings as a pre-lube before starting and also for a few minutes after shutdown to prevent "heat soak" damage to the bearings. The standby pump also takes over automatically should the main pump fail during

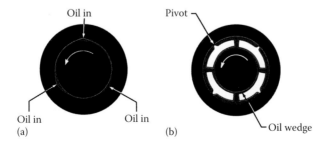

FIGURE 13.15 Antiwhip bearings—three-lobe type (a) and tilting-pad type (b). Multiplicity of fluid films suppresses the tendency for self-induced vibration of lightly loaded high-speed shafts.

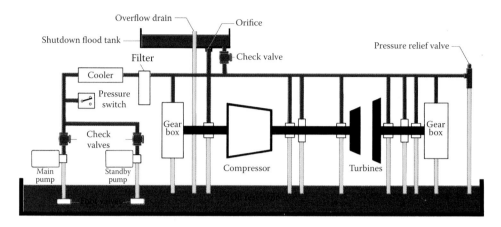

FIGURE 13.16 Lubrication system for large industrial gas turbine.

operation. Some lubrication systems, as shown in Figure 13.16, also include a flood tank to build up a reserve supply of oil during operation. This supply can flood the bearings by means of gravity for 15–20 min after system failure due to emergency power loss.

Protective devices are incorporated into the lubrication system to guard equipment against low oil supply, low oil pressure, and high oil temperature. The devices sound a warning or shut down the unit if any of these conditions occurs. In some installations, in the event of emergency power loss a pump driven by a battery-powered direct current motor is available for supplying pressurized oil on a short-term basis. Also, in some installations, an additional cooler and filter may be included. Oil flow can be directed to one filter and cooler while the other filter and cooler are cleaned during operation of the gas turbine, thus avoiding shutdown. In some installations, oil heaters are also added.

Bearing oil heater pressures are usually controlled at 15–40 psi (1.1–2.9 kP/cm^2). Hydraulic control pressures are often in the 50–150 psi (3.5–10.5 kP/cm^2) range, but much higher pressures are used in some systems.

Bearing and gear loads are generally moderate and shaft speeds are high, permitting the use of rust and oxidation inhibited type oils of "light" or "medium" viscosity.* Oil temperatures entering bearings and gears are usually in the 120–160°F (49–71°C) range, and temperature rise in these elements is 25–60°F (14–33°C). Bulk oil temperatures in the reservoir will usually be 155–200°F (68–93°C). These temperatures are well within the thermal limitations of high-quality petroleum lubricating oil of suitable type. Starting temperatures may be low, so oils of high viscosity index (such as some synthetics), which do not thicken excessively at low temperatures, are recommended. Most gas turbine manufacturers specify an ISO (International Standardization Organization) 32 viscosity oil that maintains enough viscosity at operating temperature to negate metal-to-metal bearing contact and still be pumpable by jacking pumps at startup conditions.

Base oils used in fully formulated gas turbine oils may be from API (American Petroleum Institute) group I, II, III, IV, or V, or can be a combination of base oils, depending on the specific application.

The lubricating oil is recirculated in contact with air. Portions of it are broken into small drops or mist so that intimate mixing with air occurs. Under these conditions, oil tends to oxidize, and the tendency is increased by high operating temperatures, by excessive agitation or splashing of oil, and by the presence of certain contaminants that act as catalysts. The rate of oil oxidation also depends on the ability of the oil to resist this chemical change.

---

* ISO viscosity grades 32, 46, or 68 are commonly recommended.

Oxidation is accompanied by the formation of both soluble and insoluble products and by a gradual thickening of the oil. The insoluble products may be deposited as varnish or sludge in oil passages, and their accumulation will restrict the supply of oil to bearings, gears, or hydraulic control devices. In some critical control system servo valves, even minute levels of oil degradation materials could contribute to malfunctioning of valves and loss of speed or load control. For long-time performance in reuse systems, the lubricating oil must have high chemical stability in order to resist oxidation and the formation of sludge. Gas turbines that have a common reservoir for bearings and hydraulics are more susceptible to varnish issues than typical bearing only or hydraulic only reservoirs. Hydraulic servo valves have small tolerances (typically 3 μm) that can be fouled from oxidized lubricant. Varnish-specific lube oil analysis has been developed to help turbine users manage this issue.

Water may find its way into bearing housings as a result of condensation of moisture from the atmosphere during idle periods or from other sources. The churning of oil with moisture in circulation systems tends to create emulsions. To resist emulsification, the oil should have the ability to separate quickly from water. With such an oil, any water that enters the bearing housings or circulation systems will collect at low points from which it may be drained. At the normal operating temperature of most gas turbines, water from condensation will evaporate and be removed through vents. It is also important that the oil selected have good antirust and anticorrosion characteristics.

Gas turbine oil service life is dependent on the application and service profile. Base-loaded gas turbines, with separate bearing and hydraulic reservoirs, should expect to see the charge of oil provide reliable service for two major overhaul intervals of approximately 50,000 hours per major overhaul with good maintenance and contamination control. In contrast, cyclically operated turbines or peaking turbines, with a common hydraulic and bearing reservoir, are more prone to servo valve sticking before the major overhaul period of 50,000 hours and may require a more frequent oil drain interval.

B. Aeroderivative Gas Turbines

The lubrication of aircraft gas turbines adapted for stationary service differs substantially from that of large, heavy-duty, industrial gas turbines. The lubrication systems are quite different, being much more compactly designed out of necessity. The bearings, which are rolling element types, and gears, in some instances, are "buried" within the turbines and require more intricate passageways and other arrangements for supplying and removing oil.

The bearings and contact seals run much hotter than the plain bearings and clearance seals of the industrial turbines, usually requiring the use of synthetic lubricating oil and the placing of greater emphasis on the cooling function of the oil. Because of the need for compactness, bearing and gear loading is much higher in aircraft than in stationary practice, requiring extra load-carrying ability in the lubricant.

The lubrication system for a dual, axial compressor, gas generator is shown schematically in Figure 13.17. Oil from the tank is gravity fed to boost pump D, which supplies oil at constant pressure to oil pressure pump A regardless of changes in pressure drop through the oil cooler and piping. During operation, pressure at the suction of A acts through pressure-sensing line L (1) on boost pump regulating valve G, regulating the pressure rise of oil boost pump D, and (2) on pressure operated valve N, keeping it closed. During shutdown, pressure in line L falls, permitting valve N to open and vent antisiphon loop M. Because loop M extends above oil level, this prevents leakage of oil from the tank through pumps D and A into the auxiliary gear case. Oil pressure pump A forces oil—at about 45 psi (3.2 kP/cm^2)—through strainers or filters B to main bearing and accessory drive locations. Here, the oil is jetted through calibrated orifices to parts requiring lubrication and cooling, including bearings, contact seals, and gears. The oil flows by gravity from these parts to sumps from which it is removed by individual scavenge pumps C and returned to the tank through a deaerator J. All bearing and accessory drive housings are vented by means of a breather system (not shown), which includes a rotary breather pump.

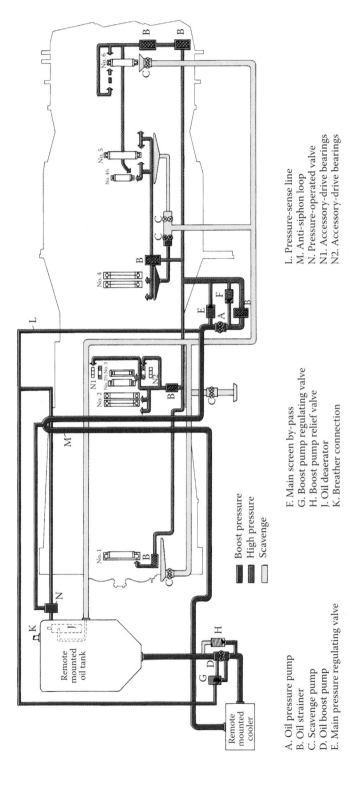

FIGURE 13.17 Lubrication system for aeroderivative gas turbine in stationary service.

A. Oil pressure pump
B. Oil strainer
C. Scavenge pump
D. Oil boost pump
E. Main pressure regulating valve
F. Main screen by-pass
G. Boost pump regulating valve
H. Boost pump relief valve
J. Oil deaerator
K. Breather connection
L. Pressure-sense line
M. Anti-siphon loop
N. Pressure-operated valve
N1. Accessory-drive bearings
N2. Accessory-drive bearings

Aeroderivative gas turbine generators have a separate lubrication system for the generator bearings.

Thermal conditions in land-based, aeroderivative gas turbines are not only more severe than in heavy-duty, industrial gas turbines, as noted previously, but are also more severe than in their airborne counterparts. In aircraft use, the turbines operate at full power only during takeoff and climb—a small fraction of the time—whereas in stationary service they may operate for long periods at full load. The lubricating oil comes in contact with hot surfaces in the 400–600°F (204–316°C) range and mixes with hot gases that leak into bearing housings. Bulk oil temperatures in the tank may be in the 160–250°F (71–121°C) range. This is a very severe condition and requires a lubricant having excellent thermal and oxidative stability to resist degradation and deposit formation. Measures that may be taken to moderate this severe thermal condition include derating engines, increasing oil reservoir sizes, and improving oil cooling equipment. In any case, it is common to idle gas turbines for several minutes before shutdown to cool them and prevent exposing the oil to high "soak back" temperatures. Synthetic ester lubricants are most commonly recommended for aeroderivative gas turbines.

C. Small Gas Turbines

Small gas turbines, as indicated in a previous section, vary considerably in design. Lubrication systems, however, are usually relatively simple versions of the type of circulation system described for large industrial turbines. Essential elements include a sump, a strainer, a shaft-driven main pump, a pressure-regulating valve, a cooler, a full-flow filter, manifold to bearings, gears, etc., and gravity return to sump. Some systems use scavenge pumps to return oil to the main sump or tank.

Most small turbines are designed to use either synthetic or petroleum lubricating oils. Turbine-type oils similar to those described for large industrial gas turbines may be used, but oils of lower viscosity may be specified for some very high speed machines or where lower ambient temperatures are encountered. In all cases, the gas turbine manufacturer specifies a suitable lubricant, and this recommendation should be followed unless specific approval is obtained for the use of some other lubricant.

BIBLIOGRAPHY

Mobil Oil Corporation. 1970. Mobil Technical Bulletin: *Gas Turbine Lubrication in Stationary Service*. Internal Mobil Oil Corporation Publication, New York.

14 Steam Turbines

In a steam turbine, hot steam at a pressure above atmospheric is expanded in the nozzles where part of its heat energy is converted to kinetic energy. This kinetic energy is then converted to mechanical energy in the turbine runner, either by the impulse principle or the reaction principle. If the nozzles are fixed and the jets directed toward movable blades, the jet's *impulse* force pushes the blades forward. If the nozzles are free to move, the *reaction* of the jets pushes against the nozzles, causing them to move in the opposite direction.

In small, purely impulse turbines, steam is expanded to exhaust pressure in a single set of stationary nozzles. As a result of this single expansion, the steam issues from the nozzles in jets of extremely high velocity. To obtain maximum power from the force of the jets' impact on a single row of moving blades, the blades must move at about half the velocity of the jets. Thus, single-stage impulse turbines operate at very high speeds. To reduce rotative speed while maintaining efficiency, the high velocity can be absorbed in more than one step, which is called *velocity compounding*.

Such a turbine (Figure 14.1) has one velocity compound stage followed by four pressure compounded stages. In the velocity compounded stage, steam is first expanded in the stationary nozzles to high velocity. This velocity is reduced in two steps through the first two rows of moving blades. The steam is then expanded again in a set of stationary nozzles and delivered to the first pressure stages. In each case, velocity is increased and pressure is decreased in the stationary nozzles. In the moving blades, velocity decreases but the pressure remains constant. The graph at the bottom of Figure 14.1 shows the changes in pressure and velocity as the steam flows through the various rows of nozzles and moving blades.

Another method of reducing rotor speed while maintaining efficiency is to decrease the velocity of the jets by dividing the drop in steam pressure into a number of stages. This is called *pressure compounding*. Refer again to Figure 14.1; the steam pressure is reduced by somewhat less than one-half in the velocity compounded stage. The steam then passes to the pressure compounded stages, where it is reduced in four steps to the final exhaust pressure. Each pressure stage consists of a row of stationary nozzles and a row of movable blades, and the whole assembly is equivalent to mounting four single-stage impulse turbines on a common shaft. The arrangement in Figure 14.1 using velocity compounding in the first stage followed by pressure compounding in the remaining stages is an approach used in many large turbines.

In reaction turbines, steam is directed into the blades on the rotor by stationary nozzles formed by blades that are designed to expand the steam enough to give it a velocity somewhat greater than that of the moving blades. (As a result of this expansion to provide steam velocity, this arrangement may be referred to as a 50% reaction turbine.) The moving blades form the walls of moving nozzles that are designed to permit further expansion of the steam and to partially reverse the direction of steam flow, which produces the reaction on the blades. The distinguishing characteristic of the reaction turbine is that a pressure drop occurs across both the moving and stationary nozzles, or blades (Figure 14.2). Normally, reaction turbines use a considerable number of rows of moving and stationary nozzles through which steam flows as its initial pressure is reduced to exhaust pressure. The pressure drop across each row of nozzles is, therefore, relatively small, and steam velocities are correspondingly moderate, permitting medium rotating speeds.

Reaction stages are usually preceded by an initial velocity compounded impulse stage, as in Figure 14.2, in which a relatively large pressure drop takes place. This results in a shorter, less costly turbine.

In the radial flow reaction, or Ljungstrom, turbine, instead of the steam flowing axially through alternating rows of fixed and moving blades, it flows radially through several rows of reaction blades. Alternate rows of blades move in opposite directions. They are fastened to two independent shafts that operate in opposite directions, each shaft driving a load.

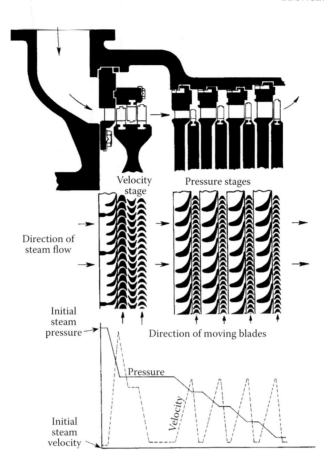

FIGURE 14.1 Turbine with impulse blading. Velocity compounding is accomplished in the first two stages by two rows of moving blades between which is placed a row of stationary blades that reverses the direction of steam flow as it passes from the first to the second row of moving blades. Other ways of accomplishing velocity compounding involves redirecting the steam jets so that they strike the same row of blades several times with progressively decreasing velocity.

After expansion in the turbine, the steam usually exhausts to a condenser, where it is condensed to provide a source of clean water for boiler feed. This simple cycle (Figure 14.3) forms the basis on which most steam power plants operate.

I. STEAM TURBINE OPERATION

Steam turbines are made in a number of different arrangements to suit the needs of various power plant or industrial installations. Turbines with a capacity of up to 40–60 MW are generally single-cylinder machines. (The terms *cylinders*, *chests*, *casings*, or *shells* are used interchangeably in this industry.) Larger units ranging in size up to 1900 MW are usually of compound type, that is, the steam is partially expanded in one cylinder, then passed to one or more additional cylinders where expansion is completed. The simple cycle shown in Figure 14.3 is water to steam to power generation, and steam to water. This forms the basis on which most steam power plants operate.

Larger high steam pressure turbines typically rotate at 1800 or 3600 rpm for 60-Hz generators and 1500 or 3000 rpm for 50-Hz generators. These turbines are typically composed of three or four separate cylinders, high pressure, medium pressure, and low pressure. The low-pressure turbine typically exhausts to a shell and tube condenser, where a vacuum is created as the steam bubbles

Steam Turbines

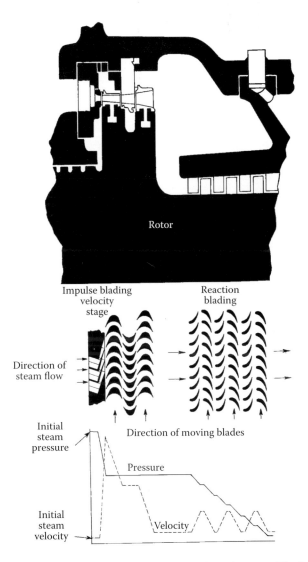

FIGURE 14.2 Reaction turbine with one velocity compounded impulse stage. The first stage of this turbine is similar to the first, velocity compounded stage of Figure 14.1. However, in the reaction blading of this turbine, both pressure and velocity decrease as the steam flows through the blades. The graph at the bottom shows the changes in pressure and velocity through the various stages.

collapse to water. Establishing and maintaining a high vacuum, targeting 29.92 in. mercury, provides optimum thermodynamic efficiencies.

A. SINGLE-CYLINDER TURBINES

Single-cylinder turbines are of either the *condensing* or *back-pressure* (noncondensing) type. These basic types and some of their subclassifications are shown in Figure 14.4.

When the steam from a turbine exhausts to a condenser, the condenser serves two purposes:

1. By maintaining a vacuum at the turbine exhaust, it increases the pressure range through which the steam expands. In this way, it materially increases the efficiency of power generation.
2. It causes the steam to condense, thus providing clean water for the boilers to reconvert into steam.

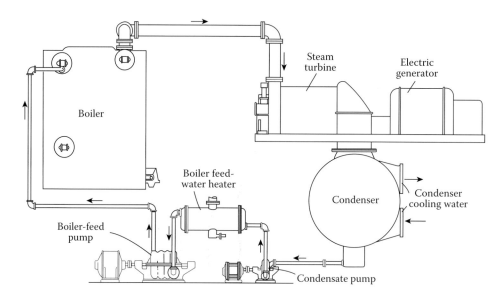

FIGURE 14.3 Simple power plant cycle. This diagram shows that the working fluid, steam and water, travels a closed loop in the typical power plant cycle.

Industrial plants frequently require steam at low to moderate pressures for process use. One of the more economical ways of generating this steam is with a *combined cycle* plant, where high-pressure steam is used to power equipment such as generators, and the exhaust steam from this equipment is used for heating or other services. The steam is generated at high pressure and, after expansion through the turbine to the pressure desired for process use, it is delivered to the process application. This permits power to be generated by the turbine without appreciably affecting the value of the steam for process use. It may be done with a back pressure turbine designed to exhaust all the steam against the pressure required for process use, or it may be done with an *automatic extraction* turbine in which part of the steam is withdrawn for process use at an intermediate stage (or stages) of the turbine and the remainder of the steam exhausted to a condenser. Such a turbine requires special governors and valves to maintain constant pressure of the exhausted steam and constant turbine speed under varying turbine load and extraction demands.

Steam can be also extracted without control from various stages of a turbine to heat boiler feedwater (*regenerative heating*). Such turbines are called *uncontrolled extraction* turbines, because the pressure at the extraction points varies with the load on the turbine.

To obtain higher efficiency, large turbines (called reheat turbines) are arranged so that after expanding part way, the steam is withdrawn, returned to the boiler, and reheated to approximately its initial temperature. It is then returned to the turbine for expansion through the final turbine stages to exhaust pressure.

High-pressure noncondensing turbines have been added to many moderate-pressure installations to increase capacity and improve efficiency. In such installations, high-pressure boilers are installed to supply steam to the noncondensing turbines, which are designed to exhaust at the pressure of the original boilers and supply steam to the original turbines. The high-pressure turbines are called *superposed* or *topping* units.

Where low-pressure steam is available from process work, it can be used to generate power by admitting it to an intermediate stage of a turbine designed for the purpose and expanding it to condenser pressure. Such a machine is a *mixed-pressure* turbine, and is another form of *combined cycle* operation.

Compound turbines have at least two cylinders or casings—a high-pressure one and a low-pressure one. To handle large volumes of low-pressure steam, the low-pressure cylinder is frequently of the double flow type. Very large turbines may have an intermediate pressure cylinder,

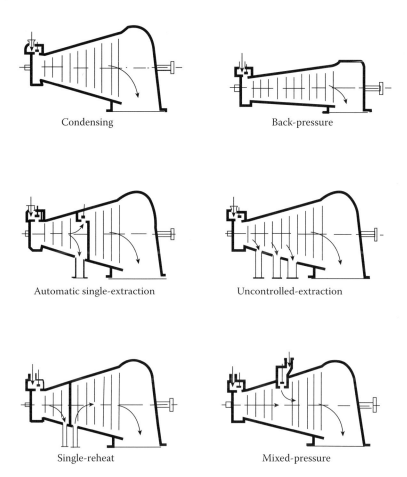

FIGURE 14.4 Single cylinder turbine types. Typical types of single cylinder turbines are illustrated. As shown, condensing turbines, as compared to back-pressure turbines, must increase more in size toward the exhaust end to handle the larger volume of low-pressure steam.

and two, three, or even four double flow, low-pressure cylinders. The cylinders may be in line using a single shaft, which is called *tandem compound*, or in parallel groups with two or more shafts, which is called *cross compound*. Reheat between the high- and intermediate-pressure stages may be used in large turbines. Steam may be returned to the boiler twice for reheating. Some of these arrangements are shown diagrammatically in Figure 14.5.

II. TURBINE CONTROL SYSTEMS

Although the trend is toward electronic speed sensing and control, all steam turbines are provided with at least two independent governors that operate to control the flow of steam. One of these operates to shut off the steam supply if the turbine speed should exceed a predetermined maximum value. It is often referred to as an emergency trip. On overspeed, it closes the main steam valve, cutting off steam from the boiler to the turbine. The other, or main, governor may operate to maintain practically constant speed, or it may be designed for variable speed operation under the control of some outside influence. Extraction, mixed pressure, and back-pressure turbines are provided with governors that control steam flow in response to a combination of speed and one or more pressures. The governors of such units are extremely complex, whereas the direct-acting speed governors on small, mechanical drive turbines are relatively simple.

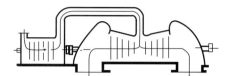
Tandem-compound, two-casing, double-flow

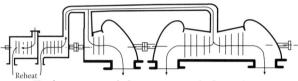
Tandem-compound, three-casing, triple-flow, reheat

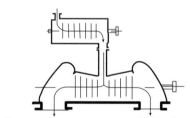
Cross-compound, two-casing, double-flow

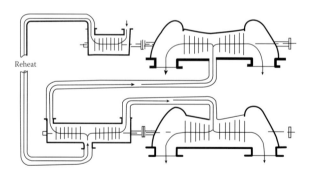
Cross-compound, four-casing, quadruple-flow, reheat

FIGURE 14.5 Some arrangements of compound turbines. Although many arrangements are used, these diagrams illustrate some of the more common ones.

A. Speed Governors

Speed governors may be divided into three principal parts:

1. Governor or speed-sensitive element
2. Linkage or force amplifying mechanism that transmits motion from the governor to the steam control valves
3. Steam control valves

Steam Turbines

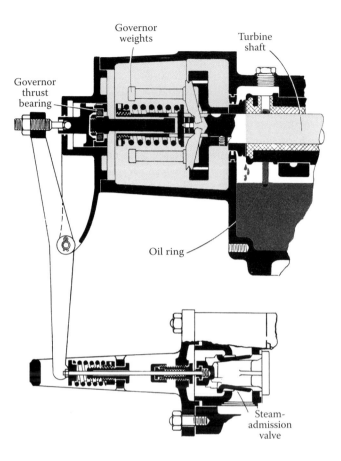

FIGURE 14.6 Mechanical speed governor. A simple arrangement such as this using a flyball governor is suitable for many small turbines.

Although most new turbine installations use electronic speed sensing and control, one commonly used speed-sensitive element is the centrifugal, or flyball, governor (Figure 14.6). Weights pivoted on opposite sides of a spindle and are revolving with it are moved outward by centrifugal force against a spring when the turbine speed increases, and inward by spring action as the turbine slows down. This action may operate the steam admission valve directly through a mechanical linkage, as shown, or it may operate the pilot valve of a hydraulic system, which admits and releases oil to opposite sides of a power piston, or on one side of a spring-loaded piston. Movement of the power piston opens or closes steam valves in order to control turbine speed.

Moderate-sized and large high-speed turbines are provided with a double-relay hydraulic system to further boost the force of the centrifugal governor and to increase the speed with which the system responds to speed changes.

A second type of speed-sensitive element is the oil impeller. Oil from a shaft-driven pump flows through a control valve to the space surrounding the governor impeller. The impeller, mounted on the turbine shaft, consists of a hollow cylindrical body with a series of tubes extended radially inward. As the oil flows inward through the tubes, it is opposed by centrifugal force and a pressure is built up that varies with the square of the turbine speed. This pressure is applied to spring-loaded bellows, which positions a pilot valve. The pilot valve, in turn, controls the flow to a hydraulic circuit that operates the steam control valves.

Newer turbines are equipped with electrical or electronic speed-sensitive devices. Signals from these devices, along with signals derived from load, initial steam pressure, and other variables, are fed to a computer where they are compared, and the appropriate signals are sent to hydraulic servo valves to adjust the steam control valves.

As indicated, the linkage between the speed-sensitive element and the steam control valves may be anything from a simple lever to an extensive hydraulic system controlled by a computer.

In small turbines, the flow of steam is controlled by a simple valve, usually of the balanced type to reduce the operating forces required. In large units, a valve for each of several groups of nozzles controls steam flow. The opening or closing of a valve cuts in or out a group of nozzles. The number of open valves, and thus the number of nozzle groups in use, is varied according to the load. The valves may be operated by a barlift arrangement, by cams, or by individual hydraulic cylinders.

Additional control valves called intercept valves are required on reheat turbines. These are placed close to the intermediate, or reheat, cylinder and are closed by a governor system if the turbine starts to speed up as a result of a sudden large load reduction. This is designed to prevent large volumes of high energy steam in the piping from getting into the high-pressure turbine exhaust, the reheat boiler, and the piping to the intermediate pressure turbine from continuing to flow and possibly cause overspeeding and emergency tripping of the turbine. In older turbines, the intercept valves were controlled by a separate governor system, but in newer machines the intercept valves are operated by the main hydraulic control system. As an additional safety measure, intercept valves are preceded by stop valves that are actuated by the main emergency overspeed governor (or other speed-sensing control system).

Automatic extraction and back-pressure turbines are provided with governors arranged to maintain constant extraction or exhaust pressure irrespective of load (within the capacity of the turbine). The pressure-sensitive element consists of a pressure transducer, and its response to pressure changes is communicated through the control system to the valves that control steam extraction, and to the speed governor that controls admission of steam to the turbine. On automatic extraction turbines, the action of the pressure- and speed-responsive elements is coordinated so that turbine speed is maintained. This may not be the case for back-pressure turbines.

III. LUBRICATED COMPONENTS

The lubrication requirements of steam turbines can be considered in terms of the parts that must be lubricated, the type of application system, the factors affecting lubrication, and the lubricant characteristics required to satisfy these requirements.

A. Lubricated Parts

The main lubricated parts of steam turbines are the bearings, both journal and thrust. Depending on the type of installation, a hydraulic control system, oil shaft seals, gears, flexible couplings, and turning gear may also require lubrication.

1. Journal Bearings

The rotor of a single-cylinder steam turbine, or of each casing of a compound turbine, is supported by two hydrodynamic journal bearings. These journal bearings are located at the ends of the rotor, outside of the cylinder. In some designs, there may be one large journal bearing between the casings that support both turbine rotors or a turbine and generator rotor (rigid coupling) instead of a separate bearing at the ends of each casing. Because of the extremely small clearances between the shaft and shaft seals and between the blading and the cylinder, the bearings must be accurately aligned and must run without any appreciable wear in order to maintain the shaft in its original position and avoid damage to shaft seals or blading.

Primarily, the loads imposed on the bearings are attributable to the weight of the rotor assembly. The bearings are conservatively proportioned so that pressures on them are moderate. The bearings are enclosed in housings and supported on spherical seats or flexible plates to reduce any angular misalignment. These horizontal split shell journal bearings are typically bimetallic, with softer metal bonded to a hard steel shell. The softer metal material is often babbitt composed of tin alloyed with copper, iron, antimony, and lead. Babbitt thickness can range from 1 mil (25.4 µm) to 100 mil (2540 µm). A mil represents one thousandth of an inch (0.001 in). The journal is normally an integral part of the turbine rotor and made from the same material steel. In a resting position, a 20-in (50.8 cm)-diameter journal bearing may have a clearance of 30 mils (760 µm). In operation, an oil wedge of 0.78–1.57 mil (20–40 µm) is formed. Babbitt creep and bearing failure can occur at temperatures above 270°F (132°C) at nominal loads of 200–1000 psi (13.8–68.9 bar). Bearing-imbedded thermocouples and/or lubricant drain temperature measurement and alarming are highly recommended and often used. Bearing metal temperature is typically 30–80°F (17–45°C) higher than the drain oil temperature. Lubricating oil is supplied through drilled passages at pressures ranging from 10 to 20 psig (0.7–1.38 bar) and temperatures ranging from 110°F to 145°F (43–63°C). A lubricating oil typically gains 20–50°F (10.7–27.7°C) between bearing inlet to bearing outlet. Oil outlet temperatures should typically not exceed 180°F (82°C), as elevated temperatures can promote premature degradation of the oil. Proper oil flow to these bearings provides fluid film for hydrodynamic lubrication and heat removal to protect the bearing from overheating. Most bearing oil outlets are equipped with a direct-read thermometer and a sight flow port for visual inspection.

The passages and grooves in turbine bearings are sized to permit the flow of considerably more oil than is required for lubrication alone. The additional oil flow is required to remove frictional heat and the heat conducted to the bearing along the shaft from the hot parts of the turbine. The flow of oil must be sufficient to cool the bearing enough to maintain it at a proper operating temperature.

When a turbine is used to drive a generator, the generator bearings are similar in design to the turbine bearings, and are normally supplied from the same system.

Large turbines are now frequently provided with "oil lifts" (jacking oil) in the journal bearings to reduce the possibility of damage to the bearings during starting and stopping, and to reduce metal-to-metal contact during turning gear operation. Oil under high pressure from a positive displacement pump is delivered to recesses in the bottoms of the bearings. The high-pressure oil lifts the shaft and floats it on a film of oil until the shaft speed is high enough to create a normal hydrodynamic film. The oil lift is also required after turbine shutdown, where the shaft is rotated for several hours or days to prevent rotor-sag and dissipate heat from the bearings.

A phenomenon that occurs in relatively lightly loaded, high-speed journal bearings, such as turbine bearings, is what is known as "oil whip" or "oil film whirl." The center of the journal of a hydrodynamic bearing ordinarily assumes a stable, eccentric position in the bearing that is determined by load, speed, and oil viscosity. Under light load and high speed, the stable position closely approaches the center of the bearing. There is a tendency, however, for the journal center to move in a more or less circular path about the stable position in a self-excited vibratory motion having a frequency of something less than one-half shaft speed. In certain cases, such as some of the relatively lightweight, high-pressure rotors of compound turbines that require large-diameter journals to transmit the torque, this whirling has been troublesome and has required the use of bearings designed especially to suppress oil whip.

Several types of bearings designed to suppress oil whip are available. Among the common types are the pressure or pressure pad bearings (Figure 14.7), the three-lobe bearing (Figure 14.8), and the tilting pad antiwhip bearing (Figure 14.9). The pressure pad bearing suppresses oil whip because oil carried into the wide groove increases in pressure when it reaches the dam at the end. This increase in pressure forces the journal downward into a more eccentric position that is more resistant to oil whip. The other types illustrated depend on the multiple oil films formed to preload the journal and minimize the tendency to whip.

2. Thrust Bearings

Theoretically, in impulse turbines, the drop in steam pressure occurs almost entirely in the stationary nozzles. The steam pressures on opposite sides of the moving blades are, therefore, approximately

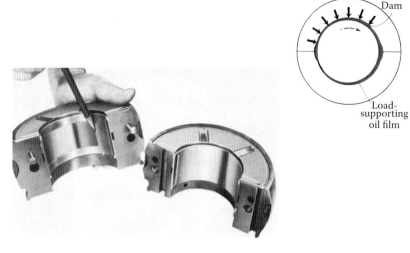

FIGURE 14.7 Pressure bearing. The wide grooves in the upper half end in a sharp dam at the point indicated. As shown in the insert, this causes a downward pressure that forces the journal into a more eccentric position that is more resistant to oil whip.

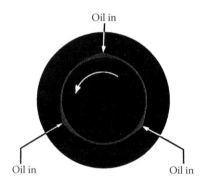

FIGURE 14.8 Three-lobe bearing. The shape of the bearing is formed by three arcs of radius somewhat greater than the radius of the journal. This has the effect of creating a separate hydrodynamic film at each lobe, and the pressure in these films tends to keep the journal in a stable position.

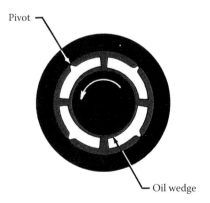

FIGURE 14.9 Tilting pad antiwhip bearing. As in the three-lobe bearing, the multiple oil films formed tend to keep the journal in a stable position.

equal, and there is little tendency for the steam to exert a thrust in the axial direction. In actual turbines, this ideal is not fully realized and there is always a thrust tending to displace the rotor.

In reaction turbines, a considerable drop in steam pressure occurs across each row of moving blades. Because the pressure at the entering side of each of the many rows of moving blades is higher than the pressure at the leaving side, the steam exerts a considerable axial thrust toward the exhaust end. Also, where rotors are stepped up in diameter, the unbalanced steam pressure acting on annular areas thus created adds to the thrust. Usually, the total thrust is balanced by means of dummy, or balancing, pistons on which the steam exerts a pressure in the opposite direction to the thrust. In double-flow elements of compound turbines, steam flows from the center to both ends so that thrust is well balanced.

Regardless of the type of turbine, thrust bearings are always provided on each shaft in order to take axial thrust and, thus, hold the rotor in correct axial position with respect to the stationary parts. Although thrust caused by the flow of steam is usually toward the low pressure end, means are always provided to prevent axial movement of the rotor in either direction. Lubricant is supplied at 10 to 20 psig (0.7 to 1.38 bar) pressure and in operation will provide a hydrodynamic film wedge of protection and remove heat. Thrust bearing and oil drain temperatures should be monitored to confirm proper lubrication.

The thrust bearings of small turbines may be babbitt-faced ends on the journal bearings, or rolling element bearings of a type designed to carry thrust loads. Medium and large turbines are always equipped with tilting pad (Figure 14.10), or tapered land (Figure 14.11) type thrust bearings.

3. Hydraulic Control Systems

As discussed earlier, medium and large turbines have hydraulic control systems to transmit the motion of the speed- or pressure-sensitive elements to the steam control valves. Two general approaches are used for these systems.

In mechanical hydraulic control systems, the operating pressure is comparatively low (less than 150 psi), and oil from the bearing lubrication system may be used safely as the hydraulic fluid. Separate pumps are provided to supply the hydraulic requirement. An emergency tripping device is provided to shut down the turbine if there is any failure in the hydraulic system.

Larger turbines now being installed are equipped with electrohydraulic control systems. In order to provide the rapid response needed for control of these units, the hydraulic systems operate at relatively high pressures, typically in the range of 1500–2000 psi.

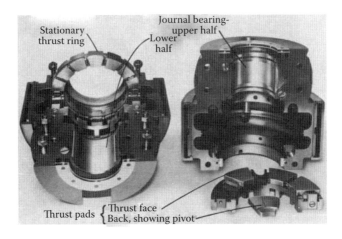

FIGURE 14.10 Combined journal and tilting pad thrust bearing. A rigid collar on the shaft is held centered between the stationary thrust ring and a second stationary thrust ring (not shown) by two rows of tilting pads.

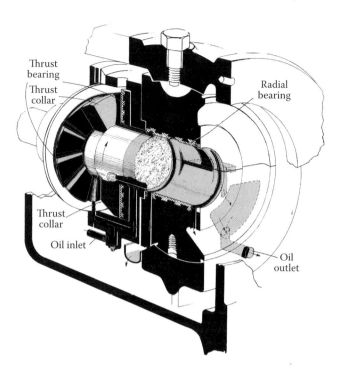

FIGURE 14.11 Tapered land thrust bearing and plain journal bearing. The thrust bearing consists of a collar on the shaft, two stationary bearing rings, one on each side of the collar. The babbitted thrust faces of the bearing rings are cut into sectors by radial grooves. About 80% of each sector is beveled to the leading radial groove, to permit the formation of wedge oil films. The unbeveled portions of the sectors absorb the thrust load when speed is too low to form hydrodynamic films.

The systems consist of an independent reservoir and two separate and independent pumping systems. The large fluid flow required for rapid response to sudden changes in load is usually provided by gas-charged accumulators.

The critical nature of the servo valves used in these systems requires that careful attention be paid to the filtration of the fluid, and strict limits on particulate contamination are usually observed. The need for precise control also necessitates that both heaters and coolers be used to maintain the temperature of the fluid and, thus, its viscosity in a narrow range. Because a leak or a break in a hydraulic line could result in a fire if the high-pressure fluid sprayed onto hot steam piping or valves, fire-resistant hydraulic fluids are widely used in these systems.

4. Oil Shaft Seals for Hydrogen-Cooled Generators

Because it is a more effective coolant than air, hydrogen is commonly used to cool medium and large generators. Shaft-mounted blowers circulate the gas through rotor and stator passages, then through liquid-cooled hydrogen coolers. Gas pressures up to 60 psi (413 kPa) are used. A further development, which has permitted increases of generator ratings over hydrogen cooling alone, is the direct liquid cooling of stator windings. Some liquid systems use transformer oil whereas others use water. Even with water-cooled stators, the interior of the generator is still filled with hydrogen.

The main connection between the type of cooling and turbine lubrication is that when hydrogen is used for cooling, some of the oil is exposed to the hydrogen. Oil shaft seals (Figure 14.12) are used to prevent the escape of the hydrogen. Turbine oil for these seals may be supplied from a separate system having its own reservoir, pumps, and so forth, or may be supplied directly from the main turbine lubricating system. In either case, before entering the reservoir, oil returning from the seals must be passed through a special tank to remove any traces of hydrogen. Otherwise, hydrogen could accumulate in the

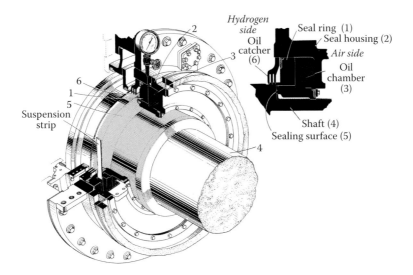

FIGURE 14.12 Shaft seal for hydrogen cooled generator. Oil under pressure in the annular chamber formed between the seal ring and housing (see inset) forces the seal ring against the shaft shoulder. From the annular chamber, the oil flows through the passage shown to the sealing face formed by a shoulder on the generator shaft.

reservoir and form an explosive mixture with air. In addition, the main turbine oil reservoir of all units driving hydrogen-cooled generators must be equipped with a vapor extractor to remove any traces of hydrogen that may be carried back by the sealing oil or the oil from the generator bearings.

5. Gear Drives

Efficient turbine speed is often higher than the operating speed of the machine being driven. This may be the case, for example, where a turbine drives a direct current generator, paper machine drives, centrifugal pumps, or other industrial machines. It is also the case where a turbine is used for ship propulsion. In these applications, reduction gears are used to connect the turbine to the driven unit.

Reduction gears used with moderate and large-sized turbines are usually precision cut, double-helical type. Double reduction gearsets are required with marine propulsion turbines, and epicyclical reduction gears are sometimes used instead of conventional gearsets. Usually, the gearsets are enclosed in a separate oil-tight casing and are connected to the turbine and driven machine through flexible couplings. Small machines may have the gear housing integral with the turbine housing and the pinion on the turbine shaft.

Reduction gears may have a circulation system that is entirely separate from the turbine system, or may be supplied from the turbine system. In the latter case, a separate pump (or pumps) is provided for the gears. Some older small-geared turbines have ring-oiled turbine bearings and splash-lubricated gears.

In marine propulsion applications, the bearing through which the propeller thrust is transmitted to the ship's hull is frequently an integral part of the gearset. It may be placed just behind the low-speed bull gear, or at the forward end of the bull gear shaft. It is usually of the tilting pad type. With a fixed-pitch propeller, the thrust bearing must be designed to operate with either direction of rotation, because turbine rotation (and propeller rotation) is reversed for the astern movement of the ship.

6. Flexible Couplings

Most modern turbo generator sets are rigidly coupled. Many other turbines, especially geared units, are connected to their loads by flexible couplings. Gear-type couplings are most commonly used.

If gear-type couplings are enclosed in a suitable housing, they may be lubricated from the turbine circulation system. A coupling of this type acts as a centrifugal separator, and means are provided for a flow-through of oil so that any separated solids will be flushed out.

Smaller couplings may be lubricated by a bath of oil carried inside the case, or with grease.

7. Turning Gear

When starting and stopping large turbines, it is necessary to turn the rotor slowly to avoid uneven heating or cooling, which could cause distortion or bowing of the shaft. This is done with a barring mechanism or turning gear. The turning gear usually consists of either an electric or hydraulic motor that is temporarily coupled to the turbine shaft through reduction gears. Rotor speed, while the turning gear is operating, is usually below 100 rpm. To provide adequate oil flow to the bearings during this low-speed operation, a separate auxiliary oil pump is usually provided. The oil coolers are used at maximum capacity to increase oil viscosity and to help maintain oil films in the bearings. If oil lifts are provided in the turbine bearings, they are also operated while the turning gear is operating.

B. Lubricant Application

One of the essential factors in reliable steam turbine operation is the provision of a lubricating system that will assure an ample supply of clean lubricant to all of the parts requiring it. The size of the turbine generally determines whether the system is simple or extensive and complex. Small turbines, such as those used to drive auxiliary equipment, are usually equipped with ring-oiled bearings, with other moving parts being lubricated by hand. Moderate-sized units driving through reduction gears may have ring-oiled bearings for the turbine and a circulation system (see Chapter 9) to supply oil to the gears and bearings in the reduction gearset. Most moderate-sized units and all large turbines are equipped with circulation systems that supply oil to all parts of the unit requiring lubrication. Separate circulation systems may be provided for the seal oil for hydrogen-cooled turbo generators and for the hydraulic control systems.

Bearing supply piping is typically surrounded by the return piping, pipe within pipe annular design. This design aids in protecting against a potential fire hazard in the event of a pipe rupture. The pipe within a pipe design will prevent the pressurized, 50–60 psig (3.4–4.1 bar), oil supply from spraying combustible lubricant toward hot steam turbine surfaces. The lube oil returns from the bearings in the external pipe and uses gravity to flow through a series of return screens into the steam turbine oil reservoir.

The turbine oil lubricant is stored in a reservoir, ranging in size up to 20,000 gal (75,700 L), which is typically located on the ground floor below the steam turbine deck. The reservoir is sized to provide suitable oil residence time to allow the separation of contaminants and air. The reservoir is kept at a slightly negative pressure, 1–2 in H_2O (250–500 Pa), by a vapor extractor to minimize potential buildup of dangerous hydrogen gas. Turbine oil at approximately 60 psig (4.1 bar) is pumped from the reservoir to the turbine bearings and is typically maintained at temperatures between 100°F and 120°F (32°C and 43°C) by shell and tube lube oil coolers. Cooling water for these heat exchangers is often supplied from cooling towers or raw water heat exchangers.

At startup, an electric motor-driven, vertical shaft, auxiliary lube jacking oil pump is used. When the turbine shaft reaches approximately 90% of the rotating speed, a turbine shaft-driven pump, located in the turbine front standard, will supply full oil flow to journal and thrust bearings, and the jacking oil pump will shut down.

C. Factors Affecting Lubrication

Steam turbines in themselves do not represent particularly severe service for petroleum-based lubricating oils. However, because of the costs and time involved in shutting down most turbines to change the oil, clean the system, and the relatively large volume of oil contained in large turbine systems, turbine operators expect extremely long service life from the turbine oil. In order to achieve long service life, oils must be carefully formulated for the specific conditions encountered in steam turbine lubrication systems.

A properly maintained turbine oil with minimum contamination and a nominal 3% annual lube oil make-up should provide reliable operation for decades.

Of the key steam turbine oil characteristics, the most important performance requirements are the ability to (1) protect against rust and corrosion, (2) provide oxidation stability, (3) maintain viscosity, and (4) shed water. Water contamination and its removal are the primary causes of premature steam turbine oil failure.

1. Circulation and Heating in the Presence of Air

The temperature of the oil in steam turbine systems is raised both by the frictional heat generated in the lubricated parts and by heat conducted along the shaft from the rotor. As the oil flows through the system, it is broken into droplets or mist, which permits greater exposure to air. System designs are usually conservative so that maximum oil temperatures are not excessive, but in long-term service some oxidation of the oil does occur. This oxidation may be further catalyzed by finely divided metal particles resulting from wear or contamination and water.

Slight oxidation of the oil is harmless. The small amounts of oxidation products formed initially can be carried in solution in the oil without noticeable effect. As oxidation continues, some of the soluble products may become insoluble, or insoluble materials may be formed. As an oil oxidizes, it should also show an increase in viscosity; in most turbine systems, however, the rate of oxidation is very slow and viscosity increases are rarely noted.

Insoluble oxidation products may be carried with the oil as it circulates. Some may then settle out as varnish or sludge on governor parts, in bearing passages, on coolers, on strainers, and in oil reservoirs. Their accumulation may interfere with the supply of oil to bearings and with governors or control of the unit. Under moderate to severe oil oxidation, there is also the potential for the oil degradation materials to plate out on bearing surfaces, reducing clearances and increasing bearing temperatures. Steam turbines that have a common reservoir for bearings and hydraulics are more susceptible to varnish issues than typical bearing-only or hydraulic-only reservoirs. Hydraulic servo valves have small tolerances (typically 3 μm) that can be fouled from oxidized lubricant. Varnish-specific lube oil analysis has been developed to help turbine users manage this issue.

Some oxidation products that are soluble in warm oil become insoluble when the oil is cooled. This can result in insulating deposits forming on cooling coils or other cool surfaces. The resultant reduction in effectiveness of cooling may cause higher oil temperatures and contribute to more rapid oxidation. These soluble and insoluble oxidation products can be reduced through specially developed filtration.

Any increase in viscosity that is within the specified International Standardization Organization (ISO) viscosity range is not necessarily harmful. However, excessive viscosity increase may reduce oil flow to bearings, increase pumping losses, and increase fluid friction and heating in bearings. All of these effects tend to increase the operating temperature of the oil and contribute to more rapid oxidation of it.

2. Contamination

Water is the most prevalent contaminant in steam turbine lubrication systems. Common sources of water contamination are as follows:

1. Gland seal steam from leaking gland shaft seals.
2. Leaking shaft seals of turbine-driven pumps.
3. Condensation from humid air in the oil reservoir and bearing pedestals.
4. Water leaks in oil coolers.
5. Steam leaks in oil heating elements (where used).

The lubrication systems of turbines that operate intermittently are more likely to become contaminated with water compared with the systems of turbines that operate continuously because of condensation buildup and transitions in gland seal operation.

When a turbine oil is agitated with water, some emulsion will form. If the oil is new and clean, the emulsion will separate readily. The water will then settle in the reservoir where it can be drawn off or removed by the purification equipment. Oxidation of the oil, or its contamination with certain types of solid materials such as rust and fly ash, may increase the tendency of the oil to emulsify, and also to stabilize emulsions after they are formed. Persistent emulsions can join with insoluble oxidation products, dirt, etc., to form sludge. The character of these sludges may vary, but if they accumulate in oil pipes, passages, and oil coolers, they can interfere with oil circulation and cause high oil and bearing temperatures.

Water in lubricating oil, combined with air that is always present, can cause the formation of common red rust and also black rust that is similar in appearance to pipe scale. Rusting may occur both on parts covered by oil and on parts above the oil level. In either case, in addition to damage to the metal surfaces, rusting is harmful for a number of reasons. Particles of rust in the oil tend to stabilize emulsions and foam, and act as catalysts that increase the rate of oil oxidation. Rust is abrasive and, when carried by the oil to the bearings, may scratch the journals and cause excessive wear. If carried into the small clearances of governor or control mechanisms, it can cause sluggish operation or, in extreme cases, sticking, overspeed, or trip the unit off line.

In some cases, water contamination in warm stagnant areas has led to the formation of microbial growth, which can block filters and control oil systems, degrade oil, and produce corrosive byproducts. A pungent smell will often accompany microbial growth.

Water concentrations in the lube oil can be reduced through centrifuge, coalescing filters, vacuum dehydration, or water traps.

Many types of solid materials can contaminate turbine oil systems. Pulverized coal, fly ash, dirt, rust, pipe scale, and metal particles are typical examples. These solid materials may enter the system in the following ways:

1. During erection and may remain even after the initial cleaning and flushing
2. Through openings in the bearing pedestals or reservoir covers
3. Carried in with water
4. Generated within the system
5. While adding make-up oil
6. When performing system maintenance or repairs

Solid contaminants contribute to deposit or sludge formation, and some tend to reduce the ability of the oil to separate from water. Some are abrasive and may be the direct cause of scoring or excessive wear. Fine metal particles may act as catalysts to promote oil oxidation.

Air is also considered a contaminant. Air entrained in the oil will cause sponginess in hydraulic controls, and may reduce the load-carrying ability of oil films. Entrained air also increases the exposure of the oil to air, and therefore increases the rate of oxidation. Excessive amounts of air in the system may lead to foaming in the reservoir or bearing housings. This could result in overflow with oil loss and unsightly and hazardous spills. In some cases, foam escaping from bearings adjacent to a generator may be drawn into the windings or onto the collector rings where it may cause breakdown of the insulation, short circuits, or arcing.

D. STEAM TURBINE OIL ADDITIVES AND CHARACTERISTICS

1. Additives

In general, a steam turbine oil is composed of 99% base oil and 1% additive. Base stocks can be selected from API base stock groups I, II, III, and IV, or in combinations of these. In formulating turbine oils, it is very important to select specific base stocks and additives and balance the formulation to provide all of the performance characteristics required to assure good turbine lubrication as well as long life. Select additives are designed to improve the performance of steam turbine oils:

- Antioxidant—often a mix of additive packages of hindered phenols and aromatic amines used to support long-term performance and reduced deposit formation.
- Rust and corrosion inhibitors—ashless acids or esters are examples of surface active additives that prevent water access to metal surfaces.
- Demulsifiers—mixed polyol esters or ethers can enhance water separation by promoting water drop coalescence enabling free water separation by gravity driven phase separation
- Defoamants—reduces surface tension so that air bubbles can be released. Silicones or polyacrylates are the two common classifications of antifoam additives.
- Metal passivator—triazoles offer protection from yellow metal corrosion and can provide some level of oxidation resistance.
- Gear wear protection, for gear-driven generators—often phosphorus based to form a protective and sacrificial layer on the gear tooth surface.

Thorough blending of these additives into the base oil is needed to ensure a complete and homogeneous blend. Blending involves precise additive dosage in the proper sequence and at a controlled temperature in blend kettles to promote complete mixing.

The first requirement of a steam turbine oil is that it must have the proper viscosity at operating temperature to provide effective lubricating films, and adequate load-carrying ability to protect against wear in heavily loaded mechanisms, and under boundary lubrication conditions. Other characteristics are concerned with providing long service life, protecting system metals, and maintaining the oil in a condition to perform its lubricating function.

2. Viscosity

In direct connected steam turbines, the main lubrication requirement is for the journal and thrust bearings. Higher viscosity oils provide a greater margin of safety in these bearings, but at the same time, increase pumping losses and friction losses owing to shearing of the lubricant films. In high-speed machines, the latter, in particular, can become an important cause of power loss and heating of the lubricant. The general practice is to use oils as low in viscosity as possible within a range that has proved suitable service. Most larger units are designed to operate on oils of ISO Viscosity Grade 32 (28.8–35.2 cSt at 40°C). Some applications require somewhat higher viscosity oils, ISO Viscosity Grade 46 (41.4–50.6 cSt at 40°C) or ISO Viscosity Grade 68 (61.2–74.8 cSt at 40°C). These grades of oil also provide excellent performance in hydraulic control systems.

Reservoir temperatures of small ring-oiled turbine bearings vary widely from one design to another depending principally on size and whether water cooling is used. The oils used for these range in viscosity from ISO Viscosity Grade 32 for the cooler units, up to as high as ISO Viscosity Grade 320 (288–352 cSt at 40°C) for the latter units.

3. Load-Carrying Ability

For geared turbines with a common circulation system supplying both the bearings and gears, higher viscosity oils or oils formulated with antiwear additives may be required to provide satisfactory lubrication of the gears. Oils of ISO Viscosity Grade 68 (61.2–74.8 cSt at 40°C) are used in many of these systems. Oils of ISO Viscosity Grade 100 (90 to 110 cSt at 40°C) are also used in some machines, particularly marine propulsion turbines. With some geared turbines, the oil is passed through a cooler immediately before entering the gears. The resultant cooler oil has a higher viscosity; thus, it provides better protection of the gears.

As noted earlier, steam turbine lubrication systems are conservatively designed. Bearing loads are moderate. Under these conditions, mineral oil lubricants of the correct viscosity normally provide adequate load-carrying ability. However, in turbines not equipped with oil lifts, boundary lubrication conditions occur in the bearings during starting and stopping. Under boundary lubrication conditions, some wear will occur unless lubricants with enhanced film strength are used.

To some extent, the increased load-carrying ability of the oil films needed during startup is provided by the higher viscosity of the cool oil. However, the additive systems in turbine oils are frequently selected to provide some improvement in film strength to ensure an additional margin of safety.

In some cases, particularly in marine propulsion and power generation installations, the need for size and weight reduction has led to the use of heavily loaded gear reducers. Performance of these gears on conventional turbine oils may not be satisfactory, and lubricants with extreme pressure properties are used. These lubricants are formulated to have the properties of good turbine oils with the addition of the extreme pressure properties; thus, they may be used throughout the turbine lubrication system. Turbine gear loading may be considered light enough not to require an extreme pressure gear lubricant.

E. Oxidation Stability

The most important characteristic of turbine oils from the standpoint of long service life is their ability to resist oxidation under the conditions encountered in the turbine lubrication system. Resistance to oxidation is important from the standpoint of retention of viscosity; resistance to the formation of sludge, deposits, and corrosive oil oxyacids; and retention of water-separating ability, foam resistance, and ability to release entrained air.

Turbine oils are usually manufactured from base oils refined from selected crudes and/or special refining processes. These oils are selected both for their natural oxidation stability and their response to oxidation inhibitors. Processing is carefully controlled and is often the most extensive applied to any base oils. Inhibitors are selected to be effective under the temperature conditions encountered in steam turbines and to provide the optimum improvement in the particular base oils used.

Oxidation life of steam turbine oils is frequently specified in terms of TOST (Turbine Oil Stability test—ASTM D943; see Chapter 3). Manufacturer specifications typically cite a minimum test life of 1000 h to reach a neutralization number of 2.0. Commercial turbine oils of ISO VG 32 typically run from 4000 to more than 10,000 h. Higher viscosity products may test out at somewhat less. Note that turbine oils represented above 10,000 h TOST are not tested according to standard ASTM D943 methodology as the upper life limit of the test is 10,000 h. Commercial steam turbine oils available today will generally give many years of service in well-maintained systems.

In addition to using the ASTM D943 for the evaluation of oxidation stability of turbine oils, the ASTM D2722 (Rotary Pressure Vessel Oxidation Test [RPVOT]) may also be used; RPVOT is discussed in Chapter 3. The RPVOT has historically been used to evaluate the oxidation stability of in-service oils to aid in maintenance planning. New oil RPVOT comparisons are not helpful as some oils have different rates of oxidation, which could result in faster RPVOT decay that may not be representative of oil in service.

1. Protection against Rusting

Unadditized mineral oils have some ability to protect against rusting of ferrous surfaces, but this ability is inadequate for the conditions encountered in steam turbine lubrication systems. Effective rust inhibitors are required in steam turbine oils.

In action, rust inhibitors "plate out" on metal surfaces, forming a film that resists displacement and penetration of water. This results in gradual depletion of the rust inhibitor. Normal make-up of new oil usually maintains an adequate level of rust inhibitor in the system.

Rust inhibitors must be carefully selected to provide adequate protection without affecting other properties of the oil, especially water-separating ability.

2. Water-Separating Ability

New, clean, highly refined mineral oils will generally resist emulsification when water is mixed with them, and any emulsion that is formed will break quickly. Certain additives, such as some rust inhibitors, some contaminants, and oxidation products, can both increase the tendency of an oil to emulsify and make any emulsion that is formed more stable. As a result, careful selection of additives is necessary if a turbine oil is to have good initial water-separating ability, and excellent oxidation stability is necessary if this water-separating ability is to be maintained in service.

3. Foam Resistance

Turbine circulation systems are, in general, designed and constructed to minimize or eliminate conditions that have been found to cause foaming. However, turbine oils may contain defoamants to reduce the tendency to foam, and the stability of any foam that does form. Because oxidation can increase the foaming tendency and also the stability of the foam, good oxidation stability is an important factor in maintaining foam resistance in service.

F. Entrained Air Release

Rapid release of entrained air is particularly important in systems supplying turbines with hydraulic governors. Excessive amounts of entrained air in these systems can cause sponginess, producing delayed or erratic response.

The rate at which entrained air is released from a mineral oil is an intrinsic characteristic of the base stock itself. It is dependent on factors such as the source of crude, and type and degree of refining. Currently, there are no known additives that will significantly improve the ability of an oil to release entrained air, but there are many additives that will degrade this ability. Thus, formulating a steam turbine oil with good air release properties is a process of selecting base oils with good air release properties, and then selecting additives that will perform the functions desired of them without seriously degrading the air release properties of the base oil.

G. Turbine Oil Compatibility Testing

Should a different turbine oil be considered as make-up to the existing charge of oil, then a compatibility test is recommended. ASTM 7155, Standard Practices for Evaluating Compatibility of Mixtures of Turbine Lubricating Oils, offers conversion guidance. Testing described in this standard practice observes changes in blended samples versus neat samples in visual inspection and selected performance tests.

H. Less Flammable Fluids

Less flammable fluids are usually used in electrohydraulic governor control systems of large steam turbines. The fluids used are most commonly based on phosphate esters or polyol esters. As noted previously, these systems are extremely sensitive to solids contamination in the fluid so considerable attention must be paid to fluid filtration.

The use of steam at high temperatures, up to as high as 1200°F (649°C), increases the possibility of fires from oil leaks in higher pressure turbine hydraulic systems. The use of these less flammable fluids are aids in mitigating this risk.

I. Maintenance Strategies

Some experienced steam turbine users opt to use lube oil analysis as a part of a planned turbine oil maintenance and rejuvenation program, sometimes called "bleed and feed." Based on periodic testing, a specified volume of steam turbine oil is removed and replaced with new oil. Based on past experience, some steam turbine users simply opt to replace 10% per year. If properly executed, this turbine oil maintenance strategy can extend turbine oil life almost indefinitely and minimize the impact of turbine oil performance decay.

BIBLIOGRAPHY

Bruce, R.W. 2012. *Handbook of Lubrication and Tribology, Volume II: Theory and Design*, Second Edition. Boca Raton, FL: CRC Press.

Mobil Oil Corporation. 1981. Steam Turbines and Their Lubrication. Internal Mobil Oil Corporation Publication, New York.

15 Hydraulic Turbines

The use of hydropower dates back to the ancient cultures of China, Greece, and the Roman Empire, which used water-powered mills for activities such as grinding grain. The first modern water turbine was developed in the mid-eighteenth century, and over the next hundred years several efficiency improvements and new designs were made. The earliest hydroelectric plants were built in the 1880s. Today, hydropower is produced in more than 150 countries and contributes approximately 19% of all electricity generated worldwide. Hydropower is considered to be a very cost-effective renewable energy source. However, there can be social and environmental impacts associated with hydropower that have to be considered.

The primary application of hydraulic turbines is to drive electric generators in central power stations. As the turbine shaft is usually rigidly coupled to the generator shaft with one set of bearings supporting both, these can be considered as one machine from a lubrication perspective. There is a wide range of sizes and operational characteristics for hydraulic turbines depending on the volume of water available and the pressure head of that water. Hydraulic turbines in commercial applications range in capacity from less than 1 MW to more than 750 MW. These units operate from as low as 40 rpm to as high as 2200 rpm but typically will operate in the range of 75–300 rpm.

Many of the locations suitable for large hydroelectric installations are in remote areas, often mountainous and with difficult access. Under these conditions, capital costs for plant construction and transmission lines are high. Therefore, in order for the power generated to be competitive in cost, plants must be designed for minimum maintenance and long service life. Generally, this has been accomplished, and the majority of hydroelectric units installed during the past 100 years are still in service.

Although many of the large hydroelectric installations are in remote locations, there are many installations in comparatively accessible locations. In recent years, the development of some of these locations has been aided by the introduction of the pumped storage concept and the use of bulb turbines.

Thermal and nuclear power plants, as well as open flume hydroelectric turbines, operate most efficiently at relatively constant loads. During off-peak periods, if the plant can be operated at or near full load, the power in excess of the load requirement is comparatively low in cost. Where a suitable location is available, this power can be used during low demand periods to pump water up to a storage reservoir. During periods of peak demand, the water is then available to drive hydroelectric generators to supplement the power available from the base load stations. In effect, pumped storage is a method of storing low-cost, off-peak power for use during peak demand periods.

Some pumped storage plants are equipped with both turbines and pumps. The generator is built as a combination motor/generator. During pumping operations, the turbine may be disconnected from the main shaft, or the turbine casing may be blown dry with compressed air so that the turbine operates without load. Similar arrangements are used with the pump. Quite a few new installations are equipped with reversible pump/turbines, and reversible motor/generators. There may be some sacrifice in efficiency with the pump/turbine combination, for example:

1. As a pump, it may be less efficient than a unit designed as a pump alone.
2. As a turbine, it may be less efficient than a unit designed as a turbine alone.

However, the lower cost of the combination machine generally offsets this loss of efficiency.

Bulb turbines (also called tubular turbines) are low head machines for what are referred to as "flow-of-stream" river applications. They can be used for relatively small, low-cost installations that can be readily blended into the surrounding country side. These machines have permitted the development of hydroelectric power in locations where installation of older types of turbines would not be practical or cost-effective.

I. TURBINE TYPES

There are several types of hydraulic turbines. They can be considered as impulse (Pelton) or reaction (pressure) types. Reaction turbines include the inward flow (Francis), diagonal flow (Deriaz), and propeller types. Bulb turbines are reaction turbines using a propeller-type runner. The choice of which type of unit to use in a particular application is a function of the pressure head and the quantity of flow available.

A. Impulse Turbines

In an impulse turbine, usually called a Pelton turbine, jets of water are directed by nozzles against shaped buckets on the rim of a wheel. The impulse force of the jets pushes the buckets on the rim of a wheel. The impulse force of the jets pushes the buckets and causes the wheel to revolve. The buckets move in the same direction as the jets of water. In order for a turbine of this type to operate efficiently, the velocity of the jets must be high; thus, a high pressure head of water is required. Pelton turbines are usually designed for pressure heads in the range of about 500–3900 ft (150–1200 m). Single units with outputs up to 200 MW have been built.

Pelton turbines are built with either a horizontal or a vertical shaft. Horizontal shaft machines are built with either one or two nozzles per runner. Typically, they are used for small to moderate-sized installations. A single runner may be connected directly to a generator, or two runners may be used, both on the same side of the generator or one on each side.

Vertical shaft Pelton turbines with four to six nozzles are now being used for larger installations. A cutaway view of an installation with four nozzles is shown in Figure 15.1. In this machine, the nozzle tips are actuated by hydraulic pistons located inside the nozzle bodies. The deflectors are actuated by a ring and lever arrangement, which in turn is actuated by hydraulic servo pistons. In many older machines, servo pistons that operate the nozzle tips are located outside the nozzles, and a mechanical linkage is used to operate the nozzle tips.

Pelton turbines are used in a number of pumped storage applications. Both horizontal and vertical shaft machines are used for this purpose. Where a horizontal shaft machine is used, the turbine is often mounted at one side of the generator/motor with the pump at the other side. Other arrangements can be used. In vertical shaft machines, the turbine is normally mounted above the pump so that the pump will operate with a positive pressure at the suction. The pump is usually coupled to the shaft through a clutch, so it can be disconnected and stopped during turbine operation. In some installations, the water is blown out of the pump, and it is left coupled to the turbine during turbine operation. The water is blown out of the turbine for pump operation. In some cases, the turbine is used to start the pump and motor and bring them up to speed. The water flow to the turbine is then shut off, and the water is blown out of the casing.

B. Reaction Turbines

In reaction turbines, the flow of water impinges on a set of curved blades which, in effect, are the nozzles. The reaction of the water on the nozzles (blades) causes them to move in the opposite direction.

In a Francis turbine (Figure 15.2), the water flows radially inward from a volute casing and is turned through 90° in the blades before flowing to the tail race outlet. In a Deriaz turbine (Figures 15.3 and 15.4), the direction of water flow is partially turned in the volute casing so that the flow is diagonally through the blades. The shaped boss then turns the water through the remaining angle to

Hydraulic Turbines

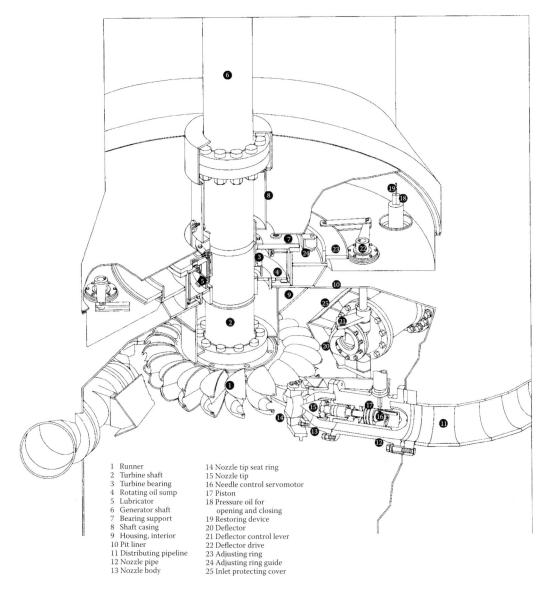

1. Runner
2. Turbine shaft
3. Turbine bearing
4. Rotating oil sump
5. Lubricator
6. Generator shaft
7. Bearing support
8. Shaft casing
9. Housing, interior
10. Pit liner
11. Distributing pipeline
12. Nozzle pipe
13. Nozzle body
14. Nozzle tip seat ring
15. Nozzle tip
16. Needle control servomotor
17. Piston
18. Pressure oil for opening and closing
19. Restoring device
20. Deflector
21. Deflector control lever
22. Deflector drive
23. Adjusting ring
24. Adjusting ring guide
25. Inlet protecting cover

FIGURE 15.1 Pelton turbine cross-sectional view and components.

direct it into the tail race outlet. In propeller turbines, the direction of flow is controlled by a volute casing or a flume so that the water flows axially through the turbine.

Francis and Deriaz turbines are intermediate to high head machines. The various types of propeller turbines (fixed blade, Kaplan, and bulb) are low head machines.

1. Francis Turbines

Francis turbines are probably the most widely applied hydraulic turbine. They are now also widely used as reversible pump/turbines. Originally used for intermediate head installations, the range of use of the Francis turbine has been extended well up into the head range that was formerly the exclusive province of impulse turbines. Francis turbines are being used at heads ranging from about 65 to 1650 ft (20–500 m), and even higher head units are under development. Outputs in the 200- to 400-MW per unit range are common, and units rated at up to 750 MW are in service.

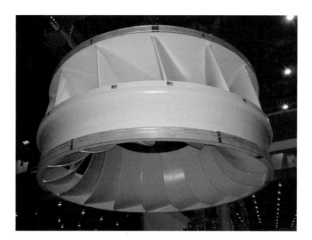

FIGURE 15.2 Francis turbine runner.

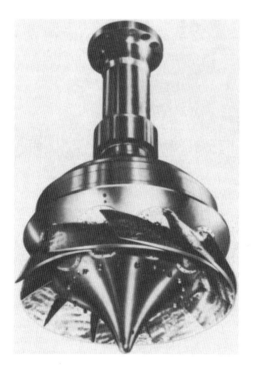

FIGURE 15.3 Deriaz turbine.

Small Francis turbines are sometimes built with horizontal shafts, but larger Francis turbines are nearly always vertical shaft machines. Water is brought in from the penstock through a spiral (volute) casing (Figure 15.5), from which it is directed into the turbine guide vanes by fixed vanes in the stay ring or speed ring (Figure 15.6). The movable guide vanes control the flow of water into the turbine to keep the turbine speed constant as the load varies. They can be operated by a single regulating ring actuated by one or two hydraulic servomotors (Figure 15.6), or by individual servomotors (Figure 15.7). With a regulating ring, the individual vanes are connected to the ring through shear pins so that if one vane is jammed by a foreign object during closing, the remainder of the vanes can still be closed to shut off the turbine.

Hydraulic Turbines

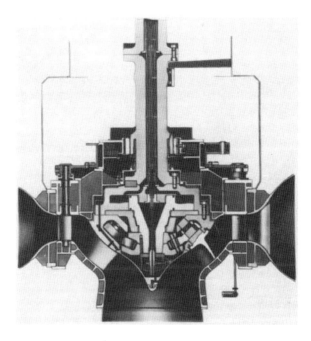

FIGURE 15.4 Cross section of Deriaz turbine.

FIGURE 15.5 Spiral casing for Francis turbine.

2. Diagonal Flow Turbines

The efficiency of Francis turbines drops off quite rapidly if the flow is less than the design value, or if the pressure head varies significantly. Where either of these conditions exists, a diagonal flow or Deriaz turbine can sometimes be used.

In the Deriaz turbine, both the guide vanes and the runner blades are adjustable (Figure 15.4). The runner blades are adjusted by a hydraulic servomotor located either in the turbine shaft or in the runner boss. The guide vanes are usually controlled by a regulating ring, similar to those used with Francis turbines. Movement of the guide vanes and runner blades is synchronized by the control system to maintain runner speed with changes in load, and keep the efficiency high as changes in the pressure head occur.

FIGURE 15.6 Assembly of Francis turbine (servomotor at top right).

FIGURE 15.7 Guide vane adjustment with individual servomotors. This shows operating mechanism for a Kaplan turbine, but is equally applicable to Francis.

Deriaz turbines are presently built for pressure heads in the range of about 60 to 425 ft (18–130 m). Designs suitable for heads up to about 650 ft. (200 m) are available. The usual construction is with a vertical shaft. Single units with ratings up to about 150 MW are in service. The larger units are reversible pump/turbines.

3. Fixed Blade Propeller Turbines

Fixed blade propeller turbines are vertical shaft machines. The blades of the runner (Figure 15.8) are cast integrally with the hub or welded to it. Because of the low heads at which propeller turbines operate, comparatively small changes in the head water or tail water level can make significant changes in the total head acting on the turbine. With a fixed blade turbine, this may cause marked

Hydraulic Turbines

FIGURE 15.8 Fixed blade propeller turbine.

changes in the efficiency of the turbine. For this reason, fixed blade propeller turbines (usually referred to simply as propeller turbines) are used only in locations where the head is fairly constant. Relatively few installations of this type of turbine exist. Outputs range up to about 40 MW per unit.

4. Kaplan Turbines

The Kaplan turbine (Figure 15.9) is the largest class of low head hydraulic turbines. The runner blades are adjustable, operated by either a hydraulic servomotor inside the runner hub, or by a

FIGURE 15.9 Kaplan turbine.

servomotor in the shaft with a mechanical linkage to the blade adjustment mechanism in the hub. The former arrangement is used for most new machines. As with Deriaz turbines, the controls for the guide vane adjustment and runner blade adjustment are synchronized to provide an optimum setting for each load and head condition. A typical Kaplan turbine installation is shown in cross section in Figure 15.10.

Kaplan turbines are built for heads from about 13 to 250 ft (4–75 m). Single units with power outputs from as little as 1 MW to units with outputs in the 175 MW range are in service.

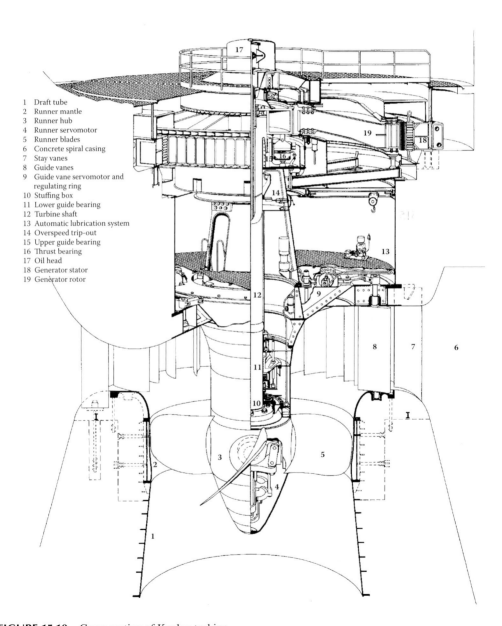

1 Draft tube
2 Runner mantle
3 Runner hub
4 Runner servomotor
5 Runner blades
6 Concrete spiral casing
7 Stay vanes
8 Guide vanes
9 Guide vane servomotor and regulating ring
10 Stuffing box
11 Lower guide bearing
12 Turbine shaft
13 Automatic lubrication system
14 Overspeed trip-out
15 Upper guide bearing
16 Thrust bearing
17 Oil head
18 Generator stator
19 Generator rotor

FIGURE 15.10 Cross section of Kaplan turbine.

Hydraulic Turbines

a. Bulb Turbines

The bulb turbine is actually a special application of the Kaplan turbine. As shown in Figure 15.11, the conventional Kaplan turbine is mounted vertically with a spiral casing carrying the water into the stay ring. From the turbine, a draft tube carries the water out to the tail race, creating a suction head on the turbine. By contrast, in bulb turbines, the turbine is mounted in a bulb-shaped section of the flume with the water flowing essentially straight through the turbine. The arrangement is much more compact, and is less costly to construct.

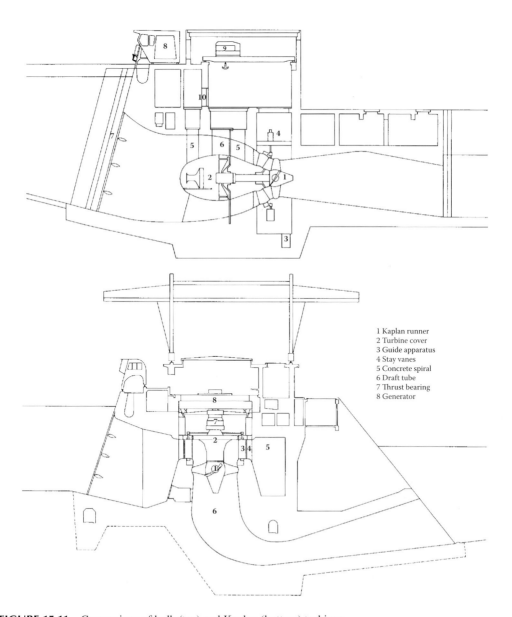

FIGURE 15.11 Comparison of bulb (top) and Kaplan (bottom) turbines.

FIGURE 15.12 Bulb turbine installation.

The shaft of bulb turbines is either horizontal or angled slightly downward toward the turbine. With a horizontal shaft, the generator is usually mounted in a bulb-shaped casing inside the flume, and is driven directly from the turbine (Figure 15.12). A mechanical drive to a generator mounted on the surface may also be used. With an angled shaft, the shaft may be extended out of the flume to drive a surface-mounted generator directly, or the generator may be mounted inside the flume.

Bulb turbines for heads up to 75 ft (23 m) are in service. Unit outputs are usually less than about 10 MW, but units up to 50 MW are in service.

b. S-Turbines

S-turbines (tubular turbines) are similar to bulb turbines except that the generator is not located in the bulb. The runner blade controls (where regulated) are contained in the bulb, but a drive shaft extends out of the Kaplan runner and through a portion of the "S"-shaped draft casing into a generator room, where it is connected to a generator. The S-turbine operates with heads of approximately 49 ft (15 m) and power outputs up to 15 MW.

II. LUBRICATED PARTS

The main parts of hydraulic turbines requiring lubrication are the turbine and generator bearings, the guide vane bearings, the control valve, governor and control system, and compressors.

A. Turbine and Generator Bearings

Horizontal shaft machines require journal bearings to support the rotating parts, including the generator armature. With the exception of Pelton turbines, thrust bearings are also required to absorb the thrust of the water acting on the runner. Vertical shaft machines require guide bearings to keep the shaft centered and aligned, and thrust bearings to carry the weight of the rotating parts and absorb the thrust of the water acting on the runner.

Vertical shaft machines have a guide bearing above the turbine, and one or two guide bearings at the upper, or generator, end of the shaft. In some cases with long shafts, an additional guide bearing may be installed about midway between the turbine guide bearing and the generator guide bearing. For reasons of accessibility, normally the thrust bearing is installed at the upper end of the shaft, either above the armature or just below it. Bearing arrangements are sometimes referred to as a *two-bearing* arrangement, wherein a combination guide and thrust bearing is located above the armature and another guide bearing is located below. In the so-called *umbrella* type, a combination thrust and guide bearing (Figure 15.13) is located below the armature, and in the *semiumbrella* type, separate guide and thrust bearings are located below the armature. Most current medium and large machines are of either the umbrella or semi-umbrella type.

1. Journal and Guide Bearings

The bearings of horizontal shaft machines are of the fluid film type, with babbitt-lined, split shells. Oil lifts to assist starting may be used in the journal bearings when the rotating parts are extremely heavy. At stabilized operating conditions, oil-lubricated journal and guide bearings will operate in the range of 140°F (60°C).

Two general types of turbine guide bearings are used in vertical shaft machines. Many older turbines are equipped with water-lubricated rubber or composition bearings. This construction minimizes sealing requirements at the top of the turbine casing, but is not satisfactory in situations where the water carries silt and other solids. As a result, most machines are built with a stuffing box at the top of the turbine, and a babbitt-lined guide bearing (Figure 15.10). Various types of seals are used in the stuffing boxes, including carbon ring packing and gland packing.

The upper guide bearings are either babbitt-lined, split cylindrical shell-type bearings (Figure 15.14) or segment bearings (Figure 15.15). The segment-type bearings are becoming increasingly popular because they permit easy adjustment of shaft alignment and bearing clearance. The segments may be crowned—that is, machined to a slightly larger radius than that of the journal plus the thickness of the oil film in order to permit easier formation of oil films. The construction shown in Figures 15.14 and 15.15, with the journal formed by an overhanging collar on the shaft, and an oil dam extending up under the collar, is now common. It eliminates the need for an oil seal below the bearing. Bearings running directly on a machined journal on the shaft, with either an oil seal below the bearing or an oil reservoir fastened to the shaft and rotating with it, are also used.

2. Thrust Bearings

Thrust bearings of horizontal shaft machines are of the tilting pad or fixed pad type. Where a reversible pump/turbine is used, tilting pad bearings designed for operation in either direction must be used. Thrust bearings for vertical shaft hydroelectric units are among the most highly developed forms of these bearings now in use. As pointed out earlier, the thrust bearing must support the weight of the rotating parts plus the hydraulic thrust of the water acting on the turbine runner. Single bearings capable of supporting loads in excess of 2000 tons (1,814,000 kg) are in service. Thrust bearings operate at the highest temperatures due to the high thrust loads. Operating temperatures will generally be in the range of 212°F (100°C).

Tilting pad thrust bearings are used on all larger machines. Some older machines are equipped with tapered land bearings. Kingsbury and Michell type bearings, in which the pads tilt on a pivot or on a rocking edge on the bottom of the pad, were used on many older machines. Most newer designs use flexible supports under the pads to permit the slight amount of tilt needed to form oil wedge films. Bearings with flexible pad supports can be run in either direction, so they are particularly suited to reversible pump/turbine units. Springs (Figure 15.16), elastomeric pads, interconnected oil pressure cylinders, and flexible metallic supports are all used. Spherical supports are also used. The bearing segments may be

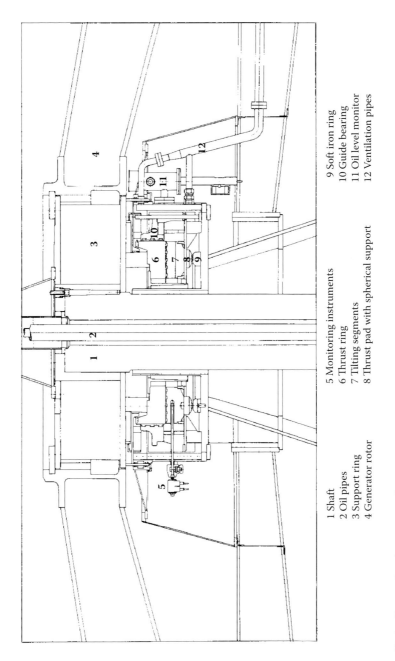

FIGURE 15.13 Umbrella type bearing construction.

1 Shaft
2 Oil pipes
3 Support ring
4 Generator rotor
5 Monitoring instruments
6 Thrust ring
7 Tilting segments
8 Thrust pad with spherical support
9 Soft iron ring
10 Guide bearing
11 Oil level monitor
12 Ventilation pipes

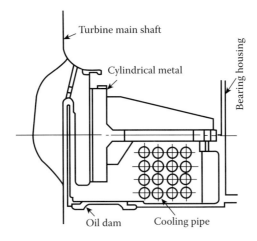

FIGURE 15.14 Structural view of cylindrical bearing for vertical shaft turbine.

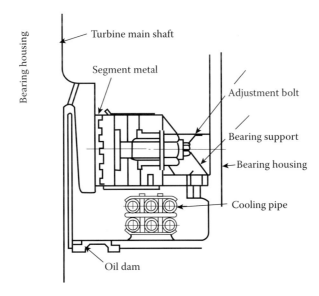

FIGURE 15.15 Cylindrical shell and segment-type guide bearings.

flat, or crowned slightly to aid in the formation of oil films. In many of the larger machines, provision is made to pump up the bearings with high pressure oil to assist starting (Figure 15.16).

B. Methods of Lubricant Application

The bearings of hydroelectric sets are either self-lubricated or supplied by a central circulation system. Circulation systems may be either unit systems, where a separate system is used for each unit in a station, or station systems, where all the units in a station are supplied from one system. In many cases, one or more of the bearings may be of the self-lubricating type, with a unit system supplying the other bearings.

In self-lubricated bearings, the oil is contained in a tank surrounding the shaft (Figures 15.13 through 15.15). Oil is lifted by grooves in the bearings, or by a ring pump on the shaft. Cooling coils can be located in the tank, or with a cylindrical shell bearing a cooling jacket may be located around the bearing shell. External cooling coils may also be used, but are generally suitable only for relatively high-speed machines where sufficient pumping force is generated to circulate the oil through the external circuit.

FIGURE 15.16 Spring supported tilting pad thrust bearing. Stationary portion showing oil lift slots.

C. Governor and Control Systems

Older hydroelectric units were equipped with mechanical hydraulic control systems with a mechanical speed-sensing device and a hydraulic system to actuate the guide vanes, and the runner blades if a Deriaz or Kaplan turbine was used. Newer machines are often equipped with electrical speed-sensing devices and electronic systems.

Older hydraulic turbine hydraulic systems generally operated at 150 psi (10.4 bar) and used the same oil as the bearing oil system. Current units operate with pressures in the 1000 psi (69 bar) range but can go as high as 2000 psi (138 bar). The hydraulic systems are now usually separate systems and require antiwear hydraulic fluids for the higher pressure systems. There is also a trend toward the use of environmentally acceptable fluids for these applications. Hydraulic pumps are driven by electric motors. Air-charged accumulators (air over oil) are used to maintain system pressure and supply the large fluid flow necessary to adjust rapidly to meet sudden changes in load. They also provide a source of fluid under pressure to shut the turbine down in the event of a failure in the system. Emergency shutdown may also be assisted by the use of closing weights on the guide vane operating mechanism, or by designing the vanes so that water pressure will close them if the hydraulic system fails.

D. Guide Vanes

The guide vanes, or wicket gates, are manufactured with an integral stem at each end that serves as the bearing journal. One bearing is used at the bottom and one or two bearings at the top. A thrust bearing may also be required. These bearings, as well as the bearings of the operating mechanism, were once grease lubricated. Centralized lubrication systems are now usually used to supply these bearings. An alternative to this are self-lubricated bearings.

E. Control Valves

In some turbines, the guide vanes are arranged to close tightly and act as the shutoff valve for the turbine. In most installations, however, separate closing devices on the water inlet are used. In the case of pump turbines, closing devices on both the inlet and outlet are used.

Hydraulic Turbines

FIGURE 15.17 Spherical valve assembly.

Closing devices include sluice valves, rotary valves, butterfly valves, and spherical valves (Figure 15.17). All are designed for hydraulic operation. Closing weights may be used for emergency shutdown. Bearings are grease lubricated.

F. Compressors

In most hydroelectric plants, compressed air is required to maintain the pressure in the hydraulic accumulators. Compressed air is also used to blow out the draft tube and turbine casing when maintenance is to be performed. Compressed air also blows out the pump or turbine when the changeover from pump to turbine operation, or vice versa, is made in pump/turbine installations. Compressed air is also used in some impulse turbine installations to keep the tail water out of the turbine when the tail water level is high. Compressors operated in hydroelectric plants are critical pieces of equipment. Air compressors can be four-stage units and operate with discharge pressures up to 1000 psi (69 bar).

III. LUBRICANT RECOMMENDATIONS

The need for extreme reliability and long service life of hydroelectric plants generally dictates that premium, long-life lubricants be used. Rust and oxidation inhibited premium turbine or circulating oils are usually used for oil applications. Viscosities usually are of International Standardization Organization (ISO) Viscosity Grade 32, 46, 68, or 100 depending on bearing design, speeds, and operating temperatures. Oils with excellent water-separating characteristics are desirable. Although startup temperatures are rarely below freezing point, the oils used must have adequate fluidity for proper circulation at those temperatures. Where oil lifts are not used for starting, oils with enhanced film strength may be desirable to provide additional protection during starting and stopping.

Hydraulic system requirements for older units were generally met with the same types of oils used for bearing lubrication. More modern high-pressure hydraulic systems have been separated from the bearing oil systems and may require antiwear-type hydraulic fluids. Good air separation properties are desirable to ensure that air picked up in the accumulators separates readily in the reservoir.

Greases used in grease-lubricated bearings require good water resistance and rust protection. They should be suitable for use in centralized lubrication systems, and have good pumpability at the lowest water temperatures. Both lithium and calcium soap grease are used. National Lubricating Grease Institute (NLGI) No. 2 consistency greases are usually used, with No. 1 consistency greases used in some extremely cold locations.

BIBLIOGRAPHY

National Geographic. Hydropower. http://environment.nationalgeographic.com/environment/global-warming/hydropower-profile/.
U.S. Geological Survey. Hydroelectric power water use. https://water.usgs.gov/edu/wuhy.html.

16 Wind Turbines

Solar energy, wind energy, geothermal energy, biomass energy, and hydroelectric energy sources are all well-known alternative energy sources. Compared to conventional energy sources such as fossil fuel or nuclear energy, these are considered renewable resources with lower carbon emissions and are better for the environment because they have much less impact on global warming. Of these alternative energy sources, wind energy has been among the fastest-growing energy sources, and the equipment used to generate electricity presents unique lubrication challenges.

Wind energy has been harnessed by man for many years, with some of the first recorded uses being to propel boats along the Nile River as early as 5000 B.C. Between 500 and 900 B.C., wind was used by the Persians to grind grain and pump water. By 1000 A.D., windmills spread north to Europe and were used by countries such as the Netherlands to drain marshes. With the advent of steel blades, there were more than 6 million windmills in service in the United States by the late 1800s. The use of a windmill to generate electricity was first developed in Scotland in 1887. Today, more than 225,000 wind turbines in 83 countries generate approximately 4% of the world's electricity, with more than 300 GW being generated. Since 1996, installed wind power capacity has had an average annual growth of ~24% and is still projected to grow 10% a year for the foreseeable future. Today, the largest wind turbines can produce up to 8 MW of power.

I. WIND TURBINE OVERVIEW

A. Wind Turbine Design

Wind is a form of solar energy and is a result of the uneven heating of the atmosphere by the sun, the irregularities of the earth's surface, and the rotation of the earth. Wind flow patterns and speeds vary considerably and are modified by bodies of water and differences in terrain. Wind turbines convert the kinetic energy of the wind into mechanical power. This mechanical power can be used to drive a generator that converts this into electricity.

Wind turbines are classified into two general types: vertical axis and horizontal axis. A vertical axis wind turbine (VAWT) machine has blades that rotate on an axis perpendicular to the ground (Figure 16.1). A horizontal axis wind turbine (HAWT) has blades that rotate on an axis parallel to the ground. There are a number of available design variations for both, and each type has certain advantages and disadvantages. The most common is the horizontal axis type with very few vertical axis machines being commercially available.

HAWTs have the main rotor shaft and electrical generator at the top of a tower, which must be turned into the wind (Figures 16.2 and 16.3). Small turbines are pointed into the wind by a simple wind vane, whereas large turbines generally use wind sensors coupled with a series of servomotor drives to adjust yaw. Yaw is the angular control of the rotation of the wind turbine housing called a nacelle. Most wind turbines have a gearbox, which turns the slow rotation of the blades into a faster rotation that is more suitable for driving an electrical generator.

Because the support tower produces turbulence behind it, the turbine rotor is usually positioned upwind of its supporting tower. Turbine blades are made stiff to prevent the blades from being pushed into the tower by high winds. Additionally, the blades are placed a considerable distance in front of the tower and are sometimes tilted forward into the wind. Larger wind turbines capable of generating about half a megawatt or more have extensive pitch control mechanisms on each blade. Generators typically require 1200–1800 rpm to operate efficiently. However, the speed of a wind rotor is usually more in the range of 10–25 rpm. In order to make up for this difference, wind

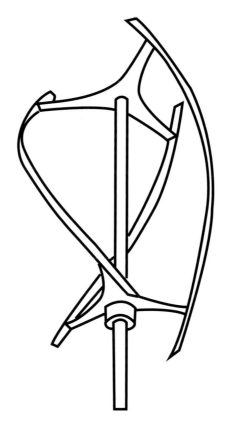

FIGURE 16.1 Vertical axis wind turbine.

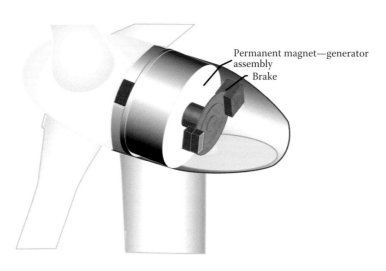

FIGURE 16.2 Permanent magnet horizontal axis wind turbine.

Wind Turbines

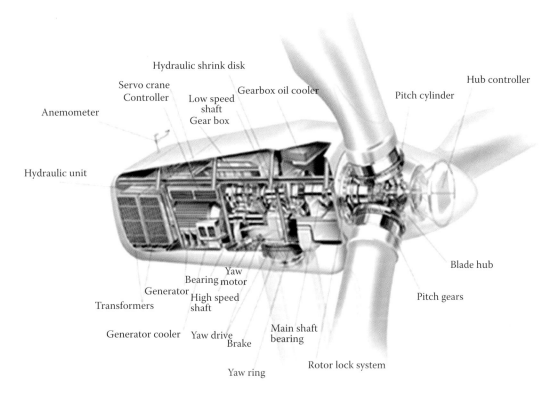

FIGURE 16.3 Horizontal axis wind turbine. The gearbox drive and other key components are identified.

turbines will usually have a gearbox transmission to increase the rotation of the generator to the speeds necessary for efficient electricity production. The gear is connected to a second high-speed shaft that, because of the gear ratio, turns at the higher speed the generator requires.

In any device, there is usually a tradeoff between complexity and maintainability. This is true for wind-powered generators. Some direct current (DC) wind turbines do not use transmissions, but instead have a direct link between the rotor and generator. These are known as *direct drive* systems (Figure 16.2). By eliminating the gearbox, the device is considerably simpler and will require less maintenance. However, in order to generate sufficient electricity with a direct drive, a larger generator is required to deliver the same power output as the AC-type wind turbines. This adds significant weight to the machine. The direct drive machines require the use of a large and expensive permanent magnet, which is usually made of rare earth elements (e.g., neodymium) that are in limited supply and is subject to geopolitical pressures, making it a strategic metal.

B. Wind Turbine Blades

Turbines used in wind farms for commercial production of electric power are usually three-bladed and pointed into the wind by computer-controlled yaw drives. Ideally, the cone tip and entire rotor assembly is pointed directly in the direction of the wind for maximum efficiency. These machines have high tip speeds of more than 300 km/h (185 mph) and are highly efficient, which contributes to good reliability. The blades are usually white in color for daytime visibility by aircraft, and range in length from 20 to 80 m (65–265 ft). The tubular steel towers can reach a height of 150 m (485 ft) or more. The blades rotate at 10–25 rpm. At 25 rpm, the tip speed can exceed 360 km/h of a longer blade. Some models operate at constant speed, but more energy can be collected by variable-speed turbines that use a solid-state power converter to interface to the transmission system. All turbines

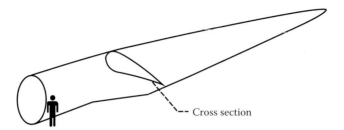

FIGURE 16.4 Wind turbine blade.

are equipped with protective features to avoid damage at high wind speeds, by feathering the blades into the wind, which ceases their rotation, supplemented by brakes.

Most modern wind turbines have only three blades on a rotor, and the total area they cover affects wind turbine performance. The use of three blades in the design is important for several reasons. Three moment arms on a common rotor mean that the main shaft will be in balance most of the time. Rotors that use the lift-principle need the wind to flow smoothly over the blade. If the blades are too close together, the turbulence from one blade can disrupt the flow of air to the blade next to it. The blades must be far enough apart so that this does not happen. The longer the length of the blades, the more of the wind's kinetic energy can be captured. The area of the blades that sweep through the air is known simply as the Wind Swept Area and can be calculated with the formula: $\pi \times radius^2$. The greater the Wind Swept Area, the more power the turbine will generate. Not all the energy of blowing wind can be harvested, because conservation of mass requires that the same mass of air exits the turbine as enters it.

Albert Betz calculated the maximum wind turbine performance, now called the "Betz limit," in 1966. Betz demonstrated his theory using a closed stream tube to document that a HAWT can only convert a maximum of 59% of the energy of wind into electricity. This optimum performance is attained when a wind turbine rotor slows the wind down by about one-third. Johannes Juul developed the "Danish Design" in 1958, which allowed alternating current (AC) for a wind turbine to be fed to the electrical power grid for the first time. This concept very quickly was adopted, and most large wind turbines today operate according to this principle.

Modern wind turbines use what is known as a lift design. The rotor blade profile is similar to that of an airplane wing and creates lift because of the differential air pressure between the flat side and the rounded side of the blade (Figure 16.4). However, because the blade is turned at an angle, the lift causes the blade to turn rather than rise. Lift-powered wind turbines have much higher power extraction than drag types and therefore are well suited for electricity generation. The rotor blades are usually made of plastic, fiberglass, or wood covered with an epoxy or urethane coating. Simple turbines with small power output (up to 2 kW) and historic windmills operate according to the principle of resistance. Vertical axis turbines have rotors that resist the wind, thus reducing wind speed. The maximum energy conversion performance of VAWTs is only about 12%, which accounts for the much greater commercial application of the horizontal axis design.

C. Wind Turbine Generator

The generator is the device inside the wind turbine that actually generates electricity. Depending on the type, it can also be referred to as a permanent magnet alternator. It takes the kinetic energy generated by the rotor and translates it into electricity. Inside the generator, coils of copper wire wound around the armature are rotated in a magnetic field to produce electricity. Generators can be designed to produce either AC or DC, and they are available in a large range of output power ratings. If the turbine is designed to produce DC current, it will have an additional component inside the housing called a rectifier, which will convert the originating AC to DC. To change from

DC to AC, a separate device called a power inverter is used. There are different requirements depending on whether or not wind turbines are tied to a power grid. In a system tied to the grid, the AC output matches that of the electrical grid itself. Most appliances in North America operate on 120 V with a 60-Hz AC frequency. In Europe and Asia, most appliances operate on 220 or 230 V with a 50-Hz AC frequency although some select countries may have differing nominal voltages.

II. GENERAL CONSIDERATIONS FOR WIND TURBINE LUBRICATION

The wind industry recognizes the importance of the lubricant as a critical component contributing to the life and reliability of the wind turbine. The use of high-performance lubricants can lead to extended component and lubricant life, reduced maintenance, maximum power generation, good ancillary system protection, all of which can be a warehouse of information about gearbox health.

A. Main Gearbox Lubrication

The main gearbox is the most critical component of a wind turbine, and it poses several unique lubrication challenges. It is important that high-performance wind turbine gear lubricants have the ability to not only provide excellent wear protection associated with all aspects of mechanical performance, but also to have the right balance to address other vital requirements including interfacial properties of the lubricant and compatibility with other parts of the wind turbine. The keys to achieving excellent equipment life comes from providing outstanding protection against bearing wear and gear wear associated with micropitting and scuffing, and providing sufficient rust and corrosion protection for the metallurgy in the gearbox.

Long oil drain intervals for wind turbine gearboxes are very common, but require retention of key performance properties of the oil. Extended oil drain intervals and filter change-out intervals result in increased production and reduced downtime. Prevention against oil aging improves productivity and reduces the overall cost of operation.

Generally, there are certain physical lubricant characteristics that most wind turbine gearbox builders and users prefer. Ideally, lubricants for the wind industry need to be light-colored and transparent in appearance, owing to the potential for accidental spills up-tower. Gear oils with low or no odor are desired in this application. Lower toxicity oils are an excellent choice where oil leakage or spillage could enter environmentally sensitive areas.

1. Industry Standards and Builder Specifications for Wind Turbine Gear Oils

Wind turbines generally have up to three different gearboxes in use: the main gearbox transmits torque from the blades and converts the slower blade speed into higher speeds for the generator to produce electricity; the pitch gear adjusts the blades according to the actual wind speed; and the yaw gearboxes rotate the nacelle into the wind. The unique demands placed on gear oils in wind turbines call for a higher level of performance above and beyond most standard industrial gear oils.

Although no standard set of performance tests for gear oils used in the wind turbine industry has been agreed to by all builders and suppliers, there are general performance and inspection features that most wind turbine and gearbox suppliers have made public. To begin with as a minimum, a gear oil needs to satisfy the requirements of the DIN (Deutsches Institut für Normung) 51517-3 standard for the classification of gear lubricants. However, because gearboxes in wind turbines represent one of the most demanding applications, the DIN standard has not been viewed as sufficient to cover all of the needs of a wind turbine gear oil. As a result, the joint effort of the IEC (International Electrotechnical Commission) and ISO (International Organization for Standardization), which consists of several wind turbine builders, gearbox builders, bearing and ancillary equipment manufacturers, insurance and certification authorities, as well as lubricant manufacturers, developed the international standard IEC 61400-4 (Design Requirements for Wind Turbine Gearboxes). This 2013 standard includes requirements and informative sections related to the lubrication of main gearboxes. Furthermore, the American Gear Manufacturers Association (AGMA) developed AGMA 6006, Standard for Design and Specification of Gearboxes for Wind Turbines, which was the basis for the IEC 61400-4 document.

These industry standards define only minimum lubricant properties. However, most of the major gearbox and wind turbine builders have developed additional in-house specifications, which should help to ensure that only the highest quality lubricants with proven performance are being used. For additional details on gear lubricant recommendations, each individual builder specification should be referenced.

2. **Viscosity and Low Temperature Requirements of Wind Turbine Gear Oils**

Wind turbine main and pitch gearbox applications can be exposed to low starting and operating temperatures. Depending on the location of the wind turbine, ambient temperatures can range from −20°F (−29°C) to above 100°F (38°C). However, wind turbine manufacturers recommend that the minimum startup temperatures be above 5°F (−15°C), which will require lubricants with low pour points. This low temperature requirement for a wind turbine gear oil is typically specified as having a pour point of −40°F (−40°C) or lower to ensure proper oil flow at startup. A more descriptive low-temperature pumpability performance measurement for these oils includes a Brookfield viscosity of 250,000 cP (centipoise) or lower at −40°F (−40°C). The low-temperature startup requirement of wind turbines compels the use of synthetic hydrocarbons such as polyalphaolefins over mineral-type lubricants. These synthetic hydrocarbon gear oils also provide excellent viscosity control at nominal operating temperature, which is needed because of the remote location of most machines.

The viscosity index of a lubricant is an important indicator of the oil's resistance to viscosity change as temperature varies. Because of the heavily loaded nature of wind turbine gearboxes, viscosity index improver additives are not used. The reason for this is that viscosity index improvers are long-chain molecules, with very high molecular weight, which are mechanically unstable under heavy loads and will shear, resulting in loss of viscosity. Hence, synthetic hydrocarbon lubricants, with a naturally high and shear stable viscosity index of 140 to 180+, are ideal for this application.

In terms of viscosity recommendations for wind turbine main gearboxes, the ISO 320 viscosity grade is used in the majority of main gearboxes around the world and is the most common viscosity specified by both gearbox and wind turbine manufacturers. There are a few instances where an ISO 220 or ISO 460 viscosity grade gear oil is used.

3. **Antiwear Performance**

a. *Micropitting Protection*

The compact designs of modern wind turbines put size and weight restrictions on the main gearbox. This high power density requires that the gear oil have high load handling capability. Case hardening of gears, which is the heat treatment process most commonly used in the wind industry, changes the physical properties of the gear and imparts wear resistance to gear teeth. Case-hardened gears are susceptible to a rather unique failure mechanism called micropitting. This failure mode is usually associated with break-in of the gears, but some of the causes include gear hardness, surface roughness, inadequate lubricant film thickness, and oil contamination. In general, micropitting can lead to more extensive gear failures.

Micropitting failure has been prevalent in the wind industry, and therefore prevention of this failure mode is a critical performance feature for gear oils. Micropitting, unlike macropitting, occurs away from the pitch line on the gear tooth at the dedendum (the radial distance from the pitch line of the gear tooth down to the root line) and the addendum (the radial distance from the pitch circle to the top of the gear). Examples of micropitting damage can be seen in Figures 16.5 and 16.6. Extremely high loads encountered in the wind turbine lead to surface fatigue that results in very small pits on the gear tooth surfaces approximating 10 μm or less in depth and width. This failure

FIGURE 16.5 Minor micropitting damage.

FIGURE 16.6 Extensive micropitting damage.

mode is also referred to as gray staining, gray frosting, peeling, or microspalling. Micropitting causes tooth profile deviations that can increase vibration and noise with the wear particles creating additional damage. It concentrates loads on smaller tooth areas, increasing the stress on gear teeth and shortening gear life.

There have been several attempts to devise a gear oil test that can predict the micropitting protection performance of a lubricant. The FZG (Forschungsstelle fuer Zahnraeder und Getriebebau) FVA 54 micropitting test has emerged as a good predictor of a lubricant's ability to resist this failure mode. The test is run in a Strama gear oil tester, using C-GF type gears, with a 40-liter circulating system to control temperature during the test. The test is usually run at 60°C (140°F) and 90°C (194°F). The test consists of three separate segments. The first segment, called the load stage phase, is conducted on one side of a test gearset that is run through a series of stage tests (Figures 16.7 through 16.9). The test gear is removed from the gearbox and evaluated for wear at the end of each 16-h stage. The point at which the profile deviation exceeds 7.5 μm of wear depth is assigned as the fail load stage (FLS) (Figure 16.7).

The second segment, the endurance phase, is run after the load stage test using the same side of the gearset. It is run at load stage 8 for 80 h. The test is then run for five successive 80-h endurance stages at load stage 10. After each 80-h segment, the gear is removed and evaluated. The test is allowed to run to the end of the last endurance stage only if the gear has not suffered more than 20 μm of wear depth. The pinion is then evaluated for profile deviation (Figure 16.7), percent surface

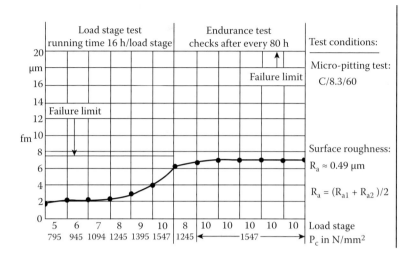

FIGURE 16.7 Report of gear face profile deviation from Forschungsstelle fuer Zahnraeder und Getriebebau (FZG) FVA 54 micropitting test.

damage (Figure 16.8), and weight loss (Figure 16.9). During the third segment of the test, the gearset is run on the opposite side of the gear faces using the load stage test segment from only stage 5 through stage 10. The pinion is then evaluated in the same manner as the first gear face for profile deviation, percent surface damage, and weight loss.

To demonstrate acceptable micropitting protection, a good wind turbine gear oil should be at least able to deliver an FLS equal to 10 or better in this test. These tests, although time-consuming (can average 10–12 weeks to complete the setup and teardown, testing, measurements, and analysis), are currently the best measurement of micropitting performance available. To date, some of the most widely accepted test institutes for the full FVA 54 test include the FZG Institute at the Technical University Munich (Forschungsstelle fuer Zahnraeder und Getriebebau—Lehrstuhl für Maschinenelemente) and the University of Bochum (Lehrstuhl für Industrie-und Fahr zeugantriebstechnik). Additionally, there are other institutes and test laboratories that can run this test.

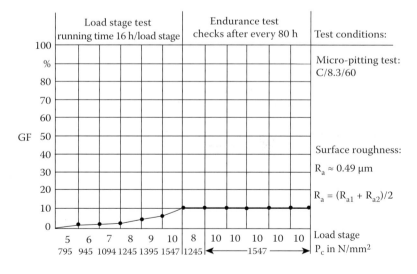

FIGURE 16.8 Report of percent gear face surface damage from Forschungsstelle fuer Zahnraeder und Getriebebau (FZG) FVA 54 micropitting test.

Wind Turbines

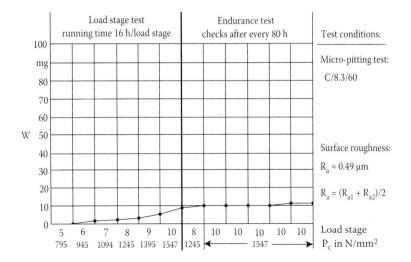

FIGURE 16.9 Report of weight loss from Forschungsstelle fuer Zahnraeder und Getriebebau (FZG) FVA 54 micropitting test.

b. Scuffing Protection

Scuffing protection for wind turbine gear oils is very important. Another common cause of failure in wind turbine gearboxes is extreme gear scuffing occurring on the high-speed pinion. The entire load from the drive train comes through the last pinion gear to drive the external generator. Although the DIN 51517-3 standard identifies FLS = 12+ as a minimum for standard gear oils in the FZG (A/8.3/90) scuffing test, most wind turbine builders require gear oils with an FLS of 14 or better. The FZG scuffing test procedure uses a test rig without a circulation system and is defined under ISO 14635-1. The scuffing test runs at the standard speed of 1500 rpm (8.3 m/s) and is augmented for wind turbines by evaluating the performance of the lubricant at double speed (3000 rpm [16.6 m/s]) or using the same FZG test rig and ISO 14635-1 procedure. In the right-hand side of Figure 16.10, a brand-new unused gear with no wear is displayed with the characteristic crosshatching necessary for evaluation in this test. On the left-hand side is a gear that has exceeded its failure stage with the characteristic scuffing damage (note the smearing and removal of metal).

c. Bearing Protection

In addition to providing micropitting protection and scuffing resistance for gears, wind turbine gear oils must also protect bearings. Wind turbine gear oils are generally required to protect the

FIGURE 16.10 Gears used in the Forschungsstelle fuer Zahnraeder und Getriebebau (FZG) scuffing test. The gear on the left indicates gear failure due to excessive scuffing wear. The gear on the right is a new gear with no wear with original crosshatching marks evident.

TABLE 16.1
Summary of FAG Schaeffler Wind Turbine Bearing Test Stages

1. Bearing Wear—High Load/Low Speed—Extreme Mixed-Friction—FE8 Cylindrical Roller thrust bearing; 100 kN, 7.5 rpm, 80°C, 80 h
2. Fatigue—Moderate Mixed Friction—FE8 Cylindrical Roller thrust bearing; 90 kN, 75 rpm, 70°C, 800 h
3. Fatigue—EHL Conditions—L11 Ten deep groove ball bearings; 8.5 kN, 9000 rpm, 700 h
4. Deposit Test/Water Added—FE8 Cylindrical Roller thrust bearing; 60 kN, 750 rpm, 100°C, 600 h

numerous sets of rolling element bearings in a gearbox. Schaeffler Technologies AG & Co. KG has developed a four-stage testing protocol (known as FAG Schaeffler Wind Turbine Bearing Test) that is generally considered by most wind turbine gearbox builders to represent the bearing performance requirements for these gear oils. The four-stage wind turbine bearing test suite includes a variety of speeds and loads that represent the types of service in a wind turbine gearbox. Stage I of the FAG Schaeffler protocol is actually the same FE8 cylindrical roller thrust bearing that is part of the DIN 51517-3 standard. This is a low-speed/high-load roller thrust element bearing test run at extreme boundary conditions, that is, at 7.5 rpm, at 80°C (176°F) for 80 h. However, the load used in the four-stage test is 100 kN (22,480 lb) for wind turbines rather than the specified 80 kN (17,984 lb) in the DIN standard. Stage II uses the same Schaeffler FE8 rig but is run at a higher speed, 75 rpm for 800 h at 70°C (158°F). The third stage uses a deep groove ball bearing in the L-11 test rig. Stage III operates under long fatigue life and elastohydrodynamic lubrication conditions. This stage of the test is run at 8.5 kN (1800 lb), 9000 rpm for 700 h at 80°C (176°F). Stage IV is a deposit test. It uses the same FE8 rig used in stages I and II, but is run at 750 rpm at 60 kN (13,488 lb) for 600 h at 100°C (212°F) with the addition of water to increase severity. Table 16.1 summarizes the test conditions. The ratings of these four tests are given in a summary report. Oils with performance scores in lower numbers are preferable, with 1.0 being the top rating.

SKF, another leading bearing manufacturer, also has a series of test protocols designed for wind turbine gear oils. SKF runs a series of bench-type tests at their laboratory facility. In one of these SKF tests, a series of bearing rollers are placed in a beaker of oil at either 100°C (212°F) or 120°C (248°F) for 8 weeks. At the end of the test, the change in viscosity of the lubricant is recorded, the appearance of the rollers in terms of sludge buildup is noted, and other visual parameters are also recorded.

4. Interfacial Properties of Gear Oils

The use of synthetic hydrocarbon gear oils in wind turbines has been proven to provide extensive protection against oil aging, while retaining the high level of oil performance needed, when long oil drain intervals are used. Because the oil drain intervals of many wind turbines tend to be 3–5 years on average (even up to 7 years in a few circumstances), use of fine oil filtration (e.g., 10 μm and below) is needed to maintain system cleanliness to enable long oil and equipment life. Lubricants used in long oil drain interval service need to retain their foam control performance after extensive filtration.

The high humidity environment under which offshore wind turbines operate can be a concern for the filterability of oil that is contaminated with some water. Several filterability tests procedures are available, but no industry standard test has been established to date.

Foam and air release characteristics for wind turbine lubricants are evaluated based on foaming characteristics in the American Society for Testing and Materials (ASTM) D892, Standard Test Method for Foaming Characteristics of Lubricating Oils. This test evaluates both the tendency of a lubricant to generate foam and characterize the stability of the foam once generated. Low numbers between 0 and 50 are considered acceptable in this test. The wind industry also stresses the performance of foam control in the Siemens/Flender foam test. This is the subject of a recently released ISO standard, ISO 12152. This method uses a small gearbox to entrain air by churning the gear oil at high speed (Figure 16.11). It evaluates the foam performance of the oil after the air is churned into the oil and assesses its ability to release air bubbles quickly. In this test, the gearbox is turned at 1500 rpm for 5 min. The standard test temperature is 20°C (68°F), but other temperatures can be applied. The gearbox is turned off, and the volume of oil increases due to entrained air and foam is recorded after 1 min of standing. Passing this test requires the oil to have less than a 15% volume increase over the baseline measurement at startup.

The wind turbine gearbox often operates under extreme corrosive environments requiring excellent performance against the rusting of steel. Rust performance is part of the DIN 51517-3 specification as defined by the ASTM D665 test method, Standard Test Method for Rust Preventing Characteristics of Inhibited Mineral Oil in the Presence of Water. An additional corrosion requirement for most wind turbine gearbox builders is a strong performance in the SKF EMCOR bearing corrosion test. Lower ratings of 0 or 1 (on scale of 0 to 5) in this test are generally required for wind turbine lubricants with both distilled water and synthetic seawater.

Besides having strong resistance to the rusting of ferrous metals, wind turbine lubricants have to be extremely resistant to corrosion of copper alloys. Strong corrosion control of copper alloys according to the ASTM D130 test, Standard Test Method for Corrosiveness to Copper from Petroleum Products by Copper Strip Test, is critical. Copper alloys are common in bearing cage materials, e.g., in a variety of pump parts.

Because of the long oil drain intervals typically used by wind turbines, the use of any zinc parts or galvanized finishing should be avoided. This is attributable to the low oxidation potential of any zinc part, as zinc will preferentially oxidize or corrode before any other metal in a machine.

A minimal level of oxidation stability performance is required for wind turbine gearbox oils. Generally, the operating temperature of the gear oil in most modern wind turbines is approximately

FIGURE 16.11 Siemens/Flender foam test rig.

70°C (158°F) or lower. This relatively low operating temperature does not significantly stress oils to the point of degradation of oxidation stability. Therefore, high levels of antioxidant inhibitor additives in the oil are not necessary. This is particularly true when synthetic lubricants that have high oxidation stability are used.

5. Material Compatibility (Seals and Paint)

Machine performance and reliability is extremely important for wind turbines. To help ensure this, many gearbox manufacturers and wind turbine suppliers have specifications for lubricant compatibility with other materials. Elastomer testing that includes fluorinated hydrocarbons and nitrile rubbers can be found in most specifications.

Nitrile rubber seals are not usually recommended for use in wind turbines because most nitrile rubbers have a short in-service life owing to the use of high levels of plasticizer. These types of seals typically do not exhibit reliable performance for more than 10,000 h, which is a little more than just 1 year of service and is far short of the gearbox oil drain interval of 3–5 years or more.

Fluorinated elastomers are generally the preferred choice of construction materials for seals and hoses. Freudenberg & Co. KG, Germany, a leading supplier of elastomers into the wind industry, has a series of specified tests for both static and dynamic elastomer–oil compatibility. For example, a static test for nitrile (NBR) rubber is 1000 h in duration at 100°C (212°F). Dynamic tests for NBR rubber are also 1000 h in duration for a gearbox with a radial lip seal filled with test oil and are run at 80°C (176°F). Fluorinated elastomers, commonly known as FKM type, are also run for 1000 h in the static test but at 130°C (266°F). The dynamic test for FKM elastomers is run at 110°C (266°F) using a radial lip seal in a gearbox. There also other varieties of FKM elastomers used for polyethylene glycol fluids.

Paint compatibility with the oil is also an important specification. Many gearbox manufacturers paint the inside surfaces as well as the outside surfaces of the machines. Therefore, compatibility with most common industrial grade paints is important. Various specifications around paint compatibility testing with oil have been developed for use with paint suppliers such as Maeder AG, Rickert GmbH & Co KG, and BASF. Moreover, sealants and thread-lock compounds, such as those supplied by Henkel-Loctite, have important compatibility considerations for wind turbine gearboxes. It is important to note that although most builders use thread-lock compounds to ensure machine integrity during the life span of the gearbox, certain types of sealants should always be avoided—in particular, sealants that contain silicone or are silicone-based. It is well known these types of sealants are incompatible with hydrocarbons of any type.

6. Gear Oil Condition Monitoring

Long service intervals are necessary for wind turbines because of their remote installation locations with limited accessibility. The highest quality lubricants are selected to provide long-term performance of gears, bearings, and all other components. To ensure flawless operation between oil change intervals, several used oil parameters are monitored particularly for the main gearbox oil. As an example, gearbox lubricants are generally saturated with water at levels of 200 ppm or less. Water levels significantly above 300 ppm indicate harmful contamination and may result in corrosion problems and/or reduced oil life.

Condition monitoring also includes the observation of a variety of wear metals to allow an assessment of the overall condition of the hardware. In general, aluminum and chromium levels above 10 ppm can indicate excessive wear of gears or bearings. Levels of iron, which is the most predominant metal in a wind turbine gearbox, will increase over time. A general guideline is that no more than 10 ppm of iron should increase per year of service of the lubricant after the running-in period. Values exceeding this level may indicate excessive wear occurring in the machine. Copper, lead, and tin are also key indicators of wear in bearings. Generally, the levels of these metals should be at 10 ppm or lower. Determination of the zinc content is also of interest, as it should not be present in a new or used gearbox. The presence of zinc in a gearbox indicates poor assembly practices or use of an incorrect part either during design or assembly. However, most builders allow zinc levels to climb approximately 10 ppm per year of service. Silicone, which is present in new oil as an antifoam additive, is generally considered acceptable at levels of 30 ppm or lower. Any increase in silicone during service is usually indicative of excessive amounts of dust or dirt contamination entering the gearbox.

The total acid number of an oil is an important indicator of aging, and it is monitored by observing the buildup of acidic oxidation products.

B. Rotor Blade Lubrication (Oil and Grease)

Wind turbines capable of generating more than 0.5 MW require continuous adjustments of the blade pitch to maximize efficiency. Most wind turbine manufacturers use a series of servomotor drives located out in the hub assembly to accomplish this. These pitch and yaw control drive motors have large reduction gear ratios that can range from 200:1 up to 1500:1 (Figure 16.12). These gearboxes are isolated from the rest of the nacelle and are usually lubricated with an ISO 320 or an ISO 460 viscosity grade synthetic lubricant.

There are alternative wind turbine designs where the pitch control operates through a control mechanism using hydraulic actuators. In this case the main oil sump and primary pump for the hydraulic system are located in the nacelle. Hydraulic oil is pumped into a dedicated void space in the center of the main gearbox shaft out into the hub assembly. The accumulators and pitch adjustment actuators are located out in the rotating hub on each individual blade. An ISO 32 viscosity grade hydraulic oil is generally used for this application. Industrial hydraulic fluids with a very high viscosity index (>140) and very good fluidity at low temperatures are used.

The large ring gear located at the bottom or base of each blade, as it is mounted in the hub assembly, is generally lubricated with grease that contains a high viscosity base oil, such as an ISO 460 viscosity grade gear oil. The grease delivery for these large ring gears is usually accomplished through the use of automatic grease dispensers that feed fresh grease on a periodic basis during the service interval.

Similar to wind turbine gear oils, there is no agreed-upon industry standard for the grease lubrication requirements for the pitch and yaw bearings (Figure 16.13). The pitch and yaw bearings are the only connections between the blade and rotor hub, and between the tower and nacelle. These bearings are highly stressed by the forces coming from the wind and the weight of the blades and nacelle. Pitch and yaw bearings do not fully rotate, but use small adjustments to rotate the blade or nacelle as needed by the direction and force of the wind. Because the pitch and yaw bearings do not rotate a full 360°, the grease is not redistributed around the bearing. For this reason, central grease dispensing systems are typically used to provide fresh grease to the bearings. The use of a

FIGURE 16.12 Pitch and yaw servomotors.

FIGURE 16.13 Blade bearing.

central grease system also simplifies maintenance and substantially extends the time between service intervals.

Large bearings that do not rotate completely may show damage (known as false brinelling) in the raceways because of oscillation and vibration under high dynamic loads. In addition, many wind turbines are located close to bodies of water so humidity (along with rain) can cause bearing corrosion. Seawater in offshore applications can be even more of a problem and can cause heavy corrosion leading to high wear rates. Because of the heavy oscillating loads and possible water ingress, a special test was codeveloped by a German bearing builder (Rothe Erde) and a university (University of Aachen—IME) to simulate the conditions seen in pitch and yaw bearing applications. This test, referred to as the Riffel or Rippling test (Figure 16.14), uses heavy loads (70 kN/15,736 lb) oscillating at a frequency of 10 Hz for a total of 1 million load cycles. During the test, a 1% saltwater solution is pumped into the test bearing. After the test is completed, the bearing is examined for scar depth and corrosion (Figure 16.15).

FIGURE 16.14 Riffel test rig.

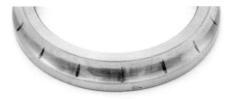

FIGURE 16.15 Bearing from Riffel test.

The blade bearings are generally lubricated with an NLGI (National Lubricating Grease Institute) 1.5 or NLGI 2 grade grease using a base oil viscosity between ISO VG 150 and 460—depending on the bearing builder and/or grease manufacturer recommendation and low-temperature torque requirement.

C. Generator Lubrication

The key lubrication points of the up-tower generator (as shown in Figure 16.3) are the rolling element bearings mounted on the housing of both sides of the stator. These bearings are lubricated with a grease containing an ISO 100 viscosity base oil, which is the same as that used in electric motors. Because of the extended maintenance of a wind turbine, these bearings are equipped with automatic grease dispensers capable of consistently delivering small amounts of grease in service. The performance of the grease is defined by the generator manufacturer based on individual in-house specifications.

D. Main Shaft Bearing Lubrication in Wind Turbines and Direct Drive Turbines

There are generally two sets of bearings supporting the main drive shaft that connect the rotor blades and hub assembly on the outside of the nacelle to the input portion of the main gearbox. These bearings are usually lubricated with a grease with an ISO VG 460 oil. The grease for the main bearing also needs to provide very good control of riffel formation. This is a phenomenon caused by vibration at zero speed, for example, when the turbine is feathered. The riffel protection performance is assessed in the Riffel test.

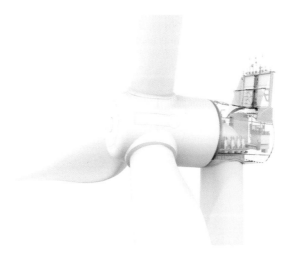

FIGURE 16.16 Direct drive wind turbine. (Courtesy of Siemens AG.)

The design life of a typical wind turbine gearbox is estimated to be 20 years. The owners of wind farms were faced in the past with some significant costs to repair or replace failed gearboxes as many gearboxes only lasted 7–10 years. A solution to this issue is to adopt a design that replaces the gearbox with a direct drive system (Figure 16.16). Technology and design improvements have made direct drive wind turbines much smaller than they used to be and much more appealing. As of 2014, Siemens was manufacturing a 6-MW unit for offshore installations. Generally, an NLGI Grade 00, ISO VG 460 grease is used for the lubrication of the bearings in direct drive systems. Because of the extended maintenance interval of a wind turbine, these bearings are equipped with automatic grease dispensers capable of consistently delivering small amounts of grease in service.

BIBLIOGRAPHY

American Wind Turbine Association (AWEA). 2013. Technical Specs of Common Wind Turbine Models. www.AWEA.org.

Barr, D., and Ethyl Petroleum Additives Ltd. 2002. Modern wind turbines: A lubrication challenge. *Machinery Lubrication Magazine*, September 2002 issue.

Bahaj, A.S., Myers, L., and James, P.A.B. 2006. Urban energy generation: Influence of micro-wind turbine output on electricity consumption in buildings. *Energy and Buildings*, 19, 154–165.

Betz, A. (1966) *Introduction to the Theory of Flow Machines*. (D. G. Randall, Trans.) Oxford: Pergamon Press.

Danish Wind Industry Association. 2003. http://www.windpower.org/en/tour/wtrb/powtrain.htm.

Deutsches Institut Fur Normung E.V. DIN 51517-3 Standard. 2014. Schmieröle CLP, Mindestanforderungen.

Krohn, S. 2001. http://www.windpower.org/tour/design/concepts.htm.

Forschungsvereinigung Antriebstechnik e.V. FVA-Informationsblatt Nr. 54 I–IV: Testverfahren zur Untersuchung des Schmierstoffeinflusses auf die Entstehung von Grauflecken bei Zahnraedern FVA-Nr. 54/7 Stand. 1993. Frankfurt.

Forschungsvereinigung Antriebstechnik e.V. FVA Information Sheet No. 243. 1995. Scuffing Test EP—-Oils, Method to Assess the Scuffing Load Capacity of Lubricants with High EP Performance Using an FZG Gear Test Rig. Frankfurt.

Gipe, P. 2004. *Wind Power Renewable Energy for Home, Farm, and Business*. White River, VT: Chelsea Green Publishing.

Global Wind Energy Council. 2015. Wind Power FAQS. http://www.gwec.net/about-winds/wind-energy-faq/.

Hansen, M.O.L. 2008. *Aerodynamics of Wind Turbines*, 2nd ed., ISBN-13: 978-1-84407-438-9 London: Earthscan.

International Standard Organization. 2005. ISO 14635-1. *Gears—FZG test procedures—Part 1: FZG test method A/8, 3/90 for relative scuffing load-carrying capacity of oils*.

Morris L. 2011. Direct drive vs. gearbox: Progress on both fronts. *Power Engineering* 115, 38.

Rocky Mountain Institute/Renewable Energy, Wind Power. 2013. Are Direct-Drive Turbines the Future of Wind Energy? www.earthtechling.com/2013/02/are-direct-drive-turbines-the-future-of-wind-energy/.

Sattler, H. 2011. Freudenberg Simmerringe GmbH und Co. KG. Static and dynamic oil compatibility tests with Freudenberg Simmerrings for the approval for FLENDER gearbox applications. (Table T 7300). FB 73 11 008.

Shotter, B.A. 1981. Micropitting: Its characteristics and implications on the test requirements of gear oils. *Performance Testing of Gear Oils and Transmission Fluids*. Institute of Petroleum, pp. 53–60, 320–323.

Steen P.C., and Jensen H.J., editors. 2009. *Wind Turbines in Denmark*. Copenhagen: Danish Energy Agency. ISBN: 978-87-7844-821-7.

Winter, H., and Oster, P. 1987. Influence of the Lubricant on Pitting and Micro Pitting (Grey Staining, Frosted Areas) Resistance of Case Carburized Gears—Test Procedures. AGMA Technical Paper 87 FTM.

Wyatt, A. 1986. *Electric Power: Challenges and Choices*. Toronto: Book Press Ltd.

17 Nuclear Power Generation

Nuclear power generation is responsible for more than 11% of the world's electricity and is growing. There are more than 430 commercial nuclear reactors located in 31 countries. At the time of this publication, about 70 new nuclear reactors were under construction with another 170 being planned with activity being greatest in China, Russia, and India. France is the country most reliant on nuclear power with 58 nuclear reactors generating ~75% of the country's electricity.

A nuclear power reactor produces and controls the release of energy from splitting the atoms of certain elements. The energy released in a nuclear reactor is used as heat to make steam to generate electricity. Nuclear reactors fall into the following categories: zero power research reactors, test reactors, special isotope production reactors, and power reactors. Basically, all nuclear reactors are similar in that they all use the fission chain reaction process to provide heat energy through the splitting (fission) of the heavy nuclei of fissionable materials. This reaction produces about 1×10^8 the energy release of burning one carbon atom of fossil fuel plus the production of extra neutrons needed to sustain the chain reaction. The fuel used is generally ^{235}U (uranium 235), ^{233}U (uranium 233), or ^{239}Pu (plutonium 239). This chapter will provide general information on reactors used in power generation, the effects of radiation on lubricants, and lubricant recommendations in nuclear power plants.

I. REACTOR TYPES

The power reactor, whose main function is to furnish energy, consists broadly of a core containing nuclear fuel, a moderator (although this is eliminated in fast neutron reactors), a cooling system, a control system, and shielding. Most commonly, pellets of uranium oxide (UO_2) are arranged in tubes to form fuel rods that are arranged into fuel assemblies in the reactor core. In practice, although they are basically similar, it is possible to design an almost endless number of different reactor types by using various combinations of fuel, coolant, and moderator. For these reasons, various countries throughout the world have pursued a particular course of design that depended on the availability of materials for construction, moderator, and fuel. For example, some European nations and Canada based their first-generation reactor designs on the use of natural uranium because of a lack of enrichment facilities. Most countries using nuclear reactors currently have the ability to produce or obtain enriched fuel.

A. Basic Reactor Systems

Among the hundreds of combinations of fuel, coolant, moderator, etc., which have theoretical possibilities as reactor systems, six basic types have been studied in research stages and have resulted in demonstration or commercial reactors.

1. Pressurized water reactor (PWR)
2. Boiling water reactor (BWR)
3. Light water graphite-moderated reactor (LWGR)
4. Fast breeder reactor (FBR), including the liquid-metal, fast-breeder reactor
5. Gas-cooled reactor (GCR)
6. High-temperature, gas-cooled reactor (HTGR)

Figure 17.1 shows the schematics for each of these reactor designs. GCRs and HTGRs are shown as the same schematic.

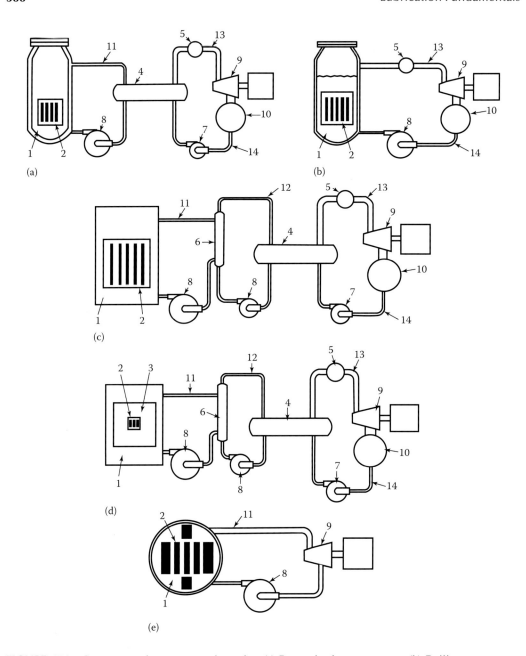

FIGURE 17.1 Common nuclear reactor schematics. (a) Pressurized-water reactor. (b) Boiling water reactor (direct cycle). (c) Sodium–graphite reactor. (d) Fast breeder reactor. (e) Gas-cooled and high temperature gas-cooled reactor. 1—Reactor; 2—core; 3—blanket; 4—boiler; 5—steam drier; 6—intermediate heat exchanger; 7—feed water pump; 8—circulating pump; 9—turbogenerator; 10—condenser; 11—primary coolant; 12—intermediate coolant; 13—steam; 14—condensate.

1. PWR

This reactor type represents the majority of commercial reactors in use. Fission heat is removed from the reactor core by ~325°C water pressurized at approximately 2000 psi to prevent boiling. Steam is generated from the secondary coolant loop system whereas the heat exchanger that cools the superheated water causes the cooler water to boil into steam. The major characteristics of the PWR reactor are:

- Light water (H_2O) is the most cost effective coolant and moderator.
- High water pressure requires a costly reactor vessel and leakproof primary coolant system.
- High-pressure, high-temperature water at rapid flow rates increases the potential for corrosion and erosion problems.
- Steam is produced at relatively low temperatures and pressures (compared with fossil-fuel boilers) and may require superheating to achieve efficiencies.
- Containment requirements are extensive because of possible high energy release in the event of a primary coolant system failure.

2. BWR

This is the second most common type of reactor in use, representing about 20% of commercial reactors. In this reactor type, fission heat is removed from the reactor by conversion of water to steam in the core. This design is similar to the PWR but uses lower water pressure (~1100 psi) so it boils in the core at ~285°C. The major characteristics of the BWR are:

- Light water is the coolant, moderator, and heat exchange medium, as in a PWR.
- Reactor vessel pressure is less than the primary circuit of the pressurized reactor.
- Heat exchangers, pumps, and auxiliary equipment requirements are reduced or eliminated.
- Has an inherent safety characteristic in that power surge causes a void formation, thus reducing the core power level.

3. LWGR

This reactor design (also known as RBMK due to its Russian origin) is found in a few commercial plants in Russia today. This unique design uses a BWR and is light water cooled with a graphite moderator. Steam is generated directly in the reactor and separated in steam drums. The cooling water is radioactive. The major characteristics of the LWGR are:

- Designed to be a large, powerful, and lower cost.
- Known for instability at low power levels.
- Operates in helium–nitrogen atmosphere.
- Two independent cooling circuits.
- Additional design changes made after the Chernobyl accident in 1986.

4. FBR

There have been several experimental and prototype fast breeder type reactors operating since the 1950s but very few commercial ones. This type of reactor is a fast neutron reactor that can produce more plutonium than it consumes. Heat from fission by fast neutrons is transferred by sodium coolant through an intermediate sodium cycle to steam boilers. No moderator is used. Neutrons escaping from the core into a blanket breed fissionable Pu-239 from fertile U-238 blanket. The major characteristics of the FBR are as follows:

- The reactor is designed to produce more fissionable material than is consumed.
- Low neutron absorption by fission products permits high fuel burnup.

- A small core with a minimum area intensifies heat transfer problems.
- Core physics, including short neutron lifetime, makes control difficult.

5. GCR

A few commercial applications for this reactor type exist in the United Kingdom. A second generation of these reactors is known as advanced gas-cooled reactors. In the GCR, heat removed from the core by gas at moderate pressure is circulated through heat exchangers that produce low- and high-pressure steam. It uses carbon dioxide gas, graphite moderator, and natural uranium fuel. The major characteristics of the GCR are the following:

- It permits low pressure coolant and relatively high reactor temperatures.
- Containment requirements are moderate and corrosion problems minimal.
- Reactor size is relatively large because of natural fuel and graphite moderator.
- Poor heat transfer characteristics of gases require high pumping requirements.
- Steam pressures and temperatures are low.
- Carbon dioxide gas is relatively cheap, safe, and easy to handle.

6. HTGR

This reactor type is not yet in commercial operation. Heat from the reactor core is carried by inert helium to the heat exchanger for generation of steam or directly to a gas turbine. The gas returns to the reactor in a closed cycle. The major characteristics of the HTGC are:

- Good efficiency can be achieved in a dual cycle with a minimum gas temperature of 760°C.
- High fuel burnup is possible, and conversion of fertile material permits lower fuel costs.
- Minimum corrosion of fuel elements will be caused by inert gas.
- High-temperature coolant minimizes the disadvantages of poor heat transfer characteristics of the gases.
- The design of fuel elements for long life is complicated by high temperatures.
- There is a limited supply of helium worldwide.

II. RADIATION EFFECTS ON PETROLEUM PRODUCTS

In general, radiation damage may be defined as any adverse change in the physical and chemical properties of a material as a result of exposure to radiation. Radiation damage is a relative term for the changes in a material that may have adverse effects on the operation of the nuclear plant. This is true of organic materials in particular; for example, the evolution of a gaseous hydrocarbon from a liquid organic material may result in an explosion hazard and an increase in liquid viscosity. Similarly, radiation of an organic fluid may result in unwanted increase in molecular size with consequent thickening or solidification of the liquid or grease. In the study of radiation damage, the concern is mainly with the adverse or undesirable changes in the lubricants that affect their ability to perform adequately in the machinery involved. It should be noted that lubricants may be able to still perform their lubrication function after reaching condemning levels as determined by conventional laboratory evaluations. This aspect is important in applications where equipment (reactor and other containment equipment) may not be accessible until such events as fuel rod changes set up on 18- to 24-month cycles. If analysis of these lubricants indicate undesirable changes in their characteristics, a judgment needs to be made on the acceptability of the lubricant to perform until scheduled outages or whether other alternatives need to be considered.

Broadly speaking, there are two mechanisms of radiolysis that must be considered in a study of the damage to organic fluids. One is the primary electronic excitation and ionization of organic molecules caused by beta (β) particles, gamma (γ) rays, and fast neutrons. The other is the capture of thermal neutrons and some fast neutrons by nuclei that would cause changes in the nuclei and the generation of secondary radiation that would result in further damage.

Two methods are used to measure radiation energy. One measure is the quantity of energy to which the materials are exposed and is called the roentgen (R); the other is the amount of energy the material absorbs and is called the *rad*. For γ radiation, the exposure unit (roentgen) is defined as the quantity of electromagnetic radiation that imparts 83.8 ergs of energy to 1 g of air.

The radiation dosage of a material is defined as an absorption of 100 ergs of energy by 1 g of material from any type of radiation. Actually, absorbed energy will vary with the type of radiation, and the effect will depend on the material exposed. For γ radiation, however, 1 rad absorbed is approximately equivalent to 1.2 R of radiation dosage. The rad is useful for comparing the equivalent energy of mixed radiation fluxes but does not distinguish between types.

From a radiation damage standpoint, 1 rad of neutron flux causes 10 times more biological damage to tissue than an equivalent absorbed energy of γ rays. For petroleum products, however, the dosage—as measured by such effects as viscosity increase—is almost equivalent for the two types. This is discussed in more detail later in this chapter.

The general levels of radiation dosage are as follows:

Dosage (R)	Effect
200–800	Lethal to humans
<5 million	Negligible to petroleum products
5–10 million	Damaging to petroleum products
>10 million	Survived by only most resistant organic structures

Based on experimental work, the damage to petroleum products may be summarized as follows:

1. Liquid petroleum products darken and acquire an acrid, oxidized odor.
2. Hydrogen content decreases and density increases.
3. Gases such as hydrogen and light hydrocarbons evolve.
4. Physical properties change, higher and lower molecular weight materials are formed, and olefin content increases.
5. Viscosity and viscosity index (VI) increase.
6. Polymerization to a solid state can occur.

It must be appreciated that the intensity of these effects or the incidence of one or more of them will depend on the amount of absorbed energy, the exact composition of the specific petroleum material, and other environmental conditions such as temperature, pressure, and gaseous composition of the atmosphere.

A. Mechanism of Radiation Damage

Organic compounds and covalent materials do not normally exist in an ionized state and therefore are highly susceptible to electronic excitation and ionization as the result of deposited energy. Covalent compounds, including the common gases, liquids, and organic materials, consist of molecules that are formed by a group of atoms held together by shared electron bonding, which yields strong exchange forces. The molecules are bound together by relatively weak van der Waal forces.

Conversely, ionic compounds, such as inorganic materials, which include salts and oxides, are already ionized (metals may be considered as being in an ionized state) and are not susceptible to further electronic excitation. Ionic compounds consist of highly electropositive and electronegative ions held together in a crystal lattice by electrostatic forces in accordance with Coulomb's law. There is no actual union of ions in the crystal to form molecules, although all crystals may be considered as being composed of large molecules of a size limited only by the capacity of the crystal to grow.

Therefore, the effect of radiation energy on nonionic compounds is to form ions, radicals, and excited species and thereby make the compounds more reactive with themselves or with the atmospheric environment. On the other hand, the effect of radiation on ionic compounds is to change the properties of the compound related to crystal structure.

B. Chemical Changes in Irradiated Materials

The physical and chemical properties of hydrocarbon fluids that make them important as lubricants change during irradiation to varying degrees based on chemical composition and the presence of additives. These changes may be traced to alteration of the chemical structure of the materials. Nuclear radiation, either directly or by secondary radiation, deposits high-level energy in the irradiated organic substance and causes ionization and molecular excitation. The ions are excited molecules that rapidly react to form free radicals, which further combine or condense (Figure 17.2).

The changes in chemical structure may be measured by various classic methods: for example, it is possible to determine the approximate number of free radicals formed by the use of scavengers such as iodine. In addition, either hydrogen or light petroleum fractions are evolved as gas. Investigations have shown that both carbon–hydrogen and carbon–carbon bonds can be broken by radiolysis. The dissociated or ionized molecules can condense, rearrange, form olefins, or other products, depending on the environment. At temperatures below 400°F (204°C), temperature effects do not seem to be significant.

Because most petroleum lubricants contain combinations of saturated and unsaturated aliphatic and aromatic compounds, the reactions of these principal hydrocarbon classes have been studied under the influence of ionizing radiation. These studies indicate, as would be suspected, that

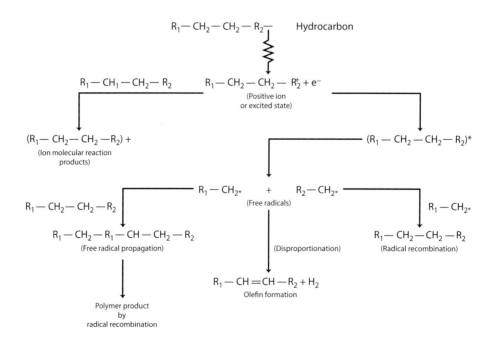

FIGURE 17.2 Radiolysis processes in hydrocarbons.

unsaturated hydrocarbons are most reactive and aromatics the least affected. Saturated compounds fall somewhere between the two extremes. Aromatic materials are highest in radiation resistance. The principal reaction is cross-linking, from which very small amounts of gas evolve.

Additional studies were made on a range of petroleum oils representative of typical paraffinic, naphthenic, and aromatic materials varying in sulfur content. The viscosity was plotted against aromatic and sulfur content (Figure 17.3). The data show that as the aromatic content increases, the viscosity increase is reduced in almost a linear relationship. The effect of sulfur content is similar but more marked. Furthermore, it was noted that radiation damage appears greatest for oils with the highest molecular weights or the highest initial viscosity.

Because it was found that naturally occurring compounds improved the radiation stability of petroleum oils and that these compounds were usually removed by refining procedures, the effect of using synthetic aromatic and sulfur as additives was studied. From these studies, it appears that a disulfide, or an alkyl selenide, provides good radiation damage protection. The disulfides prevent polymerization by a mechanism termed free radical chain stoppers. The disulfides have an advantage also of being good extreme pressure (EP) and antiwear agents but do not prevent oxidation or olefin formation.

It is well known, however, that aromatic compounds possess good thermal and radiation stability, and in the latter case protect less stable aliphatic molecules by the transfer of energy. These compounds are usually characterized by complex molecules that resonate between a number of possible electronic structures and, therefore, possess fairly stable excited energy states. The benzene rings in the aromatic structure act as a natural inhibitor by absorbing the radiation. In other words, when a paraffinic hydrocarbon absorbs energy, it is raised to an unstable state in which the energy is greater than the electronic forces that constitute the chemical bonds. The result is a bond fracture with residual free radicals. In an aromatic with an equivalent absorbed energy, the higher level is not sufficient to sever the greater electronic binding forces, and the energy is eventually liberated as heat or light. The radiation stability of aromatic petroleum extracts are, in decreasing order, polyglycols, paraffinic hydrocarbons, diesters, and silicones. The effect of aromatic compounds was

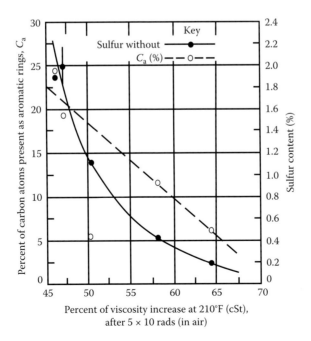

FIGURE 17.3 Radiation stability versus sulfur and aromatic content.

studied both as antirad additives to mineral oils and as pure synthetic fluids. The results are shown in Figures 17.4 and 17.5. These data show the following relationships:

1. The aromatics with bridging methylene groups between aromatic molecules are less efficient as protective agents than antirad additives with direct links between aromatic rings.
2. Long-chain alkyl groups attached to the aromatic rings make less protective agents, probably because of a difference in stability of the compound and lowering of the aromatic ring content.
3. Small amounts of a free radical inhibitor in addition to the aromatic additive substantially reduce the viscosity increase.
4. The protection afforded is not simply a direct function of aromatic content; in fact, it would appear that 40% of added aromatic material is a practical maximum. Beyond 40%, it is preferable to use a pure aromatic of suitable physical characteristics.

A study of the changes in properties and performance of conventional lube oils after irradiation shows the following:

1. Conventional antioxidant additives of the phenolic or amine type confer little radiation stability to base oils and are preferentially destroyed between 10^8 and 5×10^8 rad.
2. Didodecyl selenide, which is known to be an effective antioxidant, also has radiation protective properties. The oxidation stability is effective after an irradiation of 10^9 rad.

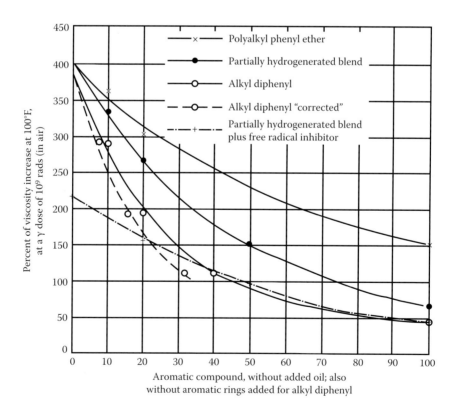

FIGURE 17.4 Radiation protection of synthetic aromatic additives.

Nuclear Power Generation

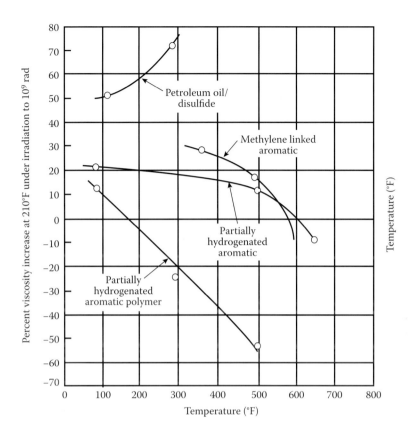

FIGURE 17.5 Radiation stability versus temperature.

3. Diester base oils, phosphate esters (antiwear additives), and halogenated EP agents produce acids at a low radiation dose.
4. Polymers such as polybutenes and polymethacrylates cleave readily and thus lose their effect as VI improvers.
5. Silicone antifoam agents are destroyed at low radiation dose.
6. In most cases, the presence of air, compared with an inert atmosphere, increases radiation damage by a factor of 1.6 to 2.3 times, as indicated by viscosity increase.

In summary, high-quality, conventional lubricating oils are suitable for radiation doses up to 10^8 rad. It should be noted that base oils that undergo severe hydrotreating or hydrocracking processes during refining may need to have sulfur- or aromatic-containing additives added to the finished product. These refining processes remove a substantial amount of sulfur and aromatic compounds. Further radiation resistance can be formulated into a good-quality petroleum oil by use of antirads such as radical scavengers or aromatic structures. These formulated oils will protect up to exposures of 10^9 rad. Above these doses and at high temperatures, synthetic-type lubricants that use partially hydrogenated aromatics blended with aromatic polymers are required. The effect of temperature at high radiation dose (10^9 rad) has been studied under a nitrogen atmosphere for these fluids (Figure 17.5).

Suitable lubricants must not only have good thermal and irradiation stability, but their wear performace must be satisfactory as well. This phase can use conventional laboratory and field testing for new lubricants. Additional special testing is also required to verify that the irradiated lubricants retain sufficient wear to protect the equipment over the expected service intervals.

1. Turbine Oil Irradiation

Studies have been conducted to compare the effects of radiation of a high aromatic API (American Petroleum Institute) Group I turbine oil (~5% aromatics) with low aromatic (<0.1%) turbine oils that use API Group II base oils. API Group II base oils have much less sulfur, nitrogen, and aromatic content than API Group I base oils. The oils in one study were exposed to 220 Mrad of gamma radiation, which is meant to simulate a loss-of-coolant accident in a nuclear plant in the containment area to evaluate how turbine oils, used in safety equipment, would perform. Equivalent results for the turbine oils with either API Group I or II base oils were obtained. The 100°C viscosities for all increased by 15–25%, the total acid number results for all did not increase to any concerning level (all <0.1), and the Rotating Vessel Pressure Oxidation Test results decreased by >90% for all oils, which represents a significant drop in the remaining oil life versus new oil, but this is considered acceptable for the desired 1 month of service required during a loss-of-coolant accident. The equivalent performance of the API Group II turbine oils is attributed partially to the aromatic structure of the oxidation inhibitors acting as a radiation inhibitor and to the inherent oxidation stability of the API Group II base oils.

2. Grease Irradiation

The damage caused by high energy radiation on greases is a dual effect. First, the radiation attacks the thickening structure and causes separation and fluidity. After this, continued irradiation results in polymerization of the base oil, resulting in thickening and gradual solidification. The precise pattern of change is dependent on the type of thickener, the gel structure, and the radiation stability of the thickener and the base oil.

In general, greases have been evaluated for radiation damage by determining the worked penetration and base oil viscosity, after irradiation, and comparing it with the original value. Greater than a 20% change is considered unacceptable. These irradiation evaluations are usually of the static type, but testing under dynamic conditions during irradiation has yielded markedly different results. Typical greases that have organic soap components of alkali earth metals, although resistant to high amounts of radiation, break down at total doses of approximately 10^8 rad. Micrographs of soap structure showed a drastic change in the normal fiber structure of the gelling agent. On the other hand, greases made with nonsoap thickeners, such as carbon black, were less affected.

The stabilization of the thickening structure under irradiation solves the problem of softening or bleeding of the base oil but will not solve the eventual solidification of the grease. This is a function of the base oil, and the solutions discussed under lubricating oil (use of antirad additives or synthetic organics as base fluids) are valid.

The mechanism of change for three greases is shown in Figure 17.6. In one case, the grease had an unstable thickener and progressively softened to fluidity. Although such a grease might protect a bearing, the problem of leakage would be substantial, and incompatibility with reactor components would be a concern. The second grease gradually decreased in penetration (solidification) after an initial increase or softening. Such a grease would cause failure in the lubricated mechanism. The third grease showed good stability with a slight softening up to 10^9 rad.

3. Radiation Stability of Grease Thickeners

The selection of the thickener of a grease designed for nuclear applications requires consideration of compatibility as well as resistance to radiation, high temperature, mechanical shear, and operating atmosphere.

Certain elements are unsuitable because their presence within or close to the reactor core could seriously affect neutron economy, or react with the fuel element cladding to cause destruction and

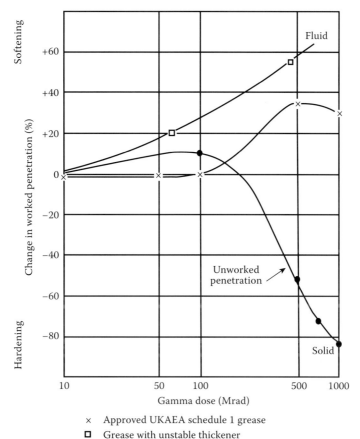

FIGURE 17.6 Effect of radiation on greases.

possible release of fission products. Accordingly, the United Kingdom Atomic Energy Authority has restricted lubricant composition on select reactors using Magnox fuel cans:

None allowed: mercury
0.1% allowed: barium, bismuth, cadmium, gallium, indium, lead, lithium, sodium, thallium, tin, zinc
1% allowed: aluminum, antimony, calcium, cerium, copper, nickel, silver, strontium, praseodymium

Under certain instances, these limits can be exceeded, when it can be shown that the metals are present in a stable compounded form and that practical compatibility tests are satisfied.

The effect of atmosphere can be illustrated by air, which has a serious oxidizing effect, especially when coupled with radiation and high temperatures. Conventional antioxidants are destroyed as noted earlier. Some of the organic modified thickeners have an antioxidant effect and perform a dual function. Hot pressurized carbon dioxide can cause rapid degeneration of conventional soap-thickened greases, presumably by carbonate formation.

In selection of a grease thickener, the compatibility of the thickener and base fluid is of paramount importance. Even an exceptionally radiation-resistant thickener, when in combination with certain base fluids, may at best yield weak gels and soften easily. For example, a satisfactory grease structure is extremely difficult to obtain by using an indanthrene pigment with a paraffinic bright stock.

Various nonsoap thickeners that form good grease structure with both mineral oil and synthetic fluid bases are available. These thickeners may be grouped as follows:

1. *Modified clays and silicas.* Typical of the modified clays are Bentone and Baragel, which are formed by a cation exchange reaction between a montmorillonite clay and a quaternary ammonium salt. This reaction produces a hydrocarbon layer on the surface of the clay—which makes it oleophilic. Finely divided silicas may be treated with silicone to render them hydrophobic, or, as with Estersil, the silica may be esterified with *n*-butyl alcohol.
2. *Dye pigments.* Organic toners or dye pigments are utilized as grease thickeners (e.g., Indanthrene).
3. *Organic thickeners.* Typical of this type are the substituted aryl ureas characterized by the diamide–carbonyl linkage that may be formed in situ by the reaction of diisocyanate with an aryl amine.

The behavior of these thickeners, when used in conjunction with a synthetic fluid, is shown in Figure 17.7. As with fluid lubricants, antiradiation compounds may be added to the grease to increase its radiation stability.

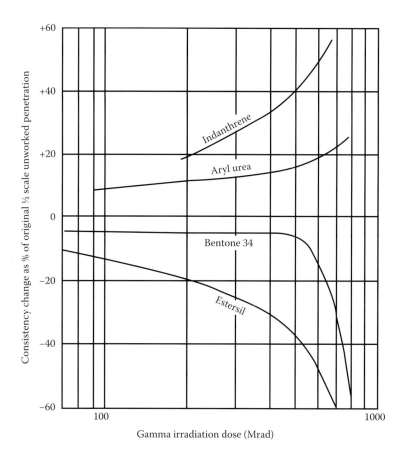

FIGURE 17.7 Effect of thickener on grease polymerization.

III. LUBRICATION RECOMMENDATIONS

Lubricants used in the nuclear industry have an additional required dimension versus lubricants used in general industrial applications. Those lubricants used in research and power reactors; fuel processing machinery; conveyors, manipulators, and cranes in irradiation facilities; viewing windows and shield doors in hot cells; and heat exchange units must also have the ability to perform in an irradiation atmosphere.

Nowhere are the operating conditions of radiation, temperature, and atmosphere more damaging than in a power reactor field. Initially, little was known of the behavior of petroleum products under irradiation, and the exact severity of the application was overestimated. This placed a design burden on nuclear power generation. As experience was gained, the original position was reconsidered. First, specific operating parameters of radiation flux, temperature, etc., were obtained, which realistically established the requirements; second, research in the radiation resistance of petroleum materials showed that conventional lubricants could withstand doses of up to 10^7 to 10^8 rad and still perform their lubrication function. New specialized lubricants were developed to withstand more than 10^9 rad.

Because the nuclear power industry is so complex and under continuous scrutiny, each plant requires separate consideration before proper lubricants and lubrication schedules can be established. Adherence to the guidance given by the local governing nuclear regulating agency is necessary, which may include qualification testing such as thermal aging, irradiation, and accident testing in addition to standard lubricant performance testing. Over time, lubricants in safety applications or areas have often been qualified for service in the equipment it was tested in, and deviation is not permitted nor advised.

Because of the criticality of equipment, lubricants that are approved by original equipment manufacturers (OEMs) should be used. Therefore, the best that can be accomplished in this chapter on lubrication recommendations is to furnish the background experience and establish general guidelines so that lubrication professionals can, after a survey of the specific conditions, recommend the best lubricants for a particular application. Lubricant formulation changes should be documented and be evaluated for acceptability.

A. General Requirements

Most lubrication professionals are familiar with the effect of factors such as speed, load, temperature, and time on the life of bearings and gears, as well as the physical and chemical properties of oils and greases exposed to these environmental conditions.

In conventional applications, the effect of speed, load, and temperature are evaluated in making recommendations. Selecting the correct lubricant and service interval is determined by evaluating the lubricant's anticipated performance under the most critical of these conditions; for example, speed may be the determining factor with antifriction bearings, and the proper grease must resist excessive softening for the service period. In most cases in a nuclear power plant, temperature is the most critical factor. Operating severity may be determined by the degree of heat and the extent of radiation exposure. The additional factor of radiation in nuclear applications affects lubricants in much the same manner as heat. Both are modes of energy and—as we have seen in a previous section on lubricants—like all organic materials, undergo major structural changes when certain thresholds of absorbed energy are reached. We know that petroleum products undergo thermal cracking and polymerization at certain temperatures and, likewise, that cleavage and cross-linking occur at certain radiation dosage thresholds.

In nuclear applications, radiation energy is expressed as flux or dose rate. The particular type of radiation and the units have already been discussed. It is sufficient to state here that this dose rate applied over a specified interval will yield the absorbed dose for an exposed material. If, for

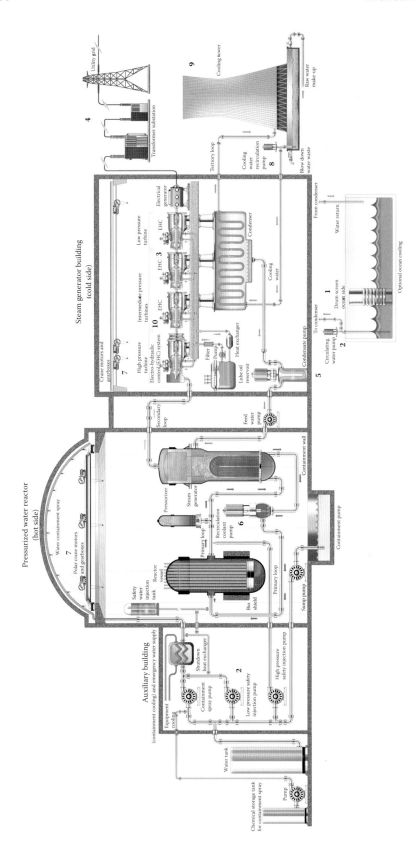

FIGURE 17.8 Typical lubricant applications in a pressurized water reactor (PWR) nuclear plant. 1—Drum screen strainer/rack rake system; 2—circulating water pumps; 3—steam generator turbines; 4—transformers; 5—condensate pumps; 6—recirculating cooling pumps; 7—overhead cranes; 8—cooling tower pumps; 9—cooling tower gearboxes; 10—turbine electric hydraulic control (EHC) systems. Other applications not shown include emergency diesel generators, air compressors, feed water pumps, building air conditioners, electric motors, and Limitorque valve controls.

example, a grease can absorb a dose of 5×10^9 rad before suffering physical or chemical changes that will render it useless as a lubricant, this grease can be used for 5000 days if the dose rate is absorbed at a rate of 1 Mrad/day or for only 5 days if the dose rate absorbed is 1000 Mrad/day.

B. Selection of Lubricants

Much has been published and extensive studies have been made that led to the correct recommendations for lubricants. In this final analysis, however, the selection of the appropriate lubricant and its application for any particular equipment must be made for each specific instance based on operating conditions and the type of bearing, gear, cylinder, etc. Nowhere is this more critical than in the lubrication of the equipment in the reactor and containment areas of the nuclear power plant. Because of the uniqueness of the designs, the severity of operating equipment, and the available relubrication or change intervals, each plant can be markedly different. The accumulated experience of lubrication professionals along with the equipment manufacturers has been helpful in numerous plants in the solution of lubrication problems and in the elimination of mechanical problems as well.

In selecting lubricants for a nuclear power plant, key personnel in the plant should be consulted. This entails consulting with the plant's equipment qualification engineer during the design phase along with the procurement engineer for commercial requirements before and after startup. In surveying plant requirements, particular attention must be paid to the radiation flux profile (some of the information is available—design basis event—developed for each plant), which has been calculated for the various parts of the plant and compared with actual surveys during operation of similar plants. As the exact radiation exposure for a lubricant in any piece of equipment may not always be known, it is advisable to select lubricants that have gone through qualification approval testing. A good general target for lubricant performance is at rates of at least 10^8 to 10^9 rad. Being conservative is a good practice even though many lubricants in nuclear plant applications are subjected to low dosage radiation on the order of 10^3 rad/h or less, which would take 12 or more years to absorb 10^8 rad. Extreme conservatism continues to be the rule in selecting lubricants in nuclear power plants. This has resulted in the use of OEM-approved, high-performance synthetic lubricants being commonplace in this industry. Figure 17.8 represents some of the typical lubricant applications present in a PWR nuclear plant.

BIBLIOGRAPHY

Nuclear Maintenance Applications Center: Radiation Stability of Modern Turbine Oils: Group II Turbine Oils Exposed to Loss-of-Coolant-Accident (LOCA) Radiation. EPRI, Palo Alto, CA: 2009. 1019593.
Sheets, M.S., and Yoshida, D.K. 1989. Final Report Class 1E Qualification of Five Mobil Oils and Five Mobil Greases. NUTECH Engineers, San Jose, CA.
Wills, J. George. 1967. *Nuclear Power Plant Technology*. New York: John Wiley & Sons.
World Nuclear Association. 2013. Plans for New Reactors Worldwide. 2013. http://www.world-nuclear.org /info/current-and-future-generation/plans-for-new-reactors-worldwide/.
World Nuclear Association. 2015. Nuclear Power in the World Today. http://www.world-nuclear.org/info /Current-and-Future-Generation/Nuclear-Power-in-the-World-Today/.

18 Compressors

Compressors are manufactured in several types and for a variety of purposes. In addition to being used to compress gas, many compressors serve as blowers or can be used as vacuum pumps. Lubrication requirements vary considerably, depending not only on the type of compressor but also on the type of gas (including any contaminants) being compressed. In general, air and gas compressors are mechanically similar so that the main difference is in the effect of the gas on the lubricant and the compressor components. The lubricant plays a role in preventing wear, sealing, minimizing viscosity dilution and additive reactions with the gas, and preventing corrosion. Refrigeration and air conditioning compressors require special consideration because of the recirculation of the refrigerant and mixing of the refrigerant with it.

Compressors are classified as either positive displacement or dynamic. The positive displacement class includes reciprocating (piston) types, several rotary types, and diaphragm types. Dynamic compressors are either of the centrifugal or axial flow type, although some mixed flow machines that combine some elements of both of these types are used.

Excluding refrigeration and air conditioning applications, in terms of the number of machines, more compressors are used to compress air for utility use than for any other purpose. Although many applications require high pressures, the vast majority of pneumatic equipment is designed for pressures between 90 and 100 psig (620–690 kPa), so most compressed air systems are designed to operate between 100 and 125 psig (690–862 kPa). These requirements are met by both portable compressors used on construction projects, in mining, and other outdoor applications and stationary compressors used to provide plant air in applications ranging from service stations to industrial plant types. In the past, these were largely the province of reciprocating compressors, but owing to various factors such as design and metallurgical improvements, increasing speed capabilities, and size reduction, large numbers of other types of compressors are now being used.

In plant air applications, the vibration associated with medium and larger reciprocating compressors generally necessitates either heavy, vibration-absorbing mountings, or special isolating mountings. The pulsation in air delivery may require pulsation dampers. Rotary and centrifugal compressors are often supplied as packaged units that can be installed with minimal mounting requirements from the vibration absorption viewpoint. Compared to reciprocating units, these packaged rotary units are more compact, offer nonpulsating flow, have fewer wearing parts, and require less maintenance.

Centrifugal and axial flow compressors deliver oil-free air. Positive displacement compressors are also available in nonlube designs, although these are not recommended for severe operating conditions such as high pressures (where the oil helps to seal), high temperatures, wet gas, or where there are corrosive elements in the gas. It should be recognized that many users of nonlube designed positive displacement compressors may use some lubricant, particularly in reciprocating cylinders, to provide longer life of components and improved sealing. Certain reactive gases, such as oxygen, require special consideration, and petroleum-based lubricants should not be used for these types of gases.

The improvements in the design of helical lobe (screw) compressors have resulted in higher pressure capability with efficiencies approaching those of reciprocating compressors. This—combined with the lower noise levels associated with rotary machines—has also increased the use of rotaries.

When considering compressor lubrication, it is necessary to recognize that compressing a gas causes its temperature to rise. The more gas is compressed, the higher will be its final temperature. When high discharge pressures are required, compression is carried out in two or more stages with the gas being cooled between stages so as to limit temperatures to reasonable levels. This also

TABLE 18.1
Effect of Staging on Discharge Temperatures[a]

Discharge Pressure Gauge		Discharge Temperature					
(psi)	(kPa)	One Stage		Two Stages		Three Stages	
		(°F)	(°C)	(°F)	(°C)	(°F)	(°C)
70	483	398	203	209	98	–	–
80	552	426	219	219	104	–	–
90	621	452	233	226	109	–	–
100	689	476	247	238	114	–	–
110	758	499	256	246	119	–	–
120	827	519	271	254	122	182	83
250	1724	–	–	326	163	225	108
500	3447	–	–	404	207	269	132

[a] Temperatures based on adiabatic compression (actual temperatures will be lower because of heat losses within system).

improves compressor efficiency and reduces power consumption. For the range of temperatures that can be reached, see Table 18.1. This table is based on adiabatic* compression of air with an intake pressure of 14.7 psia (101.3 kPa abs) and a temperature of 60°F (15.6°C) and intercooling between stages to the same temperature. These temperatures, although higher than those reached in actual practice because some of the heat is removed by cooling the cylinder walls, are indicative of the temperatures that must be considered when selecting compressor lubricants.

I. RECIPROCATING AIR AND GAS COMPRESSORS

Reciprocating compressors are used for many different purposes involving mild to extreme pressure and volume requirements. As a result, a great variety of designs are commercially available. Most reciprocating compressors are of the single- or two-stage types, with smaller numbers of multistage machines—three, four, or more stages such as Ariel compressor shown in Figure 18.1. From the lubrication point of view, single- and two-stage machines are generally similar, whereas multistage units may have somewhat different requirements, depending on pressures, temperatures, gas conditions, and the size and speeds of the pistons.

The principal parts common to all reciprocating compressors are pistons, piston rings, cylinders, valves, crankshafts, connecting rods, main and connecting rod bearings (crankpin bearings), and suitable frames that generally contain the lubrication system. In double-acting machines (which compress on both faces of the pistons), piston rods, packing glands, crossheads, and crosshead guides are required; the connecting rods are connected to the crossheads by crosshead pins. Crossheads and associated parts are also used in some multistage single-acting compressors, but the majority of single-acting compressors are of the trunk piston type, where the connecting rods are connected directly to the pistons by piston pins (wristpins). For lubrication purposes, all parts associated with the cylinders, including pistons, rings, valves, and rod packing (on double-acting machines), are considered cylinder parts (Figure 18.2). All parts associated with the driving end, including main, connecting rod, crosshead pin or wristpin bearings, crankshaft, and crosshead guides, are considered running parts or running gear. In many applications, there are two lube systems to separate the cylinder lubrication from the running gear lubrication because of substantial differences in lubricant requirements.

---

* Adiabatic compression assumes that all of the mechanical work done during compression is converted to heat in the gas, i.e., the cylinder is perfectly insulated so that no heat is lost from it.

Compressors

FIGURE 18.1 Multistage reciprocating compressor.

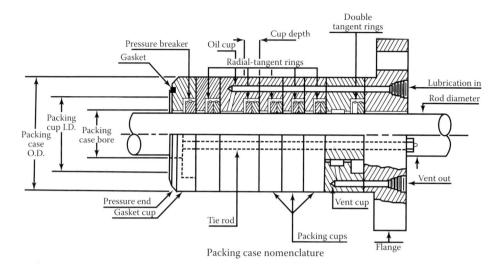

FIGURE 18.2 Mechanical rod packing design and operation.

Many reciprocating compressors are provided with cylinder cooling in order to limit the final discharge temperature to a reasonable value and to minimize power requirements. The cylinder walls and heads are cooled and, in the case of two-stage and multistage machines, the gas being compressed is cooled between stages in intercoolers. Cooling can be by air or water, but in larger machines, water cooling is usually required. In captive cooling water systems, glycol and inhibitors are used to minimize corrosion and any potential to freeze in cold operations or during shutdown periods. Frequently, the gas being cooled is further cooled in aftercoolers. In the case of air, this helps remove water and thus prevents or minimizes condensation of moisture in the air distribution system. Aftercoolers also act as separators to assist in removing oil that may be carried over from the cylinders. Air receivers are used in most large industrial systems, not only to provide a reserve to accommodate varying supply demands, but also to reduce compressor pressure pulsations, to add radiant cooling capacity, and to allow further separation of moisture and oil carryover.

A considerable amount of moisture can be condensed in intercoolers. For example, in a two-stage compressor taking in air at atmospheric pressure (70°F or 21°C and 75% relative humidity) and discharging at 120 psig (828 kPa), about 3.75 gal (14 L) of water per hour will be condensed in the intercooler for each 1000 ft^3/min (1700 m^3/h) of free air compressed. This moisture content is based on saturated air at the second-stage suction at a pressure of 50 psig (345 kPa) and 80°F. This has an influence on the lubrication of subsequent stages. Figure 18.3 can be used to calculate moisture levels condensed in intercoolers and aftercoolers for the various pressures, temperatures, and humidity conditions.

A. Methods of Lubricant Application

In reciprocating compressors, the cylinders and running gear may be lubricated from the same oil supply, or the cylinders may be lubricated separately.

1. Cylinder Lubrication

Except where cylinders are open to the crankcase, oil is generally fed directly to the cylinder walls at one or more points by means of a mechanical force feed lubricator. In a few cases, main oil feed to the cylinders is supplemented by an additional feed to the suction valve chambers (pockets). For some small-diameter, high-pressure cylinders of multistage machines, oil is fed only to the suction

Compressors

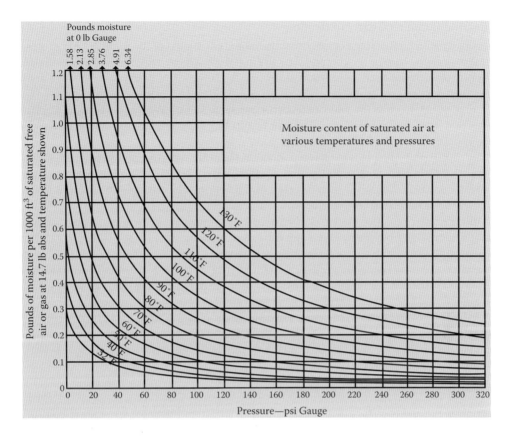

FIGURE 18.3 Vertical single-acting compressor. In this simple 3 hp machine designed for a discharge pressure of 100 psig (690 kPa), the motor and compressor are combined in a single compact unit. Lubrication is by splash system.

valve chambers. Essentially, all the oil fed to cylinders, which are not open to the crankcase, is carried out of the cylinders by the discharging gas and collects in the discharge passages, piping, and other system components such as receivers.

Cylinders that are open to the crankcase are lubricated by oil thrown from the reservoir by means of scoops or other projections on the connecting rods and cranks. Where this splash lubrication method is used, the pistons are provided with oil control rings similar to those used in automotive engines, which are designed to prevent excessive oil feeds to the cylinders.

Compressor valves require very little lubrication. Usually, the small amount of oil required spreads to the valves from the adjacent cylinder walls or is brought to them in atomized form by the air or gas stream. However, when air compressors operate under extremely moist suction conditions, it is sometimes necessary to provide supplementary lubrication by means of force feed lubricator connections to the suction valve chambers.

Valve operators, such as those used for unloader valves, may hold valves open or closed for certain types of pressure regulation systems. These generally require very little lubrication. As with suction and discharge valves, a small amount of oil is carried over from adjacent cylinder walls or is brought to the valve operators in atomized form by the gas stream.

When metallic piston rod packing is used on double-acting machines, stuffing boxes are lubricated by means of force feed lubricators. Usually, when nonmetallic packing is used, stuffing boxes are adequately lubricated by oil from the compressor cylinders. In some cases, however, mechanical force feed lubricators (or drop feed cups in older compressors) are used.

2. Bearing (Running Gear) Lubrication

In practically all reciprocating compressors, the oil for lubrication of the running gear is contained in a reservoir in the base of the compressor. Oil from the bearings, crosshead, or any cylinder open to the crankcase, drains back to the reservoir by gravity. However, a variety of methods or combinations of methods are used to deliver oil from the reservoir to the lubricated parts.

Oil may be delivered to the lubricated parts entirely by splash. Where this is done, a portion of, or projection from, one or more crankshafts or connecting rods dips into the oil and throws it up in a spray or mist that reaches all internal parts.

Many horizontal compressors have a flood system for bearing and crosshead lubrication (see Figure 9.18 in Chapter 9). Oil is lifted from the reservoir by disks on the crankshaft and is removed by scrapers. The oil is then directed to the bearings by passages, or allowed to cascade down over the crosshead bearing surfaces.

A full-pressure circulation system is often used for running gear lubrication. A positive displacement pump draws oil from the reservoir and delivers it under pressure to the main bearings (if plain) and connecting rod bearings, then through drilled passages to the wristpin bearings (bushings) and crosshead (if used). Where rolling element main bearings are used, the small controlled feed of oil required for them is commonly supplied by a drip or spray from the cylinder walls or rotating parts. Sometimes, a jet stream of oil is directed toward these bearings. In some compressors, oil is supplied under pressure to the connecting rod bearings from which it is thrown by centrifugal force to the cylinder walls. The wristpins are then lubricated by oil scraped from the cylinder walls and directed to the pin bearing surfaces by drilled passages in the pistons.

B. Single- and Two-Stage Compressors

Industrial single- and two-stage compressors range in size from fractional horsepower units to large machines of 20,000 hp (14,900 kW) or more. The smallest compressors are vertical single-acting type, usually air cooled (see Figure 18.4). Larger compressors are built in a variety of arrangements, including vertical single-acting (air- or water-cooled), vertical double-acting water cooled (Figure 18.5), and horizontal balanced opposed piston compressor (Figure 18.6). The largest machines are of the latter type. Machines may be single-cylinder or multicylinder with cylinders in tandem, opposed, or in a "V" type (Figure 18.7) or a "W" type (Figure 18.8). Assuming atmospheric conditions at the compressor suction, single-stage machines are available for pressures up to about 150 psig (1030 kPa). Two-stage machines are available for pressures up to about 1000 psig (6.9 MPa), although applications with pressures this high will generally use three or four compressor stages. Most two-stage compressors are designed for discharge pressures in the range of 80–125 psig (550–860 kPa).

1. Factors Affecting Cylinder Lubrication

The operating temperature in compressor cylinders is an important factor, both because of its effect on oil viscosity and on oil oxidation and the formation of deposits. At high temperatures, oil viscosity is reduced, so that when the operating temperature is high, higher viscosity oils may be required in order to maintain adequate lubrication films.

The thin films of oil on discharge valves, valve chambers (valve pockets), and piping are heated by contact with hot metal surfaces and are continually swept by the heated gas as it leaves the cylinders after compression. This is a severe oxidizing condition, and all compressor oils oxidize to an extent that depends on the conditions they are exposed to as well as their ability to resist this chemical change. Oil oxidation is progressive. At first, the oxidation products formed are soluble in the oil, but as oxidation progresses, these materials become insoluble and are deposited, mainly on the discharge valves and in the discharge piping, which are the hottest parts. After further baking, these deposits are converted to materials that are high in carbon content. These materials, along with contaminants that adhere and bond with them, are commonly called carbon deposits.

Compressors

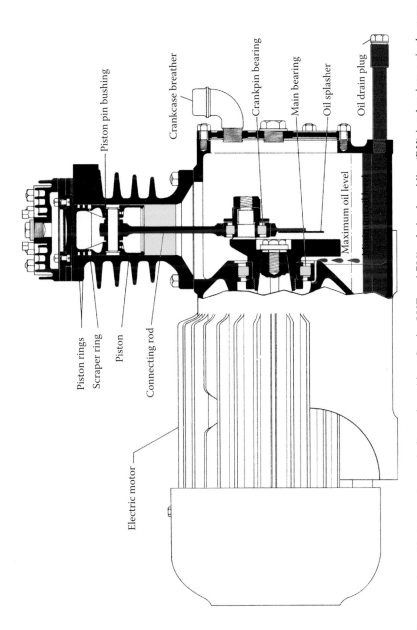

FIGURE 18.4 Moisture condensed in intercoolers and aftercoolers. Air at 0 psig, 80°F, and 70% relative humidity (RH) is taken into a single-stage compressor, compressed to 80 psig, and cooled in an aftercooler to 80°F. If the intake air had been saturated, it would have contained 1.58 lb moisture per 1000 ft³, as indicated at the left end of the 80° line. At 70% RH, it actually contained 1.58 × 0.70, or 1.11 lb moisture per 100 ft³. After compression and cooling, the air is saturated and contains 0.24 lb moisture per 100 ft³ (of free air), as read on the vertical scale by projecting to the left of the intersection of the 80° line and 80 psig lines. The rest of the moisture—1.11−0.24, or 0.87 lb per 100 ft³—has been condensed in the aftercooler.

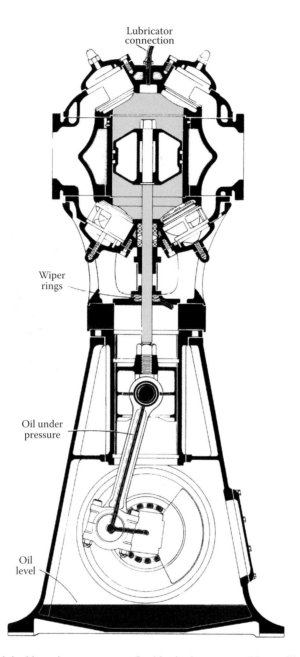

FIGURE 18.5 Vertical double-acting compressor. In this single-stage machine, rolling element main bearings are used. A pressure circulation system supplies oil to the connecting rod bearings and through drilled passages in the connecting rods to the crosshead guides. Oil draining back from the crosshead area is broken up into a fine mist, which lubricates the main bearings. A mechanical force feed lubricator supplies oil to the cylinders and wiper rings.

Compressors

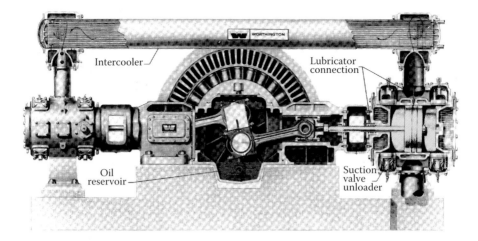

FIGURE 18.6 Horizontal balanced opposed piston compressor. This construction is used for very large capacities. There may be two to eight cylinders, each pair being arranged as shown. Oil is supplied to the cylinders and packings by a mechanical forced feed lubricator (not shown) and to the running parts by a pressure circulation system. A gear pump, driven from the crankshaft or a separate motor, delivers oil to a main distribution line and from there to the precision insert main and connecting rod bearings, the bronze crosshead pin bushings, and the crosshead slippers.

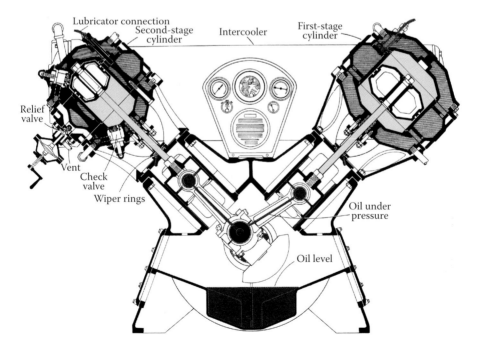

FIGURE 18.7 V-type, two-stage double-acting compressor. The cylinders are lubricated by means of a mechanical force feed lubricator that is geared to the crankshaft. Oil is delivered to the bearings through drilled passages by a pressure circulation system.

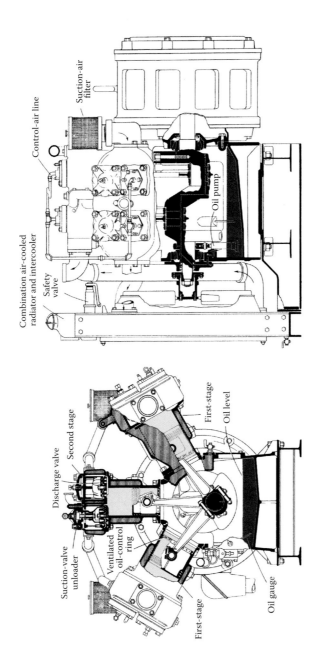

FIGURE 18.8 W-type, two-stage single-acting compressor. The first stage has two cylinders in each bank, whereas the second stage has only one cylinder for each low-pressure bank. Two stages of six cylinders form the complete compressor. The radiator contains two sections: an air cooled intercooler and a cooler for the cylinder jacket water. The connecting rod bearings are lubricated by pressure from a plunger pump. The tapered roller main bearings and cylinder walls are lubricated by oil thrown from the connecting rod bearings.

Deposits on discharge valves may interfere with valve seating and permit leakage of hot, high-pressure gas back into the cylinders. This high-temperature gas heats the gas taken in on the suction stroke so that the temperature at the start of compression is raised and the final discharge temperature is also raised. This is called "recompression" and results in efficiency losses as well as reduced flow capacity. Because the action is repeated on each stroke, the effect is cumulative, although for a constant rate of leakage, there is a tendency for the temperature to level off at some higher valve. If leakage increases, there will be a further increase in temperature.

Deposits also restrict the discharge passages and cause cylinder discharge pressures to increase and overcome the pressure drop caused by the restriction. Again, an increased discharge temperature accompanies the increased discharge pressure. Abnormally high discharge temperatures resulting from these effects result in more rapid oil oxidation and, therefore, contribute to further accumulation of deposits and temperature rise. This cycle may eventually result in a fire or explosion if it is not properly addressed.

A variety of contaminants are often present in the gas being compressed. Hard particles abrade cylinder surfaces and may interfere with ring and valve seating. Some contaminants promote oil oxidation by catalytic action, and some entrained chemicals may react directly with the oil to form deposits. Solid deposits adhere to oil-wetted surfaces and contribute to deposit buildup on discharge valves and in discharge passages. In many cases, the deposits found in compressors are largely composed of contaminants with a relatively small proportion of carbonaceous materials from oil oxidation. If these types of contaminants are found, gas filtration on the suction side of the compressor should be improved.

All of the oil fed to compressor cylinders is subjected to oxidizing conditions. As most of the oil fed to cylinders eventually leaves through the discharge valves where temperatures are highest, keeping the amount of oil fed to a minimum helps to reduce the potential for the formation of deposits on these areas as well as reduce excessive oil carryover to downstream equipment. Recommended base rates of oil feed to cylinders are shown in Table 18.2. Manufacturers of gas compressors generally recommend that oil be supplied to compressor cylinders at the rate of 1 pint for each 2,000,000 ft^2 (186,000 m^2) of area swept by the piston. The Compressed Air Institute

TABLE 18.2
Reciprocating Air Compressor Lubrication Guide (Dry Air)[a]

Cylinder Diameter (in)	Piston Displacement (ft^3/min)	Rubbing Surface (sfm)	Oil Feed per Cylinder (drops/min[b])	(pints/day)
Up to 6	Up to 65	Up to 500	0.66	0.12
6–8	65–125	500–750	1	0.18
8–10	125–225	750–1100	1.33	0.24
10–12	225–350	1100–1500	1–2	0.34
12–15	350–600	1500–2000	2–3	0.48
15–18	600–1000	2000–2600	3–4	0.62
18–24	1000–1800	2600–3600	4–5	0.86
24–30	1800–3000	3600–4800	5–6	1.15
30–36	3000–4500	4800–6000	6–8	1.44
36–42	4500–6500	6000–7500	8–10	1.80
42–48	6500–9000	7500–9000	10–12	2.16

[a] Oil feed to cylinders is in drops per minute and pints per day based on 8000 drops per pint at 75°F (24°C). To use this table for cylinder feed rates for gases other than air, multiply the feed rates shown by 3, which will provide the equivalent 1 pint for each 2 million ft^2 of swept cylinder surface. These are base starting points and may need adjustments according to gas conditions and operating parameters.

[b] The oil feed rates given are for water-filled gravity and vacuum-type site-feed lubricators. For glycerine-filled sight-feed lubricators, multiply the feed rates in drops per minute by 3 to achieve the listed pints per day.

suggests the use of 1 pint (0.47 L) of oil for each 6,000,000 ft³ (557,000 m³) of area swept by the piston in air compressors. These are general recommended starting points and may need to be adjusted (up or down) based on gas and operating conditions within the compressor.

Moisture is a factor in single-stage and multistage air compressors principally because of condensation that occurs in the cylinders during idle periods when cylinders cool below the dew point of the air remaining in them. The water formed tends to displace the oil films, exposing the metal surfaces to rusting. The amount of rust formed during any single idle period may be minor, and this may be scuffed off as soon as the compressor is started again, but in time, the process will result in excessive wear. In addition, rust tends to promote oil oxidation, and the rust particles contribute to accelerated formation of deposits. Where this potential exists, the use of oils with superior rust-inhibiting qualities and fortified with effective additives that will adhere to metal surfaces should be considered. This type of oil will help protect the metal surfaces from moisture and other liquids during these idle periods and during operation.

Oil feed to cylinders is in drops/min and pints/day based on 8000 drops per pint at 75°F (24°C). To use this chart for cylinder feed rates for gases other than air, multiply the feed rates shown by 3, which will provide the equivalent 1 pint for each 2,000,000 ft² of swept cylinder surface. These are basic starting points and may need adjustments according to gas conditions and operating parameters.

Metric equivalents:

$$1 \text{ in} = 2.54 \text{ cm}; 1 \text{ cfm} = 1.7 \text{ m}^3/\text{h}; 1 \text{ sfm} = 0.09 \text{ m}^2/\text{min}; 1 \text{ pint} = 0.47 \text{ L}$$

2. Factors Affecting Running Gear Lubrication

In general, the factors that affect compressor bearing lubrication—loads, speed, temperatures, and the presence of water and other contaminants—are moderate. The main requirement for adequate running gear lubrication is that the oil be of suitable viscosity at the operating temperature.

Much of the oil in circulation in compressor crankcases is broken up into a fine spray or mist by splash or oil thrown from the rotating parts. Thus, a large surface of the oil is exposed to the oxidizing influence of warm air and oil oxidation will occur at a rate and to an extent that depends on the operating condition and the ability of the oil to resist this chemical change. Conditions that promote oil oxidation in crankcases are mild in comparison to the oxidizing conditions in compressor cylinders, discharge valves, and discharge piping. However, the lubricant in the crankcase may remain in service for thousands of hours as opposed to cylinder lubricant that is continually replenished.

C. Multistage Reciprocating Compressors

Three-stage compressors for continuous duty are available up to 2500 psig (17 MPa). Four-, five-, six-, or seven-stage machines are available up to 60,000 psig (414 MPa) or higher. High discharge temperatures may or may not accompany these high discharge pressures. When possible, the machines are designed for compression ratios of 2.5:1 to 4.0:1 per stage, with 60°F (15.6°C) suction temperature and adequate cooling. This practice limits discharge temperatures from each stage to less than 375°F (190°C). However, compression ratios as high as 6:1 are often used, and with 60°F (15.6°C) suction temperature, the adiabatic discharge temperature is well over 400°F (204°C). Furthermore, with a suction temperature of 110°F (43°C), which is not uncommon for the later stages, the discharge temperature may approach 500°F (260°C). Even higher compression ratios, and discharge temperatures, may be encountered in compressors designed for intermittent duty.

1. Factors Affecting Lubrication

The factors of operating temperatures, contaminants, and oil feeds discussed in connection with single- and two-stage compressors also affect the lubrication of cylinders in multistage compressors.

In addition, the lubrication of multistage compressors is affected by the fact that water and oil are often entrained in the suction gas to the higher pressure stages, and by the high cylinder pressures.

Water in the form of droplets is often entrained in the air leaving the intercoolers and carried into the higher pressure stages of multistage air compressors. This water, moving at high velocity into relatively small cylinders, tends to wash the lubricant film from the cylinder walls. With an unsuitable oil in service, this results in inadequate lubrication and excessive wear, incomplete sealing against leakage, and exposure of metal surfaces to rust and corrosion.

Some of the oil carried into the intercoolers is usually entrained with the water and carried into the high-pressure cylinders. Although this oil contributes to lubrication of the succeeding stages, the results are not necessarily beneficial. This oil has already been exposed to oxidizing conditions in the lower pressure stages; thus, the total exposure is increased two or more times. As a result, these conditions could contribute to accelerated buildup of deposits even though excessively high temperatures are not involved.

High cylinder pressures acting behind the piston rings increase the rubbing pressure between the rings and cylinder walls. In addition, in trunk piston compressors, the connecting rod produces a thrust force against the cylinder wall with considerable pressure that increases with increasing compression pressure. The rubbing surfaces involved are parallel; movement between them does not tend to form thick oil films; and as pressure increases, there is a greater tendency to wipe away the thin films that are formed.

The lubrication of the running gear of multistage compressors presents essentially the same challenges as single- and two-stage machines.

2. Lubricating Oil Recommendations

As mentioned at the outset, the lubrication of compressors is not only a function of the type of compressor but also of the gas being compressed. In general, gases can be considered to be of four types: air, inert gases, hydrocarbon gases, and chemically active gases.

There are marked differences between compressor cylinder and bearing lubrication, but, in many cases, it is possible to use a single oil for the lubrication of both. In some cases, the oil required to meet the cylinder lubrication requirements may not be suitable for bearing lubrication. For example, it may be too high in viscosity, may contain special compounding that is not compatible with the materials used in some bearings, or may be of a special type that is not suitable for the extended service expected of bearing lubricating oils. Except in the case of air compressors, the following discussion pertains to the characteristics of oils for cylinder lubrication. The oils used for bearing lubrication of air compressors are usually suitable for the lubrication of bearings of compressors handling other gases. Where a problem might exist, the crankcase lubrication system is usually adequately isolated from contamination by the gas being compressed or the cylinder lubricant.

a. *Air Compressors*

The oils recommended for, and used in, air compressors vary considerably. Such factors as discharge pressure and temperature, ambient temperature, whether the air is moist or dry, and design characteristics of the machine must all be considered in the selection.

Single- and two-stage trunk piston type stationary compressors operating at moderate pressures and temperatures with dry air are generally lubricated with premium, rust-, and oxidation-inhibited oils. In portable service, these compressors may be lubricated with detergent-dispersant engine oils, typically oils for American Petroleum Institute (API) Service SJ, SL, SM, SN, CJ-4, CI-4 Plus, CI-4, and CH-4. (Note that other API service categories such as SH, CE,CF, and CG-4 are now obsolete but may still be acceptable for use.) The engine oils are also being used in many stationary compressors where the air is moist, or deposits or wear problems have been experienced with circulation or turbine-type oils. The viscosity grade used is frequently ISO VG 100 (Society of Automotive Engineers [SAE] 30), but both lower and higher viscosity oils are also used depending on ambient temperatures and machine requirements.

Under mild conditions, rust- and oxidation-inhibited turbine oils are also used as cylinder lubricants for crosshead-type compressors. Under wet conditions, compounded oils are used. The compounding may be either a fatty oil or synthetic materials. During the break-in period, either for a new or rebuilt compressor, higher viscosity oils are used and oil feed rates are increased.

In many cases where high discharge temperatures have resulted in rapid buildup of deposits on valves and in receivers with conventional lubricating oils, synthetic oils (usually diester or polyglycol based) are used. The straight synthetics offer both the ability to dissolve and to suspend potential deposit-forming materials, and offer better resistance to thermal degradation. These lubricants are usually of International Organization for Standardization (ISO) VG 100 or 150, although higher viscosity blends are used for extremely high pressure cylinders, or where severe moisture conditions are encountered.

In a number of cases where receiver fires have been experienced, fire-resistant compressor oils are used. Phosphate ester-based oils are one example of fire-resistant oil that is used in compressors as well as other applications where fire-resistant oils are desirable. For larger machines and higher pressures, or with moist air, the viscosity is usually ISO VG 100 (SAE 30) or ISO VG 150 (SAE 40). Lower viscosity grades may be used under moderate conditions with dry air or because of ambient conditions.

Where there is a need for oil-free air, reciprocating compressors, which operate without cylinder lubrication, are used. Nonlube rotaries and dynamic compressors can also be used for applications where oil-free air is required. The oil-free reciprocating compressors have polytetrafluoroethylene, carbon, or filled composition rings and rider bands, which do not require lubricants. In some newer designs, the pistons may have filled plastic composition or contain composition buttons in the piston skirts to prevent contact of the piston with the cylinder surfaces.

The running gear of crosshead compressors is lubricated normally with premium, rust- and oxidation-inhibited, circulation oils. In some cases where synthetic oil is being used as the cylinder lubricant, the same oil is recommended for the running gear with the potential of extending crankcase oil drain intervals. Seal compatibility should be validated before synthetic oils are used.

b. Inert Gas Compressors

Inert gases are those gases that do not react with the lubricating oils, and that do not condense on cylinder walls at the highest pressures reached during compression. Examples are nitrogen, carbon dioxide, carbon monoxide, helium, hydrogen, and neon. Ammonia is relatively inert but some special considerations apply to it.

The inert gases generally do not introduce any special problems, and the lubricants used for air compressors are suitable for compressors handling them. However, carbon dioxide is slightly soluble in oil and it tends to reduce the viscosity of the oil. If moisture is present, carbonic acid, which is slightly corrosive, will form. To minimize the formation of carbonic acid, the system should be kept as dry as practical. To counteract the dilution effect, higher viscosity oils than those that are normally used in air compressors are desirable.

Ammonia is usually compressed in dynamic compressors, but occasionally it may be compressed in positive displacement compressors. In the presence of moisture, it can react with some oil additives and oxidation products to form soaps. Ammonia is not compatible with antiwear compounds such as zinc dialkyl dithiophosphate (ZDDP), and oils containing these types of additives should not be used. Automotive engine oils and most antiwear-type hydraulic oils contain ZDDP. Ammonia may also dissolve in the oil to some extent, resulting in viscosity reduction. Highly refined straight mineral oils are usually used. Synthesized hydrocarbon-based lubricating oils are also used because of their low solubility for ammonia.

When compressing gases for human consumption, such as carbon dioxide for use in carbonated beverages, carryover of conventional lubricating oils into the gas stream is undesirable. Generally, medicinal white oils are required for cylinder lubrication under these circumstances. This does not apply when air is being compressed for the manufacture of oxygen for human consumption. The subsequent liquefaction of the air and distillation to separate the oxygen will leave the oxygen free of lubricating oil carryover. However, care in selection of the lubricant, controlling the rate of oil feed, and keeping the system clean is required to minimize "burnt oil" odor that can be carried over. If a conventional lubricant is used for this purpose, its effect on any catalysts should be evaluated and steps should be taken to assure that it does not contact the oxygen.

Petroleum hydrocarbons in the lungs can cause suffocation and possible pulmonary disease. Therefore, air compressors for scuba diving and other breathing air equipment should use nonlube compressors.

Under some circumstances, conventional petroleum oils cannot be used in inert gas compressors. This is the case in some process work where traces of hydrocarbons cannot be tolerated in the process gas, or some constituents of lubricating oils might poison the catalysts used in later processes. Compressors similar to those used to produce oil-free air or systems equipped with sophisticated filtration and conditioning equipment, are used where hydrocarbon carryover cannot be tolerated. Both special low sulfur, straight mineral oils and polybutenes are used where carryover of conventional oils might poison catalysts. Another example would be in paint booth applications, where carryover of silicone-type additives would create problems on surfaces to be painted (fish eyes).

In some cases, a compressor may be used alternately to compress an inert gas and a chemically active gas—for example, hydrogen and oxygen. Petroleum products form an explosive combination with oxygen. Therefore, they must not be used where oxygen is being compressed. The lubricant or lubrication system used for the compression of oxygen must also be used with the inert gas.

c. Hydrocarbon Gas Compressors

More horsepower is consumed in compressing natural gas than any other gas except air. When the volumes of other hydrocarbons that are compressed for the chemical and process industries are considered, the total horsepower consumed in compressing hydrocarbons is extremely large. Where the hydrocarbons must be kept free of lubricating oil contamination, dynamic compressors are usually used, but where high pressures are required, reciprocating compressors are used. With improved technology and the ability of some rotary compressors to achieve higher pressure and volume capacities, there is also a trend toward the use of rotaries in hydrocarbon compression for low pressure applications.

Although natural gas is mainly methane, other gases are usually present in small portions such as ethane, carbon dioxide, nitrogen, and heavier hydrocarbon gases. These heavier hydrocarbon gases are similar in many respects to the hydrocarbons that are compressed for process purposes. Occasionally, these heavier hydrocarbons can be in liquid form, which complicates the lubricant selection process.

The temperature at which a material will condense from the gaseous state to the liquid state (also the temperature at which it will pass from its liquid state to the gaseous state, i.e., its boiling point) increases with increasing pressure. Heavier hydrocarbons, with their higher boiling point, may have condensation temperatures above the cylinder wall temperature. The condensate formed under this condition will tend to wash the lubricant from the cylinder walls and dissolve in the lubricating oil resulting in viscosity reduction. Using an oil that is somewhat higher in viscosity than would be used for air under the same operating conditions can generally compensate for the dilution effect. Generally, compounded oils help to resist washing where condensed liquids are present in the cylinders. It is also usually advisable to operate with somewhat higher-than-normal cooling jacket temperatures in order to minimize condensation. This also requires the use of higher viscosity oils.

As it comes from the well, natural gas can contain sulfur compounds and is referred to as "sour" gas. Compressors handling sour gas are usually lubricated with detergent–dispersant engine oils—automotive or natural gas engine oils. These oils provide protection against the corrosive effects of sulfur. The viscosity used most frequently is ISO VG 100 and 150 (SAE 30 and 40), but if the gas is wet, that is, carrying entrained liquids, heavier oils may be used. The compressor of integral engine-compressor units is usually lubricated with the same oil used in the engine. But depending on the contaminants contained in the gas, compressor cylinders may require a different lubricant than that used in the engine crankcase.

d. Chemically Active Gas Compressors

Among the chemically active gases that must be considered most frequently are oxygen, chlorine, hydrogen chloride, sulfur dioxide, and hydrogen sulfide.

Petroleum oils should not be used with oxygen because they form explosive combinations. Oxygen compressors with metallic rings have been lubricated with soap solutions. Compressors with some types of composition rings have been lubricated with water. Compressors designed to run without lubrication are also being used. Some of the inert synthetic lubricants, such as the chlorofluorocarbons (CFCs) or fluorinated oil, can be used safely and provide good lubrication. Dry-type solid lubricants such as Teflon or graphite can be used to minimize metal-to-metal contact in this service.

Petroleum oils should not be used for the lubrication of chlorine and hydrogen chloride compressors. These gases react with the oil to form gummy sludges and deposits. If the cylinders are open to remove these deposits, rapid corrosion takes place. Compressors designed to run without lubrication are recommended. Diaphragm and nonlube rotary compressors are also used for the corrosive and reactive gases.

Sulfur dioxide dissolves in petroleum oils, reducing the viscosity. It may also form sludges by reacting with the additives in the presence of moisture, or by selective solvent action. The system must be kept dry to prevent the formation of acids. Highly refined straight mineral oils or white oils from which the sludge-forming materials have been removed, either by acid treating or severe hydroprocessing of the base stocks, are often chosen. Oil feed rates should be kept to a minimum.

Hydrogen sulfide compressors must be kept as dry as possible because hydrogen sulfide is corrosive in the presence of moisture. Compounded oils are usually used, and rust and oxidation inhibitors are considered desirable.

II. ROTARY COMPRESSORS

The five main types of rotary positive displacement compressors are straight lobe, rotary lobe, helical lobe (more commonly referred to as screw compressors), rotary vane, and liquid piston. There are many design variations available for each of these types of compressors based on application requirements. They can be single- or multiple-stage units that are designed for low pressure/high flow to relatively high pressure. As far as the lubricant coming in contact with the gas being compressed, rotary screw and rotary lobe compressors can either be dry (nonlube) or flooded lubrication. Dry lubrication is in reference to compressor rotors only. Rotary vane compressors will almost always be flooded with lubricant whereas liquid piston units will almost always be nonlube.

A. STRAIGHT LOBE COMPRESSORS

Straight lobed compressors are built with identical two- or three-lobed impellers that rotate in opposite directions inside a closely fitted casing (Figure 18.9). Timing gears outside the case drives the impellers, and these gears maintain the relative positions of the impellers. The impellers do not touch each other or the casing, and no internal lubrication is required. No compression occurs within the case because the impellers simply move the gas through. Compression occurs because of backpressure from the discharge side.

Compression ratios are low, and for this reason these machines are often referred to as blowers rather than compressors. Straight-lobed compressors are available in capacities up to about 30,000 cfm (51,000 m^3/h) and for single-stage discharge pressures up to about 25 psig (172 kPa).

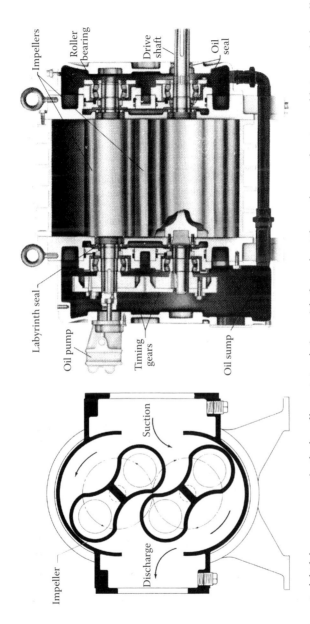

FIGURE 18.9 Rotary straight lobe compressor. As the impellers rotate, air (or gas) is drawn in at the suction opening, trapped between the impellers and the casing, then forced into the discharge area against the pressure of the system. Lubricating oil is supplied under pressure to the bearings and timing gears by means of an integral circulation system.

1. Lubricated Parts

The principal parts of straight-lobed compressor requiring lubrication are the timing gears and the shaft bearings. The bearings may be either plain or rolling element (usually roller) type. Timing gears are precision spur or helical type. The bearings and gears are generously proportioned to minimize unit loads and wear because small clearances between impellers and casings must be maintained for efficient operation. For most bearings and gearing, lubrication is by means of an integral circulation system. Some rolling element bearings at the end opposite the drive may be grease lubricated.

2. Lubricating Oil Recommendations

Lubricating oils used in straight-lobed compressors are either mineral turbine or circulating oil quality containing rust and oxidation inhibitors or synthetic lubricants where temperature extremes are encountered and long life is desired. Because the lubricant does not contact the gas being compressed in most applications, the oil generally does not require highly additized or compounded oils. Where temperatures are high or with heavy loads, antiwear oils may be used to provide additional protection for the gears. The viscosities usually recommended are ISO VG 150 for normal ambient and operating temperature and ISO VG 220 for high temperatures. Synthetic lubricants are often used where extremely high or low temperatures are encountered or where longer oil drain intervals are desired.

B. Rotary Lobe Compressors

Most rotary lobe compressors are designed for discharge pressures up to a maximum of 200 psig (1380 kPa) and outputs up to about 1100 cfm (1870 m^3/h). They are commonly used in applications similar to straight-lobed compressors such as blowers or where small amounts of compressed air may be required. The rotors (Figure 18.10) are supported by antifriction bearings, and their position, relative to each other, is controlled by timing gears. The rotary lobe compressors are available in both nonlube and flooded lube designs.

1. Lubricated Parts

The main components that require lubrication are the bearings and the timing gears. Lubrication is essential to prevent wear in the gears and bearings as this will result in loss of efficiency and

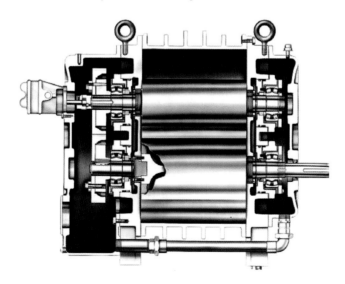

FIGURE 18.10 Rotary lobe compressor.

possibly contact of the lobes with each other or with the cases. In the flooded type, the oil also helps seal the clearances between the lobes and the cases allowing greater pressure capability, helps cool and also serves as a corrosion inhibitor in those instances where gases may contain moisture or other contaminants. In flooded applications, the lubricant introduced to the lobes goes out with the discharge gas and must be recycled back to the compressor. Generally, one common system is used for both the gears and bearings and the compressor lobes.

2. Lubricating Oil Recommendations

The lubrication requirements for the gears and bearings will generally be satisfied with an ISO VG 46 or 68 synthetic turbine- or circulating-type oil. Depending on the manufacturer and application, a different viscosity oil may be required. Oil selection will be based on the operating parameters of the compressor (nonlube or flooded) and the condition of the gas being compressed.

C. Rotary Screw Compressors

These compressors, also called helical lobe compressors, are available in single-impeller and the more common, two-impeller (rotor) design. In the two-impeller types, one common design uses a four-lobed male rotor meshing with a six-lobe female rotor (Figure 18.11). Timing gears may individually drive the rotors, or the male rotor may drive the female rotor. Gas is compressed by the action of the two meshing rotors as illustrated in Figure 18.12. The machines come in single- and multiple-stage units.

Two variations of two-impeller screw compressors are used. In the "flood lubricated" type, the oil is injected into the cylinder to absorb heat from the air or gas as it is being compressed. The oil also functions as a seal between the rotors. As oil is available in the cylinder (casing) to lubricate the rotors, these machines are now usually built without timing gears. They require an external circulation system to control the temperature of the oil and an oil removal system to remove most of the oil from the discharge air or gas (Figure 18.13).

In dry screw compressors, no oil is injected. Because the rotors are not lubricated, timing gears are required to keep the rotors from contacting each other. These compressors can be used to deliver oil-free air or gas. However, because there is no oil seal between the rotors, operating speed must be relatively high to minimize gas leakage. Water cooling of the cylinder and such features as oil-cooled impeller shafts (Figure 18.11) are usually required.

Two-impeller screw compressors are available in capacities up to about 41,200 cfm (70,000 m^3/h) and commonly used for discharge pressures in the 125 psig (860 kPa) range for single-stage units and 350 psig (2408 kPa) for two-stage units. Special application screw compressors are available for discharge pressures up to 1500 psig in compound and multiple-stage designs. Higher speed capabilities and closer tolerances in the clearance areas between rotors and cylinders are allowing increasing pressure capabilities and improved efficiencies.

The single-screw compressor (Figure 18.14) has a driven conical screw that rotates between toothed wheels. The teeth of the wheels sweep the thread cavities of the screw, compressing the air or gas as it is moved up the progressively decreasing volume of the cavity. The wheels are made of two materials:

1. A metal backing that absorbs the stresses imposed on the wheels
2. A wider plastic facing that contacts the sides of the thread cavities

Oil is injected into the case for lubrication, cooling, and sealing. As with flood lubricated, two-impeller, screw compressors, oil cooling and oil removal equipment are required as part of the compressor running gear.

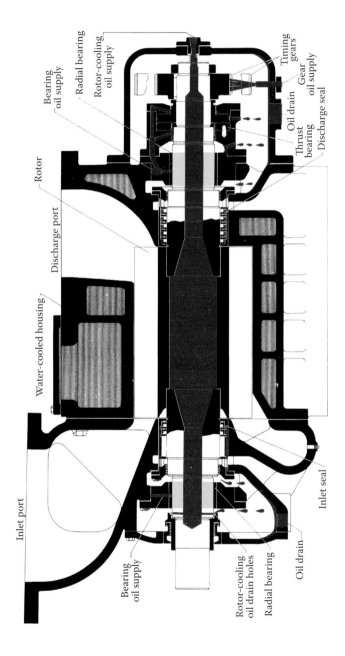

FIGURE 18.11 Two impeller rotary screw compressor. This large stationary machine has timing gears and no oil is used inside the casing. The machine is water cooled, with oil-cooled impeller shafts. Multiple carbon ring shaft seals are shown. Shaft bearings are plain, and a tilting pad thrust bearing is used.

Compressors

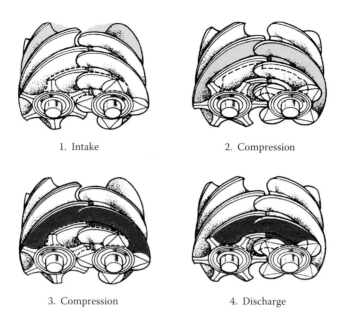

1. Intake
2. Compression
3. Compression
4. Discharge

FIGURE 18.12 Compression in a screw compressor. As the screws unmesh at the rear and below the centerline of the view labeled "intake," the expanding space created draws gas in through the inlet, filling the grooves between the screws. As the screws advance, they seal off this gas and carry it forward, compressing it against the discharge end head plate.

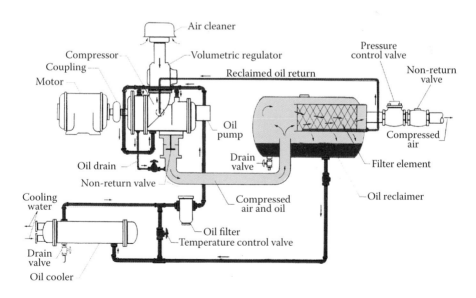

FIGURE 18.13 Lubrication system for flood-lubricated screw compressor. When the oil is cold, the temperature control valve is open, bypassing the oil cooler. As the oil warms up, the valve gradually closes and is completely closed when the oil temperature reaches about 180°F (82°C). All of the oil then passes through the oil cooler. The arrangement provides rapid warmup of the oil to minimize condensation, but limits the maximum oil temperature to provide long oil life.

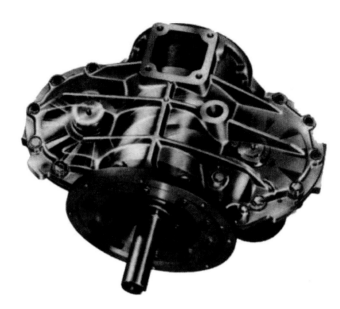

FIGURE 18.14 Single-screw compressor. The gear wheels are shown resting in the screw to indicate the manner in which they sweep the thread.

1. Lubricated Parts

The lubricated parts of dry screw compressors are the gears and bearings. Shaft bearings may be either plain or rolling element, and thrust bearings may be either tilting pad or angular contact rolling element. Gears may be either precision cut spur or helical type.

The oil lubricates the contacting surfaces of the rotors, along with the gears and bearings, in flood-lubricated machines. In both types, oil-lubricated seals may be used where gas leakage must be kept to a minimum.

2. Lubricant Recommendations

Flood-lubricated screw compressors in portable and plant air applications are lubricated by oils of ISO VG 32 (SAE 10) to ISO VG 68 (SAE 20). Heavy-duty engine oils suitable for API Service Classification SJ, SL, SM, SN, CJ-4, CI-4 Plus, CI-4, and CH-4 are used in mobile screw compressors as are automatic transmission fluids. Engine oils generally contain high levels of detergents and corrosion inhibitors that will have a tendency to pick up moisture. This characteristic may affect selection based on the need to separate water from the oil. Premium quality rust- and oxidation-inhibited oil with good demulsibility characteristics should be considered where low ambient temperatures, cyclic operation, or very humid conditions are encountered. Circulation and turbine-type oil offer improved water-separating characteristics.

Even when engine oils and automatic transmission fluids are used, problems with varnish and sludge can be encountered in severe operations. Polyalphaolefin (PAO)-based synthetic lubricants are now the most commonly used oils in these applications. Other types of synthetic lubricants such as diester, polyglycol, and synthetic blends can also be used. These synthetic oils provide a significant extension of oil drain intervals versus minerals oils, which more than offsets their higher initial costs.

Flood-lubricated screw compressors are being used for compressing gases other than air. One example is their expanded use in the field gathering process for natural gas. Owing to improvements in design that permits greater pressure and volume capabilities, as well as efficiencies closely approximating those of reciprocating compressors, there continues to be increased usage in hydrocarbon gas compression. Because of the intimate mixing of the oil with the gas, oil selection is even more critical than in reciprocating compressors. For hydrocarbon gas applications, polyglycol-based synthetics lubricants are often recommended. The polyglycols have low solubility for hydrocarbon gases, so dilution of the lubricant is minimized and good separation of the oil and gas can be obtained. Synthesized hydrocarbon-based lubricants also provide an option for lubrication of screw compressors handling hydrocarbon gases.

Dry screw compressors require only lubrication of the gears and bearings. Either synthetic or premium rust- and oxidation-inhibited circulating- and turbine-type oils of a viscosity suitable for the gears are usually used. In some cases, oils with enhanced antiwear characteristics may be desirable for added protection of the gears.

D. Sliding Vane Compressors

Sliding vane compressors are positive displacement machines where the vanes are free to move in slots in the rotor mounted eccentrically in a cylinder casing (Figure 18.15). Rotation of the rotor causes the vanes to move in and out of the rotor slots, creating pockets that increase and decrease in volume. The vanes can be held out against the case by coupling pins as in Figure 18.15, by springs or by centrifugal force alone. Where pins are used, the casing is not circular because all diameters through the axis of the rotor must be equal.

Sliding vane compressors are available in capacities exceeding 6000 cfm (10,200 m³/h) with discharge pressures of more than 100 psig (690 kPa) for single-stage units and more than 125 psig (860 kPa) for two-stage units.

Two general types of cooling are used for rotary vane compressors. Large stationary machines are water cooled. Portable and plant air machines may be air or water cooled, with the addition of oil injection into the cylinder. As with screw compressors, the latter machines may be referred to as flood lubricated, or by such terms as oil injection cooled or direct oil cooled. As with flood-lubricated screw compressors, an external system to remove the oil from the discharge gas and cool it is required.

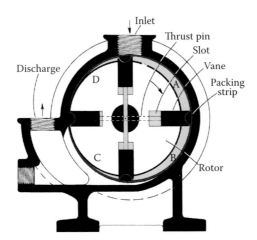

FIGURE 18.15 Rotary sliding vane compressor. The ports are located so that air (or gas) is drawn into pockets of increasing volume *A* and discharged from pockets of decreasing volume *B*. In this design, the bore of the cylinder is not circular because the total length of two vanes plus the pin, measured through the axis of the rotor, is constant.

1. Lubricated Parts

All of the sliding surfaces in the cylinder require oil lubrication to minimize friction and wear. The lubricating oil also aids in protecting internal surfaces from rust and corrosion, adds a certain degree of cooling, and helps seal clearances between vanes, rotors, cylinder walls, and heads. Shaft bearings of either plain or rolling element type are designed for oil lubrication. Where the shaft passes through the head, packing glands or seals are used, and these are supplied with oil to minimize wear and friction and to assist in preventing leakage of high pressure air or gas.

2. Lubricant Recommendations

In sliding vane compressors, some cylinder surfaces are subjected to heavy rubbing pressures as a result of the action of the vanes in their slots. The forces acting on the extended vanes tend to tilt the vanes backward, causing heavy pressure at the inner leading edges of the vanes and the trailing edges of the slots. This results in high resistance to the inward motion of the vanes and consequently, heavy pressure between the vane tips and the cylinder wall. This tends to rupture the oil films and cause wear and roughening of the surfaces involved. This process gradually diminishes efficiency. Inattention to good maintenance practices or selection of poorer quality oils could also contribute to deposit formation on vanes and in rotor slots, further restricting the free movement of the vanes in the slots.

Flood-lubricated rotary vane compressors are usually lubricated with the same types of oils recommended for flood-lubricated screw compressors. These oils give good protection against wear under the conditions encountered in the cylinders.

Large stationary rotary vane compressors differ considerably in their lubrication requirements from the flood-lubricated type. Water jackets are used to cool these machines. Cylinder lubricant is fed by force feed lubricators. Feed rates, in relation to compressor capacity, must be much higher than those for reciprocating machines. These machines tend to run cool at the inner cylinder walls, so condensation is a problem. Heavy-duty engine oils of SAE 30 or 40 grade (ISO VG 100 or 150) are widely used. Where the gas discharge temperature are above 300°F (149°C), higher viscosity oils or the use of synthetic compressor oils are recommended. Compounded compressor oils of comparable viscosity are also used.

E. Liquid Piston Compressors

In a liquid piston compressor (Figure 18.16), rotation of a multibladed rotor causes a liquid, usually water, to be thrown outward forming an annulus that rotates at rotor speed and maintains

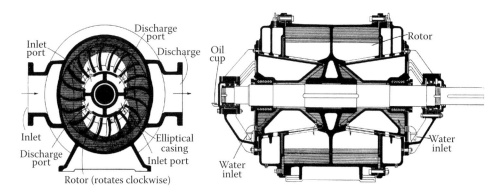

FIGURE 18.16 Rotary liquid piston compressor. The water annulus, in combination with the oval casing, forms pockets between the rotor blades that increase and decrease in volume much like those in sliding vane compressors. However, as there is no contact between the rotor blades and the casing, no internal lubrication is required. Water, or other liquid media, is fed continuously to replace the liquid phase carried out with the compressed gas.

a seal between the rotor blades and the housing. As the annulus rotates, the pockets formed between the rotor blades and the annulus increase or decrease in volume in the same manner as in a sliding vane compressor. Because of the annulus, the suction and discharge ports must be located in the cylinder heads, rather than on the outer perimeter of the cylinder. The liquid annulus eliminates the need for sliding, sealing surfaces so there are essentially no wearing surfaces inside the cylinder.

Liquid piston compressors are available in capacities exceeding 16,000 cfm (27,200 m^3/h) and for pressures up to 100 psig (690 kPa).

1. Lubricated Parts

The only parts of liquid piston compressors requiring lubrication are the shaft bearings. All models have rolling element bearings, which may be lubricated by oil bath or grease.

2. Lubricant Recommendations

Oil-lubricated bearings of liquid piston compressors are usually lubricated with premium, inhibited circulating and turbine quality oils. Viscosities are selected according to bearing requirements. Grease-lubricated bearings are lubricated with premium, ball, and roller bearing greases. NLGI Grade 2 multipurpose lithium complex greases are most often recommended.

F. DIAPHRAGM COMPRESSORS

Diaphragm compressors have been around for many years in low-pressure/low-flow applications. Many of the small units consist of a simple diaphragm that is moved back and forth in a reciprocating motion by a rod either mounted to a rotating crank or by means of an eccentric cam. With improved designs, they can be used in applications requiring very high pressures with moderate flow capacities. A common usage is in refinery operations in the production of white oils, where pressures can exceed 50,000 psig (345 MPa). Compression ratios can exceed 20:1. They are also used in the handling of corrosive or reactive gases, such as in many process industries, because diaphragms and casings are built out of exotic corrosion-resistant materials such as Inconel, Monel, and Hastelloy. The vast majority of diaphragm compressor components are made of carbon steel or stainless steel that can handle many of the chemically reactive gases.

1. Lubricated Parts

The lubricated parts consist essentially of a hydraulic system. The lubricant acts as a hydraulic oil to provide the reciprocating motion of the ram that operates the diaphragm. This ram operates in a range of 200–400 cycles/min and hydraulic pressures can exceed 2000 psig (13.8 MPa). Depending on the medium being compressed and discharge temperatures, the oil also functions to remove heat from the diaphragm. Under severe operating conditions, oil coolers may be required.

2. Lubricant Recommendations

Typically, the oil used for diaphragm-type compressors will be an ISO VG 68 or 100 antiwear-type hydraulic oil. As the oil does not come in direct contact with the gas being compressed, it is not necessary to use highly additized or compounded oils for these systems. If there is any potential for contacting the gas being compressed, special considerations will be required. If system oil temperatures are consistently high, synthesized hydrocarbon-type hydraulic oils are often used.

III. DYNAMIC COMPRESSORS

In contrast to positive displacement compressors where successive volumes of gas are confined in an enclosed space (which is then decreased in size to accomplish compression), dynamic compressors first convert energy from the prime mover into kinetic energy in the gas. This kinetic energy is then converted into pressure. As the rotors of dynamic compressors do not form a tight seal with the casings, some internal leakage occurs past the rotors. These clearances are getting much tighter and speeds much higher to allow greater efficiencies and higher pressures. The two main types of dynamic compressors are centrifugal and axial flow. Quite often, you can find both of these designs incorporated into one single shaft unit.

A. Centrifugal Compressors

In a centrifugal compressor, a multibladed impeller rotates at high speed in a casing (Figure 18.17). Air or gas trapped between the impeller blades is accelerated and thrown outward and forward in the direction of rotation. The air leaves the blade tips with increased pressure and very high velocity, and enters a diffuser ring. In the diffuser ring, because of the increasing area in the direction of flow, a reduction in velocity and a substantial increase in pressure take place. The air or gas then enters a volute casing where, again because of the increasing area in the direction of flow, a further reduction in velocity and increase in pressure take place. The outward flow of air through the impeller creates a reduced pressure at the inlet (or eye of the impeller), causing air to be drawn into the compressor.

Centrifugal compressors are particularly adapted to supplying large volumes of air (or gas) at a relatively small increase in pressure. They are inherently suited for high-speed operation—up to 60,000 rpm—and can, therefore, be directly connected or geared to high-speed driving motors or turbines. Speed-increasing drives may also be used. Centrifugal compressors are built with capacities exceeding 650,000 cfm (1,100,000 m^3/h).

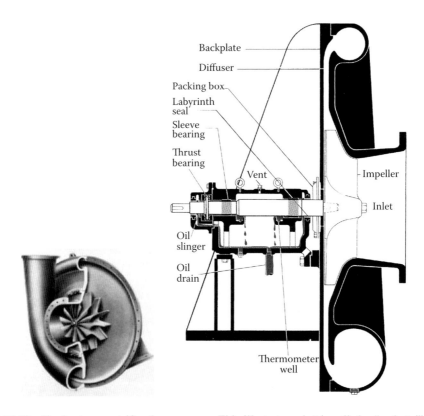

FIGURE 18.17 Single-stage centrifugal compressor. This illustrates what is called a "pedestal" machine. The impeller is overhung on a short shaft supported by two bearings mounted on a pedestal. The thrust bearing is of the fixed shoe design.

Compression ratios range up to about 5:1 with one impeller stage. A single-stage machine operating with atmospheric intake may be designed for discharge pressures as high as 60 psig (44 kPa). Multistage compressors with up to 10 stages are built. For example, a 10-stage internally geared compressor will have the final stage pinion rotating at about 50,000 rpm and be capable of discharge pressures as high as 2900 psig (20 MPa).

Centrifugal compressors are used extensively as boosters in petroleum refineries, chemical process plants, and, to a lesser extent, in natural gas pipeline operations. In booster service, air or gas is taken in at or above atmospheric pressure and discharged at still a higher pressure. Where suction pressures are high, casing and shaft seals must be designed to withstand the high pressures involved.

Centrifugal compressors are also built for plant air applications. These units are generally designed with capacities from 100 to 15,000 cfm (1700 to 25,500 m^3/h) at discharge pressures generally in the range of 125 psig (860 kPa). The best performance from a pressure ratio standpoint is about 4:1 or 5:1 in a single-stage centrifugal compressor. Two to five stages are used, with three or four stages being the most common. Pinions mounted around a large bull gear drive the impellers. In this way, each impeller is driven at its optimum speed. Where a turbine drive is used, the bull gear may be driven directly from the turbine. With an electric motor drive, the bull gear may be driven through a speed increaser. Intercooling between stages is normally used.

Centrifugal compressors discharging at low pressure usually have no provision for cooling. Higher pressure machines are cooled by water passages in the diaphragms between stages, or by injecting a volatile liquid into the gas stream. It should be noted that, because of the very high velocities encountered in dynamic compressors, any liquid (water or other) or solid contaminants in the gas stream will cause erosion of the impellers and blading.

Where the shaft passes through the casing, seals are provided. Several types are used: soft braided packing, labyrinth seals, carbon ring seals, oil film seals, and mechanical (end face) seals. The mechanical seals are similar to those used for reciprocating compressors (Figure 18.4) but designed for a rotating shaft. The type of sealing used depends on such factors as pressure, speed, gas being compressed, and the amount of leakage that can be tolerated.

1. Lubricated Parts

The bearings and gears (where used) of centrifugal compressors require lubrication, but there is no need for lubrication within the impeller casing. Oil film seals and contact as well as noncontact (gas-pressurized, close clearance labyrinth) seals are used on high-pressure compressors. Contact-type seals must be supplied with oil to provide lubrication and cooling, and enhance sealing.

With a single impeller, or group of impellers, all facing in the same direction, there is an axial thrust because of the backward force exerted by the discharging gas on the rim of the impeller. This thrust force can be balanced internally by means of a balancing drum or dummy piston, but a thrust bearing must be used to support all or part of the thrust load and keep the impeller accurately located in the casing. Thrust bearings are generally angular contact ball, collar, fixed shoe, or tilting shoe type.

Shaft bearings of high-speed units are generally plain (sleeve type), but rolling element bearings are used in some units. Plain bearings, including collar, fixed shoe, and tilting pad shaft bearings, are oil lubricated. Tilting pad shaft bearings are used on high-speed dynamic compressors because of two desirable characteristics: they are self-aligning and they exhibit good dampening. Rolling element bearings may be either oil or grease lubricated, depending on the operating parameters of the compressor and other component requirements, such as gears.

Where a compressor is driven by a step-up gearset, the drive gears and one or more of the bearings may be lubricated from the same system. In the case of packaged plant air units, the bull gear and bearings are contained in an oil-tight casing, and oil is circulated to them under pressure from a shaft-driven oil pump. Where a speed increaser is used, it is lubricated from the system that supplies the bearings.

2. Lubricant Recommendations

Because of the high speed of most centrifugal units, the oil that is most commonly used will be an ISO VG 32 synthetic or turbine quality oil containing rust and oxidation inhibitors. The high pitch line velocities on some of the gearing, particularly on the internally geared high-pressure machines, will dictate the limitations on viscosity to avoid inlet shear heating in the gear mesh. Lower or higher viscosity oil may also be used depending on compressor design and operation. Oils that protect against varnishing are important in some higher temperature applications. Additional comments on lubrication that applies to both centrifugal and axial flow machines are discussed in the subsection "Lubrication Recommendations" under "Axial Flow Compressors."

B. Axial Flow Compressors

An axial flow compressor contains alternating rows of moving and fixed blades. High velocity is imparted by the moving blades to the air (or gas) being compressed. As the air flows through the expanding passages between the fixed and moving blades, the velocity is reduced and transformed into static head, that is, pressure. A stage consists of one row of moving and one row of stationary blades (see Figure 13.8 in Chapter 13), and a relatively large number of stages—in some applications, more than 20 stages in one casing can be used. Axial flow compressors (Figure 18.18) are compact, high-speed machines suitable for handling large volumes of air at compression ratios as high as

FIGURE 18.18 Six-stage axial flow compressor. This compressor is equipped with plain journal bearings and a tilting shoe thrust bearing. Oil is supplied to the bearings by a circulation system.

30:1 in a single cylinder or casing. Machines capable of handling 1,000,000 cfm (1,700,000 m^3/h) or more have been built. They may be driven by electric motors, steam turbines, or gas turbines, the latter being fairly common in the chemical process industries where large volumes of gas are available to drive the turbine, and large volumes of compressed air or gas are required. Axial flow compressors are also used in all industrial or aircraft-type gas turbines that are used for power generation and ship propulsion except for the smaller-sized turbines.

1. Lubricated Parts

As with centrifugal compressors, the parts of axial flow compressors requiring lubrication are the shaft bearings, the thrust bearings that are used to take axial thrusts and maintain axial rotor position, and any oil-lubricated seals that may be used.

2. Lubricant Recommendations

Because the lubricants used in centrifugal and axial compressors are not exposed to the conditions inside the cylinder or casing, lubricant selection is based on the requirements of the bearings. Where a speed increaser supplied from the same lubricating oil system is used, the oil must be selected to meet the gear requirements, which are generally more severe than those of the bearings. Premium inhibited circulating- and turbine-type oils are most commonly used. Where only bearings require lubrication, oils of ISO VG 32 are usually used. Where a speed increaser is used, somewhat higher viscosity oils (typically ISO VG 46 or 68) may be required depending on the gear pitch line velocities and other requirements of the gearing. In some limited cases, lower viscosity oils may be required.

In packaged plant air units, there is considerable tendency for air to be whipped into the oil, and foaming can result. Oils with good air release properties and foam resistance are preferred. The premium long-life turbine oils in ISO VG 32 are usually recommended.

Where a dynamic compressor is driven by a gas turbine, both machines may be on the same lubrication system. The lubricant selection is then based on the requirements of the gas turbine. Both specially developed mineral oil-based and synthesized hydrocarbon-based lubricants are used. Generally, the viscosity used is ISO VG 32 with some manufacturers requiring ISO VG 46 for their turbines.

Some lubrication challenges have been experienced with dynamic compressors handling ammonia or "syngas" for ammonia production. As pointed out under reciprocating compressors, ammonia can react with some inhibitors and oxidation products to form insoluble soaps and sludges. Although it will be uncommon for the ammonia gas to contact the lubricating oil directly in dynamic compressors, there could possibly be some contamination via seal oil systems or faulty seals. The soaps and sludges formed can be thrown out in the coupling connecting the speed increaser to the compressor. This can cause coupling imbalance or eventual binding, resulting in increased vibration. The deposits can also end up in oil coolers resulting in poorer heat transfer characteristics. Deposits may also form in the speed increaser. Highly stable oils with inhibitors that are not affected by ammonia have been effective in reducing these problems. Both mineral oil- and synthesized hydrocarbon-based products are used. The viscosity used is usually ISO VG 32 or 46.

IV. REFRIGERATION AND AIR CONDITIONING COMPRESSORS

The basic principles of the refrigeration compression cycle are shown in Figure 18.19. The five essential parts basic to every system are shown: evaporator, compressor, condenser, receiver, and expansion valve (or capillary). Liquid refrigerant flows from the receiver under pressure through the expansion valve to the evaporator coils where it evaporates, absorbing heat resulting in a cooling action. The vapor is then drawn into the compressor where its pressure and temperature are raised. At the higher pressure in the discharge of the compressor, the condensing temperature of the refrigerant is higher than it would be at atmospheric pressure. When the hot, high-pressure vapor flows from the compressor to the condenser,

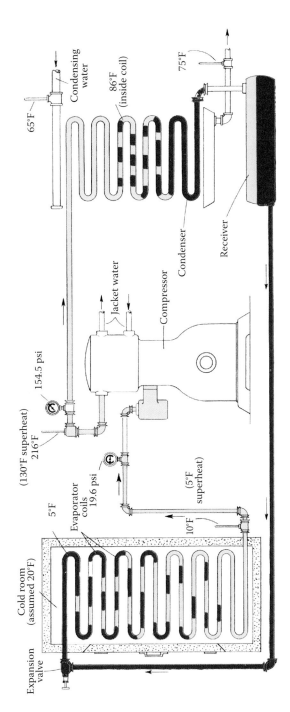

FIGURE 18.19 Basic single-stage compression refrigeration system. The elements shown here are common to all compression refrigeration systems, whether refrigeration or air conditioning.

the cooling water (air in some applications) removes enough heat from it to condense it. The heat removed from the refrigerant in the condenser is equal to the amount of heat removed from the cold room (cooling action) plus the heat resulting from the mechanical work done on the refrigerant in the compressor that is not removed by the jacket cooling of the compressor. In many commercial installations, the evaporator cools a heat transfer fluid such as brine that is then pumped through the area to be cooled. Smaller units, such as home refrigerators and freezers, room air conditioners, and automotive air conditioners have air-cooled condensers rather than water-cooled condensers.

In commercial installations, two or three stages of compression may also be used. Where system pressures or cooling capacities dictate two stages of compression is needed, two-stage compressors are used, or a combination of separate single-stage compressors. Rotary sliding vane, scroll, or rotary screw compressors are sometimes used at low to moderate pressures or for booster purposes. Multistage reciprocating compressors are used for large air conditioning installations with a trend toward the use of more scroll compressors. Reciprocating compressors are commonly used in refrigeration systems along with a small trend toward the use of the rotary vane compressor. Centrifugal compressors are also used on some commercial refrigeration systems as well as in chillers. Reciprocating, sliding vane, and scroll compressors have been used for automotive air conditioning systems with some screw and axial piston compressors also used. Some very small units such as dehumidifiers may be equipped with diaphragm-type compressors. Reciprocating compressors are used in most other applications.

Most reciprocating compressors for commercial installations are of the single-acting, trunk piston type and have closed crankcases. Because of refrigerant leakage past the pistons, the crankcases are then filled with a refrigerant atmosphere. The same is true of axial piston units used for automobile air conditioning. Crosshead and double-acting compressors have open crankcases. The majority of small- to medium-size electric motor-driven refrigeration and air conditioning units are hermetically sealed, with all of the operating parts, including the electric motor, sealed inside the unit. Evaporators may operate either dry or flooded. In dry evaporators, only refrigerant vapor is present, whereas flooded evaporators have both liquid and vapor present.

A. Factors in the Compressor Affecting Lubrication

1. Cylinder Conditions

The oil film on the cylinder walls of a reciprocating refrigeration compressor is subjected to low temperatures at the suction ports along with moderately high temperatures near the cylinder head. Because viscosity decreases with temperature, the oil near the suction ports will have considerably higher viscosity than the oil near the cylinder head. Nevertheless, the oil must spread in a thin film over the entire working surface. This is accomplished by the piston rings (or the piston itself in small compressors without piston rings) as the pistons move back and forth. The oil must distribute rapidly, but to do this it must not be too high in viscosity. On the other hand, too low a viscosity oil will not protect against wear.

Oil carried out of the cylinders to the valves and discharge piping is subjected to the temperature of the discharging refrigerant. Ordinarily, the temperature of the discharging refrigerant is not high; for example, the discharge temperature of a single-stage ammonia compressor with a compression ratio of about 5:1 should not be much in excess of 250°F (121°C). Some single-stage units operate at higher ratios and higher discharge temperatures, but in most small compressors the valve temperatures remain moderate because of the relatively large cooling area in proportion to cylinder volume. The discharge temperatures of compressors operating on fluorocarbon refrigerants are lower than equivalent machines operating on ammonia, although the compressors of automobile air conditioning systems may operate at quite high discharge temperatures.

When two or more stages of compression are used, the operating temperature in each stage will usually be lower than in single-stage machines. In rotary compressors, the discharge temperatures are also usually moderate because of low compression ratios.

2. Oxidation

In compressors with enclosed crankcases, temperatures are normally moderate and the entire machine is filled with refrigerant vapor. Very little, if any, air is present. Under these conditions, oxidation in the usual sense does not occur, although it is doubtful that it can be avoided entirely. Limited oxidation does not impair the lubricating value of an oil because the initial oxidation products formed are soluble in the oil. If oxidation progresses too far, a condition is eventually reached where some of the soluble oxidation products become insoluble when the oil is cooled. These oxidation products could then plug or restrict capillary tubing or orifices within the system.

3. Bearing System Conditions

The general requirements of the bearing systems of refrigeration compressors are similar to those of other comparable compressors. However, some special factors must be considered. In the compression of air or gases such as hydrocarbon gases, it is desirable that the oil not be miscible with the gases. In closed refrigeration systems, the oil must be somewhat miscible with the oil in order to circulate throughout the system and reach all of the components in need of lubrication.

In compressors with closed crankcases, there is very little exposure to oxygen so oxidation stability of the oil is not a major concern. However, where the same oil is used for both bearings and cylinders, as in many small units, the oil must have adequate oxidation stability for the cylinder conditions.

Where ammonia is used as the refrigerant in compressors with closed crankcases, any additives used in the oil must be types that are not affected by ammonia. If the refrigerant is soluble or partially soluble in the oil, such as the majority of the fluorocarbon refrigerants, this will dilute the oil and reduce its viscosity, which must be considered in the selection of the oil viscosity.

The motor in hermetically sealed units is completely surrounded by a mixture of refrigerant and oil. As a result, the oil must have good dielectric properties, must not affect the motor insulation, and must not react with the copper motor windings, or other system materials at elevated temperatures. Because most of these units are operated on fluorocarbon refrigerants, the dilution effect of the refrigerant on the viscosity must be considered.

In compressors where the crankcase and cylinders are completely isolated from each other, as in units having crosshead construction, the oil in the crankcase is exposed to air, and there is intimate mixing of the warm oil with air. These conditions are favorable to oxidation and require a chemically stable oil to resist oxidation.

B. Refrigeration System Factors Affecting Lubrication

If the oil being carried out of the compressor cylinder forms gummy deposits in the condenser, or if it congeals or forms waxlike deposits in the evaporator, capillary tube or expansion valve, it may seriously reduce the heat exchange capacity. Heat-insulating deposits in the evaporator make it necessary to carry a lower evaporator temperature in order to produce the required refrigeration effect. This, in turn, requires a lower evaporator pressure and increases the power required by the compressor, for a given refrigeration duty, owing to the increased pressure range through which the gas must be compressed. In addition, at the lower suction pressure, the vapor density is lower so the compressor will have to handle a greater volume of vapor, reducing refrigeration capacity. Heat-insulating deposits in the condenser increase the temperature difference between the cooling medium (water or air) and the condensing refrigerant. The resulting higher condensing temperature makes higher compression necessary and increases power consumption.

Whether heat-insulating deposits will be formed depends on the properties of the lubricating oil, the refrigerant in use, the evaporator temperature, and the equipment used in the system. The effect of some of the common refrigerants is considered separately.

1. Fluorocarbons

CFC refrigerants have been phased out for use in air conditioning and refrigeration systems because of their potential negative environmental effects on the ozone layer. As a result, more environmentally friendly non-CFC refrigerants are being used and recommended. Several alternative refrigerants have been around for many years such as ammonia, hydrocarbons, carbon dioxide, methyl chloride, and others that do not have ozone depletion potential and will be continued to be used in many applications. Non-CFC fluorocarbon refrigerants such as R-134a and blends such as R-404A, R-407C, and R-410A are the most common replacements for the CFC refrigerants such as R-22, R-11, R-12, and R-123. The use of these and other alternative refrigerants is increasing rapidly. The U.S. Environmental Protection Agency (2012) has compiled a list of acceptable refrigerants by application that can be viewed on their website, http://www.epa.gov/ozone/snap/lists/index.html.

Each of the alternative refrigerants has specific properties and operating characteristics that must be understood and handled accordingly to ensure maximum system performance as well as the safety of the people handling them and the public that is potentially exposed to them. In many systems, CFCs are and will remain in service. The Montreal Protocol banned the production of CFCs as of January 1, 1996, and hydrochlorofluorocarbons (HCFCs) were limited to production levels as of the same date; moreover, production is projected to cease by the year 2030.

Air conditioners in older automobiles, as well as many home refrigerators and air conditioners, were filled with CFC refrigerants, and some of these units are still in service. When systems containing CFCs need servicing, they must get filled with CFCs manufactured before January 1, 1996, use reclaimed CFCs from older systems, or retrofit the systems to accept one of the alternative environmentally friendly refrigerants. Gradually, all of the CFCs and HCFCs will be replaced by alternative HFC materials, as well as other gases such as isobutane, propane, and ammonia.

With the refrigerants that are miscible, or partly miscible with oil, enough of the refrigerant dissolves in the oil to depress the pour point of the oil sufficiently that congealing of the oil on evaporator surfaces does not usually occur. However, there is a temperature at which a heavy, flocculent precipitate first appears when a mixture of refrigerant and 10% of the oil is chilled. The temperature at which this occurs depends on the refrigerant, the percentage of oil in the refrigerant, and on the oil itself. Refrigeration systems using fluorocarbon refrigerants are often designed so that approximately 10% oil is present in the evaporator. In some cases, the evaporator is actually charged with this amount of oil. Under these conditions, the floc point of the oil (also known as the critical separation temperature) represents the lowest temperature that can be used with that oil.

The waxy materials that precipitate from these oil–refrigerant mixtures can also clog expansion valves and capillary control tubes, preventing their proper functioning. However, the concentration of oil in the refrigerant at the expansion valve is usually lower than in the evaporator, so the floc point is depressed below what it would be at a 10% concentration. As a result, if the oil selected has a low enough floc point for conditions in the evaporator, it will usually not cause difficulties in the expansion valve or capillaries. Difficulties in these areas attributed to mineral oils are frequently the result of minute quantities of water in the system forming ice crystals.

Oil selection can go a long way to minimizing problems related to lubrication in those systems using fluorocarbons. The use of highly refined naphthenic or paraffinic mineral oils works satisfactorily with both CFCs and HCFCs. The base stocks for these oils are usually severely hydroprocessed or acid-treated to remove wax and other undesirable materials from a refrigeration oil standpoint. For HCFCs, in addition to mineral oil being acceptable, alkyl benzene synthetic lubricants provide excellent performance. Widely used lubricant types for HFC applications include polyol esters, polyalkylene glycol, and polyvinyl ether.

2. Ammonia (R-717) and Carbon Dioxide (R-744)

Oil is slightly miscible in anhydrous ammonia and carbon dioxide. Generally, not enough of the gas dissolves in the oil to have a significant effect on the pour point of the oil. Thus, if the pour point of the oil is above the evaporator temperature, the oil will congeal on the evaporator surfaces and form an insulating film that interferes with heat flow and efficient performance of the system. To remove the oil, the evaporator must be periodically warmed in order to liquefy the oil so that it will drain from the surfaces to where it can be removed. With flooded evaporators, refrigerant flow may be so rapid that there is little or no opportunity for the oil to collect on evaporator surfaces, and the pour point of the oil may not be a major concern. Ammonia is not compatible with copper or brass and cannot be used in systems containing these metals. As with CFCs and HCFCs, ammonia works well with highly refined mineral oil. They also can use PAOs (synthesized hydrocarbons), polyalkylene glycols, and polyol esters.

3. **Hydrocarbon Refrigerants**

Isobutane (R-600A) and propane (R-218) gases are being used as replacements for CFC refrigerants in some applications. These gases are primarily used in smaller units such as hermetic household refrigerators.

4. **Sulfur Dioxide (R-764)**

Sulfur dioxide gas has a selective solvent action that, when mixed with conventionally refined lubricating oils, will result in sludge formation. It therefore requires the use of highly refined white oils or Group III base stocks with low levels of additive.

5. **Lubricating Oil Recommendations**

Table 18.3 shows general lubricant recommendations by refrigerant type. The lubricants are classified according to base oil types. The requirements of oils for refrigeration systems can be summarized as follows:

1. The oil should be of proper viscosity to distribute readily at the system's lowest temperatures yet provide adequate films to protect against wear in the cylinders and crankcases.
2. The oil should have adequate chemical stability to resist oxidation and the formation of deposits in crankcases open to the atmosphere, and to resist the deteriorating influence of high temperatures at compressor discharge.
3. In closed systems without oil separators, the oil should be miscible with the refrigerant in order to ensure that the oil will circulate through the system and return to lubricate

TABLE 18.3
Lubricating Oil Recommendations Based on Refrigerants

Refrigerant	Lubricating Oil	
	Mineral Oil[a]	Synthetic
Fluorocarbons		
CFC-11, 12, 113, 114, 500, 502	Yes	PAO, POE
HCFC-22, 123, 125, 408A(blend)	Yes	PAO, AB
HFC-134a, 143a	No	POE, PAG, PVE
Blends 404A, 407C, 410A	Yes	POE, PVE
Ammonia	Yes	PAO, PAG, POE
Carbon dioxide	Yes	PAO

Note: AB, alkylbenzene; PAG, polyalkylene glycol; PAO, polyalphaolefin; POE, polyolester; PVE, polyvinyl ether.

[a] Mineral oils are to be highly refined paraffinic or naphthenic. White oils or severely hydroprocessed base stocks should be used.

the compressor. In closed systems with separators, it is desirable that the oil not be miscible with the refrigerant in order to facilitate separation. In open crankcase systems, it is desirable that the oil not be soluble or miscible with the refrigerant so that dilution is minimized.
4. Thermal aspects require that the oil be able to withstand system temperatures without breakdown and that it does not inhibit the heat transfer characteristics of the refrigerant.
5. The oil needs to be chemically stable and not react with the refrigerant or system components. Some additives in the oil can react with the refrigerant to form deposits or sludges.
6. They must reduce friction and minimize wear.
7. They must keep the system clean and stay in service for extended intervals.

Oil viscosities recommended vary from as low as ISO VG 7 to as high as ISO VG 150.

BIBLIOGRAPHY

Mobil Oil Corporation. 1983. Compressors and Their Lubrication. Fairfax, VA.
Mobil Oil Corporation. 1971. Refrigeration Compressors and Their Lubrication. Fairfax, VA.
U.S. Environmental Protection Agency. 2012. Ozone Layer Protection—Alternatives/SNAP. http://www.epa.gov/ozone/snap/lists/index.html.

19 Lubricant Contribution to Energy Efficiency

Access to reliable and affordable energy is critical to virtually all aspects of modern life. The obvious areas include transportation, power generation, and the various energy sources used for heating and cooling our homes, public spaces, and work environments. However, a significant amount of the world's energy supply is used "behind the scenes" in manufacturing, agriculture, the Internet, and various public services, to name just a few. As a result, supplying the world's energy demand will remain a critical part of future human and societal development.

History has shown that energy demand continues to be driven by population size and economic activity. Over the past two centuries, there has been a remarkable transition that has taken place that has led to today's modern economies and lifestyles. Figure 19.1, taken from ExxonMobil's Outlook for Energy analysis (Exxon Mobil Corporation, 2014), shows the projected increase in world population and economic activity through 2040. Under current conditions, this would result in an annual world energy demand of 1200 quadrillion BTUs. However, coincident with demand, there are trends to use and produce energy more efficiently, which has the potential to reduce the 2040 worldwide energy demand by about 500 quadrillion BTUs. Thus, saving energy is critical to reducing demand and provides strong economic and social incentives to develop and implement energy-saving technologies.

In the transportation area, various regulatory bodies often mandate improvements in energy use and emission reductions. These standards can be enforced by penalties and fines aimed at providing an incentive to reach specific targets. For example, in the United States, the Corporate Average Fuel Economy (CAFE) standard will require manufacturers to reach a target of 54.5 miles/gal for its fleet of passenger and light-duty truck vehicles by 2025. In Europe, emissions such as CO, NO_x, and particulates are regulated. Table 19.1, from a 2011 European Environment Agency report, gives a broad overview of various targets set for transportation-related emissions. Some of these targets are already encompassed in existing or future standards whereas others are at the proposal stage. Exhaust emissions can be controlled by improvements in exhaust aftertreatment systems. However, emissions-based standards also act as a driving force toward more fuel-efficient engine and vehicle technologies.

Regulations in other transportation areas vary by country and type, and are often related to exhaust emissions. One exception is that fuel economy and the related emissions from aircraft have been left to market forces. Despite the fact that fuel represents a significant cost to airline operators, improvements in fuel use per passenger mile has stagnated over the past decade. Recently, the International Civil Aviation Organization has been working toward establishing a CO_2 standard for aircraft that, if enacted, may change the current status quo in the aviation arena.

Meeting future emission and energy saving targets will require significant advances in equipment, engine, and vehicle technologies. Vehicle and engine manufacturers recognize that the goals for the next decade or two will be met by using combinations of approaches rather than the emergence of a single breakthrough technology. As will be shown later, the largest efficiency gains in internal combustion engines are likely to come from improvements in thermodynamic efficiency and engine operation. For example, engine downsizing coupled with turbocharging and the adoption of gasoline direct injection (GDI) are trends common between multiple vehicle manufacturers. These technologies are already making valuable contributions to improving engine efficiency. In parallel, increasing the use of hybrid systems and new transmissions including six-speed and higher and continuously variable transmissions (CVTs) will further improve the overall use of the energy

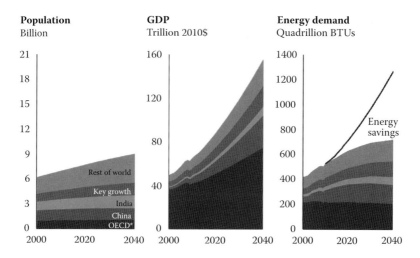

FIGURE 19.1 Projected world population, economic and energy demand through 2040. Single line indicates demand in the absence of future energy savings. *Mexico and Turkey are included in key growth countries.

TABLE 19.1
Summary of Goals for Transport-Related Emission Standards

Transport Segment	Target(s)	Date(s)
All transport greenhouse gas (GHG) emissions (excluding international maritime)	20% lower versus 2008	2030
	60% lower versus 1990	2050
New passenger vehicles	130 g CO_2/km	2015–2017
	95 g CO_2/km	2020
New light vans/commercial vehicles	175 g CO_2/km	2014–2017
	147 g CO_2/km	2020
New maritime ship efficiency via the energy efficiency design index	10% lower versus 2011	2015
	15–20% lower versus 2011	2020
	30% lower versus 2011	2025

output from the engine. Although these transformations are not directly related to friction reduction, they will have an effect on the lubrication systems and fluid requirements. In fact, in some cases the lubricant may be a key factor in enabling the successful implementation of a technology.

Direct reduction in friction losses will come from multiple sources. Improvements in design and manufacturing technology will result in systems that have lower overall friction losses. Advances in materials and surface coatings may also contribute to lowering surface friction. Lubricants will need to continue to evolve to keep pace with these technology developments and to contribute toward improving the efficiency of systems that both produce and consume energy. In this chapter, the major mechanisms of friction loss in lubricated systems, and how they influence various types of machine elements are examined.

I. FRICTION LOSS MECHANISMS

There are two major sources of friction loss within a lubricated contact. The first is attributed to shearing of the lubricant as it passes through the contact. The second results from the energy dissipated when surface asperities collide and slide over one another within the contact. In addition

Lubricant Contribution to Energy Efficiency

to the losses within the contact, there are a number of losses that occur outside the contact. For example, churning of the oil by rotating parts can adsorb measurable amounts of energy. Similarly, in circulating systems, it is necessary to supply power to push the lubricant around the system and through any restrictions such as valves and filters. Other energy loss mechanisms include any momentum transfer that occurs when the fluid velocity changes after impact with a moving surface and windage losses that can be associated with the force required to move components through the air and any oil mists.

In most mechanical systems, the total energy loss comprises a combination of the different sources. Furthermore, each of the moving components will contribute varying amounts to the combined total. Energy losses are thus highly dependent on system design, the lubricant properties, and operating variables.

A. Viscosity—Shear Losses

In the definition of viscosity, the term "resistance to flow" is often used. More formally, as shown in Figure 19.2, viscosity is the variable that relates the rate at which the fluid is sheared with the stress that is developed within the fluid that is resisting the shearing motion. For a lubricant filling the gap between two surfaces, the shear rate is determined by the distance between the surfaces and the speed of one surface relative to the other. The shear stress is transmitted across the liquid–solid boundary and, over a given area, will generate a force that resists the direction of motion. This can also be considered as the force required to maintain a given speed and becomes an energy loss when summed up over any distance.

These definitions form the basis for all the viscous-related energy losses in lubricated contacts. These simple relationships become more complicated when other lubricant properties such as shear thinning and pressure–viscosity effects are considered. In addition, in most systems, the separation of the surfaces also depends on speed and lubricant viscosity among other variables that modify the shear rate. However, despite these factors, reducing a lubricant's viscosity is commonly used as a means to reduce friction.

B. Boundary Friction

In previous chapters, we have seen that the generation of fluid films depends on many factors. In hydrodynamic bearings, the thickness of the film and friction losses are related to the fluid's viscosity as well as the speed and load applied to the bearing. Figure 19.3 shows an example of how friction can vary with these parameters. This is often referred to as the Stribeck curve, named after Richard Stribeck, who studied the friction performance of journal bearings in Germany during the late 1800s and early 1900s.

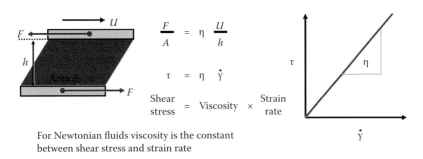

For Newtonian fluids viscosity is the constant between shear stress and strain rate

Units: poise (centipoise, cP), Pa-s

FIGURE 19.2 Definition of viscosity and relationship to shear stress.

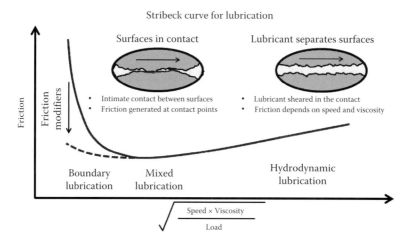

FIGURE 19.3 Stribeck curve showing relationship between speed, viscosity, load, and friction for hydrodynamic bearings.

The Stribeck curve provides an example of how a lubricated contact can transition between different regimes of lubrication. At high speed, low load, and higher viscosity, the lubricant film will be relatively thick and can fully separate the two surfaces. In this regime, located on the right-hand side of Figure 19.3, the viscous forces provide the main contribution to the friction loss in a hydrodynamic bearing. As viscosity increases, the viscous friction losses also increase. If speed or viscosity is reduced, or load is increased, the lubricant film gets thinner as we move to the left on the Stribeck curve. At some point, the highest points on each of the surfaces begin to make contact with one another. At each one of these small contacts, a tiny amount of localized friction force is generated as the surfaces slide past each other. Within the contact, there may be many of these contacts depending on the relative roughness of the two surfaces and the thickness of the lubricant film. The total friction generated in the contact is thus a combination of the viscous forces and the friction generated at the asperity contacts. Because both hydrodynamic and surface effects govern behavior, this is referred to as the mixed lubrication regime. If the speed or viscosity is reduced or load is increased, the lubricant film generated by hydrodynamic effects is reduced even further. This results in a greater number of asperity contacts. In this regime, surface friction effects dominate, and the term "boundary lubrication" is used. As the friction generated at surface contacts is generally larger than the friction generated by shearing the lubricant, the total friction increases, which results in the upturn seen in the left-hand side of the Stribeck curve.

The Stribeck curve was originally applied to hydrodynamic bearings. However, most machine elements and even entire systems such as internal combustion engines will exhibit similar behavior. The precise shape of the curve and the conditions under which the transition occurs between the different regimes depends on many factors. For example, machine design will influence how well hydrodynamic films are formed, and material selection and surface roughness will affect how much surface contact occurs and how much surface friction is generated. Thus, each machine element exhibits its own unique Stribeck curve.

Given that the Stribeck curve exhibits a minimum, it is tempting to think that minimizing friction is merely a matter of selecting a lubricant that provides an optimum film thickness. In practice, this is not a simple task. Most machines operate under variable conditions, including start–stop cycles. Even when they are running, they often operate over a wide range of speed, load, and temperature conditions—all of which moves the system up and down the Stribeck curve. Furthermore, different machine elements in a system may be working in different parts of their cycle compared with companion components. An operating engine is a good example of this as each cylinder places

Lubricant Contribution to Energy Efficiency

loads on the engine connecting rod bearings at different crankshaft angles. Thus, even if we are able to optimize the friction losses at a given speed and load condition for one component, it is unlikely that this represents an overall optimum for the entire machine.

Another factor that influences how well an optimum film thickness can be attained is durability. In addition to generating surface friction, the surface contact that occurs under the mixed and boundary lubrication regimes can also lead to increased wear. Although this can be controlled using effective lubricant additive chemistry, the selection of lubricant viscosity is often restricted by a desire to reduce the risk of unacceptably high wear. As an example, if an oil for a given component is selected that has an operating viscosity that minimizes the friction loss, it is likely that the system would operate deep into the boundary lubrication regime during any transitions to higher loads or temperatures. In practice, equipment design and selection of a lubricant is a compromise between obtaining the best overall efficiency while managing the risk of adverse wear. However, as design and engineering models become more sophisticated, the margin for error can be reduced, allowing machines to operate much closer to energy efficiency optimized conditions.

In the mixed and boundary regime, the friction between the moving surfaces is generated by different mechanisms and behaves differently than the friction generated via fluid shearing. As most practical engineering surfaces have some amount of roughness, the contact between opposing surfaces only occurs at the points where high spots or asperities on each surface interact with one another. In fact, the actual area of contact is typically only a small percentage of the so-called nominal contact and is determined by the microgeometry of the surfaces. Figure 19.4 shows this effect for two nominally flat surfaces that are in contact with one another.

Under these conditions, the relationship between the real area of contact and the pressure at the asperity contacts can be defined as

$$A = W/P \qquad (19.1)$$

where
A = total real area of contact
W = applied load
P = average pressure within the asperity contacts

It is noteworthy that Equation 19.1 suggests that the real area of contact does not depend on the nominal or macro scale contact at all. Although there are some limits, it is true that the real area

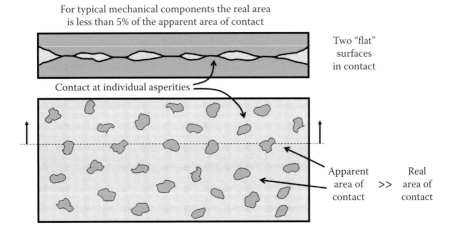

FIGURE 19.4 Contact between two nominally flat surfaces occurs at discrete contact points where the relative high points of each surface interact.

FIGURE 19.5 Junction between two opposing asperities being sheared by relative sliding motion.

of contact depends mostly on material properties and the nature of the rough surfaces. This is the origin for some of the earliest laws of friction.

Figure 19.5 shows an individual asperity contact. If a junction is formed between the two surfaces, a shear strength value can be defined that must be exceeded in order to break it.

The force required to break the junction is simply stated as

$$f = \tau a \qquad (19.2)$$

where
 f = force required to break a single junction
 a = contact area of a single junction
 τ = shear strength of the junction

In order for the two surfaces to slide over one another, it is necessary to continuously break the junctions that are formed as asperities slide over one another. The total force is simply the sum of all the individual asperity forces and is directly linked to the total area of contact, as shown in Equation 19.3:

$$F = \tau A, \qquad (19.3)$$

where F is the total force.

The friction coefficient is defined as the force required to move one surface relative to another divided by the load keeping them in contact as given in Equation 19.4:

$$\mu = F/W \qquad (19.4)$$

where μ is the friction coefficient.

Substituting for F and W gives

$$\mu = \frac{\tau A}{PA} = \frac{\tau}{P} \qquad (19.5)$$

Although the relationship described in Equation 19.5 is a simplification of a much more complex process, it does reflect several important characteristics. First, as mentioned earlier, the coefficient of friction is independent of the nominal area of contact. Second, the shear strength of the junction between the two surfaces has a direct impact on the coefficient of friction. This is heavily influenced by such factors as material properties and surface chemistry. The latter is often modified by the presence of any films formed by a liquid lubricant. Third, the pressure that exists at the contacting asperities governs the coefficient of friction. This is controlled by a combination of the applied loads, the material properties, and the surface topography.

The preceding simple analysis goes a long way toward explaining the three laws of friction published by Guillaume Amontons in 1699:

1. The force of friction is directly proportional to the applied load.
2. The force of friction is independent of the apparent area of contact.
3. Kinetic friction is independent of the sliding velocity.

Lubricant Contribution to Energy Efficiency

The first and second laws are a direct result of the relationship between load and the real area of contact. The random nature of rough surface topography ensures that, on average, a linear relationship exists between load and real area of contact. Roughness and material properties define how much real surface contact is required to support an applied load and does not depend on the macro scale apparent area of contact. The third law is also known as Coulomb's law and is based on the assumption that once it is moving, the real area of contact remains constant despite the speed of the surfaces.

These laws were originally developed to explain dry friction. The presence of a lubricant alters the observed friction as it introduces different mechanisms and effects. For example, friction in the presence of a liquid lubricant most often results in friction varying with speed. This occurs because, as stated earlier, the viscous contribution depends on speed and the hydrodynamic effect will change the separation of surfaces and hence the amount of surface contact.

C. Friction Modifiers Used in Lubricants

In the previous discussion, it has been assumed that the friction developed between contacting asperities is related to the shear stress required to break the junction. If this property can be controlled, it is possible to change the overall friction generated within a contact. Lubricants often contain friction modifiers that are intended to control boundary friction by a number of different mechanisms. Chapter 9, which focused on lubricating oils, provides a brief explanation of friction modifiers under the description of antiwear additives. The following discussion gives greater expansion on the definition and purpose of friction modifiers.

Friction modifiers used in lubricants falls into a broad class of surface active additives. They can be subdivided into three basic types. The first type consists of a group of compounds that preferentially adsorb onto the surfaces. The second type consists of materials that chemically react with the surface, creating a new layer or surface. The third group consists of solid materials that are dispersed in the oil that, once entrained in the contact, can alter the dynamics of the asperity interactions. In each case, the purpose of these additives is to modify the number and properties of junctions formed between contacting asperities.

1. Surface Adsorbing Friction Modifiers

The first type of friction modifier is typically based on organic polar molecules that preferentially adsorb onto the surfaces. Examples include glycerol mono-oleate, partial esters, and fatty amides. It is generally understood that these additives work by forming an assembled layer or even multiple layers on the surface owing to a combination of hydrogen bonding with the surface and intermolecular forces acting between adjacent attached molecules. Figure 19.6 shows how polar molecules

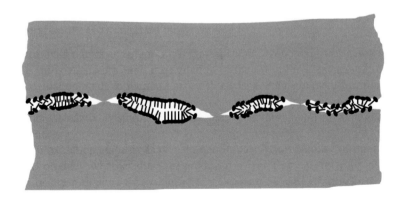

FIGURE 19.6 Attachment of polar molecules to the surfaces within a lubricated contact.

might be assembled on two contacting surfaces. Under relatively mild conditions, this layer can help reduce the occurrence of direct surface interactions, thereby reducing the amount of boundary friction occurring within the contact.

2. Chemically Reactive Friction Modifiers

The second class of friction modifier additives is typically based on a metal that chemically reacts with the surface creating a new layer or surface. Molybdenum (Mo) is the most widely used metal although others such as tungsten can also be used. The metal is often combined with other elements such as sulfur and oxygen, which, when coupled with alkyl chains, makes the molecule soluble in lubricant base stocks. In order for this class of additive to perform its function, it is often necessary for a direct reaction with the surface to occur. For oil-soluble molybdenum-containing additives, the friction-reducing effect is associated with the formation of molybdenum disulfide (MoS_2) on the surface. The layered structure of MoS_2 resembles graphite and provides a plane along which it is easy for one layer to slide over its neighbors. These easily sheared layers that form in the junctions between asperity contacts reduce the shear stress required for asperities to slide past on another, resulting in lower friction. It has been observed that the formation of MoS_2 occurs only within the rubbing track of the surface. The precise nature of the chemical and physical processes leading to the formation of reacted films remains an area of study, but the use of such additives provides an effective means of reducing friction.

Figure 19.7 shows the effect of adding the two different types of soluble friction modifiers to a polyalphaolefin (PAO) base stock. The test configuration was a steel ball rubbed against a flat steel specimen. The friction force is measured as the temperature is ramped up in a controlled fashion. The results from this simple test demonstrate some key features of these types of friction modifiers. At the beginning of the test, all three cases exhibit a similar friction response. In the case of the organic friction modifier, the friction is maintained at a relative constant level throughout the test. This compares with the rapid rise in friction that occurs using the PAO base stock alone. Although

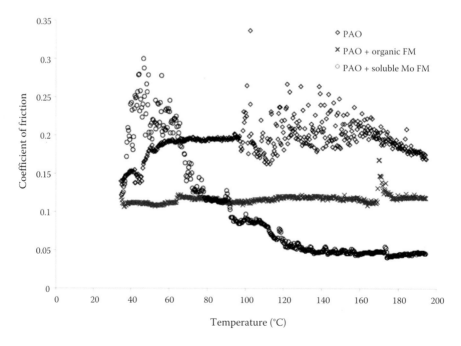

FIGURE 19.7 Friction test results from a reciprocating ball on a flat surface experiment showing the effect of organic and molybdenum friction modifiers as a function of temperature.

difficult to observe directly, it can be assumed that the organic friction modifier has either reduced the occurrence of asperity contacts or alternatively reduced the friction within them. The results also suggest that the formation of an adsorbed layer occurs relatively quickly and at relatively low temperatures. In contrast, the soluble Mo friction modifier initially results in a higher friction coefficient. As the experiment progresses, the friction reaches a peak and thereafter progressively reduces to a value considerably lower either than the base stock alone or with the organic friction modifier. It appears that the formation of an effective film occurs over time and may require reaching an activation temperature before a friction-reducing surface film is formed.

The results in Figure 19.7 represent an ideal case. In formulated lubricants, friction modifiers have to compete with other surface active additives in order to effectively form a layer on the surface. Similarly, other additives may form reacted films, interrupting the process of producing a friction-reducing film. Both types of additive exhibit operational limits in terms of temperature and load over which they are effective. In general, the purely organic friction modifiers are effective under relatively moderate conditions, but the films can be easily removed during the rubbing process. Reacted films are often more tenacious and work under more extreme conditions, but these have limits and require a surface that they can readily react with (e.g., ferrous metal surface). Despite such restrictions, use of these types of friction modifiers offers an effective means of reducing the boundary friction losses in a lubricated contact.

3. Solid Dispersion Friction Modifiers

The third class of additive are dispersed solid materials. This class also includes MoS_2 along with other materials such as graphite. These solid additives are not soluble in the lubricant base stock and require dispersion. Once entrained in the contact area, these solids can form a third body barrier between contacting asperities. If the solid particles contain an easily sheared plane, such as the case with MoS_2 and graphite, then it serves as a means of substituting the high friction of direct asperity contact with a lower friction preferred shear plane. These systems look attractive because they do not rely on a specific chemical reaction or adsorption mechanism to work. However, they do possess some significant disadvantages that limit their application. The first of these is stability in the oil. There is a strong tendency for the solids to agglomerate and separate out from the liquid phase. This creates problems in storage and during any equipment downtime. In addition, entrainment into the contact is not guaranteed, and under many conditions, the friction reduction benefits are not apparent. In some cases, these issues can be partially resolved using much smaller distributions of particle size. In recent years, there has been a rapid growth in interest in the use of dispersed nanomaterials. These include both carbon-based structures such as carbon nanotubes and graphene (thin sheets of graphite) and noncarbon materials such as tungsten carbide. The use of nanodispersed materials has been shown to reduce friction. However, the current lack of information on health risks for handling some nanomaterials has restricted commercialization in mainstream liquid lubricants. This situation may change in the future.

D. Friction in Concentrated Contacts

Contacts between solid bodies can be divided into two broad classes, as shown in Figure 19.8. Conformal contacts are often formed between two solids where the shape and size of the two opposing surfaces are relatively similar to one another. Examples include the various types of hydrodynamic journal bearings and bushings commonly found in many mechanical systems. Conformal contacts are also found in situations where one or both of the materials are highly elastic. Examples include vehicle tire–road surface contacts and elastomeric seals against rotating shafts. Nonconforming or concentrated contacts occur when the two surfaces have dissimilar geometries such as when a ball or cylinder is loaded against a flat surface. This important class of contact is found in many mechanical systems including rolling element bearings, gears, and engine valve train systems.

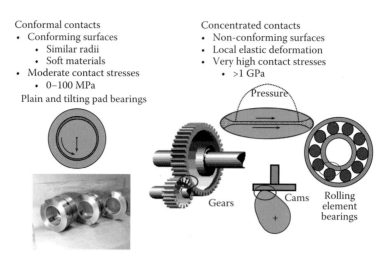

FIGURE 19.8 Comparison of conformal and nonconformal (concentrated) contacts and examples of where they are found.

The behavior of a lubricant within a concentrated contact can be very different than when it passes through a more conformal contact. The film formation process and generation of friction forces are governed by the properties and response of the lubricant at the much higher pressures than exist within conformal contacts. Figure 19.9 shows a cross section view of a lubricated concentrated contact. These are often referred to as elastohydrodynamic lubrication (shortened to EHL or EHD) contacts.

The distinguishing feature of nonconforming contacts is the high pressure that exists between the two surfaces. The point of contact between the surfaces occurs over a tiny footprint. Elastic strains result in localized conformity between the surfaces that spreads the load over a finite area. This contact region is often referred to as the Hertzian contact region after Heinrich Hertz, who studied the contact between elastic solids in the late 1800s. Despite the local conformity, the normal pressures generated in the contact region are very high and often exceed 1 GPa (1×10^9 N/m²).

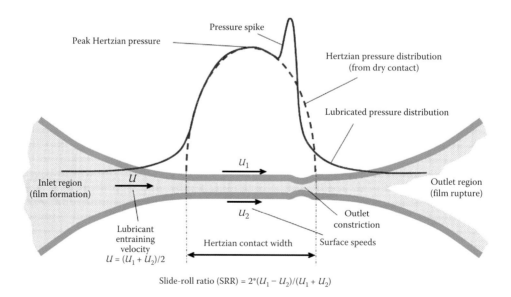

FIGURE 19.9 Cross section view of a concentrated elastohydrodynamic lubrication (EHL) contact.

Under normal hydrodynamic conditions, it would be extremely challenging for a fluid to enter into this high-pressure region to generate a separating lubricant film. However, many fluids, including the hydrocarbons used as lubricants, exhibit an increase in viscosity when subjected to increasing pressure. In the inlet region, it is the interplay between the increasing pressure, increasing fluid viscosity, and the hydrodynamic lubrication mechanism that allows a fluid film to be generated between the two surfaces and gives rise to the EHL name associated with this lubrication regime.

The formation of the film depends on how the viscosity of the lubricant responds to pressure. The relationship between the lubricant viscosity and pressure is often represented by a simple relationship:

$$\eta = e^{\alpha P} \tag{19.6}$$

where
η = dynamic viscosity
α = pressure–viscosity coefficient
P = local pressure

The pressure–viscosity coefficient of a fluid is determined by its molecular structure. For lubricants, the behavior is often dominated by the properties of the base stock. However, blending of additives and combinations of different base stocks all influence the pressure–viscosity effect.

The friction generated in a well-lubricated EHL contact, in which there is little surface–surface interaction, is due to the shearing of the lubricant within the high-pressure Hertzian contact zone. Figure 19.10 shows measured values of viscosity versus pressure for a low-viscosity PAO. Viscosity is plotted on a log scale for which Equation 19.6 would produce a straight line. Clearly, the fluid response departs from this simple model. Equation 19.6 remains useful for helping predict the formation of a lubricant film because the inlet region, which dominates the film-forming process, is at relatively modest pressure. However, serious errors occur if this relationship is applied in an attempt to predict fluid viscosity losses in the highly loaded portion of the contact. More accurate models exist that can produce a much better fit to the typical viscosity–pressure response. However, in order to predict friction response, it is necessary to use them as part of a complex computer simulation of the entire EHL contact.

A more direct approach is to physically measure the friction generated in a well-controlled EHL experiment or test. The method is often referred to as a Traction or EHL Traction measurement. It is

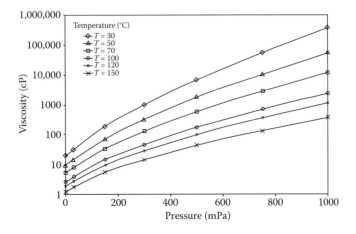

FIGURE 19.10 Results from viscosity measurements of a polyalphaolefin (PAO) conducted at different temperature and pressures.

based on reproducing an EHL contact in which the speed of both contacting surfaces is controlled separately. Popular configurations include a ball loaded against the flat surface of a disk and rollers and/or disks loaded against each other.

Figure 19.11 shows a typical configuration of a ball on flat disk style EHL traction measuring device. An EHL contact is reproduced between the ball and the flat surface of the disk. It is important that the individual speeds of the ball and disk, the load applied to the ball, and the test temperature are all well controlled. In order to measure only the fluid contribution to the friction in the contact, very smooth balls and disks are used. It is also essential to use speeds and fluid viscosities that provide good separation between the two surfaces. A typical lubricant traction curve is obtained by selecting an average entraining speed as defined in Figure 19.11 and selectively adjusting the ball and disk speeds to keep this constant while varying the sliding speed. When the speed of both surfaces are equal, corresponding to a slide-roll ratio (SRR, as defined in Figure 19.11) of 0, there is no relative sliding between the surfaces. Consequently, there is no shearing action on the fluid and there is no sliding friction generated in the contact. The speed of either the ball or the disk is then sequentially increased whereas the speed of the other component is reduced by equal amounts to obtain different SRRs. This generates a traction (i.e., friction) force within the contact that is measured by a suitable transducer. The traction coefficient is calculated by dividing the traction force by the load applied to ball.

Unlike boundary friction, the EHL traction coefficient varies with temperature, applied load, the sliding speed, and the size and shape of the contact zone. In addition, other factors such as shear-related heating along with material and fluid thermal properties influence the results. As such, unlike viscosity, traction measured under EHL conditions should not be considered a fundamental property of the fluid alone. Nevertheless, the measurement of fluid traction properties has proven to be an extremely useful tool for identifying and developing lubricants aimed at reducing friction and improving energy efficiency for systems operating under EHL conditions.

Figure 19.12 shows a series of traction curves generated at different peak contact pressures for a PAO lubricant. The results clearly demonstrate the influence of pressure on the traction coefficient. For reference, several typical examples of the range of contact pressures for different machine elements are included. Similarly, traction is most often observed to decrease as temperature is increased. Measurements such as these can be used to identify new materials exhibiting low traction, study blend effects, and characterize the performance of fully formulated lubricants.

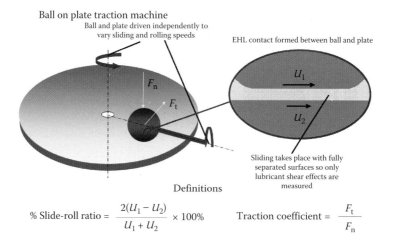

FIGURE 19.11 Typical configuration of a ball on plate EHL traction measuring system. The disk and ball are loaded against each other and their speeds are independently controlled to allow testing under different rolling and sliding speeds. The definitions of slide-roll ratio (SRR) and traction coefficient are also shown.

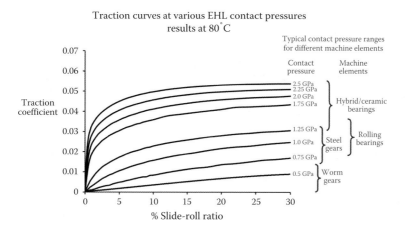

FIGURE 19.12 A series of measured EHL traction curves for a polyalphaolefin (PAO) fluid having a nominal kinematic viscosity of 100 cSt at 100°C. The results from 0.5 to 1.25 GPa were measured using steel components and the higher pressure measurements were obtained using a tungsten carbide ball and disk combination. The figure also shows typical peak pressure ranges for different types of machine elements.

Figure 19.13 shows the relative traction performance for a series of common lubricant types. Conventionally refined mineral oils are used in a wide variety of lubricants. Their traction behavior results from the complex mix of different molecular species that are present and are not optimized for low EHL traction performance. The use of base stocks from more complex refining processes and the use of synthetic base stocks can provide the means to achieve significant reductions in EHL traction. However, fluids that produce a low traction response typically also possess lower pressure–viscosity coefficients. This can result in slightly thinner EHL films when compared with a higher traction fluid running under identical conditions. Understanding the balance between these properties, and how they may affect performance in different machine elements, is important.

This section was a review of the fundamentals involved in the generation of different sources of friction within lubricated contacts. In operating machines, there are multiple components that generate friction via a number of different mechanisms. For example, in an internal combustion engine,

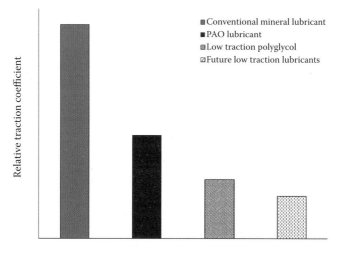

FIGURE 19.13 The relative EHL traction performance for common lubricant classes. New lubricant base stocks may become available to further reduce traction.

there are tens of individual component types and hundreds of different contacts formed between different moving surfaces. The friction losses are the sum of all these individual losses, and it can be challenging to determine which loss mechanism is dominant. In the following sections, different machine elements and systems will be examined as well as how reducing friction will help improve efficiency in these systems.

II. HYDRODYNAMIC FLUID FILMS

The various configurations and applications of hydrodynamic bearings have been reviewed in Chapter 8. They are available in a wide range of sizes, can be designed to support very high loads, and can operate over a very large speed range. They are most often used in any situation where it is necessary to support a rotating part and can be configured to resist radial loads, axial loads, and combinations of both. This section covers friction losses for some examples of industrial applications. The losses associated with hydrodynamic bearings in internal combustion engines will be reviewed in a later section.

A. Hydrodynamic Friction in Bearings in Industrial Applications

Bearing design and lubricant selection are often aimed at generating a fluid film that is capable of fully separating the bearing surfaces. For industrial applications, particularly large equipment, the design life is typically decades and the replacement cost is high. Consequently, the first consideration is ensuring durability rather than minimizing friction losses. Referring to Figure 19.3, this is equivalent to operating toward the right-hand side of the Stribeck curve—well away from the minimum point. However, the advent of more sophisticated design tools, smoother surfaces, and improved manufacturing accuracy have made it possible to more closely define desirable operating limits. This has allowed designers to become less conservative and has opened the door to operate nearer the minimum on the Stribeck curve. The result is that in many applications, and for new equipment designs, there is a trend toward using lower viscosity lubricants.

Despite being designed to operate in the hydrodynamic regime, it is almost impossible to avoid transitions into the boundary lubrication regime. The simple start–stop cycle guarantees a transition through the regimes. Moreover, any transient loads may momentarily result in some surface contact and boundary lubrication. These conditions are often short lived. Hence, the contribution toward total friction from boundary lubrication is typically small compared with the viscous losses. Also, many bearings use a combination of steel surfaces loaded against a softer, more compliant surface coating such as babbitt. Although recent results suggest that some friction modifiers can reduce boundary friction in hydrodynamic bearings, they are not really optimized for these materials, and their use in common industrial lubricants is rare.

The friction losses in fluid film bearings can be estimated using the hydrodynamic theory. The Reynolds equation shown in Equation 19.7 describes the relationship between fluid flow and pressure generated in the gap between bearing surfaces and balances them with the external applied force.

$$\frac{d}{dx}\left(\frac{h^3}{\mu}\frac{dp}{dx}\right) = 6U\frac{dh}{dx} \tag{19.7}$$

where
- x = measured in the flow direction of the bearing
- h = local thickness of the lubricant film
- U = speed of the bearing
- p = local pressure within the lubricant
- μ = lubricant viscosity

This one-dimensional form of the Reynolds equation works well for the so-called long bearings. Long bearings are those where the dimension in the axial direction is similar to or greater than the bearing diameter. The length/diameter ratio (L/d) is a nondimensional parameter often used to define bearing geometry. Equation 19.7 ignores any fluid flow in the axial or z direction. In short bearings, the side flow of fluid out of the bearing cannot be ignored and requires the inclusion of a second term, as shown in Equation 19.8

$$\frac{d}{dx}\left(\frac{h^3}{\mu}\frac{dp}{dx}\right) + \frac{d}{dz}\left(\frac{h^3}{\mu}\frac{dp}{dz}\right) = 6U\frac{dh}{dx} \tag{19.8}$$

Reasonable analytical approximations can be made for short bearings. However, closed-form solutions do not exist for the more general case, and it is necessary to solve Equations 19.7 and 19.8 numerically. Once a numerical solution is achieved, it is possible to calculate the friction loss in the bearing. At each point in the bearing, the fluid viscosity and local average shear rate is known. Using the relationships in Figure 19.2, it is possible to calculate a local shear stress that can be summed up over the entire bearing surface to calculate the total force required to keep the shaft rotating. This can also be expressed as a bearing torque and is a direct measure of the friction losses.

It is often not practical to arrange a full numerical solution for all bearings in a complete system. Fortunately, there is a wealth of published results from authors who have modeled different bearing configurations. One of the earliest, which came from Raimondi and Boyd (1958), provides response curves expressed in terms of nondimensional groups that allows their application to a wide range of bearings and operating conditions. Figure 19.14 shows their results for friction for bearings with different arcs (β) and $L/D = 1$. This, along with similar charts, forms the basis for bearing design

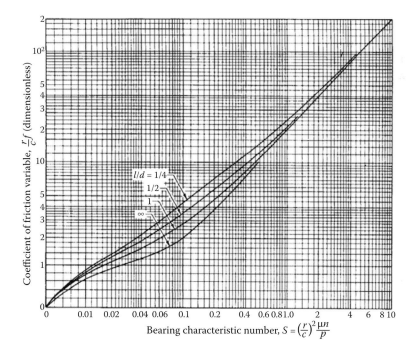

FIGURE 19.14 Results obtained by Raimondi and Boyd for the friction generated in bearings of different arc length β. The bearing characteristic number includes the bearing radius/clearance (r/c) parameter. (Courtesy of Society of Tribologists and Lubrication Engineers.)

and, coupled with the Stribeck curve, can be used to optimize bearing design and lubricant selection to minimize the viscous losses.

The approach just described has been available for many decades and provides a means to achieve at least an approximate optimization for the balance between durability and energy efficiency. However, within the model, there are assumptions that ignore effects that may help achieve even better optimizations. The simplest bearing models, based on Equations 19.7 and 19.8, assume that the lubricant is Newtonian (i.e., viscosity does not vary with shear rate) and also ignores the heating effect as the fluid is sheared in the contact. This localized heating changes the local fluid viscosity that modifies local pressure generation and flow. More sophisticated models simultaneously solve the Reynolds and heat energy equations and use viscosity models that account for the effects of local temperature and shear. These models require highly detailed knowledge of the design and material properties of the bearing and lubricant. They are typically used by bearing manufacturers and are not readily accessed by equipment users. However, the trend is toward increasing use of such tools to more fully explore the advantages of lubricants with different viscosity profiles and thermal properties such as different types of synthetic lubricants.

B. Measuring Bearing Friction

The friction coefficient for most hydrodynamic bearings is relatively small. However, as the amount of energy being transferred by the rotating shaft can be extremely large, particularly in some of the high-capacity industrial applications, even a small reduction in friction coefficient can add up to a substantial improvement in total energy consumed. Also, it is typical that multiple bearings are used, and the benefits are cumulative.

It is extremely difficult to obtain accurate measurements of bearing friction in operating equipment. Daily fluctuations in operating conditions and temperatures can lead to changes of a similar order of magnitude to the measured friction. Furthermore, the physical methods for direct measurement of bearing torque are not usually practical. Indirect methods such as monitoring fuel or energy use or the temperature rise due to friction can work if monitored over a sufficient period to reduce uncertainty. The measurement of temperature rise can either be on the component itself or as an increase in oil temperature between the bearing inlet and outlet. These methods can be useful if only a qualitative comparison between different bearings or lubricant is required. However, converting the data to an absolute measurement of friction loss is not easily achieved.

Under carefully controlled laboratory conditions, it is possible to obtain direct measurements of bearing friction. This requires running a fully loaded bearing under conditions closely matching field operation. This results in a requirement to apply and support high loads and a means to drive a rotating shaft. It is also necessary to have an oil circulation system and a means of heating and cooling the oil to maintain a target temperature. These combined requirements usually result in a relatively large and complex test stand. Some commercial test systems are available, but often it is necessary to develop a custom test unit.

One configuration used to measure bearing torque is shown in Figure 19.15. In this case, the reaction torque generated in the bearing is measured directly via a mechanical connection and load cells. In order to reduce any friction effects in the measurement system, the entire bearing unit is supported on a hydrostatic air bearing system capable of supporting the applied loads. Instrumentation is used to measure and control variables including oil temperature, flow rates, and pressures.

In bearing applications using the soft babbitt material, surface temperature can be a limiting factor. As temperature increases, the yield stress of the babbitt can reduce to a point where it begins to flow even under well-lubricated conditions. Thermocouples imbedded in the surface of the bearing are often used to map out local temperatures and can reveal details on how the fluid is behaving within the contact. Despite the complexity, such instrumentation is worthwhile and can be an aid in determining operating limits and associated lubricant properties.

Lubricant Contribution to Energy Efficiency

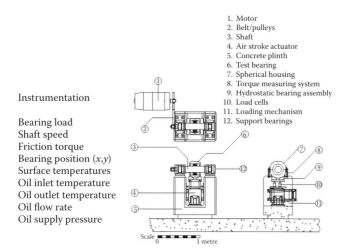

FIGURE 19.15 Hydrodynamic bearing test machine schematic. This machine is capable of measuring bearing friction torque.

The use of advanced synthetic lubricants that exhibit improved viscosity characteristics can yield advantages in bearing applications where the temperature varies. As shown in Figure 19.16, it is possible to match viscosity for two different lubricants at a single temperature. This can be selected to be a normal or critical operating condition. For a high viscosity index (VI) lubricant, the viscosity will be lower than the comparative lubricant for all temperatures below the reference temperature. This yields lower viscous and churning losses. The compromise is that the film thickness will be lower than the comparative case. However, the lubricant film is thicker than that obtained at the reference condition, ensuring that no adverse wear will occur. Departures in operation resulting in temperatures above the critical condition will result in thinner films for both oils. However, the effect is minimized for the higher VI lubricant.

Figure 19.17 shows test results comparing a conventionally refined, 95 VI mineral oil with a lower ISO (International Organization for Standardization) viscosity grade PAO synthetic oil with a higher VI. The results were obtained from carefully controlled tests on a thrust-loaded tilting pad bearing, which was run under a wide range of speeds and load conditions. The results compare the

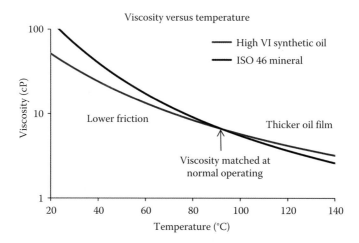

FIGURE 19.16 Viscosity versus temperature curves demonstrating how high viscosity index (VI) lubricants can reduce friction at lower temperature conditions.

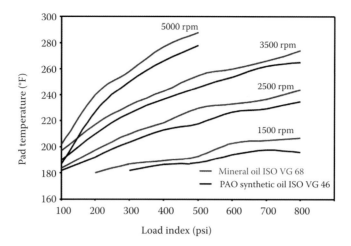

FIGURE 19.17 Temperature test results from a tilting pad thrust bearing. This compares the maximum pad temperature for a conventionally refined mineral oil (viscosity index [VI] ~95) with a lower viscosity polyalphaolefin (PAO) oil with a higher VI. Maximum pad temperatures were obtained using thermocouples imbedded in the tilting pad surfaces. (Courtesy of Kingsbury, Inc.)

maximum temperature measured on the surface of one of the tilting pads. Under all conditions, the lower ISO viscosity grade PAO lubricant produced lower surface temperatures. Direct bearing friction measurements were not made. However, because the thermal properties of the two oils are similar, it can be assumed that the reduced temperatures were due to lower viscous friction-induced heating effects of the PAO lubricant.

A similar series of results have been obtained for plain journal bearings tested using the apparatus shown in Figure 19.15. The absolute friction and power losses were measured for a series of bearings of different L/d ratios and clearance ratios. Figure 19.18 shows the results for one clearance ratio at all three L/d ratios tested. The contour maps were plotted using the measured data and show the reduction in power loss associated with replacing a conventionally refined lubricant with a lubricant that has a much higher VI. The nominal test temperature was set to be coincident with the point at which the viscosity curves cross, similar to that shown in Figure 19.16. Consequently, the lubricant film thickness generated by both oils was similar under the same conditions. The results show that under many conditions there is a benefit to using higher VI oils even when the nominal operating viscosity is constant. The improvements vary depending on conditions, but increase for long (i.e., high L/d) bearings. For lower L/d bearings, the improvements are less pronounced and in some conditions are even negative.

The exact mechanism behind the friction reduction is not fully understood. It may be in part due to pressure–viscosity effects. Alternatively, reduced viscous losses for the higher VI oil in the cooler unloaded regions of the bearing may account for the differences. This is an early example of where efficiency gains from lubricants may come from in the future. Although hydrodynamic bearings are relatively efficient machine elements, it is clear that there remain opportunities to improve efficiency. In the short term, improved maintenance practices and a higher awareness of appropriate lubricant selection can result in higher efficiencies. Over the long term, improved designs and development of new lubricants are likely to result in even more efficiency improvements.

C. FRICTION IN ROLLING ELEMENT BEARINGS

As their name implies, rolling element bearings use components that are designed to roll against one another rather than slide. Reducing the amount of sliding helps minimize friction and the term

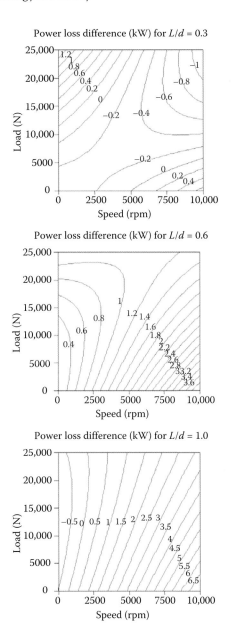

FIGURE 19.18 Test results showing the reduction in power losses obtained in different L/d ratio journal bearings at different speeds and loads when a high viscosity index (VI) oil replaces a lower VI oil.

antifriction bearing is often used. Despite their inherently low friction characteristics, there are sources of friction loss that can be further reduced to help minimize energy losses.

Rolling element bearings are available in a variety of different configurations. The two main classes are ball bearings and roller bearings. Chapter 8 shows the many configurations available in each of these groups. Their size can range between the micro scale examples of which would fit on the head of a pin to huge custom-manufactured bearings used in large-scale industrial and mining equipment. Their application range is also very large, and rolling element bearings can be designed to operate over a wide range of speed, load, temperature, and environmental conditions.

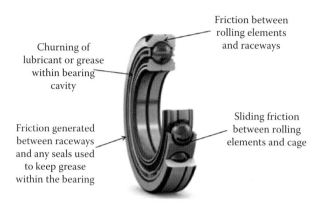

FIGURE 19.19 Major sources of friction generated in a rolling element bearing.

There are wide ranges of lubricants used in rolling element bearings. The class of lubricant is most often determined by the type of equipment the bearing is installed in. For industrial applications, where a liquid lubricant is used, the viscosity can range from low viscosity hydraulic fluids and spindle oils up to the very high viscosity lubricants used in open gear systems. However, majority of rolling element bearings are lubricated by greases and are found in large numbers in everyday household applications such as motorized kitchen equipment, pumps, and electric motors.

In spite of the wide range of configurations, applications and types of lubricants used, the energy loss mechanisms are similar. Figure 19.19 identifies some of the sources. The relative contribution of each varies with type, application, and operating conditions.

1. EHL in Rolling Element Bearings

The contacts formed between the rolling elements and the raceways operate under EHL conditions as described earlier in this chapter. Although rolling element bearings are designed to operate under rolling conditions, there is always some relative sliding or slip that occurs in the contact. In bearings using spherical rolling elements, the sliding that occurs within the contact is determined by geometry. Figure 19.20 shows the distribution of sliding for an EHL contact formed within a deep

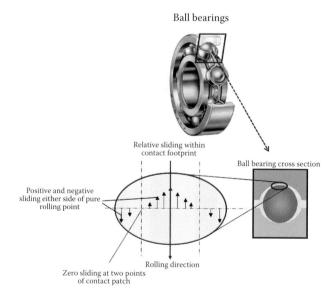

FIGURE 19.20 Regions of relative sliding that occurs within a deep groove ball bearing EHL contact.

groove ball bearing when running under nominal pure rolling conditions (i.e., no gross slip occurring between the balls and raceways). The bearing geometry only allows absolutely pure rolling to occur at two positions within the rolling element–race contact. Away from these locations, there is relative sliding that occurs between the two surfaces in both positive and negative directions. This effect occurs for all ball and spherical roller bearings. If gross slip occurs, that motion would be superimposed on top of the distribution shown in Figure 19.20.

In more complex geometries, such as angular contact bearings, the rolling elements may exhibit another source of sliding. Depending on the loading, the rolling elements can be forced to spin in a direction that is off-axis compared with their rolling direction. This adds a spin component within the contact that adds another relative sliding motion.

Irrespective of the direction and type, any relative sliding within the EHL contact represents a source of energy loss. Depending on the relative thickness of the EHL film and roughness of the two surfaces, the losses may be attributable to fluid shearing, boundary friction, or a combination of both. As described in Chapter 8, the ratio of EHL film thickness and combined surface roughness, as shown in Equations 19.9 and 19.10, can help determine which friction mechanisms are occurring within the contact. Rolling element bearing manufacturers often use an alternative value called the kappa (κ) value. This is a ratio of viscosities where the operating lubricant viscosity is divided by a reference viscosity calculated to give a nominal specific film thickness of one. Kappa and specific film numbers can be similar except that the former ignores the effect of pressure–viscosity effects. This can lead to significant differences, particularly when synthetic lubricants are used.

$$\lambda = h/\sigma \qquad (19.9)$$

and

$$\sigma = \sqrt{\sigma_1^2 + \sigma_2^2} \qquad (19.10)$$

where
λ = (lambda value) specific film thickness
h = EHL film thickness
σ_1 and σ_2 = root-mean-square surface roughness values for the two surfaces

For rolling element bearings, $\lambda < 1$ is assumed to be boundary lubrication, $1 < \lambda < 4$ mixed lubrication, and $\lambda > 4$ full EHL. True operating conditions may be higher than calculated because of the running-in effects that often reduces the surface roughness. Under full film conditions, the traction properties (Figure 19.12) of the oil are the most influential factors in determining the losses with the contact. Figure 19.13 shows the relative traction performance for a range of common lubricants. Lower traction synthetic oils generate less friction and heat and result in energy savings. At lower λ conditions, surface friction contributes to energy losses and can be influenced by friction-modifying additives.

2. Other Sources of Friction and Churning Losses in Rolling Element Bearings

The sources of friction in cylindrical roller bearings are a little different. Figure 19.21 shows the roller–race contact points for a cylindrical roller bearing. In the absence of gross slip, the line contact formed between the roller and raceway operates under pure rolling conditions, which results in very low friction losses. However, drag on the rollers from other friction sources will introduce small amounts of sliding. The contact between the end of the roller and the retaining rib on the raceway operates under high sliding and can be a source of friction loss. Friction is increased if the bearing is subjected to axial loads that force the contact to occur. This is a particularly demanding

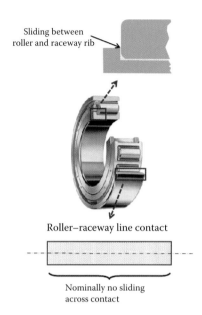

FIGURE 19.21 Roller–raceway contact points in a cylindrical roller bearing.

area to lubricate well and is sensitive to any geometric imperfections in the rib or end of the roller. Avoiding misalignment and use of high-quality bearings can help reduce friction.

The interaction of the cage and rolling elements is another source of friction that is common in most bearings configurations. Cage design varies by bearing type and materials used. Pressed steel cages are common but various other materials such as polymeric are also used. The rolling element–cage contact interaction is pure sliding and the friction generated depends on cage design, materials, and bearing type. Advice on selection of cage type for reducing friction should be obtained from the manufacturer.

Lubricant churning is also an energy loss mechanism present in rolling element bearings. The losses depend on operating speed, lubricant viscosity, and amount of oil in the bearing cavity. Unlike hydrodynamic bearings, it is unnecessary to fill the bearing cavity with lubricant. Adequate lubrication can occur with a very small volume of lubricant present. Lubricant flow rates are often determined by cooling requirements rather than supplying enough oil to form lubricant films. Churning losses in rolling element bearings are often small. However, it should be remembered that many rolling element bearings are used in equipment where other components such as gear wheels and pinions will contribute significant churning losses to the system.

Grease-lubricated bearings represent a special case. The grease remains in the bearing cavity, and the grease thickener contributes to the overall properties of the lubricant. In addition to the grease properties, the amount of grease used in the bearing cavity has a large impact on energy losses. An overfilled bearing will create high churning losses and higher running temperatures. However, too little grease may result in oil starvation. The optimum grease charge depends on many factors, and recommendations from grease and bearing suppliers should be followed.

Recently, both bearing and lubricant manufacturers have been focusing on reducing friction losses in rolling element bearings. New greases based on advanced synthetic base stocks and different thickener soaps are becoming more widely used. Bearings using new materials and coatings, improved surface finishes, friction-reducing cage designs and materials, and even advances in seal technology all contribute to reducing energy-sapping friction. We can expect a continuous improvement in bearing technology aimed at reducing their contribution to machine and system energy losses.

Lubricant Contribution to Energy Efficiency

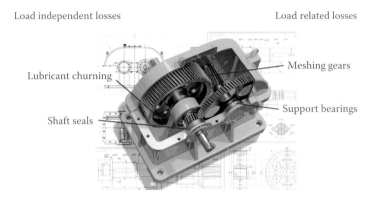

FIGURE 19.22 Sources of friction losses in a typical enclosed gearbox.

D. Friction in Gears and Gearboxes

Gears are used to transmit motion between different components in a system. Their primary functions are to change the relative speed, to alter direction of movement, or both. Unlike bearings, which are generally used to support loads while allowing rotation, gears are often part of the load and energy flow path within a system. Hence, it is easy to relate the efficiency of a gear system directly to a loss of energy available as output. The many different types of gears have been covered in Chapter 8. The two most basic forms are enclosed gears, in which all components are housed within a single casing, and open gears, which operate without any casing.

Figure 19.22 shows a parallel shaft gearbox and lists some of the sources of friction loss. These can be characterized as losses that will vary with the load transmitted through the gearbox and those that are essentially independent of load. In some cases, such as lubricant churning, the losses vary with speed.

1. Friction between Gear Teeth

Most gear teeth are based on an involute profile that ensures a constant rotational speed throughout the meshing cycle. Figures 8.53 and 8.54 show a typical meshing sequence that occurs between two gear teeth. Although there are some geometric differences between all the different gearbox configurations, the basic action is similar. A key feature is that the amount of sliding between the gear teeth varies as the contact progresses across the two contact surfaces. For most gear contact cycles, the sliding will start at a maximum value, reduce to 0 at the point where the contact occurs at the pitch line, and then increases again as the contact point progresses until the gear teeth are no longer in mesh. An EHL film is formed in the concentrated contact between the gear teeth in the same manner described earlier in this chapter. Between meshing gears, there is significant shearing of the highly pressurized oil as it passes through the contact. Figure 19.23 shows how the various stages of sliding map onto example lubricant EHL traction curves. The amount of sliding and values of traction will depend on gear design and operating conditions. However, the contact spends a significant amount of the meshing period on the high sliding parts of the traction curve, and the overall meshing losses can be equated to the shaded area in Figure 19.23. Reducing traction has the additional benefit of reducing losses in the rolling element bearings that are used to support the gear shafts. The cumulative effect of reducing traction often results in substantial improvements in overall gearbox efficiency, and there are continuous efforts to develop lubricants with lower traction properties.

Similar to rolling element bearings, the specific film thickness between gear teeth provides an estimate of the mechanism of lubrication. Because of manufacturing and machining complexities, gear finishes are often rougher than found on rolling element bearing surfaces. Therefore, many gears operate under nominal mixed or boundary lubrication conditions. Friction modifiers may help reduce meshing friction in some cases, although those based on surface adsorption mechanisms are

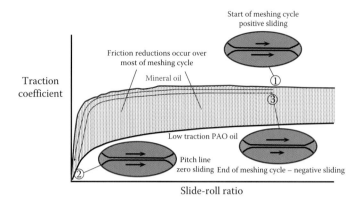

FIGURE 19.23 The sliding within a gear EHL contact progresses throughout the meshing cycle. The shaded area indicates how low traction oils can reduce friction over most of the meshing cycle.

unlikely to survive the combination of high contact stresses and aggressive sliding that occurs in gear contacts.

Parallel shaft gears are remarkably efficient. Under full load, each mesh typically loses less that 1% of the power transmitted across the interface. The efficiency of a complete gearbox is often in the high 90% range, particularly at high loads where the impact of churning losses represents a lower fraction of the overall losses.

Direct measurement of gear efficiency requires specialized equipment. Accurate measurement of efficiency requires that the difference between power input and output needs to be quantified. Because parallel shaft gear efficiency is high, the result is buried in a small difference between two large numbers. The use of very accurate load and speed transducers can provide useful results, but any signal or response noise can limit the overall accuracy. Another approach is to use the four-square design, in which two sets of gears are loaded against one another. In this case, the power is not transmitted outside of the system but load is recirculated within a loop formed between the two sets of gears. It is then possible to directly measure the force or torque required to rotate one of the gear shafts, which provides an estimate of the friction force that is being generated in the two gear systems. This is obviously a more complex testing setup and is generally limited to specialized test facilities.

Gearboxes that have offset shafts such as hypoid gears have a much higher sliding component between the meshing gears. The efficiency of these types of gears is often significantly less than that of parallel or regular bevel gears. Rear axle gearboxes used on commercial vehicles and some light duty trucks and passenger vehicles are examples where this hypoid gear configuration is used. Dynamometers are often used to measure the overall efficiency of automotive axle units.

Worm gears (see Chapter 8) are another class of gears that exhibits very high sliding. The most popular configuration is a steel gear, or worm, that acts against a bronze wheel. The worm can be placed either above (worm over) or below (worm under) the wheel and, in some cases, the wheel can be steel. Worm gears have the advantage in that very high gear ratios can be obtained in a single gear mesh. This design results in a small and simple unit compared with either a multiple shaft or planetary system that would be necessary to achieve the same gear ratio using conventional steel gears. The penalty is that worm gears are typically considerably less efficient in part owing to the high sliding friction that occurs in the worm and wheel teeth contacts as shown in Figure 19.24.

The contact pressure between the steel worm teeth and bronze wheel teeth is relatively modest compared with steel-on-steel contacts. Despite this, the lubricant pressure–viscosity effect still plays a dominant role, and worm gear efficiency can be significantly improved using lubricants with low EHL traction. Figure 19.25 shows the measured efficiency using oils exhibiting different traction properties. Comparing these results with the traction performance shown in Figure 19.13,

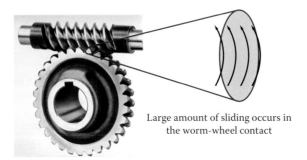

Large amount of sliding occurs in the worm-wheel contact

FIGURE 19.24 Worm over wheel configuration for a steel worm meshing with a bronze wheel. The high sliding that occurs in the gear contact is a source of friction loss for these types of gear.

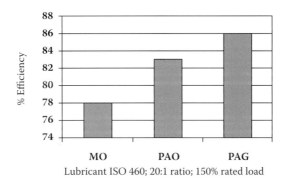

Lubricant ISO 460; 20:1 ratio; 150% rated load

FIGURE 19.25 Efficiency measurement results from tests conducted on a worm gearbox using lubricants with different EHL traction properties. EHL, elastohydrodynamic lubrication; MO, mineral oil; PAG, polyalkylene glycol synthetic; PAO, polyalphaolefin synthetic.

there is a significant efficiency gain in worm gears for low traction polyalkylene glycol (PAG)-based lubricants. PAG lubricants are finding increasing use in worm gear systems because of this advantage. Many of these worm gears are filled for life and are not exposed to some of the incompatibility risks that PAG lubricants have with hydrocarbon-based lubricants such as those using mineral or PAO base stocks. For this reason, PAGs, despite their low traction and potential efficiency benefits, are not routinely used in other gear systems where exposure to other fluids is more likely.

The efficiency improvements highlighted in Figure 19.25 yield several additional advantages. The reduced friction losses can often lead to lower operating temperatures. This can help improve oil life and, in systems that are thermally limited, can allow the unit to operate at a higher load condition. This has led to the possibility of using smaller worm gear units for a given application.

2. Churning Losses in Gears

Enclosed gears operate with a lubricant contained within a sump. The spinning shafts and gears churn the oil, which can be a significant loss mechanism. For example, light truck and passenger vehicle gears and axle systems spend much of their time under partial or low load operating conditions. This may also coincide with higher speeds, such as motor or freeway driving. Such conditions result in increasing the relative contribution of churning losses. When tested over a typical automotive use cycle, the churning losses in a vehicle driveline system often provide a similar impact to efficiency as do friction losses in the gear contacts.

One measure that can be used to reduce the losses is to scavenge the oil out of the sump and use a return system to inject the oil where it is needed. This design is sometimes used in high-end, niche applications such as motorsports but results in greater system complexity as well as high engineering and packaging costs. A more cost-effective approach is to use lower viscosity/higher VI lubricants. This must be balanced against the possible durability risks associated with the thinner lubricant films generated in the gear contacts. However, improved gear finishing and manufacturing techniques are already enabling reductions in recommended lubricant viscosity for many systems. This trend is likely to continue as a means of improving operating efficiency of many gear systems.

It is difficult to generalize how much churning losses contribute to industrial, marine, and other gear systems. Systems that operate at high load for much of their duty cycle will have a lower proportion of churning losses compared to systems operating under mixed duty cycles. Nevertheless, the same mechanisms apply, and we can expect a trend toward lower viscosity lubricants for those systems in which a reduction in churning losses will yield a tangible efficiency improvement.

III. FRICTION LOSSES IN HYDRAULIC SYSTEMS

Chapter 7 describes the basic concepts of hydraulic systems and their various configurations. Hydraulic systems transmit force and movement through a fluid medium usually referred to as a hydraulic fluid. Hydraulics represents one of the largest and broadest classes of industrial systems and can be found in mobile equipment, manufacturing, process industries, and all classes of transportation vehicles. Because of the number and ubiquitous nature of hydraulic systems, any gains in efficiency can have a significant overall impact on a country's overall emissions and energy use. Recently, there has been an increased effort to develop more energy-efficient hydraulic systems that will likely result in improved designs and higher-performing fluids.

A. Sources of Friction in Hydraulic Systems

There are many examples of hydraulic circuits shown in Chapter 7. Most will consist of one or more pumps, and may contain various junctions, valves and filters, a fluid reservoir, and the hydraulic actuators. Energy losses can occur throughout the system but can be broadly separated into losses at the pump and losses in the circuit.

Losses within the circuit will occur as a result of fluid flowing through the various pipes and components within the system. In pipes, the losses are directly related to fluid viscosity. Careful selection of piping and configuration can help reduce these losses, and there is now an increased emphasis on system design that can help reduce these losses. These flow-related losses result in a reduction in pressure within the fluid that reduces the amount of force and power available as output. Other system components can also create pressure reductions, some related to viscosity and others due to generation of turbulence and other flow effects. These losses depend on system design and fluid properties. Routinely, the hydraulic fluid properties at operating temperature and ambient pressure are used to estimate flow-related losses. Reducing fluid viscosity will reduce flow-related losses. However, system leakage and pump viscosity requirements often set minimum viscosity requirements. At typical hydraulic system pressures, there is a small pressure–viscosity effect. However, as system pressures increase and a greater variety of fluid options are considered, there may be some opportunities to improve overall flow efficiency by careful fluid selection and better system optimization.

The pump is the other major source of hydraulic system losses. Figure 19.26 shows an axial piston pump and highlights the two major loss mechanisms: volumetric efficiency and mechanical efficiency. Mechanical efficiency is simply a function of all the friction losses that occur between the pump's moving parts. These occur for all pump types and will be related to both viscous and boundary friction losses. The combined mechanical losses will most often be reduced as viscosity is reduced. However, this improvement is limited to when the boundary friction contribution becomes dominant as lubricant films become thinner.

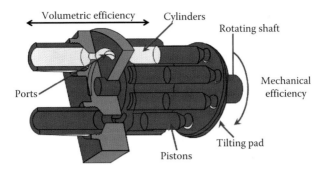

FIGURE 19.26 Losses in an axial piston pump can be divided into volumetric efficiency and mechanical efficiency. These losses also occur for other pump configurations.

Losses that affect volumetric efficiency occur during the pressurization of the hydraulic fluid. In any pump, there are leakage paths that allow the fluid to escape from the pressurized side of the pump back to the unpressurized part. The energy that has been used to pressurize the leaked fluid is lost and results in a reduction in pump efficiency. Fluid compressibility also has an impact on volumetric efficiency.

Figure 19.27 shows a typical pump efficiency curve as a function of fluid viscosity. The combination of mechanical and volumetric losses result in a peak in the overall efficiency and an optimum viscosity range. The shape of the curve and the position of the optimum range depend on pump design, manufacturing tolerances, surface finishes, and materials.

1. Hydraulic Fluid Selection

For optimum efficiency, it is necessary to select a hydraulic fluid where the viscosity coincides with the ideal range shown in Figure 19.27. If a high VI fluid is used, the effect of temperature variation can be minimized because it will remain closer to the optimum condition. Furthermore, when coupled with low friction and traction properties, it can result in significant efficiency improvements when compared with conventional fluids. Figure 19.28 shows an example of the gains measured during tests performed on a hydraulic vane pump. Replacing an ISO 46/100 VI conventional hydraulic fluid with an advanced high VI oil with low EHL traction yielded efficiency gains of between 3.5% to more than 6%, depending on operating conditions. As noted before, efficiency gains will depend on many factors, but it is clear that measurable improvements are well within reach by careful fluid selection. Some care needs to be exercised when evaluating high VI fluids. One method of gaining

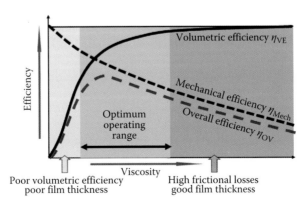

FIGURE 19.27 Pump efficiency versus fluid viscosity curve showing combination of volumetric and mechanical efficiency effects.

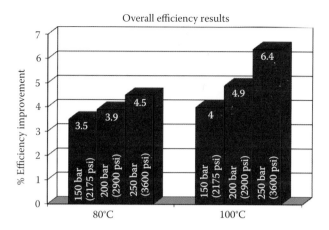

FIGURE 19.28 Results from vane pump testing showing measured efficiency gains obtained by replacing a conventional ISO 46 hydraulic fluid with 100 viscosity index (VI) with an advanced high VI hydraulic fluid with low EHL traction.

high VI is to use polymeric viscosity modifier additives. Some of these polymeric viscosity modifiers are based on high molecular weight materials, which are very effective at increasing VI but have poor mechanical stability. This will result in a reduction of bulk viscosity, because the polymer is broken apart as it is exposed to the high shear zones within the hydraulic circuit. High VI fluids that are shear stable offer the best solution for achieving efficiency gains while maintaining fluid performance throughout their useful life.

Controlled dynamometer and field tests have shown that the efficiency benefits shown in Figure 19.28 can yield similar fuel economy benefits in mobile equipment applications. These are significant improvements that provide both operational cost savings and related emission reductions. Given the potential impact on global emissions and energy use, continued improvements in hydraulic system efficiency from future developments in new system components and fluids can be expected.

IV. VEHICLE AND INTERNAL COMBUSTION ENGINE EFFICIENCY

Vehicle fuel economy and emissions receive ongoing attention from regulators worldwide. Reduction in fossil fuel dependence and reducing greenhouse gas emissions are the key drivers that are likely to remain in place for the future. The impact of CAFE standards in the United States and the emissions-based standards used in Europe have already been mentioned. These are just two examples of a global network of standards already in place. The common factor is that all of them are likely to set ever more stringent targets on vehicle emissions and fuel use.

A. Energy Use in Vehicles

Although all forms of motorized transportation consume fossil fuels, passenger cars receive the greatest amount of attention from regulators and the society at large. Although there are examples of all-electric passenger vehicles and use of other energy sources, the internal combustion engine powers the vast majority of passenger vehicles. It is widely understood that, although vehicle architecture and design changes will occur, the internal combustion engine will continue to be used in the foreseeable future as the primary source of energy conversion in passenger vehicles and trucks.

The internal combustion engine converts the chemical energy of fuel into mechanical work through the process of combustion. Most engines are based on either the Otto cycle for gasoline or the diesel cycle for diesel engines. Considering the vehicle as a whole, there are a large number of

Lubricant Contribution to Energy Efficiency

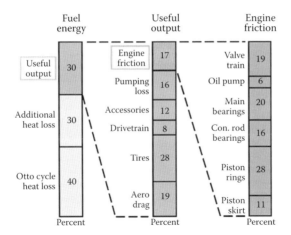

FIGURE 19.29 Example distribution of fuel energy use in a gasoline engine-powered passenger vehicle.

energy conversion and power transfer steps that lie in the path between the fuel and the ultimate propulsion of the vehicle. Figure 19.29 shows how the fuel energy is used to propel a gasoline-powered passenger vehicle. The analysis is based on literature data and was calculated for a fully warmed up vehicle traveling at 40 miles/h (64 km/h). The relative contributions vary based on vehicle type and operating conditions, and will be different for more modern vehicles than those included in the analysis. However, the major sources of losses are common to most vehicle types. Similar estimates are available in various public sources from government agencies and National Research Laboratory reports.

A relatively small amount of the available fuel energy is turned into usable output from the engine. The thermodynamic cycle and heat losses together form the greatest proportion. These are unavoidable and are dictated by the physics of the cycle or the need to maintain temperatures in a range at which the engine materials can operate. However, they represent areas where large gains can be made in fuel economy and will drive many of the engine and drivetrain changes that will occur in the future. The remaining available output is further distributed into powering various systems and overcoming multiple loss mechanisms. The losses due to internal friction within the engine is a relatively small amount of this segment and is further distributed between the different components of the engine.

In order to achieve the aggressive efficiency and emission targets, it will be necessary for vehicle manufacturers to capture the cumulative benefits of reducing losses from multiple sources. The majority of these will not be aimed directly at engine friction reduction. However, many will affect the lubricant and lubrication systems and may require a different mix of performance properties for future lubricant products.

B. Trends in Automotive Design Impacting Efficiency

One of the biggest sources of energy loss is the inefficiency of the thermodynamic cycle. For passenger vehicles, this is further compounded by the fact that, for much of the time, the engine is forced to operate under inefficient partial throttle conditions. Significant fuel economy gains can be made by improving thermodynamic efficiency and increasing the time spent under more advantageous engine operating conditions.

1. Engine Trends

A common trend among multiple passenger vehicle manufacturers is turbo downsizing. This enables a smaller engine to deliver similar performance to a larger, naturally aspirated counterpart.

For example, the common V6 engine configuration can be replaced by turbocharged four-cylinder engine while still maintaining the performance that consumers have come to expect. Turbo downsizing results in multiple benefits that account for its popularity. The increased compression ratio improves thermodynamic efficiency and recycles some of the energy contained in the exhaust system that would otherwise be lost. In addition, the sweet spot for efficient operation can be increased, resulting in an increased proportion of time operating under advantageous conditions. Despite the addition of a turbocharger unit, the overall size and weight of the power plant are reduced and can yield further improvements in aerodynamic and vehicle weight-related losses. The system is more complex and the turbo unit that operates at high speeds and temperatures places a significant additional demand on the lubricant. Good oxidation and deposit control are essential properties for engine oils used in turbocharged engines.

GDI is a fueling strategy that is becoming increasingly popular and is often coupled with turbo downsizing. As its name implies, in GDI engines the fuel is injected directly into each cylinder rather than into a manifold upstream of the cylinder. It requires the use of individually controlled high-pressure injectors. The main benefit is derived from the ability to precisely control fueling using the engine management system. Careful mapping of fuel injection amount and timing against engine operating conditions provides considerable fuel economy benefits. Furthermore, some of the partial throttle problems are avoided because power is controlled via fuel management rather than throttling the air/fuel mixture before it enters the cylinder. Direct injection has been used in diesel engines for many years, and the gasoline version shares with them the problem of soot generation. Soot can be generated because of the uneven burning of the fuel owing to the incomplete mixing of the fuel and air within the cylinder. The soot can find its way into the engine oil with concerns being expressed on the potential effect on wear of some engine parts.

There are a number of approaches that are aimed at varying the exhaust and air intake strategies. In a conventional engine, the valves that control these processes are driven by cam fixed shafts. This sets the sequencing and amount of valve opening and closing to a fixed pattern for all engine operating conditions. A number of systems are designed to provide a means to change valve behavior based on engine operation and include variable valve timing, continuously variable valve timing, variable valve lift (VVL), and valve phasing as examples. Each of these can affect the valve operation in different ways but all aim to provide an extra degree of control of valve operation during different phases of engine operation. This can reduce air pumping losses, help maintain temperatures, manage available power, and reduce emissions. In some cases, it can even adjust the thermodynamic cycle. An example of this is conversion from the normal Otto cycle to the Atkinson cycle that is achieved when the inlet timing is extended, allowing some of the intake air to return into the manifold. Adjustable valve systems require more complex cam and associated drive systems, which are frequently controlled through hydraulically operated mechanisms. This introduces a requirement that the engine oil also needs to perform as a hydraulic fluid with management of air entrainment and separation becoming important attributes.

2. Automotive Transmission and Powertrain Trends

These trends all deal with improving the conversion of fuel to available power output. Managing the power more effectively is another complementary approach. One method is to enable more freedom to operate the engine in its optimum range. This can be achieved by providing more gear ratio options between the engine and drive wheels. In passenger vehicles, there has been a steady increase in the number of forward gear ratios in both manual and automatic gearboxes. Nine-speed automatic transmissions are already in use, and even more ratios may be available in the future. Fuel economy gains of 6–16% have been claimed when compared with conventional four-speed options. Automated manual transmission (AMT) and dual-clutch transmission (DCT) offer alternatives to the use of a torque converter and can yield additional efficiency gains.

The ultimate control over engine speed is to use a CVT. Those systems based on drive belts have been available for some time but have torque limits that confine their use to low- and moderate-power output engines. Infinitely variable transmissions (IVT), based on passing power across EHL films, have shown some initial promise. Various configurations using toroidal components have been researched and included the possibility of a geared neutral that would remove the need for a torque converter. Interest in these systems has waned recently, which may be attributable to the difficulty in simultaneously optimizing a fluid to possess both the high EHL traction properties required to transmit power across an interface with the need to have reasonable VI characteristics for low-temperature fluidity. This is a good example of a new component technology that required a parallel development in fluid technology. IVTs are in use for nonpassenger vehicle applications, but at this point the CVT and AMT options appear to have the advantage for passenger vehicles.

3. Other Vehicle Trends

Another method of altering engine duty cycles is use of hybrid systems. The simplest of these is a simple stop–start system that shuts the engine down when the vehicle is at rest and restarts it when necessary. This saves fuel on city driving cycles for some added complexity. A more robust starting system is required, and the engine components must now go through about a factor of 10 times more stop–start cycles. In some cases, this has led to the use of polymer coatings on bearings to help reduce wear. Stop–start engine protection may become a desired feature for future engine oils.

Further hybridization involves coupling the engine with another energy source and/or energy storage system. These take various forms using mechanical, hydraulic, and battery-based systems. Battery-based systems are currently the most common hybrid energy sources used for passenger vehicles. These battery-based hybridization configurations seek to capture and recycle the energy lost through braking. Some allow for precharging the battery, which adds to the onboard available energy at the start of a journey. Electric hybrid vehicles have seen rapid growth and will no doubt continue to be part of the new vehicle mix. However, it remains a relatively small amount of the total new vehicle output and may be limited because of their increased initial cost compared with the actual improvements in fuel economy.

Duty cycles for most commercial hybrid vehicles appear to be relatively moderate. At this stage, it does not appear that any special new lubricant performance features will be required. However, some of the trends discussed in the next section to help reduce engine friction can be equally applied to engines used in hybrid vehicles.

The remaining areas for vehicle-related efficiency gains are unlikely to influence engine oil technology. Improvements in tire and aerodynamic losses clearly fall into this category. Also, many of the ancillary components, such as the water pump and generator, will switch to an on-demand approach rather than a direct coupling to engine speed via a drive belt. This approach may affect the choice of grease technology for use in the various bearings in these components, but this is an open question at present.

C. Engine Friction Reduction

Despite its relatively modest contribution to the overall distribution of losses, there is considerable ongoing effort to reduce engine friction. Simply reducing engine speed reduces the viscous churning loss. This can be achieved by using more advanced transmission systems as well as the use of turbocharging to some extent. Use of lower viscosity oils also reduces churning and other viscous losses over the entire speed range. There has been a continuous shift toward lower viscosity grade engine oils for passenger vehicles. The once popular 10W grades were originally supplanted by 5W and, more recently, by 0W grades. As an example, Society of Automotive Engineers (SAE) 0W-20 is now a commonly recommended viscosity for engine oils for new gasoline-powered vehicles. The SAE J300 engine oil viscosity specification has been recently extended to include lower viscosity grades down to 0W8. Table 3.1 shows the complete list of SAE viscosity grades and their viscosity limits.

Reduction in viscosity has been less dramatic for commercial vehicle engine lubricants. SAE 15W-40 remains a mainstream viscosity grade. However, there is clear evidence that fuel economy for commercial vehicle engines can be reduced by using lower viscosity oils such as SAE 10W-30 or 5W-30. Figure 19.30 shows the results from a series of engine tests conducted using different viscosity commercial vehicle engine oils. The fuel economy gain over a reference SAE 15W-40 viscosity oil is plotted against the high temperature, high shear viscosity. Over the range tested, there is a progressive improvement in fuel economy as viscosity is decreased. Maintaining engine durability when lower viscosity oils are used remains the primary concern for original engine manufacturers and operators. However, it is anticipated that future commercial engine oils will also follow a downward viscosity trend in order to capture the potential fuel economy gains.

Most engine systems will exhibit improved fuel economy when using lower viscosity lubricants. Figure 19.31 (from Tellier et al., 2013) shows another example. Testing of a General Electric VGF F18GL natural gas engine revealed statistically significant gains of 1.1% and 1.3% for SAE 30 and 20 oils, respectively, compared with a SAE 40 oil that is in current use.

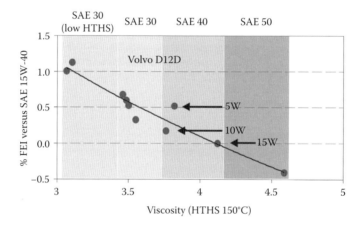

FIGURE 19.30 Commercial vehicle engine test results conducted on different viscosity engine oils. Efficiency is plotted against the measured American Society for Testing and Materials (ASTM) D5481 high temperature, high shear (HTHS) viscosity.

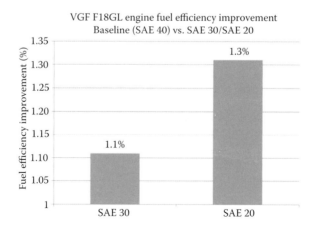

FIGURE 19.31 Fuel efficiency increase experienced in a General Electric VGF F18GL natural gas engine when using lower viscosity Society of Automotive Engineers (SAE) 30 and 20 gas engine oils compared to the reference SAE 40 viscosity oil.

Lubricant Contribution to Energy Efficiency   471

The practical limit for viscosity reduction will be different depending on the hardware and operational cycle. As shown in the Stribeck curve (Figure 19.3), there is a point at which lower viscosity will result in an increase of boundary friction that limits further friction reduction. Also, other limiting factors, such as the volatility of an engine oil, may set the lowest practical viscosity that can be used.

The addition of friction modifiers can help reduce the impact of boundary friction and extend the lower limit for improving fuel economy using low viscosity lubricants. The piston–cylinder liner contacts and engine valve train components are examples of engine components that can benefit from use of surface active friction modifiers. Another approach is use of alternative materials and coatings, which will be discussed later.

D. Measuring Fuel Economy

For passenger vehicles, there are a number of standardized methods for quantifying fuel economy effects. The Sequence VIE engine test is the method that is proposed for use in the new ILSAC (International Lubricant Standardization and Approval Committee) GF6 engine oil specification under development in the United States. The test method is fully described in ASTM (American Society for Testing and Materials) D7589. It uses a standardized V6 engine and requires reference runs to account for any drift in engine performance and to establish a baseline to calculate the fuel economy performance. The fuel consumption is measured twice in six-stage test sequences covering a range of different operating conditions. In the first test, sequence results are obtained for the new oil. This is followed by an in-engine aging step that simulates approximately 6500 miles of use on the oil. The same six-stage test sequence is then run on the aged lubricant. The fuel economy performance for the new and aged oil sequences is then calculated separately using different weighting factors for each of the stages. Engine oils will need to meet viscosity grade-dependent limits in order to pass and qualify for the ILSAC GF6 designation.

In Europe, the New European Driving Cycle is used to quantify vehicle emissions. The test method uses a complete vehicle that is run on either a flat road or a stationary roller test system capable of simulating aerodynamic drag. The test is started using a cold vehicle (20–30°C), which is then driven through a sequence of four urban driving cycles followed by an extra urban driving cycle. The test sequence captures the warming up effects associated with a relatively cold start and has been shown to favor use of lower viscosity lubricants. Since the test was originally conceived, vehicle power and driving styles have changed, and there have been concerns raised on the ability of the procedure to reproduce results relevant to current vehicle use. A harmonized world light vehicles test procedure (WLTP) is under development as a likely replacement.

Measuring fuel economy for commercial vehicles and their engines provides a different set of challenges. Engines can be bench-tested and vehicles may be placed in chassis dynamometers. However, their complexity and cost are much higher compared with passenger vehicles. The SAE J1321 fuel economy test provides standardized cycles for testing heavy-duty trucks. The procedure can be conducted on either a test track or an open road. Multiple repeated tests are necessary to obtain results that are statistically relevant.

These test methods are based on testing a complete engine or vehicle system. This is useful for qualifying an engine oil against a reference engine oil or quantifying overall fuel economy performance. However, as shown in Figure 19.29, the friction response of the complete engine or vehicle system is the sum of multiple friction sources occurring in different components. Because the mechanisms and conditions differ, it is very difficult to gain an understanding of how friction is being impacted at the individual component level. This is particularly the case for lubricant effects because a change in lubricant composition and performance will likely have different impacts on each of the components.

A breakdown of the distribution of engine friction can be obtained using one of two methods. The first is to instrument an engine to isolate the various friction forces. Examples include use of strain gauges on piston connecting rods or use of floating cylinder liners to extract the instantaneous friction between piston and cylinder liner. This can provide real-time data, but any inertia forces must be factored out in order to quantify the friction-related forces. Moreover, it is extremely difficult to measure more than a few components at a time. Another approach is to use motored friction testing. In this case, the engine is not fired but is driven by an external motor system. The torque required to run the engine at different speeds is then a measure of the internal friction. If the engine is fully configured, it is important to remove the pumping losses associated with drawing the air into the engine and pumping it out. This can be achieved using an in-cylinder pressure transducer to map out the pressure response over a complete engine cycle. Once the baseline friction response is set for the complete engine, it is possible to sequentially remove engine component groups such as the cylinder head, piston and connecting rods, and crankshaft. The contribution of each group is then subtracted from the previous set of results. This method can quantify the contribution of friction for different sets of engine components to help quantify the breakdown shown in Figure 19.29. It is possible to run such tests over a range of conditions while using multiple lubricants. A drawback of this method is that the engine is not fully loaded and the temperature distribution of a fired engine is not reproduced.

An improved understanding of engine friction breakdown is useful to both the engine designer and those looking to reduce the lubricant-related friction. At the most basic level, it is possible to identify the components that contribute the most to overall friction losses and how the distribution varies for different operating conditions. From the lubrication perspective, it can help determine which lubricant properties are most important to reduce overall losses.

Engine tests are both costly and complicated to set up and run. Detailed studies are more readily accomplished using bench tests designed to simulate different engine components. Reciprocating piston rings loaded against cylinder liners, single-engine bearing tests, and isolated cam shaft tests are common configurations. Results obtained from a single-engine bearing test show a clear transition from the hydrodynamic into the boundary lubrication regime. Because the response is isolated to a single-engine component, these types of test are easier to interpret compared with a full-engine test. They also allow testing of experimental lubricant compositions that would be impossible to run in full-engine tests because of the requirements associated with running an oil in a full engine. In this example, a shift in response is attributable to the use of shear thinning polymeric viscosity modifiers.

Engine component tests do not always simulate the true engine environment. However, they do use the correct materials and are a step closer than many of the common and standardized test procedures, and represent a good compromise between simpler tests and full-engine testing. Because they are less costly and time-consuming to run, they often enable larger parametric studies than is practical in full-engine tests. Furthermore, the improved control over conditions and measurement can provide more repeatable results. Engine component tests are another useful tool to help identify ways to reduce friction in engines.

Given the strong economic and regulatory incentives for reducing friction in engines and other vehicle components, a continuous evolution of engine and vehicle technology is expected. These trends are already having a significant impact on engine lubricant composition and performance, and future lubricants will need to continue to meet evolving demands.

E. Use of Materials and Surfaces to Improve Efficiency

The drive to improve efficiency will affect the design of future systems. In parallel, it can also lead to the introduction of new materials. Improved strength to weight and inherently lower surface friction are two drivers for new technology. Furthermore, simply reducing surface finish can enable the use of lower viscosity lubricants. Taking this a step further, there is considerable work on imposing specific surface features on the surface to enhance lubrication. Some of these approaches are even beginning to take experiences learned from the natural world and have been coined as "bio-inspired or bio-mimicking tribology."

Most common lubricants have been designed and optimized over the years to work on ferrous-based materials. The antiwear additive zinc dithiophosphate (ZnDTP), which is commonly used in engine oils, is a good example. ZnDTP reacts with iron surfaces to form a multilayered film that is very effective in providing wear protection under thin film conditions. If one or both surfaces are nonferrous, these reacted tribofilms will be different or even nonexistent.

The selection of a new material is governed by multiple considerations. For example, the use of aluminum engine blocks is largely driven by weight reduction and improved heat transfer. Although both of these provide potential energy-saving benefits, it introduces the problem of lubricating an aluminum interface. Various approaches can be used, but most involve a modification of the surface layer. This example highlights an important issue related to material selection. It can be difficult to find materials that optimize both bulk and the surface properties that affect friction and wear processes. In many cases, the best solution is to use a surface coating.

The use of surface coatings is rapidly becoming more commonplace. Coatings that were once used in niche applications, such as motorsports, are now finding increasing use in mainstream products. Reduction in friction and wear are the two main drivers. There is a large range of coatings that can be used. Diamond-like carbon (DLC) coatings is one class that has been receiving considerable attention because of their low friction and, in some cases, high hardness that can result in improved wear. However, research has illuminated the fact that friction and wear performance of coatings is highly influenced by lubricant composition and performance. Consequently, there has been a rapid rise in the amount of research exploring the interaction of lubricant components on DLC coatings.

Polymer coatings are another example of surface treatment. As noted earlier, polymer-coated bearing surfaces are being used to help protect against frequent stop–start cycles. The use of polyether ether ketone coatings on piston skirts has been shown to reduce friction. Polymer coatings introduce yet another class of materials that the lubricant not only has to be compatible with, but also must enhance performance of or at least have a neutral impact.

Engines, in particular, are composed of many component systems and many different interfaces. In the search for improved efficiency, coating and material selections will no doubt vary between manufacturers and even between different components in the same engine. If this trend continues, it makes for an even more complex environment in which the lubricant must operate. Currently, these considerations are not included in industry specifications. However, even if they are not included in industry-wide specifications, it is likely that in the future individual manufacturers will be including different material-based performance requirements on future lubricants.

BIBLIOGRAPHY

ASTM D7589-15. Standard Test Method for Measurement of Effects of Automotive Engine Oils on Fuel Economy of Passenger Cars and Light-Duty Trucks in Sequence VID Spark Ignition Engine. ASTM International. West Conshohocken, PA, 2015. www.astm.org.

European Environment Agency (EEA). 2011. Laying the Foundations for Greener Transport, TERM 2011: Transport indicators tracking progress towards environmental targets in Europe. EEA Report No. 7/2011.

Exxon Mobil Corporation. 2014. Outlook for Energy: A View to 2040. http://corporate.exxonmobil.com/en/energy/energy-outlook (updated annually).

Raimondi, A. A., and Boyd, J. 1958. A solution for the finite bearing and its application to analysis and design: III. *ASLE Transactions* 1(1), 194–209.

Tellier, K., Donahue, R. J., Murphy, R., and Zurlo, J. R. 2013. Energy Efficient Gas Engine Lubrication. Paper No. 118. ExxonMobil Research and Engineering. Paulsboro, NJ.

20 Handling, Storing, and Dispensing Lubricants

ExxonMobil's lubricants are quality products made to exacting standards. Their quality levels are designed to provide effective and economical performance when they are used as recommended in the applications for which they are intended. ExxonMobil takes every practical precaution in product storage and handling to ensure that the products are maintained on specification through delivery to the customer. After delivery, it becomes the responsibility of the user to exercise proper care in handling, storing, and dispensing of the lubricants. This is necessary to protect the quality built into each lubricant so that it can deliver optimum performance in the use for which it is intended, and to maintain product identification and any precautionary labeling that may exist. A good lubrication program should also include steps to ensure proper handling and disposal of used lubricants.

The first steps in achieving optimum performance from quality oils and greases are proper handling, storing, and dispensing. These are necessary for two primary reasons: first to preserve the integrity of the products; and second to preserve identification and any precautionary labeling. It is poor practice to buy high-quality lubricants and then permit degradation prior to their use through contamination or deterioration. Poor storage and handling practices also increase the risk of misapplication when the identification on the containers has become illegible through improper handling, or the products have been transferred to inadequately or improperly marked containers. Proper handling, storing, and dispensing are also important for plant and personnel safety, to protect against health hazards, and to minimize the risk of environmental contamination. Most petroleum products are combustible and require protection against sources of ignition. They are not generally health hazards, but it must be recognized that excessive exposure to them can be undesirable and should be avoided. Good hygienic practices should be exercised when contact has occurred. Finally, contamination or leakage produces waste that must be disposed of, aggravating the disposal problems and environmental concerns.

Sound procedures and the correct dispensing equipment will minimize lubricant contamination, which can damage machinery. Pumps, oil cans, grease guns, measures, funnels, and other dispensing equipment must be kept clean at all times and covered when not in use. Where operating conditions justify them, centralized dispensing or lubrication systems that keep the lubricants in closed systems and, therefore, protected against contamination, are highly recommended. Systems of this type are available to handle many types of oils and greases. There are other advantages: lubrication servicing generally can be performed faster, which results in less waste; integral metering devices can supply important consumption data; and cost-effective bulk deliveries can be used.

Deterioration of lubricants can result from exposure to heat or cold, intermixing of brands or types, prolonged storage, chemical reaction with fumes or vapor, entrance of dust and abrasive particles, and water contamination.

Relatively simple precautions and procedures in the handling, storing, and dispensing of lubricants and associated petroleum products can achieve significant economic and operating benefits. In most cases, the benefits of investing in lubricant dispensing equipment greatly exceed the purchase price.

Economic benefits are based mainly on the elimination of waste due to the following preventable factors:

1. Leakage or spills from damaged or improperly closed containers
2. Contamination due to exposure of lubricants to dust, metal particles, fumes, and moisture

3. Deterioration caused by storage in excessively hot or cold locations
4. Deterioration due to prolonged storage
5. Residual oil or grease left in containers at the time of disposal or return
6. Mixing different brands or types of lubricants that are incompatible
7. Leaks, spills, and drips when filling a reservoir or lubricating a machine

Operating benefits, which also are reflected in dollar savings, include the following:

1. Reduction of machine problems attributable to the lubricant resulting in fewer downtime occurrences.
2. Reduced material handling time. It has been estimated that labor costs for lubricant application can be as high as eight times the cost of the lubricant applied.
3. Better housekeeping. Oil or grease spilled on floors is a major safety and fire hazard.

Although the information contained in this chapter provides general suggestions on good practices, it is the responsibility of the purchaser to identify and adhere to all government, local, and legally mandated regulations. Areas such as plant safety, handling and storing of flammable or combustible materials, any special precautions to address health and safety issues, fire prevention and protection, ventilation, and disposal of wastes need to be an integral part of the lubrication program. The recommendations and suggestions in this chapter are believed to be consistent with the standards of the U.S. Federal regulations. However, it must be recognized that in many cases, more stringent local standards may apply. These should always be checked so that conformity with them can be established and maintained.

I. HAZARDOUS CHEMICAL LABELING FOR LUBRICANTS

Many countries are adopting new hazardous chemical labeling requirements in alignment with the United Nations' Globally Harmonized System of Classification and Labelling of Chemicals (GHS; United Nations, 2011). The intent of these requirements is to help ensure improved quality and consistency in the global classification and labeling of all chemical substances, mixtures, and preparations that includes lubricants.

The primary purpose of GHS is to enhance protection of human health and the environment by reducing differences among the various systems in use around the world. It is expected that a common platform will provide more consistent framework of information on the safe handling and use of hazardous chemicals and enhance worker comprehension, allowing them to avoid injuries and illnesses related to exposures to hazardous chemicals.

The GHS standard requires that information about chemical hazards be conveyed on Safety Data Sheets (SDSs) and labels using visual notations to alert the user, providing immediate recognition of the hazards. SDSs, which must accompany hazardous chemicals, are the more complete resource for details regarding hazardous chemicals. Labels must provide instructions on how to handle the chemical so that chemical users are informed about how to protect themselves. The label provides information to the workers on the specific hazardous chemical. Although labels provide important information for anyone who handles, uses, stores, and transports hazardous chemicals, they are limited by design in the amount of information they can provide.

GHS uses nine pictograms; however, not all countries adopting GHS choose to require all of them. For example, at this time U.S. Occupational Safety and Health Administration (OSHA) only enforces the use of eight. The environmental pictogram is not mandatory, but may be used to provide additional information. Figure 20.1 shows the symbol for each pictogram, the written name for each pictogram, and the hazards associated with each of the pictograms. Most of the symbols are already used for transportation, and many chemical users may be familiar with them.

Handling, Storing, and Dispensing Lubricants

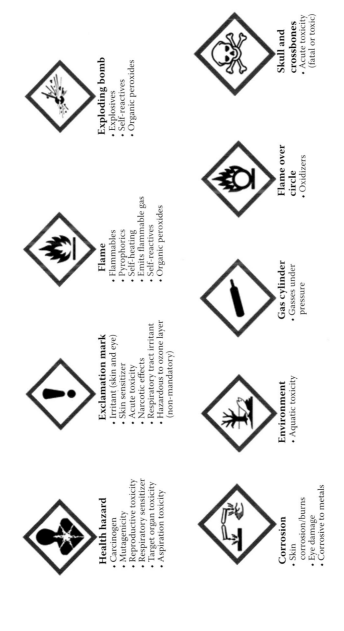

FIGURE 20.1 GHS pictograms and hazards. GHS, UN Globally Harmonized System of Classification and Labelling of Chemicals.

II. HANDLING

Handling is defined here as those operations involved in the receipt of the supply of lubricants and their transfer to in-plant storage. The type of handling involved depends on the form of receipt of the lubricants either in packages, minibins, or in bulk.

A. PACKAGED PRODUCTS

All shipments of oils, greases, and associated petroleum products in containers up to and including 208-L or 55-gal oil drums and 180-kg or 400-lb grease drums are considered to be packaged products. In some cases, shipments of various petroleum products in intermediate bulk containers (IBCs; 1040 L or 275 gal) are considered to be packaged products as well.

The size, shape, and weight of the most popular standard lubricant containers are shown in Figure 20.2. To aid in planning storage space, racks, or shelves, the dimensions of the containers illustrated and cartons of smaller packages are shown. Most containers are made of sheet steel or plastic in thicknesses and constructions suitable for their contents, the intended service, and applicable freight regulations. Grease gun cartridges, and nonmetal lubricant containers that are typically used for smaller packages are made of plastic or spirally wound fiber board with an oil-or grease-proof liner and metal ends.

Each container is sealed with appropriate covers, gaskets, lids, bungs, caps, and seals. In addition, each container is stamped, embossed, stenciled, lithographed, or labeled with the brand name of the lubricant it contains and appropriate safety information. Most lubricants are delivered to the user by truck or freight car in packages or bulk (in tanks and minibins).

Drums can be unloaded without damage from trucks or freight cars, and it is typically done with mechanical aids such as fork trucks, hydraulic lifts, and hoists. Before unloading, the brakes of the truck should be set firmly, and if any possibility of movement exists, the wheels to the truck or freight car bed should be blocked.

Under no conditions should drums or pails be dropped to the ground or onto a cushion such as a rubber tire. Doing so may burst seams, with consequent leakage losses, and possible contamination of the contents. Leakage may also constitute an environmental, safety, or fire hazard. In addition, the empty drums may have a refundable deposit or resale value if they are not damaged. Packaged lubricants can be safely and easily unloaded with a forklift truck and transported directly to storage. Generally, this unloading must be accomplished at a loading dock because it is usually necessary to drive the lift truck onto the truck or freight car to complete the unloading. Where a lift truck is to be driven onto a truck or freight car, all applicable safety regulations relative to such operations must be adhered to. Where a loading dock is not available, use of the truck's hydraulic lift gate is recommended, as shown in Figure 20.3, to safely unload packaged lubricants.

1. Moving to Storage

After unloading, drums can be moved safely to the storage area by properly equipped forklift trucks, either on pallets or held in specially equipped fork jaws (Figure 20.4). If fork trucks are not available, the drums can be handled and moved with barrel trucks or drum handlers (Figure 20.5). Drums should be strapped or hooked to the frame of a barrel truck and unloaded carefully.

When rolling drums, care should be taken to avoid hard objects that are higher than the drum body clearance and that might puncture the shell and result in leakage or contamination of the contents as well as potential injury to personnel handling the heavy drums.

Caution. When rolling drums, maintain firm control to prevent the possibility of a drum running away and injuring personnel, damaging machinery, or damaging the drum itself.

Smaller containers are usually shipped on pallets. They should be handled with the same care as the larger containers.

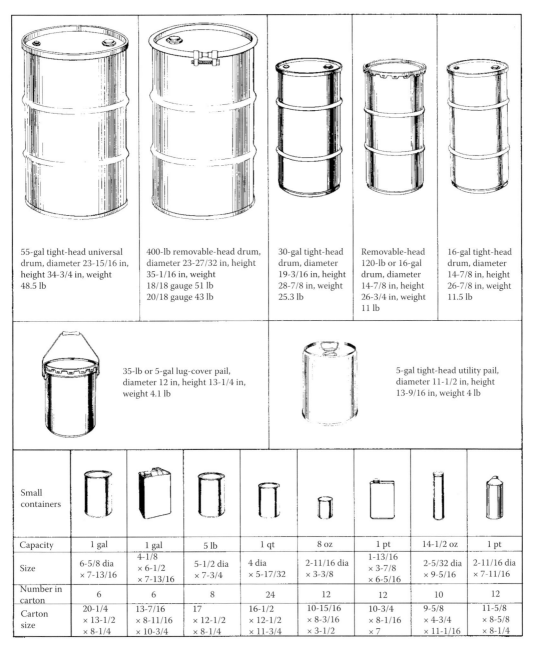

FIGURE 20.2 Typical lubricant containers (not drawn to scale). Standard containers that are commonly used for ExxonMobil products, with their outside dimensions in inches shown. Carton sizes for smaller packages are also given. Note that a 55-gal or 400-lb drum on its end requires about 4 ft² of floor space, and on its side, about 6 1/2 ft². The 14 1/2-oz package shown is a grease cartridge for loading hand guns. The 1-pt container at the lower right is a spray can (aerosol) for ease of application of specific products. Metric unit equivalents: 1 in = 25.4 mm; 1 lb = 0.45 kg; 1 gal = 3.785 L; 1 oz weight = 28.35 g; 1 fluid oz = 29.57 ml; 1 ft² = 0.093 m².

FIGURE 20.3 Hydraulic lift gate. Many trucks used for delivering packaged products are now equipped with hydraulic lift gates that can be used to lower the packages to the ground or a loading dock.

FIGURE 20.4 Lift truck for handling drums. The hydraulically actuated arms on this truck will clamp and lift four drums. Other models are available that will clamp one or two drums at a time, and rotators can be added to permit tilting or inverting the drums.

Smaller containers of lubricants are usually packed in fiberboard cartons. These should be unloaded and moved to storage with care. The cartons should be left sealed until the product is required for use.

B. Bulk Products

The advantages of bulk delivery and bulk storage of lubricants have resulted in the increased use of this method of operation in locations where the quantities of lubricants used are sufficient to justify

Handling, Storing, and Dispensing Lubricants

FIGURE 20.5 Manual drum handler. Hand-operated hydraulic systems clamp the drum then lift it for transporting.

the installation costs. The term "bulk" in this context refers not only to deliveries in tank cars, tank trucks, tank wagons, railcars, and special grease transporters, but also to deliveries in any containers substantially larger than the conventional drums. Under the latter category are a considerable number of special bulk bins—IBC or jumbo drums (minibins) that will transport on the order of 275 gal (1040 L, also listed as 1000 L) or more of oil or 3680 lb (1670 kg) of grease. Bulk bins (IBC) are usually offloaded and left on the customer's premises while the product they contain is being used, but the lubricant may also be pumped off into on-site bulk storage tanks, or supply tanks for central lubrication or dispensing systems (Figure 20.6).

Bulk handling and storage systems offer both economic and operating benefits to the lubricant user. Economic benefits include the following:

1. **Reduced handling costs**—When delivery is made directly into permanent storage tanks, handling costs are reduced markedly compared to those for drums or smaller packages.
2. **Reduced floor space requirements**—Permanent tanks occupy much less floor space than is required for an equal volume of oil in drums.
3. **Reduced contamination hazards**—Bulk storage tanks are usually filled directly from the tank car or tank truck through tight fill connections. The exposure of the lubricant to contamination is greatly reduced compared to handling and dispensing the same volume of packaged product.
4. **Reduced residual waste**—The user is charged only for the amount of product delivered into a bulk tank. Practice varies with bulk bins, but in many cases the user is allowed credit for residual material in the bin, or charged for only the amount added to the bin when it is

FIGURE 20.6 Bin for grease or oil. This type of bin, shown here stacked ready to be moved into a plant, is used most widely by ExxonMobil, but other designs with different capacities may be used in some areas.

refilled. As the amount of residual oil left in drums is usually on the order of about 1 gal or 4 L, and the amount of grease left in drums may range up to 20 lb (9 kg), the amount of lost lubricant attributable to residual waste with these packages is significant.

5. **Simplification of inventory control**—The use of bulk storage facilities sharply reduces the number of individual items to be ordered, tallied, routed, and processed for return or salvage.
6. **Reduced container deposit losses**—Damaged drums or containers that cannot be returned for other reasons such as excessive or unacceptable materials left in them result in lost drum deposit value and/or additional disposal costs.
7. **Lower lubricant purchasing costs**—When purchasing larger quantities of lubricant in bulk versus drum, it is very common to expect a lower purchasing cost for the lubricant.

The operating benefits to be achieved through the use of bulk facilities include the following:

1. **Simplification of handling**—Less handing time means that personnel are freed for other tasks.
2. **Constant availability of needed lubricants**—Time spent waiting for placement and opening of new drums and installation of pumps is minimized.
3. **Reduced contamination hazards**—Minimizing contamination means fewer lubrication problems on the manufacturing floor.
4. **Lubricants can be made available at strategic points**—Pumps and pipelines can be used to move lubricants from permanent tanks to locations where repetitive lubrication operations are performed.
5. **Improved personnel and facility safety**—Any reduction in handling operations reduces exposure to conditions that could result in injuries to personnel, plant, or environment.

These benefits can be achieved to their fullest extent only when the bulk system is properly designed and installed, and cooperation between the supplier and user results in uncontaminated products being delivered to the plant at the time that they are required. In determining the feasibility of installing bulk capabilities, other factors such as compliance with all government and local requirements need to be considered. Insurance requirements are another consideration.

1. Unloading

Before the receipt of bulk deliveries of lubricants, whether they are in tank cars, tank wagons, or special grease transporters, certain precautions should always be observed. The storage tanks should be gauged to ensure that there is sufficient capacity available for the scheduled delivery. Empty tanks should be inspected and flushed or cleaned if necessary. In the case where large tanks must be entered for inspection or manual cleaning, applicable safety rules should always be observed.

Before the delivery is unloaded, a check should be made to ensure that the correct fill pipe is being used, the valves are set correctly, and any crossover valves between storage tanks are locked shut. Best practices include using properly labeled or color-coded hoses or connectors (or use symbols instead of colors to address personnel with color blindness). In addition, use of specific connector fittings to only hook up with the proper mating connector can avoid unwanted cross contamination. It is also strongly advised to take a midtank sample from the delivery container for a quality assurance reference. Each type of bulk delivery requires several additional special precautions, which are outlined in the following sections.

2. Tank Cars and Tank Wagons

Only trained employees should unload tank cars of flammable or combustible liquids. The car brakes should be set, wheels blocked, and "stop" signs and safety cones set out. Before attaching and unloading connection, the dome cover should be opened and the bottom outlet valve checked for leakage. All hoses, pumps, valves, and other connections should be checked for cleanliness. After unloading is complete, the hose should be disconnected, the dome cover closed, and the valve closed immediately. Caps or other protective covers should be reinstalled on hoses and other connections.

3. Special Bulk Grease Vehicles

Although tank trucks have been used successfully to carry bulk grease, cleaning problems are considerable and tend to discourage the use of this type of vehicle. A more satisfactory approach has been the use of specially designed and constructed bulk grease vehicles. As an example, some vehicles are capable of carrying 38,900 lb (17,645 kg) of grease in two 19,000-lb (8618 kg) compartments. The tanks are fully insulated for heat retention in long hauls. For excessively long hauls or for deliveries made in cold temperatures, the bulk grease vehicle tanks should be equipped with heating capabilities to facilitate offloading.

These units are designed primarily for use in refilling bulk bins or the main storage tanks in centralized grease-dispensing systems in plants with large volume requirements for specific greases. They are usually used for transporting pumpable greases in the consistency range up to and including NLGI (National Lubricating Grease Institute) No. 2. Each bulk grease delivery requires careful study and engineering to assure satisfactory operation.

III. STORING

Proper storage of lubricants and associated products requires that they be protected from sources of contamination, from degradation due to excessive heat or cold, and that their identification be maintained. Also important are the ease with which the products can be moved in and out of storage, and the ability to operate on a "first in, first out" basis. Increasingly important in the selection, location, and operation of petroleum product storage facilities are applicable fire, safety, and insurance requirements.

Flammability and combustibility classification of lubricants is mainly based on the flash point of the lubricant and is controlled by governmental or local regulations. The U.S. National Fire Protection Association (NFPA), in its flammable liquids code NFPA 30,* defines flammable liquid as any liquid having a flash point below 100°F (37.8°C) and having a vapor pressure not exceeding 40 psia (2068 mm Hg) at 100°F (37.8°C), and classifies it as a Class I liquid.

---

* The definitions contained here are those of NFPA 30 (current revision in 2015 edition).

Flammable liquids (Class I) are further subdivided as follows:

Class IA includes those liquids having flash points below 73°F (22.8°C) and having a boiling point below 100°F (37.8°C).
Class IB includes liquids having flash points below 73°F (22.8°C) and having a boiling point at or above 100°F (37.8°C).
Class IC includes liquids having flash points at or above 73°F (22.8°C) and below 100°F (37.8°C).

A combustible liquid is defined as any liquid having a flash point at or above 100°F (37.8°C). Combustible liquids are divided into two classes (Class II and Class III):

Class II liquids include those with flash points at or above 100°F (37.8°C) and below 140°F (60°C).
Class III liquids include those with flash points at or above 140°F (60°C). They are divided into two subclasses, as follows:
Class IIIA liquids include those with flash points at or above 140°F (60°C) and below 200°F (93.3°C).
Class IIIB liquids include those with flash points at or above 200°F (93.3°C). This class of liquids, because of their lower flammability, does not require the special handling precautions that apply to the more flammable materials.

When a combustible liquid is heated for use near its flash point, it must be handled in accordance with the next lower class of liquids.*

Most lubricants, because of their relatively high flash points, under these regulations are Class IIIB combustible liquids. Select low-viscosity industrial lubricants (e.g., some low-viscosity spindle oils) may be classified as Class IIIA. Petroleum solvents that are commonly used for cleaning parts and equipment are typically considered Class II combustible liquids. However, depending on the conditions of storage and use, they may require handling as Class IC flammable liquids.

Suggestions included in the section on oil houses, size, and arrangement are believed to be consistent with the U.S. OSHA standards in effect as of the date of publication, which are similar to those of the NFPA. It is important to comply with all government and local regulations.

A. Packaged Products

Packaged lubricants and associated products should be stored in a warehouse or in an oil house. Outdoor storage should be avoided whenever possible. The best storage area for lubricants is a well-arranged, properly constructed, and conveniently located oil house. Warehouse storage is desirable when the oil house lacks the space needed to stock the complete lubricant inventory required.

1. Outdoor Storage

Storing lubricant drums or other containers out-of-doors is a poor practice. The hazards of this type of storage include the following:

1. Identifying drum markings or labels may fade and become unreadable under the combined action of rain, sun, wind, and airborne dirt. Rusting can obliterate drum markings. When the markings become so deteriorated that the drum contents cannot be identified, it may be necessary to discard the material or take a sample and perform a laboratory analysis to determine the nature of the contents. This is a costly and time-consuming procedure. On

---

* U.S. OSHA regulations, which apply to any workplace having employees, requires that when a combustible liquid is heated for use to within 30°F (16.7°C) of its flash point, it must be handled in accordance with the requirements of the next lower class of liquids.

the other hand, accidental use of the wrong lubricant as a result of incorrect identification can cause severe damage to equipment.

2. Seams of containers may be weakened owing to alternating periods of heat and cold, which causes the metal to expand and contract cyclically. The net result may be a loss of the contents by leakage or contamination if the drums are subsequently subjected to rough handling or improper storage.
3. Water may get into the drum around the bungs and contaminate the contents. A drum standing on end with the bungs up can collect rainwater or condensed atmospheric moisture inside the chime (Figure 20.7). This water can be gradually drawn in around the bungs by the breathing of the drum as the ambient temperature rises and falls. This can occur even with the bungs drawn tight and the tamper-proof seals in place.
4. Dirt and rust that accumulate inside the chime and around the bungs may contaminate the contents when the drum is finally opened for use.
5. Extreme heat or cold can change the physical properties of some products and render them useless. Water may separate from soluble oils, or invert emulsion and fatty oils may congeal in compounded oils. Emulsifiers, in wax emulsions, may separate and cause irreversible degradation. Cold lubricants dispense slowly and consequently may need to be warmed up before use, increasing the possibility of further damage due to overheating. Cold lubricants dumped into operating systems may cause temporary operating problems because of the substantial difference in viscosity and thermal effects.
6. Contaminating rust can develop inside a container if water leaks in from any source.
7. Sunlight may darken oil if it is exposed directly especially to lubricants stored in IBCs made of plastic. The ultraviolet radiation from the sun may form undesirable chromophores (quinones and benzaldehydes) and deplete antioxidant additives, which will accelerate oil oxidation and shorten oil life. The chromophores are degradation or rearrangement products of the light-sensitive antioxidant in the lubricant formulation. Antioxidant additives may deplete in order to protect the lubricant from gross oxidation. Chromophoric materials may be intensely colored and can cause the lubricant to darken and/or change color prematurely. The color may become yellow at low concentration levels and brown at higher concentrations. The best prevention is to always avoid exposure to direct sunlight.
8. Maintaining proper storage temperature is important to maintaining oil properties. Very high temperatures may accelerate oil oxidation, whereas very cold temperatures may cause

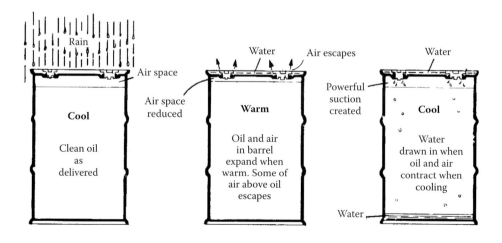

FIGURE 20.7 Entrance of moisture due to expansion and contraction (breathing) in an upright drum.

some additives to drop out. Generally, most oils and greases should be stored in temperatures no higher than 45°C (113°F). Some oils, such as metal working fluids, require a minimum storage and handling temperature above 41°F (5°C) to protect from freezing.

Of course, any of these conditions that result in irreparable damage to the drums can mean that the resale value of the drums is lost and the drums become a scrap disposal problem. To minimize the harmful effects of unavoidable outdoor storage, a few simple precautions and procedures can be very helpful.

As a general rule, lubricants in drums should never be stored outdoors. When drums must be stored outside, a temporary shelter (Figure 20.8) should be set up to protect them from rain or snow.

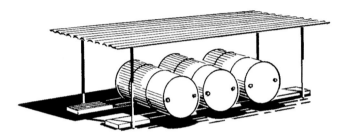

FIGURE 20.8 Outdoor drum shelter.

FIGURE 20.9 Metal drum covers. Weights have been placed on these covers to reduce the possibility that the covers will be blown off by high winds.

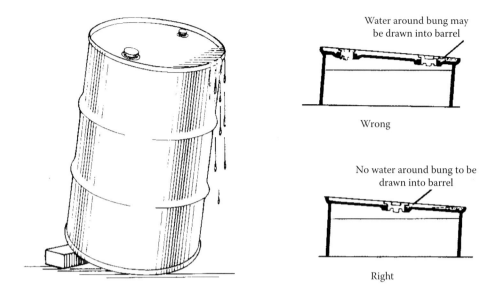

FIGURE 20.10 Correct method of blocking up a drum.

Drums should be laid on their sides with the bungs approximately horizontal as shown. In this position, the bungs are below the level of the contents so that breathing of water or moisture is greatly reduced, and water cannot collect inside the chime. For maximum protection, the drums should be stood on end (as long as product identification is visible) with the bung ends down on a well-drained surface. Where all of these approaches are impractical, drum covers such as those shown in Figure 20.9 can be used. Regardless of the position drums are stored in, they should always be placed on blocks (Figure 20.10) or racks several inches above the ground to avoid moisture damage.

Drums that must be stored outdoors with the bung end up before use should be cleaned carefully to eliminate the hazard of collected rust, scale, or dirt falling into the drum contents.

The drum storage area should be kept clean and free of debris that might present a fire hazard.

2. Warehouse Storage

Mechanical handling equipment is needed for efficient movement of drums into and out of a plant warehouse. Hand- or power-operated forklift trucks or stackers (Figure 20.11) are widely used for this purpose. They offer the advantage of a single handling operation from warehouse storage to the oil house or point of use. A chain block or trolley with a proper drum sling mounted on an I-beam bridge (Figure 20.12) can be used to move drums in and out of storage. This type of equipment does not require the aisle space needed to maneuver a forklift. Mechanical handling equipment is closely regulated by safety regulatory agencies. A thorough study is required to assure that the system and equipment meet all requirements.

Racks and shelving should provide adequate protection for all containers. Aisle space should be adequate for maneuvering whatever type of mechanical handling equipment is used. Each type of lubricant stocked should not be allowed to block access to the older stacks, so that a "first in, first out" basis can be maintained. This will minimize the hazard of deterioration owing to excessive storage time. A wide variety of racks and shelving (Figure 20.13), both assembled and in components for assembly on site, are commercially available.

When lubricants and other petroleum products are to be stored in a plant warehouse, location and rack construction should be considered with respect to applicable insurance, fire, and safety regulations. The material storage location should also be considered on the basis of receiving and

FIGURE 20.11 Drum stacker. The lifting arms slip under a drum resting on the floor. The drum can then be lifted and moved into the storage rack. This unit is battery powered, but manual models are also available.

dispensing convenience. In addition, the warehouse design should prevent humid and/or dirty environments by using proper ventilation, sealing, etc. To prevent safety incidents or product tampering, facilities should have some controlled access to the lubricant storage areas.

B. Oil Houses

1. Function

The function of an oil house in an industrial plant is to provide a central point for the intermediate storage and day-to-day dispensing of lubricants, cutting fluids, and other related materials needed to lubricate and maintain the plant's production equipment. Where soluble cutting fluids are used, the oil house is a typical location where emulsions are prepared. In many cases, the oil house is configured with the ability to purify or recondition used or contaminated oils.

The oil house should be equipped to properly maintain and clean all the equipment used in daily lubrication work. This includes pumps, grease guns, oil cans, portable dispensing units, strainers filter elements, and grease gun fillers. In addition, storage space should be provided for small containers of products, guns, cans, and other equipment necessary to properly dispense and apply lubricants. Facilities should have some controlled access to the lubricant oil houses to prevent safety incidents or product tampering.

Handling, Storing, and Dispensing Lubricants

FIGURE 20.12 Chain block and trolley. A hand or power operated chain block mounted on an I-beam bridge can be used to move drums into and out of storage racks. The racks shown are constructed of 2-in pipe and standard fittings. The pairs of flat strips supporting the upper drums can be removed to permit access to the drums below.

FIGURE 20.13 Storage racks. Racks of this type can be shop built or obtained commercially in various heights and widths. Drums can be conveniently moved into and out of this rack with the stacker shown in Figure 20.11.

The administrative function of the oil house should include the maintenance of machine lubrication charts and inventory records, conformance with established lubrication service schedules, and good housekeeping practices.

2. Facilities

The simplicity or complexity of oil house facilities will depend largely on the size of the plant and the comprehensiveness of its lubrication program. In general, a well-equipped oil house will contain adequate stocks of the following items:

1. Drum racks, either of the rocker type or tiered with labels to identify lubricant type
2. Oil and grease transfer pumps
3. Oil dispensing filtered ventilation systems
4. Labeled grease gun fillers with proper connector contamination caps
5. Labeled grease guns
6. Sealable oil transfer cans or dispensing containers
7. Portable equipment such as oil wagons, lubrication carts, sump drainers, catch pans, and power grease guns
8. Maintenance supplies such as lint-free wiping rags, cleaning materials, filters, spare equipment lubricant applicators, sight glasses, and grease fittings
9. Containers of absorbent materials for cleaning up oil spills
10. Grounding straps for use with combustible materials

In addition, batch purification equipment to recondition used oils may be desirable, soluble oil mixing equipment may be required, and solvent tanks for cleaning parts and lubrication equipment generally are required.

When plant usage of a specific lubricant warrants it, a permanent bulk tank may be installed in the oil house.

3. Size and Arrangement

These factors are also largely determined by plant size and lubrication service requirements. The total rack storage needed can be calculated by determining the quantity of each type of oil, grease, solvent, etc., which will need to be stocked in the oil house, and then converting this to rack dimensions by means of the container dimensions shown in Figure 20.2. The amount of space needed for miscellaneous equipment (solvent tanks, mixing equipment, cabinets for lubrication service personnel, portable dispensing equipment, etc.) must also be considered.

Individual lockers for lubrication service personnel can contribute to good housekeeping, safety, and general oil house efficiency. Lockers of steel construction are durable and easy to keep clean. They should be of sufficient size to store normal individual equipment, with adjustable shelving to permit maximum space utilization (Figure 20.14). This type of storage cabinet should be ventilated to prevent accumulation of vapors or fumes. If safety cans of cleaning solvents are to be stored in them, the lockers must be of an approved design.* Used wiping rags or waste should never be stored in these lockers, but should be placed in an approved disposal container.

Adequate cleaning capabilities is an oil house essential. Dispensing equipment (grease guns and the like) must be cleaned regularly for proper functioning. Airborne dirt and dust collect quickly on oil-wetted surfaces, so oil containers and transfer equipment should be cleaned regularly to remove these contaminants. Solvent tanks should be large enough to handle the largest equipment used. In general, baths of nontoxic safety solvents are permitted by safety regulations

---

* Where a reference is made to an "approved design" or "approved type" of container, locker, etc., the intent is an item that has been tested by and carries the label of one of the recognized testing organizations.

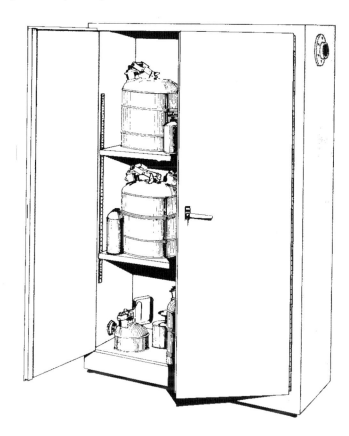

FIGURE 20.14 Individual locker. Adequate space is provided for the lubrication serviceman's equipment. The locker should be provided with top and bottom ventilation to dissipate fumes and vapors. Where flammables are to be stored in the lockers, a safety locker of approved design must be used.

provided that the tanks have properly designed covers of the automatic self-closing type, and ventilation is adequate. Two tanks, one for cleaning and one for rinsing, will assure proper cleaning of equipment.

One oil house layout (Figure 20.15) is offered as a guide to planning. Note that the sliding doors (1 and 6) are wide enough to admit the largest expected forklift load. The doors shown in this layout are self-closing fire doors, which may not always be required by local fire regulations, but generally offer the advantage of increased safety. The portable equipment storage area (13) provides ample maneuvering space for lift trucks to remove and replace the oil and grease drums (2 and 12). The open layout permits easy, unimpeded access from any part of the oil house to the door (6) leading to the plant. Local or higher level codes may require provisions for secondary containment (e.g., diking) of spills.

4. Optimum Utilization of Manpower

An oil house should be located centrally with respect to the lubrication service activity while minimizing the intrusion of foreign contamination. A study of service requirements based on the total travel of lubrication service personnel from the oil house to their work can help to determine the most efficient location for the oil house. Travel distances from the warehouse, unloading dock, or other storage facilities should also be taken into account. In multibuilding plants, it may be advantageous to build a separate oil house in a central location within the plant area. In such cases, an oil warehouse and the oil house may be combined in one building, with consequent savings in handling and storage costs.

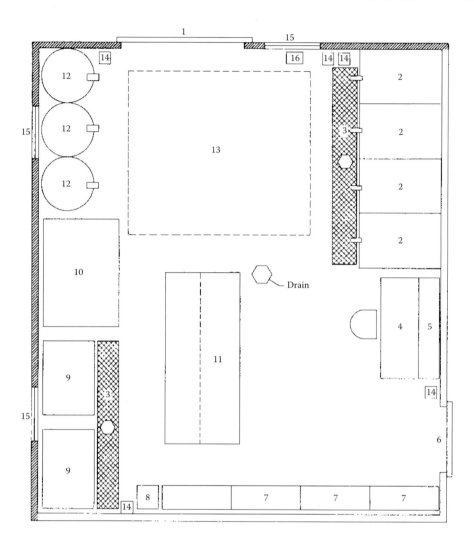

FIGURE 20.15 Oil house layout. Some of the features of this layout are as follows: 1 and 6, self-closing fire doors wide enough for passage of lift trucks or other material handling equipment; 2, drum racks; 3, grating and drain; 4, desk; 5, filing and record racks; 7, individual lockers or storage cabinets; 8, waste disposal container; 9, solvent cleaning tanks; 10, purification equipment, soluble oil mixing equipment, or other special equipment; 11, cabinets and racks for equipment, supplies, and small containers; 12, grease drums with pumps; 13, parking area for oil wagons, etc.; 14, fire extinguishers; 15, ventilators; 16, container of sawdust or other oil absorbent.

5. Housekeeping

Orderliness and cleanliness in the oil house are essential. In an orderly storage arrangement, the chances of mistaken identification of products or applications are greatly reduced. Regular cleaning schedules should be set up and maintained. Each container or piece of equipment should bear a label showing clearly the product for which it is used. This label should provide enough information to enable personnel to correctly identify the desired product in the original containers. For example, identifying an oil container as "Mobil SHC" would not be adequate identification because the same plant might use several viscosity grades or types of Mobil SHC Oils. The best practice uses full product names or specific product coding. Color coding is frequently suggested for this

Handling, Storing, and Dispensing Lubricants

identification, but it must be remembered that a considerable number of people are color-blind, which can defeat a color-coding system. Labels should be replaced as frequently as necessary to maintain legibility. Figure 20.16 provides an example of lubricants grouped by type or application and viscosity.

Every piece of equipment used in the oil house should have a space reserved for it and should be in its place when not actually in use. A chart or list of these locations posted near the product storage racks will facilitate location of needed items. A best practice includes having samples in clear, transparent plastic bottles for comparison so people can become familiar with the proper color of each lubricant. This will help prevent product misapplication as well as provide indication when the oil in service oil has been significantly degraded or contaminated.

Observance of cleanliness and an orderly routine will be reflected in the attitude and efficiency of the lubrication personnel. Their increased sense of pride and responsibility will have a direct bearing on the lubrication service in any plant. It will also contribute to improved safety against fire and personal injuries. In this respect, it is often considered to be of such importance that it receives major emphasis from compliance officers and some regulatory agencies.

Note: Although comparatively few lubricants and associated products require precautionary labeling, in those few cases where it is required, similar precautionary labels should be affixed to all equipment used for those products. Dispensing and application equipment used for products requiring precautionary labeling should not be used for any other products unless thoroughly cleaned first.

6. Safety and Fire Prevention

Warning signs should be posted in every oil house to alert personnel to the presence of combustible materials. If flammables are used or stored, or Class II combustibles are used in such a way that they must be treated as flammables; then the applicable warning signs for these hazards should also be posted. Standard warning placards complying with all applicable local safety regulations (OSHA or other) relative to flammables or combustibles are generally available. "No Smoking" signs, in red, should be prominently displayed with the no smoking rule rigorously enforced.

Handheld fire extinguishers and automatic fire suppression systems are essential for oil house safety. Hand-operated fire extinguishers should be mounted at strategic points throughout the oil house, particularly near those areas where cleaning tanks are located. They should be inspected periodically (at least as frequently as required by applicable fire regulations) to ensure that they are in satisfactory operating condition.

Fire extinguishing methods for flammable or combustible liquid fires include the following:

1. Suppression of vapor by foam
2. Cooling below the flash point by water spray or fog (not a direct stream, which could spread the fire!)
3. Excluding oxygen, or reducing it with carbon dioxide (CO_2), to a level insufficient to support combustion
4. Interrupting the chemical chain reaction of the flame with dry chemical agents or a liquefied gas agent

All personnel employed in the oil house should be thoroughly instructed in the location and use of the fire extinguishing equipment.

Rags, paper, or other solid materials that have been soaked in flammable or combustible liquids should be placed in an approved type of disposal container with a self-closing cover. The container should be emptied at the end of each shift and the contents removed to a safe location for reconditioning or incineration. All spills should be cleaned up promptly. If an absorbent is used, it

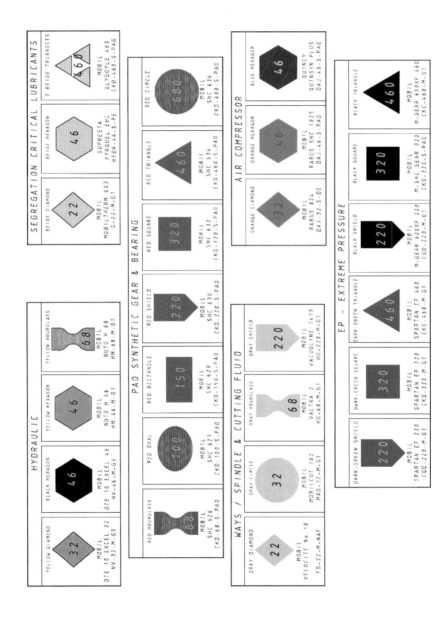

FIGURE 20.16 Oil house lubricant labeling. Note different symbols, shapes, and colors are used to differentiate lubricants. The ISO (International Organization for Standardization) viscosity grade is shown in the symbol as is the name of the color (to address any color blindness).

Handling, Storing, and Dispensing Lubricants

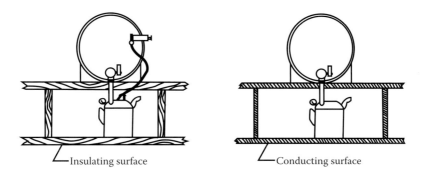

FIGURE 20.17 Bonding for control of static electricity. At the left, if either container is on an insulating surface, a bond wire should be used. At the right, if both are on the same conducting surface, an adequate bond can be obtained by keeping the faucet or nozzle in contact with the container being filled.

should be swept up promptly, placed in an approved disposal container, and removed to a suitable disposal area.

Ventilation equipment should be kept in first class operating condition at all times. Without good ventilation, vapors could collect within the oil house and be a hazard to both personnel safety and health and act as a fire hazard. The solvents used for cleaning parts and equipment in an oil house may present fire and health hazards. A few simple precautions can minimize these hazards:

1. Use only nontoxic solvents such as mineral spirits with flash points above 100°F (37.8°C).
2. Static electricity generated by the flow of products such as solvents can cause sparking, which can be a source of ignition. This is not generally a problem with products with flash points above 100°F (37.8°C), but simple precautions will eliminate the hazard. Drums containing solvents should preferably be grounded during opening and while installing faucets or pumps. If the drum is not resting on a grounded, conducting surface, an adequate ground can be obtained by clipping a wire between the drum chime and an appropriate ground such as a water pipe. When filling small containers from a solvent drum, a bond between the drum and container should be made. If both the drum and container are resting on the same conducting surface, adequate bonding will be obtained if the faucet or nozzle is kept in contact with the container; otherwise, a bond wire clipped between the container and the drum should be used (Figure 20.17).
3. Use only approved-type safety containers (Figure 20.18) for transferring solvents from drums or bulk storage to the cleaning tanks or point of use.
4. Prolonged inhalation of solvent fumes can cause headaches, dizziness, or nausea. Good ventilation around areas where solvents are used should be provided, preferably in the form of an exhaust system with hoods.
5. Keep covers on the solvent tanks when they are not in use. If an exhaust system with hoods is used over the solvent tanks, make sure that the exhaust system is operating before the covers of the tanks are opened.

In cases where inadequate ventilation conflicts with safety regulations, or problems of disposing of dirty solvents make it impossible to clean with petroleum solvents inside the oil house, arrangements should be made for cleaning in an area where the requirements can be met such as in a painting department or a general cleaning area.

FIGURE 20.18 Safety container. The container is self-closing with a liquid tight seal. The container must have an appropriate and approved testing agency label such as U/L or F/M label.

7. Security

Experience indicates that the oil house should be kept closed and locked when not in use by lubrication personnel. This will prevent confusion that may arise when unauthorized or uninstructed personnel enter and obtain their own choice of lubricants from open containers. The use of an improper lubricant may lead to equipment damage and could impact production.

C. Lubricant Deterioration in Storage

Lubricants can deteriorate in storage, usually as a result of one of the following causes:

1. Contamination (most frequently with water)
2. Exposure to excessively high temperatures
3. Exposure to low temperatures
4. Long-term storage where the storage interval has exceeded the lubricant's shelf life

Some contaminated or deteriorated products can be reconditioned for use, whereas others must be degraded to inferior uses, destroyed, or otherwise disposed of. The decision as to which course of action to follow is dependent on such factors as the amount of product involved with its value. This needs to be evaluated relative to the cost of reconditioning or salvaging, the type and amount of contaminants present, the degree of deterioration that has occurred, and the effect of the contamination or deterioration on the functional characteristics of the lubricant as well as environmental, health, and safety issues. Some of these considerations are discussed at more length in the following paragraphs.

1. Water Contamination

In many cases, water contamination is relatively easy to identify, although a quantitative indication of the amount present usually requires analysis via an on-site water test kit or laboratory. Frequently, a quantitative determination is not required, and the presence of water can be visually detected by haze or by suspended or free water. Where doubt exists, a simple "crackle" test—in which a small amount of the oil is placed in a shallow dish and heated to or past the boiling point of water on a hot plate—provides a fairly positive identification of water. When water contamination has been identified, steps that can be taken depending on the type of lubricant and its use.

2. High-Temperature Deterioration

a. Greases

Some oil separation (bleeding) may occur in select greases when they are stored for prolonged periods, particularly in a high-temperature environment. Slight bleeding may be normal. The separated oil will accumulate on the surface of the product. The amount of oil separated should be considered in deciding whether the grease can be used satisfactorily. Where the amount of oil separated is relatively small, it can be mixed back into solution or poured or skimmed off. Where excessive oil separation has occurred, it is probable that the bulk of the grease has changed sufficiently to be unsuitable for its intended use, so it must be disposed of.

Evaporation of the original water content of a water-stabilized calcium soap grease can cause separation of the product into oil and soap phases. However, this usually occurs only at quite high storage temperatures. Where it occurs, the grease usually will be rendered unfit for service.

b. Lubricating Oils

Most premium grades of hydraulic, process, circulation, and engine oils are not affected by even prolonged storage at temperatures below 113°F (45°C). Prolonged storage at higher temperatures—direct container contact with furnaces, steam lines, exposure to sunlight directly in outdoor storage, etc.—may cause some darkening because of oxidation. When in doubt as to the ability of the oil to perform satisfactorily, a sample should be sent to the laboratory for analysis and evaluation. This assumes that there is a sufficient amount of product involved to justify the cost of the analysis before using the product.

c. Products with Volatile or Aqueous Components

This group of products includes diluent-type open gear lubricants, solvents, naphthas, conventional soluble oils, wax emulsions, emulsion-type fire-resistant fluids, and water–glycol fire-resistant fluids.

Long storage at elevated temperatures of diluent-type open gear lubricants can cause evaporation of the diluent. If a significant proportion of the diluent is lost by evaporation, the product will be difficult to apply.

Caution: The diluent used in some types of open gear lubricants may be classified as a health hazard on inhalation, and precautionary labeling is required. Containers of such materials should be kept tightly closed when not in use, and stored in a cool place.

Volatile products, such as solvents and naphthas, may suffer loss by evaporation. Usually, the quality of the remaining product is not affected other than ease of application. If the products are in sealed containers, high temperatures can cause distortion or bursting.

Some products that contain significant amounts of water, such as fire-resistant water-in-oil (invert) emulsions and water–glycol fluids, are usually unharmed by small amounts of water evaporation. Because invert emulsions lose viscosity, and loss of water and fire-resistant characteristics are dependent on the water content, caution needs to be exercised to assure appropriate water content in these fluids. Loss of water can be addressed by adding water (following the manufacturer's instructions) as needed. Agitation is usually necessary when water additions are made to these products.

3. Low-Temperature Deterioration

Below-freezing temperatures normally will not affect the quality of most fuels, solvents, naphthas, and conventional lubricating oils and greases. The major difficulty associated with storage outdoors or in unheated areas during cold weather is with dispensing of products that are not intended for low temperature service. Some oils may suffer from some of the additives dropping out or settling. Some additives such as silicon-type defoamant can be resuspended in the oil by means of agitation at room temperature, whereas others such as active sulfur metalworking fluids cannot be brought back into the solution once the additive has dropped out. When possible, containers should be brought indoors and warmed before attempting to dispense the product.

Products that contain significant amounts of water should not be exposed to temperatures below 40°F (4°C). Certain fire-resistant invert emulsions are formulated to resist physical and chemical changes brought about by a small number of freeze–thaw cycles, but if they are excessive, the invert emulsions can sustain damage.

Repeated freezing or long-term exposure to freezing temperatures may destroy the emulsification properties of conventional soluble oils. Usually, there is no change in the appearance of the product, although it may have a cloudy cast, and the product must be disposed.

Some oils, when subjected to repeated fluctuations of a few degrees above and below the pour point, may undergo an increase in their pour point (pour point reversion) by 15–30°F (8–17°C). As a result, dispensing may be extremely difficult even when the ambient temperature is above the specified pour point of the product. Oils that contain pour point depressants or those that have relatively high wax content (e.g., steam cylinder oils) are the most prone to pour point reversion. Although this problem occurs relatively infrequently, products that may exhibit this phenomenon should be stored at temperatures above their pour points.

A product that has undergone pour point reversion may return to its normal pour point when stored for a time at normal room temperature. Heavy steam cylinder oils may require 100°F (37.8°C) or higher storage temperature for reconditioning.

4. Long-Term Storage

Long-term storage (several years from the manufacturing date) at moderate temperatures has little effect on most premium lubricating oils, hydraulic fluids, process oils, and waxes. However, some products may deteriorate and become unsuitable for use if stored longer than 3 months to a year from the date of manufacture and packaging. The nominal shelf life of most lubricating oils and greases is typically at least 60 months when stored in the original sealed container in a sheltered environment. An approximate guide to the maximum recommended storage times for some shorter-term storage products is shown in Table 20.1. Products stored in excess of the times

TABLE 20.1
Typical Shelf Life of Shorter Storage Life Products

Product	Shelf Life (months)
Gas engine oils	36
Refrigeration compressor oils	24–36
Water-soluble metal working fluids	12
Water–glycol fire-resistant fluids	24
Invert emulsion-type fire-resistant fluids	6
Custom blended soluble oils	3
Wax emulsions	6

Note: Shelf life begins from the date of packing. The above is only a guideline and is dependent on the lubricant formulation. Products may be requalified for an additional storage period through laboratory testing.

listed may still be of acceptable quality for use, but a confirming laboratory analysis should be conducted to ensure that the product meets the chemical and physical characteristics of the original products. This testing can be used to requalify the product for an additional storage period. Where storage time and conditions for a product are extremely critical, these facts are generally indicated on the package.

There is a fundamental difference between product life in storage and product life in service. When a lubricant is in service, it is exposed to dynamic conditions (e.g., circulation, splashing, churning), and its suitability for continued use becomes a function of other factors.

IV. DISPENSING

Dispensing includes withdrawal of the products from the oil house or other storage, transfer to the point of use, and application at the point of use. The point of use is defined as the bearing, sump, reservoir, friction surface, system, etc., where the product will perform a lubrication function. Packaged products may be dispensed directly from the original containers, or be transferred to various types of dispensing or application equipment. Bulk products may be transferred by means of portable equipment to the point of use, but frequently are dispensed directly from the tanks or bins through permanently installed piping systems to automatic application devices, or automatic or manually operated dispensing points. Where lubricants and related products are dispensed by methods other than completely closed systems, the basic requirements for effective dispensing are as follows:

1. Containers or devices used to move lubricants and related products should be kept clean at all times.
2. Each container or device should be clearly labeled for a particular product and used only for that product.
3. The device used for introduction of a product to a point of final use should be carefully cleaned before the filling operation is started; this includes grease fittings, filler pipes and the area around them, filler screens, filler holes, and quick disconnect fittings. Sumps and reservoirs should be thoroughly cleaned and flushed before filling the first time, and should be checked when being refilled and cleaned as necessary.

The most common dispensing equipment used in industrial manufacturing or process plants includes the following, in sizes, types, and quantities suitable to specific needs:

a. In the oil house
   Faucets
   Grease gun fillers
   Grease paddles
   Highboys/modular oil bins
b. From oil house to machine
   Lubrication carts and wagons
   Oil cans
   Safety cans
   Portable greasing equipment
   Portable grease gun fillers (where grease cartridges are not used)
   Oil wagons
   Sump pumps
   Catch pans
   Rags and funnels
   Tools to gain access to reservoirs and/or drain systems

FIGURE 20.19 Rocker type drum rack. The hook on the handle engages the top of the drum chime while the pads slip under the bottom chime.

A. IN THE OIL HOUSE

1. Faucets

Faucets are used to dispense oils, solvents, and other fluids from storage tanks or drums to containers for use in the manufacturing area. They are available in different sizes for fast- or slow-flowing fluids to fit either the 2-in or 3/4-in drum opening. The faucet can be inserted in the opening, and the drum lifted by a crane, lift truck, or chain hoist to a rack that will hold it in a horizontal position while the fluid is dispensed. An alternate method is to use one of the many rocker drum racks (Figure 20.19) to tip the drum onto its side and support it during dispensing. Racks of this type with casters permit easy positioning of the drum on the rack.

Drum faucets are commercially available in steel, brass, stainless steel and plastic to suit specific fluid needs (noncorrosive, nonsparking, etc.). Some types provide a padlock hasp to prevent inadvertent or unauthorized opening. Faucets with a self-closing feature are generally preferred because they minimize spillage if the faucet is accidentally opened. Faucets equipped with flame arresters are available for use with flammable liquids.

Flexible metal hose extensions may be attached to some faucets to contact the lip or edge of the container being filled to reduce the risk of static electricity discharge.

2. Transfer Pumps

Pumps are used to transfer oils, greases, solvents, coolants, etc., from drums or tanks to other containers. In the manufacturing area, they may be used to fill reservoirs or lubrication equipment directly from drums or lubrication wagons.

A pump that can be inserted in the bung opening of a drum is desirable for dispensing oil. These pumps are of positive displacement design and can be obtained to deliver measured quantities of lubricants. The simplest types are hand-operated. An excellent type of hand-operated pump has a closable return (Figure 20.20) actuated by a spring, to return drippage to the drum without the danger of contamination that would occur through an exposed drain. Air-operated drum pumps (Figure 20.21) and electrically operated pumps are also available.

Drum pumps are also available for dispensing semifluid or soft greases that are packed in closed-head drums. These pumps can be fitted into the 2-in bung opening, and deliver 20 lb (9 kg) or more

Handling, Storing, and Dispensing Lubricants

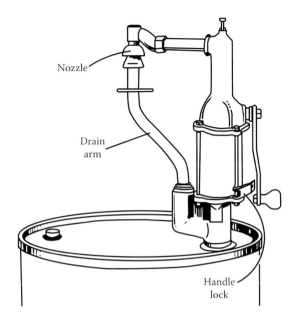

FIGURE 20.20 Drum pump with return.

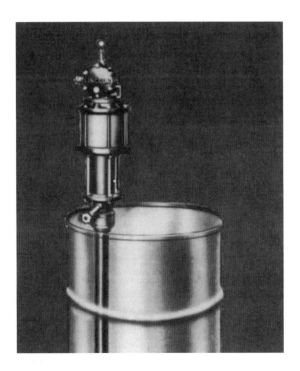

FIGURE 20.21 Air-operated oil pump.

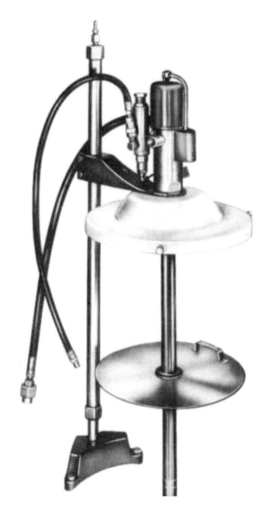

FIGURE 20.22 Air-operated grease pump and hoist. The pump is equipped with a follower plate in order to ensure feeding of relatively stiff greases and to minimize waste.

of grease per minute and are capable of high-pressure discharge. They are used for transferring grease to smaller servicing equipment such as grease gun fillers and portable greasing equipment.

Stiffer greases are packed in open head drums, and pumps for these products are usually mounted in the center of a drum cover that clamps on the top of the drum. This type of pump arrangement is generally equipped with follower plates that push the grease toward the suction and keep the grease from adhering to the sides of the drums (Figure 20.22). Figure 20.22 also illustrates a hoist arrangement that can be used to remove and install pumps rapidly and easily with a minimum of lubricant contamination. Most pumps can handle greases of the NLGI 2 and 3 consistency, depending on the type of grease.

3. Grease Gun Fillers

Where grease cartridge guns are not used, filling is generally done using grease gun fillers. These can be used to fill directly from the original container or from a reservoir in the gun filler. Models are available to fit various containers (including 35 lb [16 kg] pails, 120 lb [54 kg] and 400 lb [182 kg] open head drums). They may be manually (Figure 20.23) or electrically operated.

Handling, Storing, and Dispensing Lubricants

FIGURE 20.23 Hand-operated grease gun filler. This unit fits a 120-lb open head drum and is equipped with a follower plate. Similar units fitting 35-lb pails or 400-lb drums are available.

Where grease guns, fillers, and other grease dispensing equipment are to be filled with a grease that is too stiff to pump with a transfer pump, a grease paddle normally must be used. Preferably, the paddle should be made of metal or plastic to prevent contamination that might occur with a wooden paddle (i.e., splinters or wood chips). After filling is completed, the cover of the grease container should be replaced securely to prevent contamination.

4. Highboys

Highboys or modular, racked oil bins with permanently mounted pumps are frequently used for oil storage and dispensing. They have the advantage of neat appearance, compactness, and sturdiness, but they also suffer from certain disadvantages in that they are difficult to clean and require an additional product transfer, with risk of contamination, to fill them. The usual highboy design has no drain for cleaning, so there is a tendency for foreign matter to collect at the bottom of the tank. When new oil is added to the tank, the resulting agitation may cause this foreign material to be suspended in the oil. As a result, any oil drawn off before this can settle may contain harmful contaminants. When configuring these types of systems, a best practice is to use a dedicated pump(s) for each oil or oil type to reduce the chance of comingling lubricants and ensuring that each bin is configured with an air breather.

Where highboys must be used because of space limitations, they should be equipped to permit easy draining and cleaning.

B. From Oil House to Machine

Moving lubricants from the oil house to machinery, where they are to be used, is a critical phase of dispensing that justifies the same care as handling in the oil house. The problem again is one of preventing contamination and confusion of products. It is usually further complicated by the necessity for transporting a variety of lubricants in different types of containers or lubrication equipment. This phase of dispensing problems is essentially a matter of choosing containers of equipment that can be handled economically with a minimum risk of contamination and confusion.

As a general rule, the most desirable containers are those that can be filled in the oil house and then emptied directly into the machine, or used to lubricate the machine. Each product should have its own container or piece of dispensing equipment that should be clearly marked for that product. Such equipment should not be considered interchangeable, unless they are emptied and thoroughly cleaned before being used for another product. It is also a good practice to consult your lubricant supplier to generate a list of compatible products that can be intermixed for noncritical applications and products that cannot be mixed because of incompatibility issues or potential changes in critical physical and chemical balances.

Caution: Galvanized containers or piping should not be used for transporting oil. Many of the industrial oils used today contain additives that can react with the galvanized zinc to form metallic soaps. These soaps may thicken the fluid and clog small orifices, oil passages, wicks, etc. The zinc may also carry over into circulating oils and cause interpretation problems with oil analysis results.

1. Oil Dispensing Containers

Although it is outnumbered by automatic oiling devices ranging from simple bottle oilers to complex centralized oiling systems, the common oil can is still widely used to carry oil to machines. Its chief advantages are traditional acceptance and low cost, combined with the need for an easy method of applying small quantities of oil to open bearings. Its chief disadvantages are higher labor costs per unit of oil dispensed, the increased hazard of lubricant contamination as compared with closed automatic systems, and the generally inferior performance of bearings designed for hand oiling. When hand oil cans must be used, thorough precautions should be taken to maintain lubricant cleanliness.

The simplest hand oil can (sometimes called a pistol oiler) is the diaphragm type. A more practical and desirable oil can is the positive delivery type that delivers a definite quantity of oil from any position (Figure 20.24). The trigger actuates a simple pump incorporated in the can, and an adjustment permits varying the quantity of oil delivered with each stroke.

Where larger quantities of oil than can be delivered with an oil can are required, special containers are used. Open pails and cans invite contamination and are not a good solution. A practical container for quantities of oil that can be hand carried is the safety can (Figure 20.25). Various styles are available. They can be obtained in many capacities with self-closing spouts and fill covers. Some are provided with removable spouts of flexible metal (or plastic) hose to permit filling of less accessible reservoirs. Such cans are sturdy, easily cleaned, and closed against contamination. Self-closing safety cans of approved design should always be used for transporting flammable materials.

Plastic containers of oils in 1 quart (1 L) to 5 gal (20 L) sizes are used in some plants. Their advantages include light weight, low cost, and freedom from rust or corrosion. However, some

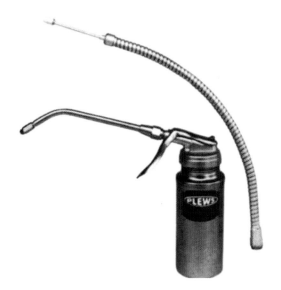

FIGURE 20.24 Pistol oiler. The flexible spout can be used to access difficult-to-reach application areas.

FIGURE 20.25 Large oil containers.

plastics are affected by oils or by the additives in some oils, so it is always necessary to check with the manufacturer to be sure that the plastic used is compatible with the fluid to be carried.

2. Portable Oil Dispensers

A wide variety of equipment is available to transfer quantities of oil from the oil house to the point of use. Included are bucket pumps (Figure 20.26) that can be used to transport a few gallons of liquid lubricant and discharge it directly into the machine sump or reservoir. Where larger quantities must be transferred, wheeled dollies (Figure 20.27) that can transport a standard 16 gal (or 60 L) or 55 gal (or 208 L) drum are available. The drum can be equipped with a manual, air, or electrically operated transfer pump so that the lubricant can be pumped directly into the machine. Even larger quantities can be transported in special oil carts (Figure 20.28), which may also be equipped with a manual or power-operated pump.

FIGURE 20.26 Lever-operated bucket pump. This unit holds 30 lb or 15 kg of a lubricant such as a gear lubricant. Extension on the base is to step on and hold the unit firmly while it is being operated.

FIGURE 20.27 Drum dolly. With the wheels raised, a drum can be rolled onto the platform of the dolly. The wheels can then be lowered and the drum moved to the point of use where product can be dispensed with a transfer pump.

Handling, Storing, and Dispensing Lubricants

FIGURE 20.28 Oil cart. The drain on the right facilitates cleaning the tank.

3. Portable Grease Equipment

Small quantities of grease (3–24 oz or 85–680 g) can be dispensed from hand grease guns. These are available in push-, screw-, lever- (Figure 20.29), and air-operated types. Some of the lever- and air-operated guns can be loaded by suction, with a gun filler, by cartridge, or by any combination of the three. Where small quantities of grease at high pressure are required, special lever-operated guns, sometimes called booster guns, are available.

Couplers and coupling adapters can be fitted to grease guns to enable their use with many various types of fittings. Extension hoses can be attached to many guns to facilitate lubrication of difficult-to-reach fittings.

Before loading a grease gun, it should be checked for cleanliness and cleaned if necessary. When using a grease gun filler, the fittings on both the filler and the gun should be wiped clean with a lint-free cloth. Hand loading should be avoided whenever possible. If it is necessary to load from an open grease container, check the surface of the grease for dirt and other contamination and if necessary, scrape off the top layer to remove any contaminants and hardened grease. Make sure that the grease is of the correct thickener type to assure compatibility and performance. Use a clean metal

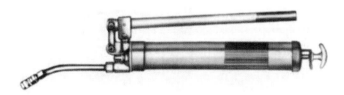

FIGURE 20.29 Lever-operated grease gun. Models are available to deliver various pressures. Capacity for this model is 24 oz (680 g), and it can be loaded by suction, by paddle, or with a grease gun filler. Similar guns of smaller size are available for cartridge loading.

or plastic paddle to transfer the grease from the container to the gun, or load the gun by suction filling. Before suction filling, carefully wipe off the surfaces of the gun that will be submerged in the grease. Grease cartridges should be wiped to remove dust before loading into the gun. The metal top of the cartridge should be pierced or torn off carefully to avoid the possibility of metal or other materials falling into the grease.

Grease guns should be marked clearly to indicate the type of grease they are used for. A gun should be used for only one type of grease to avoid problems that might arise because of incompatibility of different brands or types of grease. If it is necessary to change the type of grease to be used, the gun should be thoroughly cleaned in solvent and dried before refilling. After filling, grease should be pumped through the nozzle to flush out any remainder of the previously used product and any traces of solvent.

Where more of a particular grease is required than can be provided by a single filling of a gun, several alternatives are available. A gun filler on a pail or drum of grease, or a portable gun filler (Figure 20.30) can be taken to the location where lubrication is being performed. Manual bucket pumps on standard 35-lb or 18-kg pails can be used. A variety of air and electric power guns are also available. Most of this equipment will handle semifluid and soft greases without difficulty. Some fibrous and firmer greases may require a follower plate arrangement to ensure that the grease slumps to the pump inlet. Very firm greases may require an arrangement that provides positive priming, as shown in Figure 20.31, to ensure feeding to the pump inlet.

Scheduled calibration of grease equipment is a best practice. Technicians can ensure that the correct volume and pressure of grease are achieved by routing calibration. Simple calibration of a manual grease gun includes dispensing a set weight of grease on a small scale and recording the number of "shots" or strokes. Technicians can check the volume and pressure of pneumatic greasing equipment using a grease flow test kit (Figure 20.32).

FIGURE 20.30 Portable grease gun filler. The 30-lb or 15-L tank on this unit can be filled in the oil house. The unit can then be taken to the plant to refill guns as required.

FIGURE 20.31 Electric power gun and cutaway. The cutaway view shows the helix arm and worm gear, which ensure feeding of firmer greases to the pump inlet.

FIGURE 20.32 Grease flow test kit. For calibration of greasing equipment. Available from ExxonMobil.

4. Lubrication Carts and Wagons

The lubrication program of a plant may be so organized that personnel are regularly assigned to service distant parts of the plant where they must apply a number of different oils and/or greases to various types of machinery. It is important in these cases to supply each person with a practical means for safely transporting the necessary supplies. Such equipment may be either purchased or built. It may be elaborate or simple, depending on the variety of lubricants used and on the machines to which the lubricants are to be applied. A simple, practical lubrication wagon may be nothing more than a shop cart with sufficient room for needed supplies. Space is needed for oil containers, grease guns, hand oil cans, grease pails or portable gun fillers, and a supply of miscellaneous necessities such as spare grease fittings, tools, replacement cartridges, and clean lint-free cloths for cleaning lubrication equipment and application points. Lubrication instructions for the machinery

FIGURE 20.33 Commercial lubrication cart. Space is provided for gun fillers, fluid lubricants, grease guns, and miscellaneous supplies.

to be lubricated should also be carried. Such a cart is flexible and permits a quick change in the type of lubricants carried.

More elaborate lubrication carts can be obtained from the suppliers of lubrication equipment. The type shown in Figure 20.33 provides space for portable grease gun fillers, grease guns, and a number of containers of fluid lubricants. This type can handle mainly fluid lubricants, but space is available to carry miscellaneous supplies. If there are provisions for practical and easy cleaning of the tanks, these carts are very satisfactory.

C. Closed System Dispensing

A number of types of lubricant dispensing and application systems can be considered essentially closed systems, in which the lubricant is exposed to the minimum possibility of contamination. These systems are not truly closed in that the supply tank or reservoir must be charged with lubricant periodically. The lubricant supplied to these applications can be dispensed at the point of use through a quick disconnect fitting or metering device, substantially reducing the risk of contamination compared to the dispensing methods already discussed. Some of the more common types of these systems are discussed in the following sections.

D. Central Dispensing Systems

Although lubricants can be dispensed from bulk bins or bulk tanks into containers, oil wagons, grease gun fillers, etc., transfer by some form of piping system is often justified. Such systems are usually custom designed and built where large volumes of lubricants are to be handled, but smaller systems may be assembled from equipment modules available from lubrication equipment suppliers. Electric, hydraulic, or air-operated transfer pumps—similar to those discussed earlier for dispensing from drums that will handle greases or oils—are available to fit tanks or bulk bins. Where lines extend beyond the capacity of the tank-mounted pump, booster pumps can be installed at suitable points along the lines.

Dispensing nozzles of the type shown in Figure 20.34 can be used to control dispensing in systems of this type, or in assembly line filling of equipment such as gear units. The dial can be preset

Handling, Storing, and Dispensing Lubricants

FIGURE 20.34 Metering control nozzle. The dial can be preset to deliver from 1 to 60 quarts of oil then shut off flow. A counter provides a record of the total amount of lubricant dispensed.

to deliver from 1 to 60 quarts (liters) and can be shut off automatically. The meter also records the total amount of oil dispensed, up to 9999 gals.

E. MAINTENANCE AND SERVICE

Comparatively little maintenance is required for central dispensing systems. Proper cleanliness precautions should be observed when filling storage tanks or reservoirs. An adequate quantity of lubricant should always be maintained in the tank so the transfer pump will not suck in air that might cause binding or locking. Pump pressures should be checked periodically, and lines should be inspected for leaks or damage.

Again, as stated earlier, any industrial plant lubrication program should meet all governmental and local codes and regulations that address all health, safety, and environmental factors.

BIBLIOGRAPHY

Mobil Oil Corporation. 1990. Mobil Technical Bulletin: *Handling, Storing, and Dispensing Industrial Lubricants*. Fairfax, VA.

National Fire Prevention Association. 2015. NFPA 30. Flammable and Combustible Liquids Code. http://www.nfpa.org/codes-and-standards/document-information-pages?mode=code&code=30.

United Nations. 2011. *Globally Harmonized System of Classification and Labelling of Chemicals (GHS)*. New York: UN.

United States Department of Labor. 2012. Occupational Safety and Health Administration – OSHA 29 CFR Part 1910.106. Flammable Liquids. https://www.osha.gov/pls/oshaweb/owadisp.show_document?p_table= STANDARDS&p_id = 9752.

21 Practices for Lubricant Conservation and Machinery Reliability

Conservation of natural resources and protection of the environment have become common goals in our society today. Depending on current plant practices, these goals can be accomplished generally with minimal cost impact in the areas of lubricant use and disposal. Control of lubricants within the plant to prevent the generation of oily wastes is just one such operation. Any lubricant that becomes a waste increases operating costs from the standpoint of material purchases, waste disposal, and environmental protection issues. A plant lubricant may become a waste product and, as such, may require reclamation or disposal under proper control to prevent pollution for the following reasons:

1. Contamination with water, other lubricants, foreign matter, dirt/dust, process materials, wear metals (from the lubricated equipment) or dilution
2. Degradation during use and lubricant shelf life expiration because of improper storage, aging and depletion of additives, oil oxidation and varnish formation, change in viscosity, or loss of lubricity
3. Escape from the system through leaks, spills, line breakage, faulty gaskets or excessive foaming; overlubrication in all-loss systems; carry-off on products involved in the manufacturing process

One key element to preventing any lubricant from becoming a waste product lies in the selection, storage, and handling (handling, storage, and dispensing are discussed in Chapter 20) of these products in the plant, from receipt as new materials to disposal of used materials.

With the increased emphasis on preventing pollution of ground and surface waters, strict regulations have been enacted covering the composition of industrial effluent to both surface waters and municipal sewage plants. Regulations covering plant effluent and stream water quality vary from area to area, and—depending on the jurisdictional control of the particular body of surface water, groundwater systems, or sewage plant authority—may be under the control of various local or country legislation groups. The first step for anyone concerned with the problem of preventing pollution from industrial sources is to understand the effluent, stream, groundwater, or sewage standards governing the particular situation.

Before discharging any fluid waste into a waterway, it is essential to comply with the existing regulatory requirements at that particular location. In general, as far as oil content is concerned, the effluent should be free of visible floating oil (one such guideline is to keep the oil content below 15 mg/L; however, different local regulations may exist). This level assumes a dilution effect in the receiving stream. For example (considering differences in local regulations), if the stream is to be used for municipal water supply, its oil content would need to be kept below 0.2 mg/L according to one local guideline. If, on the other hand, the receiving body of water is being used for recreational, agricultural, or industrial uses, oil content of up to 10 mg/L may be acceptable. By the same token, in order not to interfere with the planned use of the stream, effluent limits generally must also be met for color, odor, turbidity, dissolved oxygen, heavy metals (e.g., lead and mercury), phenols, phosphates, suspended solids, etc.

As it is impossible to cover the myriad of various regulations here, this chapter will instead focus on the general practices aimed at reducing the generation of lubricant wastes while maximizing their useful life.

I. OVERVIEW OF IN-PLANT LUBRICANT HANDLING

Proper in-plant handling consists of efficiently using petroleum lubricants to prevent them from prematurely becoming waste products. Furthermore, this chapter will cover in general terms the proper disposal of spent lubricants in such a way that they do not become a detriment to the environment.

Based on 2012 figures, world lubricant demand exceeded 40.5 million metric tons. The industrial sector consumed a significant portion of these lubricants as hydraulic fluids, process oils, and metalworking fluids. These fluids require recycling or reclamation for continued beneficial use to conserve our natural resources or, when no longer suitable, disposal in an approved manner to prevent environmental damage.

ExxonMobil is committed to addressing the challenge of sustainable development, which includes balancing growth with environmental protection so that future generations are not compromised by actions taken today. Our field engineers apply the same philosophy when working with lubricant customers and end users. Sound in-plant handling practices and purification of lubricants deliver quantifiable economic and environmental benefits.

An initial step toward efficient lubricant usage and used oil disposal should be part of an overall lubrication program such as that offered by ExxonMobil. Part of this program would include expert engineering services to ensure proper product selection, maximization of lubricant life, minimization of leaks, beneficial machine maintenance, optimum oil drain intervals, and improved handling and storage practices to prevent contamination and spills. Planning, implementing, and maintaining a sound lubrication program requires time and financial resources; however, the benefits outweigh the costs and can be measured in reduced lubricant consumption, healthier equipment, and improved machine availability.

Each industrial plant has unique challenges and limitations that require specific solutions to minimize used oil generation and waste oil disposal in an effort to control pollution. The following guidelines are applicable to most lubricant users:

1. Select the right lubricants, including hydraulic fluids, gear lubricants, metalworking fluids, coolants, and crankcase oils, to obtain the longest service life possible.
2. Establish a program for good preventive maintenance to keep equipment in good operating condition (key element).
3. Set up good housekeeping and contamination control procedures.
4. Where feasible, use a plantwide, multimachine circulation system to replace small, single-machine reservoirs.
5. Provide oil purification equipment for circulation systems to ensure optimum use of lubricants where practical.
6. Identify the nature and sources of waste generation and disposal problems.
7. Attack the disposal problem at the source, not at the plant effluent stage.
8. Know the appropriate regulatory agencies and follow local regulations.
9. Keep metalworking fluids, solvents, and lubricant streams separate and distinct from each other to reduce complexity in recycling or reclamation.
10. Maintain separate sewer systems for sanitary waste, process water, and storm drainage.
11. Do not use dilution as the solution to pollution.
12. Concentrate before you treat for final disposal.
13. Use effective and cost-efficient lubricant practices when commissioning new or rebuilt machinery.

Benefits of a sound lubrication conservation program exceed the up-front costs of implementation. The potential benefits include:

1. Economic		Less lubricant purchases
		Reduced application costs
		Increased machinery availability
		Reduction in lubricant inventory costs
		Increased production capacity
		Reduced maintenance costs
2. Safety		Reduced potential for injury to personnel
		Reduced potential of fires from poor housekeeping
		Lower insurance costs
3. Environmental		Avoidance of high remediation costs
		Fewer fines
		Improved public opinion

II. PRODUCT SELECTION

Many factors enter into the selection of the proper product depending on its use as a lubricant, hydraulic fluid, or metalworking fluid. Speed, load, and temperature must be considered in the selection of gear and bearing lubricants; types of metals and severity of operation in machining; and the type of engine, speed, and fuel in crankcase and cylinder lubrication of stationary diesels, gas engines, and gas turbines. These factors and their effects on product selection are well known and usually recognized.

There are other factors that will minimize waste oil disposal problems in a plant. These factors, which should be considered in the lubricant selection process, are long oil service life, the ability to help control leakage, the ability to minimize harmful contamination effects, compatibility with other products, value as a by-product, and ease of disposal.

ExxonMobil field engineers work with end users to recommend the fewest number of correct lubricants. Product consolidation is a key element in any effective lubrication program as is the use of long-life lubricants. Best in-plant practices include the establishment of engineering or lubricant standards for newly installed or purchased equipment. These standards focus on the use of currently inventoried oils or greases to minimize the number unique oils and greases the plant must store, maintain, and ultimately dispose of after use.

A. LONG SERVICE LIFE

One important characteristic of a lubricant that will do much to minimize the used oil disposal issues is long lubricant service life. The longer the interval between drains or oil changes in any machine, the less waste oil generated and fewer shutdowns required for oil changes. Long service life is the result of many factors including machine and operating conditions, maintenance practices, proper selection of the lubricant, and product quality. Product characteristics inherent in the base oil or additive package that enhance long life are the following:

1. Chemical stability to ensure minimum buildup of oxidation products
2. Resistance to changes in viscosity
3. Control of acidity and deposit formation
4. Additive robustness and retention to ensure the needed film strength

5. Detergency to prevent deposits and keep system components clean
6. Viscosity–temperature characteristics
7. Ability to maintain those product qualities that impact performance

All of these items lead to the selection of the highest quality product to achieve the longest service life. In addition, premium product selection is usually the most economical when the total cost of ownership is considered. Although the initial material price may be higher, the cost of application, maintenance, and disposal in man-hours is sharply decreased and the overall cost of lubrication is lower. For example, synthetic lubricants typically cost more to purchase but can achieve substantially longer service life, resulting in fewer oil drains, increased production capacity of equipment, and less waste disposal. High-performance synthetic lubricants often have proven to be the most desirable and economical solution for conservation.

B. Compatibility with Other Products

Where the possibility exists of one product mixing with another during use, such as the hydraulic fluid in a machine tool with metalworking fluid, the products used for both fluids should be selected so that cross contamination will minimize the negative effects on performance. For example, in the instance mentioned, if the metalworking fluid could be used as the hydraulic fluid, frequent oil change-outs or disposal may not be required because of any admixture. Other opportunities will become apparent in a survey of the types of lubricants used in the plant and the product availability from lubricant suppliers. Even at the disposal stage, compatibility of used oil will permit consolidation without forming complex chemical mixtures with unique disposal challenges.

C. Value as By-Product

As most lubricants will ultimately require disposal at some point, consideration should be given at the selection stage as to the value of the used oil as a by-product. Many lubricants, after satisfying their primary function, depending on local regulations, can be used for less demanding service such as a fuel or feedstock material for reclamation or rerefining. The ability of a lubricant to serve again as a by-product after initial use should be considered in the selection of a new product, both from the point of view of conservation of natural resources and that of waste disposal. In some cases, the cost of disposal of a select few lubricants may be more than the original purchase price of those lubricants. Knowing this upfront allows for better control of their use and planning for eventual disposal costs.

D. Ease of Disposal

The last factor to be considered in lubricant selection is the ease of disposal once it exceeds its useful life. Ease of disposal is related to the lubricant type or formulation. For example, oil-in-water and water-in-oil emulsions differ in the ease with which the emulsions can be broken. Also, a few select products have been developed (e.g., certain metalworking fluids) that may be disposed of in municipal sewage systems (assuming no undesirable contamination). Always check with local authorities and suppliers first. Another example of products that generally present fewer problems as pollutants are the environmentally aware lubricants that are typically readily biodegradable with low toxicity. In general, the value of a by-product increases with the ease of disposal.

After selecting the correct lubricant, users should refine their handling, storage, and dispensing program with a focus on contamination control and waste minimization.

III. IN-SERVICE HANDLING

Once a lubricant is charged in a system and until it requires replacement, much can be done to prevent the escape of the lubricant from the system or its premature drainage because of degradation. The prevention or minimization of leaks, spills, or drips, which can complicate the disposal of lubricants, will also reduce the potential for accidental pollution of plant effluent. A correctly designed and maintained lubrication system coupled with in-service oil purification ensures maximum lubricant life.

A. Reuse versus All-Loss Systems

There are many ways of applying lubricants to bearings, gears, machine ways, cutting tools, etc., varying from hand oilers, grease guns, through bottle and wick oilers, and splash and mist feed, to the most sophisticated circulation systems and centralized bulk-fed systems. Some are all-loss systems in which the lubricant is used in a once-through application. In certain cases, the lubricant is consumed in the process of lubrication, such as in a two-cycle engine where the lubricating oil is mixed with the fuel, is carried off on the product or in exhaust gases, or is collected for disposal. An enclosed lubrication system, such as in gear cases, large engines, turbines, paper machine drying systems, or machine tools, continually reuses the same lubricant until its lubricating characteristics change or are degraded. Enclosed lubrication systems are recommended where practical, both from the standpoint of conservation of products and minimization of pollution. These systems are more economical in terms of costs of material, application labor, housekeeping, and disposal. When these systems are coupled with purification systems, they actually furnish better conditioned lubricant throughout the use period to the bearings, gears, and other lubricated components of the machine.

The advantages of enclosed lubrication systems include ease of adjustment or lubricant feed rates for numerous points (e.g., multiple bearings or multiple cylinders) that might have different oil flow demands. Enclosed systems can also help to prevent contamination of the lubricating oil or cross contamination with metalworking fluids, process oils, and process material.

Oil mist lubrication systems can be an excellent method of lubrication for specific machine applications while also being quite economical in lubricant consumption. Unless carefully controlled, oil mist lubrication systems may cause fogging and condensation on floors and machine exteriors, whereas excessive oil mist can lead to concerns regarding exposure through inhalation. Proper installation, care, and maintenance are needed to ensure proper oil feed rates without any excess lubrication.

Air over oil (air–oil) systems are an alternative to oil mist lubrication, drip oilers, or brush-type oilers and can replace grease for select applications. These systems inject oil into the air stream to lubricate the component. Advantages include minimal oil fog compared to oil mist systems and a reduction in consumption compared to drip and brush oilers. A disadvantage includes their all-loss aspect.

B. Prevention of Leaks, Spills, and Drips

Any loss of lubricant from a system or machine component, not called for in the design and for which a collection system is not provided, complicates the disposal problem. It has been estimated that leakage from circulating systems, including hydraulic systems, reaches approximately more than 100 million gal (379 million L) a year, which would involve more than 5 million man-hours in labor to resupply the system with make-up oil. With proper leakage control, it is estimated that more than 70% of this could be reduced or eliminated. In addition to saving material and labor, the benefits that accrue from proper leakage control include increased production; decreased machine downtime; prevention of cross contamination of other lubricant or coolant systems; elimination

TABLE 21.1
Losses by Oil Leaks (This Is the Purchasing Cost Impact of Lubricant Leakage)

Leakage[a]	Loss in 1 Day		Loss in 1 Month		Loss in 1 Year	
	Gal (L)	Value[b] (U.S. Dollars)	Gal (L)	Value[b] (U.S. Dollars)	Gal (L)	Value[b] (U.S. Dollars)
One drop in 10 s	0.11 (0.42)	0.79	3.37 (12.74)	23.60	40 (151)	280
One drop in 5 s	0.22 (0.85)	1.58	6.75 (25.52)	47.25	81 (306)	567
One drop per second	1.12 (4.25)	7.88	33.75 (128)	236.20	405 (1530)	2835
Three drops per second	3.75 (14.17)	26.25	112.50 (425)	787.50	1350 (5103)	9450
Stream breaks into drops	24 (90.72)	168	720 (2722)	5040	8640 (32,659)	60,480

[a] Drops are approximately 11/64 in (4.37 mm) in diameter.
[b] Based on an approximation of $7.00/gal ($1.85/L).

of safety hazards to plant personnel; and prevention of accidental pollution of the plant's storm, process, or sanitary waters. Also, leaks, spills, and drips seriously complicate the disposal problem because of collection difficulties. Any collected waste oil will usually have a lower by-product value during disposal because of contamination. Furthermore, when cross contaminated, it often forms tight emulsions or complex mixtures with other liquids that are not easily separated. The cost impact (oil only) of oil leaks can be significant, even for seemingly small drips, as seen in the data in Table 21.1.

System leakage control on any machine requires attention to two general classes of joints through which fluid may be lost: moving joints (dynamic) and static joints. Moving joints include rod or ram packing, seals for valve stems, pump and fluid motor shaft seals, and in some instances, piston seals. Static joints include transmission lines, pipes, tubing, hoses, fittings, couplings, gaskets and seals and packing for manifolds, flanges, cylinder heads, and equipment ports (hydraulic pump and motor ports).

Little or no leakage will occur past newly installed seals or packing for moving joints if correct materials and installation procedures have been used. Some of the causes of leakage in moving joints include improper installation (resulting in seal and packing damage), misalignment, wrong seal used, and rough or scarred finishes on rods or shafts. However, both internal leakage past pistons, vanes, valves, etc., and external leakage past rod, shaft, valve stem packing, housings, etc., may be expected to develop in time even under normal service conditions. Internal leakage may cause problems in machine operation and loss of lubricant through consumption in the machine. External leakage may cause problems with oil loss and environmental pollution as noted earlier. Only a properly planned and executed preventive maintenance program will find and correct such leakages.

Leakage of oil from static type joints may be the result of one or more of several causes:

1. Use of unsuitable type of joint or transmission lines
2. Lack of care in preparing joints—machining, threading, cutting, etc.
3. Lack of care in making up joints
4. Faulty installation such that joints are loosened by excessive vibration or ruptured by excessive strain
5. Severe system characteristics that subject the joints and lines to peak surges of pressure due to "water hammer"

6. Improper torquing of bolts or fasteners
7. Incompatibility of joint materials with lubricant (takes time to show up)
8. Use of threaded pipe for high-pressure lubrication systems

In addition to slow leakage, a combination of these items may contribute to fatigue failure, line breakage, and loss of large quantities of lubricant and hydraulic fluid.

Several general measures to reduce leakage are as follows:

1. Use joint-type seals and packing material for installation and maintenance work that have proved satisfactory in service. A best practice is to avoid using Teflon tape as a thread or joint sealant.
2. Train maintenance personnel in principles of proper installation of joints, seals, and packing and maintain surveillance to ensure proper execution of accepted procedures.
3. Minimize the number of connections and make all lines and connections accessible for checking and maintenance.
4. Design installations to avoid excess vibration, mechanical strain, twisting, and bending.
5. Locate and protect tubing, valves, etc., from damage by shop trucks, heavy work pieces, or materials handling equipment.
6. Protect finely finished surfaces in contact with seals and packing from abrasive and mechanical damage.

One other area that should be given consideration for avoiding spills is the proper dispensing and draining of lubricant from machine reservoirs. The dispensing of new lubricants has already been discussed in Chapter 20, and the drainage of waste oils is addressed later in this chapter.

C. Elimination of Contamination

It is well known that the largest single cause of waste lubricant generation is due to contamination. It is estimated that 70% of the oils that are rendered unfit for service are attributable to contamination. This contamination covers a wide range of materials including foreign matter, degradation products formed from the lubricant during use, and cross contamination through admixture with other fluids. Contamination reduces the life of the oil, shortens component life, and increases disposal complexity. In almost all cases, proactive contamination control is more cost-effective than cleaning, filtering, or removing the contaminants from oil.

1. Central Reservoir Maintenance

Reservoirs for circulating lubricating systems, which include hydraulic and metalworking fluid systems, may be one or a combination of three types: integral with the machine and located in the base, separate from the machine but individually associated with it, and centrally located reservoirs serving multiple machines. Whether a central system is economically justified depends on many factors, among which are the size of the facility, the number of different grades or brands of fluid in use, and the facility layout. Obviously, in any facility with multimachine lubrication requirements, a central system can be better justified if, at most, two multipurpose lubricants are handled.

All reservoir types can become contaminated in a similar fashion, and preventive methods apply equally to each. Dirt, water, dust, lint, and other foreign substances contribute to the formation of emulsions, sludges, deposit, and rust when present in a circulating lubrication system. These materials also detract from the performance of the lubricant, accelerate lubricant degradation, and increase the potential for wear or loss of performance of the system components. These contaminants should either be removed or the lubricant in the reservoir should be drained and replaced. Some contaminants present in the waste oil can limit its value as a by-product for use as fuel or for reclamation. The following lists of numerous contaminants that are sometimes present in circulating systems:

- Air
- Assembly lubes
- Cleaning solvents or solutions
- Coal dust
- Dirt
- Drawing compounds
- Dust
- Gasket sealants and materials
- Grease from pump bearings
- Lint from cleaning rags or waste
- Metal chips
- Metalworking fluids
- Oil-absorbent material
- Packing and seal fragments
- Paint flakes
- Persistent emulsions
- Pipe scale
- Pipe threading compounds
- Rust particles
- Rust preventives
- Water
- Way lubricants
- Wear particles (metal)
- Weld spatter
- Wrong oil

Considering each of these individually may suggest sources of contamination and means to prevent their entry into the system. Contamination prevention of foreign matter involves good system design, good maintenance, and good housekeeping practices. The following general recommendations will help in controlling reservoir contamination by foreign matter (Figure 21.1):

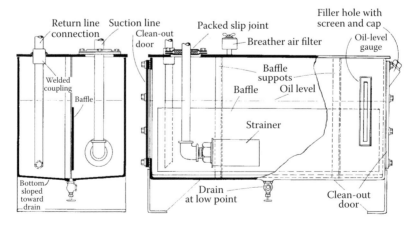

FIGURE 21.1 Contamination control in a central reservoir.

1. Removable-type reservoir covers should fit well, be gasketed, and be tightly bolted on.
2. Clearance holes in the reservoir cover for suction and drain lines should be sealed, preferably by a compressible gasket and bolted flange-type retainer.
3. The oil filler hole should be equipped with a fine mesh screen and dust tight cover.
4. The breather hole should be provided with an air filter and be regularly maintained. This should be the only point in the system to equalize vacuums and pressures or the only pathway for air to enter or exit the oil system.
5. The suction or oil inlet should be equipped with a strainer to prevent large debris and other foreign matter from entering the system and should be inspected regularly.
6. Use a magnetic pickup near the bottom of the reservoir, a magnetic dipstick, or a magnetic drain plug to help reduce abrasive ferrous particles from being circulated, which can damage machine components.

2. Cross Contamination

Where a machine uses separate systems and lubricants for hydraulic operation, bearing and gear lubrication, and machine tool cooling, every effort of design and operation must be exerted to prevent contamination. For example, machine tool hydraulic systems are often subject to contamination with water-soluble chemical coolants or oil emulsions. These can cause deposits on valves, emulsion formation, and foaming in the hydraulic system because of poor oxidation resistance or chemically active agents and fatty acids. Similarly, in the case of machine tool operation, metal chips may be allowed to pile up around metalworking fluid drains. Such piles act as dams and may cause pools of metalworking fluid to overflow onto the ways and contaminate the way lubricating system. The seals of rotating shafts in lubricating systems may allow cross contamination of the metalworking fluid with lubricating oil, which is commonly referred to as tramp oil.

3. Proper System Operation

Another type of contamination that can severely impact the performance of the lubricant is caused by degradation products formed in the lubricant as a result of system operating conditions. For example, excessively high operating temperatures can cause oxidation of the oil. The oxidation products formed could result in changes in viscosity, an increase in acidity, and the formation of varnish and deposits. High system oil temperature may be the result of improper selection of the lubricant, low oil levels, inadequate system capacity, poor cooling, system malfunction, or a poor original design. Although most premium lubricants have high oxidation stability, proper system design, maintenance, and operation will complement this inherent characteristic. Adequate lubricant system capacity or use of oil coolers will help control the bulk temperature during recirculation.

In addition to adequate temperature control to maximize oil life and prevent harmful effects of oxidization, equipment owners or designers should be attentive to the following oil temperatures: reservoir (bulk oil), oil inlet (oil to the application), and oil outlet (oil exiting the application). Calculating or estimating the target oil viscosity and the required oil film thickness at the application(s) is critical to maximizing component life. Lubricant temperature and flow rate at the application (oil inlet) point is as important as the bulk oil temperature in the reservoir and must be considered when making changes or upgrades to oil cooler circuits and reservoirs.

IV. IN-SERVICE LUBRICANT PURIFICATION

In circulating systems, the lubricant, hydraulic fluid, or metalworking fluid should be kept as free of contaminants as possible. This can be accomplished by preventing entry of contaminants into the system and also by in-service purification during operation. Lubricant purification, depending on the nature and extent of the contaminants, may consist of continuous bypass or full-flow treatment.

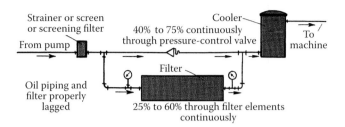

FIGURE 21.2 Continuous bypass purification.

In many instances, a combination of the two methods is incorporated into the circulating system. In addition, critical large-capacity systems may use portable filtration units when adding oil to the system or for periodic contaminant removal. Other options include permanently connecting independent lubricant purification units to large reservoirs or using separate batch units to purify or reclaim drained lubricants for reuse.

A. Continuous Bypass Purification

In the continuous bypass purification system (Figure 21.2), a portion of the oil or coolant delivered by the pump is diverted continuously from the main line for purification. The cleaned oil or coolant is then returned to the system. The remaining unpurified stream is delivered through the main line to the system. This system of purification is reasonably effective when the rate of contamination is low.

The bypass is limited to applications where continuous purification of 5–10% of the total pump discharge is sufficient to keep the entire charge in good condition for a satisfactory period.

A throttling orifice to prevent too great a withdrawal of bypass fluid limits flow to the bypass filter. Gradual clogging of the elements reduces the flow of dirty oil through the filter. This condition causes an increase in pressure on a gauge on the filter between the orifice and the filter elements. When this gauge shows a predetermined increase in pressure, the filter elements should be cleaned or replaced.

B. Continuous Full-Flow Purification

In a continuous full-flow system (Figure 21.3), the entire volume of lubricating oil or coolant is forced through a filter before passing through the cooler to the lubricating system. The oil may pass through the cooler ahead of the filter if the cooled oil is still at a good filtering viscosity.

Clogging of the filter decreases the flow of the fluid through the elements. This condition is indicated when gauges on the inlet and outlet of the filter show an increasing drop in pressure across the filter elements. Before this drop becomes excessive, a relief valve in the filter opens and bypasses unpurified fluid around the elements to maintain a full supply of fluid to the system. Before this condition happens, it is recommended that the filter elements be replaced. Filters installed in parallel

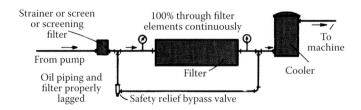

FIGURE 21.3 Continuous full-flow purification.

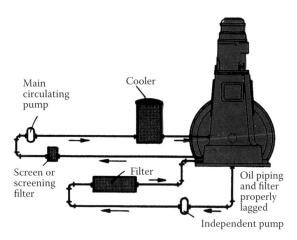

FIGURE 21.4 Continuous independent purification.

with the required valving permits the replacement of elements in one filter at a time, without shutdown. This should be done whenever the pressure differential across a filter reaches the maximum value recommended by the filter or machine manufacturer.

C. Continuous Independent Purification

Lubricating oil or coolant is sometimes purified continuously via a system entirely independent of the main circulating system. Used oil is drawn from the sump by an independent pump, delivered to a centrifuge (purifier) or fine filter (polishing filter), or both, and after purification, is returned to the circulating system (Figure 21.4).

Although this purification system would operate normally during circulation system operation, it can be used when the circulation system is shut down, and can be shut down without circulation system interruption, at any time to replace the filter elements or clean the purifier whenever the pressure drop across the unit dictates.

D. Periodic Batch Purification

Periodically, the entire charge of lubricating oil or coolant is removed from the system for purification. The oil or coolant may be allowed to settle, then be reheated and passed through a centrifuge, reclaimer, or other types of purification equipment. Typically, this type of batch purification is used in large capacity systems such as utility turbine systems where a single charge of oil can exceed 5000 gal (19,000 L) and have an expected life beyond 20 years. For example, during a major turbine overhaul or during a scheduled downtime, the entire batch of oil may be removed from the system and placed in a holding tank for batch purification. This oil is replaced in the system with new oil or oil that has been previously purified. The oil removed from the system can then be batch-purified on a leisurely basis.

E. Full-Flow and Bypass

In such a purification system, the dirty fluid from the machine passes through the purification system at a full flow rate, and a portion of the stream delivered to the machine is diverted and returned for further purification. The bypass stream is designed so that when the main stream is shut down the purification system may continue to operate. Such a system can provide for gross filtration of

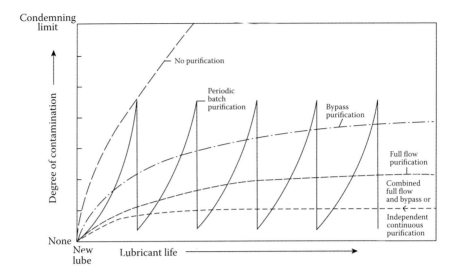

FIGURE 21.5 Effectiveness of various purification methods.

the full-flow stream and fine filtration of the bypass stream at a slower flow rate with the combination yielding a cleaner fluid. In addition to fine bypass filtration, the bypass system can be equipped with coalescing filter elements, which can be used to remove small amounts of moisture from the fluid.

A comparison of the four methods (Figure 21.5)—full-flow, continuous bypass, combination full-flow and bypass, and batch purification—shows that the batch method yields the cleanest fluid at the least cost but requires downtime and has a widely fluctuating efficiency. The combination method, which can function when the system is either operating or shut down and can use more stringent purification methods on the bypass portion without sacrificing operating flow rates, yields the highest overall efficiency.

Many industrial oils can be retained in service for long periods before draining by use of in-system or portable purification systems. Other contaminated oils can be reused after independent purification; they can be returned to the system provided they are clean, they retain their original performance characteristics, and the additives have not been removed by the purification process. If the fluid is no longer suitable for its original use, purified lubricants may be used in noncritical applications and all-loss systems such as hand oiling of chains, bar oil for chainsaws, or for flushing purposes. Before reuse, always check the base properties of purified lubricants. Lubricants cross contaminated with metalworking fluids or vice versa that cannot be adequately purified for their original purpose should be drained and used as a heating oil, be rerefined if applicable, or disposed of in another way.

V. PURIFICATION METHODS

The purification methods most commonly used in industrial plants are settling, size filtration, centrifuging, and clay or depth filtration. It is noteworthy that conventional fine filtration will not remove significant quantities of additives because most of them are soluble in the oil. This applies to the majority of industrial and automotive-type lubricants. Special care should be taken in certain instances where very fine filtration (<3 μm) with high (>100) beta (β) ratios is used; some additive can be removed from highly additized oils such as engine and transmission oils or compounded oils such as steam cylinder oils. The contaminants and additives removed and the various purification methods are outlined in Table 21.2.

Practices for Lubricant Conservation and Machinery Reliability

TABLE 21.2
Contaminants and Additives Removed by Purification

Materials	Settling	Size Filtration	Clay Filtration	Centrifuging	Reclaiming	Rerefining
Water	Yes[a]	_[b]	_[b]	Yes	Yes	Yes
Solids	Yes[c]	Yes[c]	Yes[c]	Yes	Yes[c]	Yes
Oxy products	No	No	Yes	No	No	Yes
Fuels, solvents	No	No	No	No	Yes	Yes
Oil additives	No	No	Yes	_[d]	_[d]	Yes

[a] Except water held in tight emulsions.
[b] A little water can be absorbed or held back by the filter material.
[c] Solids will be removed down to the filtration limit (rating).
[d] Some additive may be removed along with the water.

Settling and size filtration is often used on the machine or in the circulating system. A central reservoir of proper design will permit settling and may be equipped with size filtration. Clay filtration and centrifuging may be incorporated into the system as separate units. Reclamation systems had historically been batch operations conducted on the drained lubricant, but enhancements now make them more portable allowing them to be rotated from system to system similar to portable filter units.

A. Settling

Water and heavy solid contaminants will separate by gravity whenever oil is held in a quiescent state for a suitable period. For separation of solids, water, and oil, the system as shown in Figure 21.6 can be fabricated. Moderate heating of the oil from 160°F to 180°F (70–80°C) by a steam coil or low watt density electric immersion heater will decrease the viscosity of the oil and accelerate the settling process. Care must be exercised with selection of heating elements to assure that surface temperatures are not so high that they actually cause oxidation or thermal cracking (heating element surface temperatures below 225°F or 107°C are desirable).

In the dual tanks used in batch purification (Figure 21.6), preliminary straining of gross debris and particulates is achieved in the upper tank. Used oil is introduced into one of the duplicate settling tanks well below the level of any clarified or partially clarified oil present. The used oil should be added slowly to prevent agitation of the settled contaminants. The floating suction permits withdrawing clean oil from the surface above a minimum level of water and sludge concentrate in the conical bottom.

B. Filtration

Filtration or straining may be used either as full-flow, continuous bypass, or a combination of these two methods as discussed earlier. Straining by metal screen or woven wire in cartridges or plates removes only the largest particles, is low in initial cost, and allows maximum flow rates almost independent of fluid viscosity. Conventional filtration will not remove water or oil-soluble contaminants but as discussed, coalescing-type filters can be used where removal of small amounts of water is desired. Any filtration system that removes water can also remove some of the polar rust inhibitors that attach themselves to the water. Surface-, edge-, and depth-type filters allow a variation in the size of particle removal with the surface- and depth-types suitable for the finer particles. These can be high in both initial and operating costs and are adaptable to single or multimachine systems. When the system is properly designed, little or no downtime is required for their servicing.

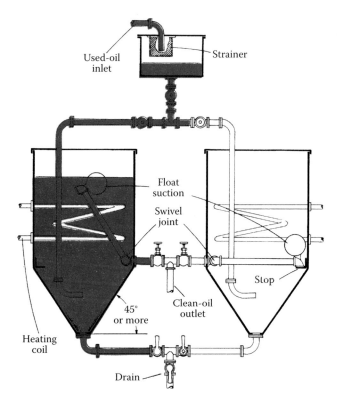

FIGURE 21.6 Settling tanks for batch purification.

C. Size Filtration

Cloth, stainless steel (reusable filters), metal edge, paper edge, surface-type strainers, or depth-type filters are used to remove particles from the oil.

Wire cloth strainers are used to remove only the larger particles—50 µm and larger. Metal-edge type filters (Figure 21.7) clean by forcing the oil to pass through a series of edge openings formed by a stack of metal wheels or between the turns of metal ribbons. The majority of these units will filter out particles of 90 µm; some will filter particles as small as 40 µm. Paper-edge filters use elements of impregnated cellulose disks or ribbons. Impregnated cellulose disks will filter particles from 0.5 to 10 µm; ribbon elements will filter down to approximately 40 µm size particles. Depth-type filters (Figure 21.8) use Fuller's earth, felt waste, or rolled cellulose fibers in replaceable cartridges and can filter out particles from 1 to 75 µm in size. Surface-type filters use replaceable resin-treated paper, closely intertwined fiberglass strands (or other synthetic fibers), or woven cloth. The paper elements are available to filter out particles down to about 1 µm, and the cloth down to about 40 µm.

1. Depth-Type Filters

These are probably the most versatile of industrial-type filters. The depth-type filters can be equipped with cartridges using a variety of filter media. Commercially available cartridges include the following:

1. An adsorbent-type filter containing Fuller's earth in a woven cloth container. Other materials include activated alumina and clay. These materials are recommended for removal of soluble contaminants such as acids, asphaltenes, gums, resins, colloidal particles, and fine solids. It will also remove polar-type additives and is not recommended for

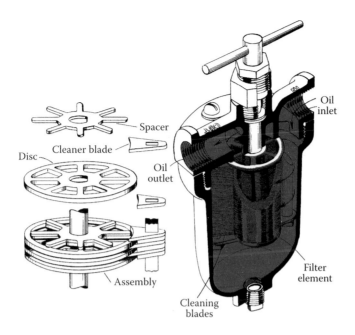

FIGURE 21.7 Edge-type oil filter.

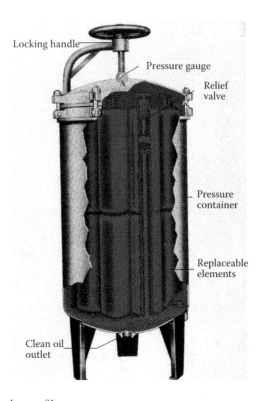

FIGURE 21.8 Typical depth-type filter.

general industrial lubricants unless the additive is replaced following purification or the oil is used for purposes not requiring the removed additives. Although Fuller's earth will remove small amounts of water (approximately 1 quart or 0.95 L of water per element), it is not recommended where gross water contamination is prevalent. Typical applications for this type of filter medium are quench oils, transformer oils, and vacuum pump oils. It is also recommended for filtering certain synthetic fluids such as phosphate esters and silicones.
2. A cellulose-type filter element, usually a combination of cotton waste and wood fibers such as redwood and excelsior, is usually recommended for removal of large amounts of gross solids. It will not remove additives or water; the cellulose type retains its resiliency in the presence of water and water emulsions and usually permits higher flow rates. Both types are recommended for filtering engine oils, hydraulic fluids, turbine oils, fuel oils, and lubricating oils.
3. Resin-impregnated, accordion-pleated paper filter cartridges are recommended for medium contaminant loads, high flow rates, and low pressure drops such as required by full-flow oil filtration. The cartridges are available for particle removal in the 1- to 20-μm size range. These are recommended for most industrial and automotive oil applications.

2. Dense Media Filters

These filters are effective when in bypass service to remove particulate contamination not captured by full flow filters in systems with stringent oil cleanliness targets (Figure 21.9). The potential benefits of this type of filtration include extended oil drain intervals and increased component life. These filtration units can be installed in individual pieces of equipment by tapping into the hydraulic pressure line, bleeding off a small amount of oil, and returning back to the system reservoir. The elements will filter down to 1 μm, absolutely removing contaminants including wear metals and some moisture.

3. Clay Filtration

Certain clays, such as Fuller's earth, adsorb oil oxidation products, and the depth of clay will screen out fine particles. Fuller's earth may be contained in cartridge-type (Figure 21.10) units or may be mixed with the oil and filtered through size filtration units. The major disadvantage of clay filtration is that certain oil additives may be removed with the contaminants, which is not ideal for most additized lubricants. In practice, clay filtration of lubricants is mainly limited to removing degradation products from phosphate ester electrohydraulic controlled fluids used in power generation turbines.

FIGURE 21.9 Dense media bypass filteration units for hydraulic systems. Filter element is shown at bottom right.

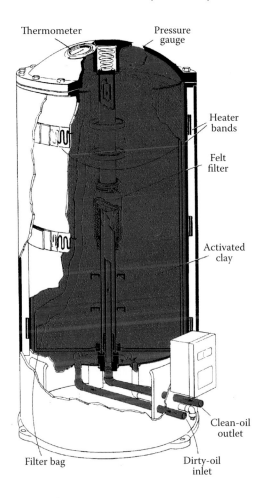

FIGURE 21.10 Activated clay-type filter.

D. MULTIPURPOSE PURIFIERS

Filtration, settling, and free water removal are often combined in a single unit, as shown in Figure 21.11. These units have capacities for treating from 50 to 2600 gal/h (189–9840 L/h). The oil conditioner is a three-compartment unit that provides dry clean oil by

1. Removing free water and coarse solids via horizontal wire screen plates in the precipitation compartment (Figure 21.11a)
2. Removing suspended solids via vertical, cloth-covered, leaf-type filter elements in the gravity filtration compartment (Figure 21.11b)
3. Polishing the oil and stripping it from moisture vapor (good for small amounts of moisture) via cellulose filter cartridges in the storage and polishing compartment (Figure 21.11c)

The leaf-type filter elements may be removed individually without shutting the system down. It has micronic selectivity down to less than 1 μm particle size and will not remove rust or oxidation inhibitors except small amounts of polar compounds that may be attached to any moisture that is removed.

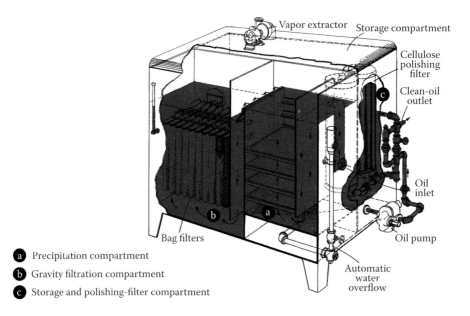

- **a** Precipitation compartment
- **b** Gravity filtration compartment
- **c** Storage and polishing-filter compartment

FIGURE 21.11 Combined filtration and settling purifier.

E. Centrifugation

Used lubricating oils can also be purified by centrifugation, a process for separating materials of different densities whether they be two liquids, a liquid and a solid, or two liquids and a solid. If the liquids are mutually soluble, the centrifugation process will not be able to separate these materials. A centrifugal force is produced by any moving mass that is compelled to depart from the rectilinear path that it tends to follow; the force being exerted in the direction away from the center of the curvature path. The centrifugal force is directly proportional to the mass or specific gravity of a material, and the higher the specific gravity, the further it will travel from the center of rotation. The centrifuge is a machine designed to subject material held in it, or being passed through it, to relatively high centrifugal force (500 to more than 10,000 g's), with means of collecting the separated materials. A batch centrifuge holds a given quantity of material and must be stopped periodically to discharge the solids and clarified liquid, and must also be recharged. Continuous centrifuges accept a steady stream of material, applies a centrifugal force, and continuously discharges the separated components. A centrifugal separator or purifier handles a continuous stream of mixed, nonsoluble liquids and effects their separation. A basket centrifuge is designed to hold a mass of material, usually a mixture of solid and liquid, and by subjecting the mass to centrifugal force, separates the liquid that passes through the basket walls leaving the solid behind.

Centrifugation by disk and tubular centrifuges, operating at high forces, are excellent for removal of (at high rates) the relatively small particles and separation of fluids of close but differing densities. The initial investment and operating costs are relatively high, and a fairly large floor space is required for the units with large volume capacity. Smaller volume capacity units are also available that can be rotated from machine to machine in industrial applications similar to portable filtering units.

F. Centrifugal Oil Filters (Separators)

Initially designed for internal combustion engines, centrifugal filters have expanded their use into industrial applications. These units are generally small and suited for systems with oil capacities up to about 100 gal (378 L) and can develop centrifugal forces of about 2000 g's. With these forces, they have the capability of removing particles down to about 0.5 μm. They are small units that

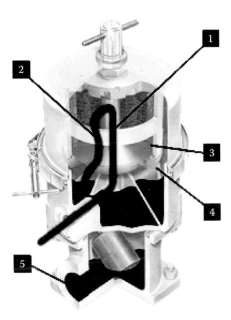

FIGURE 21.12 Spinner II® centrifugal oil filter. (1) Dirty oil enters separation chamber through hollow spindle. (2) Oil passes through spinning rotor where centrifugal force 2000 times greater than gravity separates contaminants from oil. (3) Contaminants accumulate on the rotor surface as solid cake. (4) Clean oil exits through opposing, twin nozzles that power the centrifuge up to 4000 rpm. (5) Clean oil returns to the sump/reservoir from the level control base. (Image courtesy of T.F. Hudgins Inc.)

handle about 2 gal/min (7.5 L/min) in a side stream and are driven by oil pressure and flow that spins their rotors up to about 6000 rpm. Because they are designed to discharge clean oil directly to the crankcase or reservoir, a simple valve can shut them down for cleaning without shutting down the equipment (Figure 21.12). Clean-out generally involves the removal of the sludges built up on the walls of the unit. They can remove free water and solids along with minor amounts of degradation products but do not remove additive materials other than the minor amounts that have reacted with the contaminants that are removed.

G. Sludge and Varnish Removal

In lubrication and hydraulic systems, the oil can degrade over time. Heat, contamination, and additive depletion can lead to this degradation. If the degradation or contamination continues over time, sludge and varnish deposits can form in the lubrication/hydraulic system. These deposits can lay down in important equipment areas with tight tolerances such as hydraulic control valves or on lube oil piping, heat exchangers, and reservoirs. The deposits can lead to erratic equipment performance, poor heat transfer/oil flow, and further acceleration of lubricant degradation—all of which costs the industry millions of dollars every year in downtime, premature equipment wear, and shortened lubricant life.

Figure 21.13 shows the difference between a clean hydraulic spool valve and one that has varnish deposits. This varnish can result in the valve spool sticking and poor equipment operation. Besides varnish deposits, sludge can also be deposited within systems. An example of sludge buildup on a filter compared to a clean filter that used a premium quality hydraulic oil is illustrated in Figure 21.14. Varnish and sludge deposits can be prevented by using premium-quality lubricants with special additives that promote "keep clean" performance.

Companies can successfully remove sludge and deposits using several techniques (discussed in the following subsections).

FIGURE 21.13 Varnish deposits on a servo valve spool.

FIGURE 21.14 Sludge deposits on a filter. The top filter is laden with sludge from degraded lubricant versus the clean filter at the bottom which used a premium quality hydraulic oil with keep clean performance.

1. Electrostatic Precipitation Filtration

The composition of lubricant degradation products can consist of very fine particles that are too small to catch on traditional filters. These fine particles can be squeezed through traditional filter media because of their nature. Electrostatic precipitation filters can provide a means to capture some of these fine particles. These filters remove contaminants from lube oil by use of an induced electrostatic charge. Electrostatic precipitators subject the fluid and varnish particles to an electric field. This electric field causes the particles to agglomerate and are either collected on an oppositely charged surface or are captured by the filter media. The filtration technique can work well under the right conditions. The nature of the contaminants, oil flow, and oil temperature through the filter can all influence its effectiveness.

2. Resin Filtration

Resin filtration is another means of removing lube oil degradation products. Filters with resin remove polar by-products of oil degradation by attracting and absorbing the material. Sized and operated properly, this technique has proven successful in removing varnish-forming materials in lube oil.

3. High-Velocity Chemical Flushes

Varnish or carbonaceous material can adhere to equipment surfaces in cases where filtration methods have a difficult time removing these lube oil degradation products. To clean these components typically requires a combination of high-velocity flushing with use of a specialty chemical cleaner/solvent/detergent added to the lubricant.

To avoid equipment damage, this flushing should take place during scheduled downtime using the proper equipment and techniques. In addition, some of the specialty chemical cleaners may have poor interaction with the in-service oils and are therefore required to be completely removed before returning the equipment to normal service. Additional flushing with fresh lubricant and draining and refilling may be needed to remove this chemical cleaner mixture from the system.

VI. RECLAMATION OF LUBRICATING OILS

Typically, reclamation units have the capability of removing solids, water, volatile solvents, and fuel dilution products from lubricating oil but will not remove any significant levels of oil additives or degradation materials. If the oil to be reclaimed is not severely degraded or contaminated with high flash point materials, it can be reused in its original application or else downgraded to a less severe application. Rerefining, on the other hand, strips all additives and degradation materials from the lubricating oil and essentially produces a rerefined base stock, which must be readditized to obtain the desired characteristics. The process and units used are similar to those used in a conventional refinery. On-site rerefining is not appropriate for individual industrial plants because of scale, complexity of additive blending, local environmental regulations, and economic considerations.

A. Reclamation Units (Oil Conditioners)

In addition to settling and filtration as shown in Figure 21.11, if heating elements and a vacuum chamber were added, this would constitute a reclamation unit (sometimes called an oil conditioner or vacuum dehydrator). This unit may be used to remove water and light ends usually due to contamination from solvents and other volatile materials. Reclamation may be either by batch or continuous operation. Units are available in various sizes from those that are portable, capable of handling from as little as 6 gal/h (22.7 L/h), to large stationary units with capacities as high as 3000 gal/h (11,355 L/h). The portable units can be rotated from one machine to another similar to filter carts.

In a reclamation unit, the contaminated oil is generally prefiltered to remove the bulk of large solids and then passed through heaters to attain the desired temperature. When the correct temperature (~180°F or 80°C) is reached, the oil is routed through filters into a vacuum chamber where much of the water and volatile materials (solvents, fuels, etc.) are removed. The high temperature lowers the viscosity of the oil, which aids in separation of the water and volatile materials in the vacuum chamber and helps increase the final filterability after leaving the vacuum chamber. The filtering units are generally cascading units starting with large particle size removal capability (40 µm) down to 5 µm or less in the final stage. This cascading filtration improves the efficiency of filtration and reduces filter replacement costs. The clean oil is then passed through an initial heating section to reclaim the heat for heating of the dirty oil entering the unit, and then the cooled oil is transferred to the operating unit or a holding tank. Reclamation units do not remove additives other than those polar materials that become attached to the water molecules.

VII. WASTE OIL COLLECTION AND ROUTING

We have discussed the selection, in-service handling, and purification of lubricants for industrial and commercial applications to help minimize contamination and assure long equipment and lubricant life. Eventually, all lubricants will become waste oil and must be drained from the system, collected into waste storage tanks, and prepared for one of several methods of final disposal. The fact that waste lubricants can be defined as hazardous materials by some local environmental regulations can add complexity and cost to final disposal. The following general considerations are taken into account when it comes to disposal of lubricants:

1. Handling of waste oil so that it will have the optimum stability for disposal in order to generate the highest value as a by-product.
2. The need to be compliant with local environmental regulations regarding waste oil collection and disposal. This includes preventing the waste oil from polluting any in-plant or nearby process water, storm drain, and sewer systems as well as any groundwater and surface waters outside the plant.

This can be accomplished by using proper collection and routing systems. An example of an intermediate-size multimachine collection and reclamation system is shown in Figure 21.15. A simplified alternative system may consist of a combination of a sump tank cleaner and centrifuge for renewing and purifying used oil containing sludge, chips, and other solid contaminants similar to a portable reclamation unit. Most of these types of units have their own pumping and discharge system, although some units will work on vacuum to pull the oil into the unit and, by means of reversing valves, will use air pressure to discharge the oil back to the sump. Waste oil can then be carted back to holding tanks or drums for final disposal.

Whatever the size of the system for the collection and routing of the waste oil, there are certain principles that should be followed:

1. Separate collection systems should be used for each type of lubricant, and separate storage should be provided until final disposal. Neat oil and lubricants should be kept separate from emulsions, chemical coolants, etc.
2. Lubricating oils should be kept separate on the basis of the degree of contamination even though the type of oil may be the same. A relatively clean hydraulic fluid should not be mixed with waste metalworking fluids or dirty lubricants. The value of the waste hydraulic fluid as a by-product is considerably lower when mixed with metalworking fluids or coolants.
3. Reservoirs should be drained immediately after shutdown while the oil is still at operating temperature and solids are still in suspension. Proper safety precautions must be exercised if the oil is excessively hot. Certain oxidation products also become insoluble in cold oil and would precipitate out, remain in the reservoir, and contaminate a new charge of oil.
4. Every precaution should be made to prevent the contamination of process water, storm drains, and sewer systems by waste oil streams. This is particularly a consideration where floor drains exist and spills of lubricating oil occur. It is also possible that leaky heat exchangers or oil coolers can cause contamination of the process stream or the cooling water being treated. It is common practice to maintain oil pressures through heat exchangers at a higher level than water pressure, to prevent the lubricant from contamination with water in the event of leaks.
5. Troughs, gratings, and drains in the open floor that are meant to catch leaks, spills, and drips are not desirable because of the high potential for further contamination and admixture with noncompatible products, making separation or reclamation very difficult.

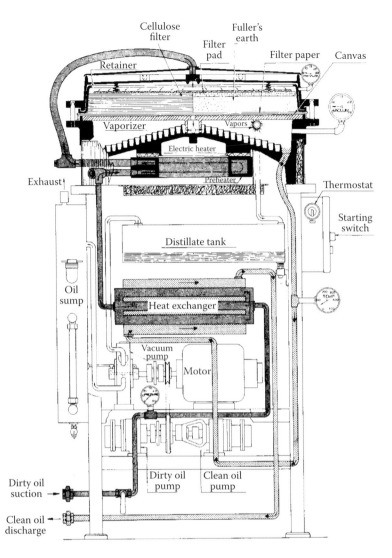

FIGURE 21.15 Multimachine collection system.

Whether individual machine collection systems or plantwide systems are used will be decided on the basis of economics and feasibility. However, the plantwide system is desirable in most cases. Individual machine systems can be serviced with commercially available oil carts, similar to those shown in Figure 20.28 in Chapter 20.

If it is not feasible to connect the individual machines to a waste oil collection and reclamation system, consideration should be given to a safe system to carry waste oil back to the waste oil storage area. If any type of containers (e.g., drums, minibins, intermediate bulk containers) are used for temporary waste oil storage, they must be clearly marked and kept closed. The waste oil area should be carefully designed so that waste oil will not leak or contaminate the floor and surrounding area. It should be isolated from ordinary water floor drains or washing stations. Where it is impossible to have separate oil drainage systems, the floor drains in appropriate areas should be equipped with oil traps. Additionally, storing waste oil in containers that can be confused with new lubricants should be avoided, so as to prevent misapplication or contamination.

FIGURE 21.16 Example of debris found on a strainer after flushing a lubrication system.

VIII. EQUIPMENT COMMISSIONING AND FLUSHING

As important as keeping the lubricant clean while in service, a plant should also ensure proper lubricant cleanliness upon the commissioning of a new or rebuilt piece of equipment. Each year, industry loses millions of dollars in lost revenue and increased maintenance costs because of inadequate lubrication system and equipment component flushing. Figure 21.16 illustrates the type of debris that is present and can be removed by flushing during equipment commissioning. If not removed, this contamination can lead to severe equipment damage and poor equipment functionality.

The key elements supporting the goal of achieving the successful flushing of a lubrication system include:

- Understand the nature of the material that should be removed from the lube system.
- Establish clear goals for the flushing.
- Develop and implement written flushing procedures. This includes conducting safety risk assessments, selecting the appropriate flushing fluid, specifying all equipment needed to conduct the flushing, use of proper flushing techniques, and addressing any compatibility issues between the equipment, the lube system, and the flushing fluid.

As an example, for turbine applications, plant personnel can use the ASTM (American Society for Testing and Materials) International D6439-11, Standard Guide for Cleaning, Flushing, and Purification of Steam, Gas, and Hydroelectric Turbine Lubrication Systems, as a guide in developing executable flushing plans for lubricating systems.

IX. FINAL DISPOSAL

Lubricants including metalworking fluids, hydraulic fluids, and gear oils eventually reach a state where they must be disposed of as waste oil. The method of final disposal will depend on the type of oil, contaminants, and condition of the waste oil. If local regulations allow, there are several generally accepted methods that permit disposal with adequate safeguards to prevent water and air pollution, including the following:

1. Reclaiming the heat value by using as a fuel supplement
2. Rerefining the waste oil to produce a base oil for lubricants
3. Incineration
4. Use of coal or petroleum coke spray, or density control spray for specific processes where appropriate

One of the more feasible methods of waste oil disposal is to burn them as fuel; however, local regulations need to be reviewed before pursuing this option. Testing of the candidate waste oil is recommended to assure that certain hazardous materials do not exist in sufficient quantities to violate clean air standards. Burning waste oils as a fuel is highly recommended if it is an approved method of final disposal for the following reasons:

1. It is the most widely applicable method from the standpoint of type of waste oil and location of generating source.
2. If the burning and heat recovery can be done at the site of the waste oil generation, it is sometimes considered recycling by local regulations.
3. There are practically no volume limitations on the amount of waste oil that may be handled by this method. It is feasible for the large, medium, and smaller users of petroleum lubricants.
4. It turns what could be a liability (disposal of waste oil) into an asset. Often, contract waste oil haulers charge to remove waste oils. A significant economic benefit can be realized by using waste oil as a fuel supplement to offset high fuel prices and to take advantage of the high BTU content of lubricating oils.

Furthermore, there is an ever-increasing demand for petroleum products (gaseous and liquid) as an energy source. Therefore, any secondary use of waste oil, subsequent to maximizing its use in its primary function, will recover additional value as a petroleum product and conserve natural resources. Burning of used lubricants as fuel is the simplest of disposal methods.

Although practically all liquid petroleum products may be oxidized into carbon dioxide and water by burning, not all of them are suitable for use as fuel or for mixing with conventional fuel. Some, because of contamination with large amounts of water, low flash petroleum products, excessive sediment, or contamination with hazardous materials, lose their value as fuel and must be incinerated. This clearly demonstrates the need for proper in-plant handling of used oil to maximize the value of waste oil during its eventual disposal.

BIBLIOGRAPHY

ASTM D6439-11. 2011. Standard Guide for Cleaning, Flushing, and Purification of Steam, Gas, and Hydroelectric Turbine Lubrication Systems, ASTM International, West Conshoken, PA.
Mobil Oil Corporation. 1971. Mobil Technical Bulletin: In-Plant Handling to Control Waste Oils. New York.

22 In-Service Lubricant Analysis

In addition to selecting the most appropriate lubricant for an application, it is important to monitor and maintain the condition of the lubricant in order to achieve the expected life of the equipment and the lubricant. One tool that successful maintenance professionals use to accomplish this is in-service lubricant analysis, which is also often called used oil analysis.

In-service lubricant analysis was pioneered in the 1940s and 1950s in the railroad and aerospace industries and became a more mainstream maintenance tool for general industry in the 1970s. The use and evolution of used oil analysis continues to grow and plays a key role in today's maintenance programs. Used oil analysis has value regardless of the type of maintenance program in use—whether it is a reactive, preventive, predictive, or a reliability-centered program.

A well-managed used oil analysis program will maximize both equipment life and oil life while reducing maintenance costs and unscheduled downtime resulting in increased productivity. The annual savings stemming from the benefits that can be realized with a successful used oil analysis program has been well documented. This program can easily provide a return of 10 to 100 times more than the expenditures associated with it.

When properly used, used oil analysis can detect changes in equipment wear rates or oil condition that provide an early warning to maintenance professionals so that they can investigate and take corrective action if necessary. Although a single oil analysis result can provide useful information, the real value is in trending used oil analysis data for a specific piece of equipment over time. Used oil analysis can help predict or prevent premature equipment failures due to accelerated wear; however, it is not a good predictor of random catastrophic equipment failures that are usually unrelated to equipment condition (e.g., a broken or fractured shaft, coupling, bearing).

Most used oil analysis programs are centered on securing a representative oil sample from the oil sump of a piece of equipment and transporting that sample to a laboratory for analysis and interpretation. Another growing trend that continues to evolve as oil and instrument technology advances is on-site oil analysis. In some cases, on-site lubricant analysis can provide substantial benefits to an operation by having access to fast results to take immediate corrective action. This on-site analysis is available in two forms today: (1) taking a sample and manually analyzing it on-site with laboratory-quality or simulated/correlated testing instruments, or (2) using in-line equipment condition monitoring instruments and sensors for real-time analysis. The criticality of the equipment and the need to understand the oil and equipment condition at a moment's notice will determine whether on-site analysis is most appropriate for the type of maintenance program being used. Obviously, sophistication of the maintenance program, available manpower resources, and cost will factor into the decision on whether to use on-site analysis. Today, it is difficult to replace the full breadth of precision testing available at a laboratory and hence its popularity. However, online condition monitoring is expected to play a major role in the future for many maintenance professionals.

I. ESTABLISHING AN IN-SERVICE LUBRICANT ANALYSIS PROGRAM

To extract the desired value from a used oil analysis program, there are several factors that should be considered long before the selection of a lubricant analysis provider or implementation of a program. Besides determining an annual budget for a used oil analysis program, the amount of involvement in performing, managing, and condition-based decision making should be established at an early stage.

A. Elements of a Successful In-Service Lubricant Analysis Program

Based on decades of experience, the key elements of a successful in-service lubricant analysis program should include the following:

1. *Obtain management commitment*—Having management sponsorship and leadership is critical particularly in understanding the goals and value of the program and ensuring that the necessary resources are available. Without this, many well-intentioned oil analysis programs will fall short of extracting the desired value and often will be abandoned over time.
2. *Define the desired goals and metrics*—Besides having a good understanding of the purpose of the program, it is important to establish numerical goals that fit into the overall maintenance program goals. Specific goals around overall maintenance costs, equipment uptime/reliability, lubricant consumption, parts/equipment replacement costs, and productivity improvement, are just some examples to consider.
3. *Establish program accountability*—A key to any successful program is to ensure that accountability/ownership of the program is well known and accepted, and that roles and responsibilities of all involved parties are well defined.
4. *Train and educate involved personnel*—Initial and regular training are an essential part of program execution. There are specialized skills that are vital to the program that need to be developed for all parties involved; otherwise, any data received could be of questionable or limited value.
5. *Respond to the analysis results*—This is the pivotal focus of the entire program. If no action is taken when early warning or critical conditions are identified, then there is little purpose to continuing the program.
6. *Measure program results versus goals*—Annual reviews are a best practice in evaluating the success of the program, documenting resultant savings, identifying continuous improvement opportunities, and establishing goals for the next year.

The oil analysis provider, the lubricant supplier, or the oil analysis consultant should have significant experience in establishing a successful oil analysis program and should be able assist their client with any of the above elements. They can be a tremendous asset in helping to achieve the desired benefits of the program.

B. Selecting an In-Service Lubricant Analysis Provider

There is much more involved in selecting a used oil analysis provider than just considering the cost of the analysis or the perceived quality of the oil analysis laboratory. A number of important selection criteria and considerations to use in comparing potential used oil analysis providers are highlighted in Table 22.1.

Ideally, the potential used oil analysis provider should be able to adequately demonstrate its capability for each of the key selection criteria areas and meet desired goals. There are cases where a used oil laboratory itself may not be able to meet all of these selection criteria and additional resources such as the oil supplier, original equipment manufacturer (OEM), information technology solution provider, or expert oil analysis consultant may be required to supplement the support needed for the used oil analysis program.

Those discerning maintenance personnel who desire to have a top-tier used oil analysis program, or have very critical equipment to protect, or expect to send to a laboratory hundreds or thousands of samples per year, may want to consider an on-site visit to those prospective used oil analysis laboratories to assist them in their selection process. They may want to also consider undertaking a pilot

TABLE 22.1
General Guidelines and Considerations Associated with Selecting a Used Oil Analysis Provider

Selection Criteria	Key Considerations
Application appropriate test slates	Determine whether the offered test slates are specific to each application and whether the individual tests will be able to adequately assess equipment condition, oil contamination, and oil condition.
Laboratory quality program	Verify an active quality program is in place with process for continuous improvement. Should meet or exceed requirements of ISO 17025.
Oil analysis results quality	Demonstration of consistent and reliable analysis results. Regular participation and willingness to share results from an ASTM Interlaboratory crosscheck or equivalent round robin program. Evaluate frequency of instrument calibration and comparison versus a known standard sample and skill level of laboratory personnel.
Sample turnaround time	Determine laboratory commitment to analyze samples and share results for most standard test slates within 48 hours of sample receipt. Confirm ability to handle occasional rush/priority samples.
Laboratory location/logistics	Validate provider's location as to whether sample shipping time is acceptable. Mail stop size/location and customs treatment can possibly delay samples getting to select laboratory sites.
Cost of analysis	Determine total cost of oil analysis to include all costs associated with the analyses, sample kits, shipping, and any other add-on costs. Conduct overall value analysis with like-to-like comparisons on test slates, service capability and other selection criteria.
Oil analysis results interpretation	Evaluate depth and breadth of results interpretation. Determine how warning limits are determined, applied, and updated. Understand level of detail/sophistication of corrective action and troubleshooting recommendations for both the oil and equipment. On-site availability to assist with troubleshooting of equipment.
Information management	Review access to and completeness for all sample point data, ease of use of system, relevance of data analysis and customization capability, and general fit/linkage into overall maintenance program. Overall IT support capability.
Ease of doing business	Assess overall order fulfillment process, appropriate sample kit materials/mailers, sample label bar coding capability for preregistered sample information, ready access to knowledgeable people and information.
Customer experience	Ready access to a customer reference list and ability to contact select customers to review their experience and overall level of satisfaction.
Training and support	Online access to training materials and on-site support in establishing and maintaining the program as needed.
Business continuity plans	Evaluate provider's backup plans to continue service in the event of an extended laboratory or IT system outage. Review overall financial stability of provider and any dependency on key positions (e.g., owner/manager, IT developer, chief chemist).

program or trial period with the used oil analysis provider before making a final commitment to ensure that their needs and goals can be met. A good value analysis and experience will ultimately determine the best fit.

II. USED OIL ANALYSIS PROGRAM STARTUP RECOMMENDATIONS

Proper planning is vital to the successful startup and continued operation of a used oil analysis program. A well-designed used oil analysis program should include predefined sample points, oil sampling intervals, and a repeatable oil sampling procedure. As with any activity, safety must be considered first.

A. Sample Point Selection

Before any oil sample is taken, it is necessary to predetermine which equipment and what sample points on the equipment to include in the oil analysis program. Each maintenance professional should evaluate this up-front, and the selection of sample points can vary by facility, depending on such factors as the desired goals of the program, the type of maintenance program in use, and resource availability. The following are various considerations to take into account when going through the sample point selection process:

- Criticality of the equipment to production
- Cost of downtime associated with the specific equipment
- Expense of the equipment, cost of repair and parts replacement, and availability of parts replacement
- Expected life of the equipment
- Operating and environmental severity the equipment is exposed to
- Capacity of the oil sump (users may want to establish a minimum threshold)
- Access and safety associated with securing an oil sample

Once the sample points are selected, a best practice is to preregister each sample point with the laboratory before sending in a sample for analysis. This will allow complete and accurate registration of equipment information, which is essential to interpretation of oil analysis results and comparison with similar equipment. Having the laboratory register sample point information off of the label is very inefficient and is subject to many transcription errors along the way. The more modern oil analysis programs have online equipment registration capability.

B. Oil Sample Intervals

Establishing oil sample intervals for each sample point is an important facet of any good used oil analysis program. The real value that comes from the program is in the trending of oil analysis results, and this begins with determining oil sample intervals and regularly adhering to these intervals. This provides a credible historical trend of machine performance. Some of the more progressive used oil analysis providers include sample interval fields in the online sample point registration that will automatically send timely reminders when an oil sample is due.

A good general guidance is to follow OEM recommendations when it comes to establishing oil sample intervals. This is especially important during any warranty period. In the absence of OEM recommendations, Table 22.2 provides several general guidelines on sample frequency for various types of equipment, which are based on past experience.

More frequent initial sample intervals might be appropriate for equipment that may be nearing or exceeding operating design parameters such as load or speed, or may be operating under severe environments where extreme heat or a greater potential for contamination ingress is present. More frequent sample intervals should also be considered when determining the optimum oil drain period. For most engines (except very large engines like those seen in gas transmission, marine service, or stationary power), it is quite common for the sample interval to coincide with the oil drain interval. Less frequent oil sample intervals could be considered for equipment that are inactive for extended periods or for less critical equipment. As a trend develops over time and very little change is seen in the oil analysis results, then sample intervals may be further extended. Additionally, oil type (e.g., synthetic versus mineral) and oil age could influence sample intervals. Lastly, the oil analysis result itself could be an indicator to increase the frequency of sample intervals as higher-than-expected wear, high contamination levels, or deteriorating oil conditions progress.

TABLE 22.2
General Guidelines for Oil Sampling Intervals

Application	Sample Point	Sample Frequency
Off-highway	Diesel engine	250–500 h
Off-highway	Wheel motor	250–500 h
Off-highway	Differential/gear	500–1000 h
Off-highway	Hydraulic system	500–1000 h
Off-highway	Transmission	500–1000 h
Off-highway	Final drive	500–1000 h
On-highway	Diesel engine	15,000–25,000 miles or 24,000–40,000 km
On-highway	Transmission	25,000–50,000 miles of 40,000–80,000 km
Industrial	Landfill gas engine	250 h
Industrial	Natural gas engine	500 h
Industrial	Paper machine lube system	1–3 months
Industrial	Turbine	1–3 months
Industrial	Compressor	3–6 months
Industrial	Gear drive	3–6 months
Industrial	Hydraulic system	3–6 months

C. Taking a Representative Used Oil Sample

One of the most important steps to ensuring accurate analysis results and gaining the most benefit from these results is to regularly adhere to correct oil sampling practices. The failure to collect a representative used oil sample can have serious consequences stemming from the misinterpretation of analysis results. Typically, these erroneous results mean that an undesirable condition goes undetected or costly corrective action is taken where it is not needed. All too frequently this extremely important step does not receive the attention it truly deserves.

Oil analysis providers or lubricant suppliers can provide training and guidance for taking representative oil samples from equipment. Having a documented procedure and training maintenance personnel on using these procedures is a recommended initial step.

1. Recommended Sampling Procedure

The following is a summary of best practices for taking representative used oil samples:

- Always use a clean, dry container to draw oil samples.
- Take oil samples at a consistent interval and location.
- Sample when the machine is running and at normal operating temperature.
- Ensure a safe work environment for taking samples; otherwise, draw samples as soon as possible after the equipment has been shut down.
- Install and use properly placed sample valves if possible. For circulation systems, this should be downstream of the filter or in the main return line.
- Samples taken downstream of the lube oil pump through a sample port should use dedicated stainless steel tubing.
- Clean the area around the sample point, then fill the sample bottle to the neck and discard this oil as bottle flush. Then for the final sample, fill the same bottle up to its neck (do not fill the bottle to the very top of the lid to avoid leakage during sample shipment). For samples that will receive a particle count cleanliness measurement, the best practice is to

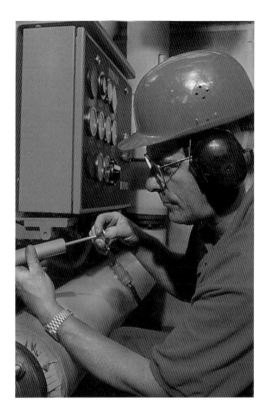

FIGURE 22.1 Use of a sample vacuum pump to extract a used oil sample.

fill the final bottle two times and discard this flush oil from the reservoir before collecting the final sample.
- If using a sample vacuum pump (Figure 22.1), first draw the sample and discard the flush oil from the flush bottle. Then connect the new, clean, and dry sample bottle and fill the sample bottle to the neck and discard this as flush, then collect your final representative sample again filling the bottle to the neck. For samples that will receive a particle count cleanliness measurement, the best practice is to fill the final bottle two times with flush oil from the reservoir, then collect the final sample.
- Ensure that samples are taken at least 6–12 inches (15–30 cm) from the bottom of the reservoir to avoid introducing any settled contaminants such as dirt, water, and sludge.
- Record key sample details (date sampled, oil name and age, equipment hours/mileage), and ensure that this information is included on the sample label.
- Ship sample in approved packaging to oil analysis provider within 24 hours.

2. **Visual Inspection of the Sample**

A considerable amount of information can be gathered by simply looking at the oil sample, and it is quite helpful to use clear plastic sample bottles for oil samples intended for this purpose. Inspecting each sample carefully before submitting it to the laboratory for analysis is highly recommended to validate whether a representative sample was taken and whether some form of corrective action should be taken immediately.

Sample clarity is an excellent indicator of contamination. A lubricant in good condition is generally clear and bright. The exception is engine oil, which can turn very dark to black owing to discoloration from contaminants such as soot, partially burned fuel, or combustion by-products. Sample haziness or cloudiness can be an indication that contaminants such as water, water-based coolants, refrigerants, or an incompatible lubricant is present. Excessive air entrainment in the sample can also cause a hazy appearance. In some cases, free water can be seen at the bottom of the sample, which is usually an indication of severe water contamination in the sample or possibly an unrepresentative oil sample. Note that an oil sample stored in a cold environment may appear hazy because of dissolved water, but when warmed up to a normal indoor temperature it may appear clear.

Sediment and particulate matter found at the bottom of the sample bottle can tell even more of the story. Nonmagnetic sediment in an otherwise clear and bright oil sample may suggest that dirt, dust, wear metals other than iron, or sand contamination is present. Ferrous particles could indicate rust or a severe wear condition. Passing a magnet under the sample bottle can help to identify the type of particle.

It is highly recommended to take corrective action before sending a sample to the laboratory if contamination (water, dirt, metal) is visible. Figure 22.2 is an example of a visual inspection that identifies contaminants. Visible particles or water in a sample reflect the possibility of abnormal equipment condition. In most systems, having visible contamination in the oil is undesirable and could indicate that equipment damage may be occurring, which should prompt an immediate call for action. Under these conditions, waiting to take action until the analysis results come back from the laboratory may only exacerbate the situation and further shorten equipment life. Another sample should be taken if there is any question whether this was a representative oil sample to confirm the findings. Corrective action might include draining or removing the water, filtering the oil, or—in extreme contamination cases—first draining the oil, then flushing the system, and finally refilling with new lubricant. The cause of contamination ingress should be addressed and equipment condition should also be evaluated at this time.

FIGURE 22.2 Example of visible contaminants in four used oil samples. Left to right: separate water layer at bottom, hazy appearance owing to dispersed water, heavy particulate matter adhering to bottom of bottle, sludge and sediment.

III. IN-SERVICE LUBRICANT ANALYSIS TESTING

As mentioned previously, having the test slate identified is recommended at the startup of the used oil analysis program before any samples have been secured and submitted to the laboratory. Again, either the used oil analysis provider or lubricant supplier can provide guidance here.

The testing of used oil samples often differs from the tests performed on new oil. Typically, tests used for new oil conform to the American Society for Testing and Materials (ASTM), International Standardization Organization (ISO), or other industry test standards. Tests conducted for used oil frequently deviate from established industry test standard procedures, but many still correlate back to these industry tests. There are a variety of reasons for this deviation. Most industry tests are designed for new oil and may not give representative or meaningful results for used oil. Some select tests designed specifically for used oil often give more relevant results. Limited sample quantity, use of less toxic solvents, and the use of automation in the laboratory have all had an influence in the adoption of modified test methods. Sample testing time and cost are other factors that have prompted used oil analysis providers to modify industry standard test procedures. All of this can make it difficult when comparing results from one used oil analysis laboratory to another because the test methods used are different, although there should be some degree of correlation back to industry test standards.

A. COMMON USED OIL ANALYSIS TESTS

There are a multitude of used oil analysis tests that are used depending on the type of equipment or application, the lubricant in use, the severity of the application, and the price of the analysis. There are often multiple test methods or procedures available to analyze the same physical or chemical property of the oil or contaminant (e.g., water). In some cases, quicker and lower-cost tests are used as a screener or conditional test before a more expensive and complex test is conducted. It is good practice to understand the test methods being used by the laboratory to have a good idea of their application and relevance as well as any limitations. Table 22.3 is a general summary chart used to describe some of the more common tests utilized for routine used oil analysis.

Other more advanced or specialized used oil analysis tests may be used depending on the lubricant in use and application. A short list of some of these other used oil analysis tests include: Flash Point, Insolubles, Ferrography, Chlorine, Ultra Centrifuge Rating, Membrane Patch Colorimetry, Linear Sweep Voltammetry, Rust, Demulsibility, Foam, Air Release, Rotating Pressure Vessel Oxidation Test, Color, and Initial pH (ipH). As is the case with the more common routine oil analysis tests, there may be a variety of different test methods available for each test.

1. Grease Analysis

In certain applications, testing in-service grease can provide insights to abnormal equipment conditions. These applications include large, slow-turning bearings on wind turbines, mining shovels, side tank agitators, paper machine rolls, and steel mill working rolls. Typically, tests include wear/contamination detection using spectrometric analysis and contamination analysis using infrared analysis. In some cases, testing for grease drop point, penetration, and other grease characteristics may provide value.

To ensure meaningful results and analysis, obtaining representative grease samples is important and can prove challenging. Some devices such as sampling syringes can aid in this effort.

B. APPLICATION SPECIFIC USED OIL ANALYSIS TEST SLATES

Having the right combination of tests in a test slate is critical to making an overall assessment of the condition of the oil, the condition of the equipment, and the type and amount of oil contamination. This involves using a test slate that selects the most relevant tests for the lubricant in service and the application. Caution is advised as some used oil analysis providers may offer the same test slate irrespective of the application or lubricant with the premise that having more data is better.

TABLE 22.3
Common Tests Used for Routine Used Oil Analysis

Test	Purpose and Method(s)
Viscosity	Indicates a fluid's resistance to flow at a given temperature (typically measured at 40°C or 100°C and reported in centistokes). Results indicate either a physical change in the lubricant or contamination by other fluids.
Oxidation	Signals the deterioration of the oil due to thermal breakdown and aging causing a physical change in the oil. Typically measured using differential infrared analysis and reported in absorbance per centimeter (Abs/cm).
Nitration	Signals the deterioration of the oil due to nitrogen oxides forming during engine combustion or rapid compresssion of entrained air. This physical change in the oil is a precursor to varnish and sludge. Typically measured using differential infrared analysis and also reported in Abs/cm.
Total acid number (TAN)	Measures the oil's acidity representing a chemical change in the oil. Indicates degradation of the oil in service leading to deposit formation. Typically, ASTM D-664 is used/referenced—results reported in milligrams of potassium hydroxide per gram of test oil neturalized (mg KOH/g).
Total base number (TBN)	Measures the oil's ability to neutralize harmful acidic compounds produced in the combustion process. Typically, ASTM D-2986 or D-4739 is used/referenced—results reported in milligrams of potassium hydroxide per gram of test oil neturalized (mg KOH/g).
Water	Determines the amount of water contamination—reported in either % or parts per million (ppm). Water causes chemical problems for the oil and physical problems for equipment. Various tests with different levels of precision are available to measure free, bound, and/or dispersed water.
Glycol	Indicates the presence of coolant (antifreeze) contamination in engine oil coming from the elements of ethylene glycol. Various tests with different levels of precision from positive/negative to % volume are available. Coolant inhibitor metals often are a good source to identify glycol contamination.
Fuel dilution	Indicates the presence of fuel (diesel or gasoline) contamination in the oil often originating from faulty fuel injectors or engine blow-by. Typically reported in % volume. Various tests with different levels of precision are available with gas chromatography being the most effective and used.
Fuel soot	Measures the amount of insoluble carbon soot in the oil that is the by-product of incomplete engine combustion—reported in % weight. Several methods commonly used include thermogravimetric analyzer (TGA), Soot Index using IR analysis correlated to TGA, and a fuel soot meter.
Particle count	Physical measurement of the size and amount of particles per 1 ml of oil. Indicates the relative cleanliness of the oil and filtration efficiency. Various test equipment/methods are used with results expressed using ISO 4406 identifying the number of particles >4, >6, and >14 µm that correlate to an ISO cleanliness code rating.
Particle quantifier (PQ) index	Measures the mass of ferromagnetic wear particles/debris in the oil irrespective of the size of the ferrous particles. Results are an index reported in unitless whole numbers. Effective tool for detecting large ferrous wear particles and trending all ferrous particles. Excellent complement to spectrometric metals analysis.
Spectrometric metals analysis	Provides elemental metals analysis of up to 22 metals in the oil representing wear metals, additive metals, and contaminant metals. Measures submicron metals up to 6–10 µm in size—results reported in ppm or wt.%. Various instruments/methods in use with inductively coupled plasma (ICP) among the most common and precise.

This may not necessarily represent best practice as extraneous or less relevant tests or data can be more costly, more confusing, and be a distraction. Consulting with the used oil analysis provider, lubricant supplier, and/or primary OEMs of the equipment in place is advisable to ensure that the most appropriate test slate is available and selected.

Table 22.4 shows a list of recommended used oil analysis test slates for engines and automotive applications. The diesel engine test slate refers to distillate fueled engines. For diesel engines using residual (bunker) fuel such as those used in some marine and stationary power applications, a slightly different test slate including the detection of asphaltenes is suggested. The landfill gas engine test slate would also address engines that generate power from other gases besides landfill including sewage, biogas, coal mine gas, and other special gases. The automotive powertrain test slate would include applications such as transmissions, differentials, final drives, planetary drives, and axles.

Table 22.5 is a list of recommended used oil analysis test slates for industrial applications. Specialized tests and test slates may be warranted for specialized or more critical industrial applications. For some heavier viscosity products, such as open gear lubes, it may be more appropriate to analyze the viscosity at 100°C. Testing for oxidation may be difficult for some synthetic lubricants because of infrared band interference of some components, and in these cases Total Acid Number (TAN) may be a more appropriate test.

TABLE 22.4
Recommended Used Oil Analysis Test Slates for Engines and Automotive Applications

Test	Diesel and Gasoline Engine	Dual Fueled Engine	Natural Gas Engine	Landfill Gas Engine	Automotive Powertrain
Viscosity at 100°C	×	×	×	×	×
Oxidation	×	×	×	×	×
Nitration		×	×	×	
Total acid number (TAN)			×	×	
Total base number (TBN)	×	×	×	×	
Water	×	×	×	×	×
Glycol	×	×	×	×	×
Fuel dilution	×	×			
Fuel soot	×	×			
Particle count					×
Particle quantifier (PQ) index	×	×	×	×	×
Chlorine				×	
Spectrometric metals analysis	×	×	×	×	×
Aluminum	×	×	×	×	×
Barium	×	×	×	×	×
Boron	×	×	×	×	×
Calcium	×	×	×	×	×
Chromium	×	×	×	×	×
Copper	×	×	×	×	×
Iron	×	×	×	×	×
Lead	×	×	×	×	×
Magnesium	×	×	×	×	×
Molybdenum	×	×	×	×	×
Nickel	×	×	×	×	×
Phosphorus	×	×	×	×	×
Potassium	×	×	×	×	×
Silicon	×	×	×	×	×
Silver	×	×	×	×	
Sodium	×	×	×	×	×
Tin	×	×	×	×	×
Titanium		×	×	×	
Vanadium	×	×	×	×	
Zinc	×	×	×	×	×

TABLE 22.5
Recommended Used Oil Analysis Test Slates for Industrial Applications

Test	Hydraulic	Circulating Systems	Gear Drives	Steam Turbine	Gas Turbine	Compressors	Paper Machines
Viscosity at 40°C	×	×	×	×	×	×	×
Oxidation	×	×	×	×	×	×	×
Total acid number (TAN)				×	×		
Water	×	×	×	×	×	×	×
Glycol						×	
Particle count	×	×	×	×	×	×	×
Particle quantifier (PQ) index	×	×	×	×	×	×	×
Spectrometric metals analysis	×	×	×	×	×	×	×
Aluminum	×	×	×	×	×	×	×
Barium	×	×	×	×	×	×	×
Boron	×	×	×	×	×	×	×
Calcium	×	×	×	×	×	×	×
Chromium	×	×	×	×	×	×	×
Copper	×	×	×	×	×	×	×
Iron	×	×	×	×	×	×	×
Lead	×	×	×	×	×	×	×
Magnesium	×	×	×	×	×	×	×
Molybdenum	×	×	×	×	×	×	×
Nickel	×	×	×	×	×	×	×
Phosphorus	×	×	×	×	×	×	×
Potassium	×	×	×	×	×	×	×
Silicon	×	×	×	×	×	×	×
Silver				×	×		
Sodium	×	×	×	×	×	×	×
Tin	×	×	×	×	×	×	×
Titanium				×	×		
Vanadium				×	×		
Zinc	×	×	×	×	×	×	×

C. Test Precision

When comparing test results from an identical sample that was split and either analyzed in the same laboratory or in another laboratory brings test precision into consideration. Test *precision* is the magnitude of variability that can be expected between test results from a particular test method, either from the same laboratory or from different laboratories. All test methods, no matter how good, have systematic and random errors. The precision of a particular test tells you how large this variability could be.

The *precision* of a test does not tell you how accurate the test method is, as *accuracy* is the closeness of agreement between an individual measured value and the value considered to be the *true value*. The *true value* is an accepted reference value or conventional true value. An accepted reference value is usually obtained from a certified reference material. Figure 22.3 describes the relation of varying precision, accuracy, and trueness.

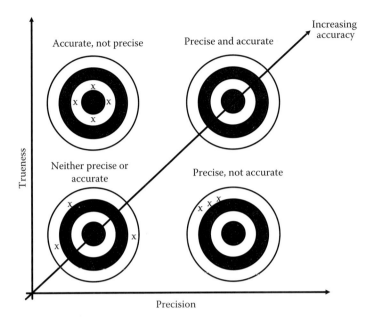

FIGURE 22.3 Examples of varying precision, accuracy, and trueness.

Within precision, there are a number of definitions:

1. Repeatability (r)—The difference between successive results obtained by the same operator in the same laboratory with the same apparatus under constant operating conditions on identical test material
   Note: Repeatability is typically used to check the "sameness" of two successive results.
2. Reproducibility (R)—The difference between two single and independent results obtained by different operators working in different laboratories on nominally identical test material
3. Site repeatability (r')—The difference between successive results obtained by different operators in the same laboratory with the similar apparatus under constant operating conditions on identical test material (normally between r and R)

Most industry standard test methods will have a different precision statement that will show whether the method has good or poor precision. Therefore, one needs to be aware of what this is to make a decision on whether or not two test results are considered similar. For many used oil test methods used in a particular laboratory that do not adhere strictly to industry standard test procedures, this precision statement may not have been defined and is therefore not available.

Below are some example precision statements for some industry standard tests:

Viscosity, ASTM D445
Repeatability—1 case in 20 exceeds 0.26%.
Reproducibility—1 case in 20 exceeds 0.76%.
Total Base Number (TBN), ASTM D2896
Repeatability—1 case in 20 exceeds 6.2%.
Reproducibility—1 case in 20 exceeds 16.2%.

TBN, ASTM D4739
Repeatability—1 case in 20 exceeds $0.22X^{0.47}$, where X is the average of two separate readings.
Reproducibility—1 case in 20 exceeds $1.53X^{0.47}$, where X is the average of two separate readings.

TAN, ASTM D664
Repeatability—1 case in 20 exceeds $0.117X$, where X is the average of two separate readings.
Reproducibility—1 case in twenty exceeds $0.44X$, where X is the average of two separate readings.

IV. INTERPRETATION OF USED OIL ANALYSIS RESULTS

After assuming that a representative used oil analysis sample has been taken and submitted to the laboratory where the most appropriate tests for the specific product application have been performed, the next most critical phase is the interpretation of test results. A good interpretation of used oil analysis results should include four key elements:

1. Overall condition of the sample results
2. Comparison and rating of each test against established condemning limits
3. Identification of the likely cause of any abnormal result
4. Corrective action recommendations to address any abnormal results

A. OVERALL SAMPLE RESULT CONDITION

A best-in-class program should include sample result reports that summarize the overall condition of the sample. This is typically done using either stoplight colors (green, yellow, or red) or words (e.g., acceptable/satisfactory or unacceptable/unsatisfactory/alert) or a combination. Ideally, the overall condition should be based on a summary of equipment condition, lubricant condition, and the degree of lubricant contamination, which are all based on the evaluation of the results of each laboratory test. Figure 22.4 provides an example of a used oil analysis report.

B. CONDEMNING LIMITS USED TO EVALUATE SAMPLE RESULT CONDITION

Most used oil analysis programs use absolute predefined condemning limit alarms to help the user identify when an abnormal lubricant and/or equipment condition exists. An absolute alarm means that if a specific oil analysis test (e.g., iron, water, viscosity) exceeds a predefined condemning limit, then an alarm is triggered. Typically, an oil analysis program has a lower level (less severe) condemning limit assigned to a test that triggers a Caution or Borderline alarm and also has a higher level (most severe) condemning limit assigned to a test that triggers an Alert or Unsatisfactory alarm. At each alarm level, it is common to have comments on the cause of the condition and also recommended corrective actions as appropriate.

There is no standard being utilized in the used oil analysis industry for establishing predefined condemning limits for each test. There have been various guidelines published, but each used oil analysis provider ultimately selects their own predefined condemning limits. Some may use OEM limits when available, whereas others select what they feel is most appropriate. To determine predefined limits, the more advanced used oil analysis providers use a combination of statistical analysis of historical data including correlation to wear, new oil specification data and composition knowledge, correlation with field application knowledge and experience, OEM or industry limits,

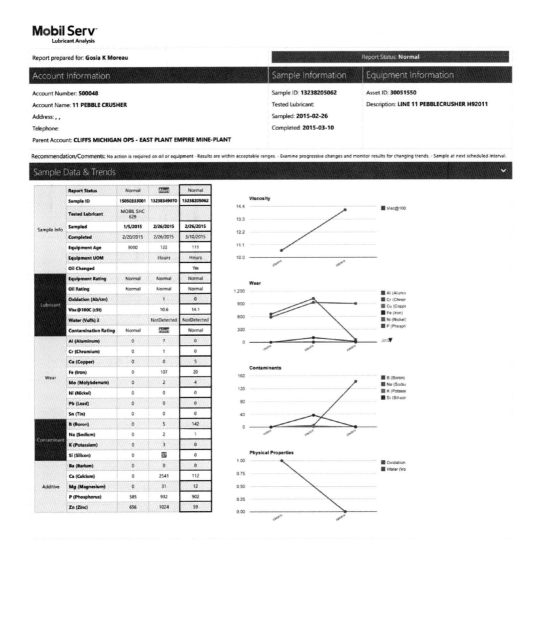

FIGURE 22.4 Used oil analysis sample report displaying overall sample condition.

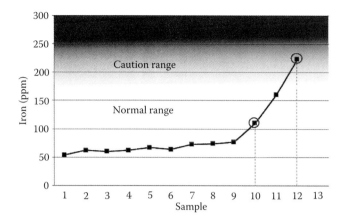

FIGURE 22.5 Examples of predefined absolute and trend analysis oil limits. Traditional predefined absolute used oil alarm limits would not highlight an abnormal result for iron until sample #12 when it surpasses the Caution alarm range of 175 ppm. A properly configured trend analysis alarm would have been triggered at sample #10, because this result significantly deviates from the past nine sample results, indicating that a change is occurring that should be investigated. Sample #11 would also trigger a trend analysis alarm because of the significant change versus the previous 10 samples.

and test precision. Limits should be based on both the lubricant and application and not just the lubricant. Some used oil analysis providers also have the flexibility to allow the more sophisticated users to define their own absolute limits based on their own equipment knowledge or history, operating conditions, warranty considerations, and overall maintenance goals. Another type of condemning limit occasionally used is based on the rate of change over time such as wear rate per mile or per hour (i.e., ppm/mile or ppm/h).

With recent improvements in analytics, oil analysis result alarming took a major step forward. Oil analysis reporting has the capability to not only use the traditional absolute predefined alarm methods, but now incorporates trend alarming. To generate the alarm, the first generation of trend alarm programs typically have calculated the mean and standard deviation of the historic data, then looks at the last data point and determines if the result from the calculated mean is higher or lower than a certain number times the standard deviation. The newest advanced generation of trend alarming uses complex algorithms that are tailored to individual tests and the typical results generated from these tests, and then applies unique statistical methods for that test result. Figure 22.5 illustrates the use of both predefined absolute limits and trend analysis limits. The key to this sophisticated form of trend analysis is knowledge of the equipment being evaluated, the test method and its precision, and the meaning of the test results all being supported by advanced knowledge of applied statistics. This added capability will help maximize the effectiveness of a user's oil analysis program, which should help optimize both equipment and lubricant life and costs.

C. Diagnosing the Cause of Abnormal Oil Analysis Results

A key element of oil analysis interpretation is the identification of the source and cause for an abnormal oil analysis result. Experienced oil analysis providers should be able to assist in this effort, and many provide this kind of information on the oil analysis report. Besides understanding the actual test method and its relation to an alarm condition, knowledge of the equipment, its operation, and the lubricant in service is critical. Tables 22.6 and 22.7 highlight some of the typical sources of equipment wear for automotive and industrial equipment.

TABLE 22.6
Typical Sources of Wear for Automotive Equipment

	Diesel and Gasoline Engine	Natural/Landfill Gas Engine	Automotive Powertrain
Aluminum (Al)	Pistons, bearings, blocks, housings, bushings, blowers, thrust bearings	Pistons, bearings, blocks, housings, bushings, blowers, thrust bearings	Pumps, clutch, thrust washers, bushings, torque converter, impeller
Chromium (Cr)	Rings, roller/taper bearings, liners, exhaust valves	Rings, roller/taper bearings, liners, exhaust valves	Roller/taper bearings
Copper (Cu)	Wrist pin bushings, bearings, cam bushings, oil cooler, valve-train bushings, thrust washers, governor, oil pump	Wrist pin bushings, bearings, cam bushings, oil cooler, valve-train bushings, thrust washers, governor, oil pump	Clutches, steering disks, bushings, thrust washers, oil cooler
Iron (Fe)	Cylinders, block, gears, crankshaft, wrist pins, rings, camshaft, valve train, oil pump, liners, rust	Cylinders, block, gears, crankshaft, wrist pins, rings, camshaft, valve train, oil pump, liners	Gears, disks, housing, bearings, brake bands, shift spools, pumps, PTO, shafts
Lead (Pb)	Bearings	Bearings	
Silver (Ag)	Bearings, end wrist pin bushing	Bearings	Bearings
Tin (Sn)	Pistons, bearing overlay, bushings	Pistons, bearing, bushings	

TABLE 22.7
Typical Sources of Wear for Industrial Equipment

	Gas and Steam Turbine	Hydraulic/ Circulating System	Compressor	Gear Drive	Paper Machine
Aluminum (Al)		Pump motor housing, cylinder gland	Rotors, pistons, bearings, thrust washers, block housing	Thrust washers, oil pump, bushings	
Chromium (Cr)		Rods, spools, roller/taper bearings	Rings, roller/taper bearings	Roller/taper bearings	Bearings
Copper (Cu)	Bearings, oil cooler	Pump thrust plates, pump pistons, cylinder glands, guides, bushing, oil cooler	Wear plates, bushings, wrist-pin bushings, bearings (recips.), thrust washers	Thrust washers, bushings, oil cooler	Bearings cages, bushings, oil cooler
Iron (Fe)	Bearings	Pump vanes, gears, pistons, cylinder bores, rods, bearings, pump housing	Camshaft, block, housing, bearings, shafts, oil pump, rings, cylinder	Gears, bearings, shaft	Bearings, gears, housings
Lead (Pb)	Bearings	Bearings	Bearings		Bearings
Silver (Ag)	Bearings	Bearings	Bearings	Bearings	
Tin (Sn)	Bearings	Bearings	Pistons, bearing, bushings		Bearings
Titanium (Ti)	Bearings, turbine blades				

TABLE 22.8
Typical Causes and Remedies for Abnormal Lubricant Conditions

	Potential Cause	Suggested Remedy
High total acid number (TAN)	High sulfur fuel, overheating, excessive blow-by, overextended drain intervals, improper oil	Evaluate oil drain interval, confirm type of oil in service, check for overheating of oil or equipment in service, check for severe operating conditions, drain oil
Low total base number (TBN)	Overheating, overextended oil drain, improper oil in service, high sulfur fuel	Evaluate oil drain intervals, verify "new" oil base number, verify oil type in service, change oil, test fuel quality service
Nitration	Improper scavenge, low operating temperature, defective seals, improper air/fuel ratio, abnormal blow-by	Increase operating temperature, check crankcase venting hoses and valves, ensure proper air–fuel mixture, perform compression check or cylinder leak-down test
Oxidation	Overheating, overextended oil drain, improper oil in service, combustion by-products/blow-by	Use oil with oxidation inhibitor additives, shorten oil drain intervals, check operating temperatures, check fuel quality, evaluate equipment use versus design, evaluate operating conditions
High viscosity	Contamination soot/solids, incomplete combustion, oxidation degradation, leaking head gasket, extended oil drain, high operating temperatures, improper oil grade	Check air/fuel ratio, check for incorrect oil grade, inspect internal seals, check operating temperatures, check for leaky injectors, check for loose crossover fuel lines, evaluate operating conditions
Low viscosity	Additive shear, fuel dilution, improper oil grade	Check air/fuel ratio, check for incorrect oil grade, inspect internal seals, check operating temperatures, check for leaky injectors, check for loose crossover fuel lines, evaluate operating conditions

D. Corrective Action Recommendation Related to Abnormal Oil Analysis Results

Once an abnormal wear condition on a piece of equipment is identified, most maintenance professionals understand how to correct the condition, whether it be to repair or replace the affected components, or just closely monitor the situation for future changes. When it comes to further understanding an abnormal lubricant condition or severe lubricant contamination event, a used oil analysis provider can provide deeper insights into the cause and potential remedy. Specific cause and corrective action comments should be an integral part of an oil analysis report—this is indicative of the quality of the analysis interpretation and experience level of the analysis provider. The best-in-class oil analysis providers will have application and lubrication experts available to further discuss the meaning of the analysis report along with assisting in troubleshooting the identified problem area. Tables 22.8 and 22.9 provide some insight into common causes and potential remedies of abnormal lubricant condition and lubricant contamination.

E. Corrective Action Recommendations Correlating to Other Predictive Maintenance Techniques

Combining in-service lubricant analysis with other predictive maintenance techniques can help optimize an equipment reliability program. Using measurements from vibration analysis (including ultrasonics), infrared temperature, current, and pressure, etc., can create a more complete picture of equipment condition. Using the combined data not only addresses the symptom but often can identify and address the root cause(s).

TABLE 22.9
Typical Causes and Remedies for Abnormal Contamination Conditions

Contaminant	Potential Cause	Suggested Remedy
Fuel dilution	Extended idling, stop and go driving, defective injectors, leaking fuel pump or lines, incomplete combustion, incorrect timing	Check fuel lines, check cylinder temperatures, worn rings, leaking injectors, seals, and pumps, examine driving or operating conditions, check timing, avoid prolonged idling, check quality of fuel, repair or replace worn parts
Fuel soot	Improper air/fuel ratio, improper injector adjustment, poor quality fuel, incomplete combustion, low compressions, worn engine parts/rings	Ensure injectors are working properly, check air induction/filters, extended oil drain intervals, check compression, avoid excessive idling, inspect driving/operating conditions, check fuel quality
Insolubles (solids)	Extended oil drain interval, environmental debris, wear debris, oxidation by-products, leaking or dirty filters, fuel soot	Drain oil, flush system, change operating environment, reduce oil drain interval, change filters
High particle count	Defective breather, environmental debris, water contamination, dirty filters, poor make-up oil procedure, entrained air, worn seals	Filter new oil, evaluate service techniques, inspect/replace oil filters, inspect/replace breather, high pressure system flush, evaluate operating conditions
High particle quantifier (PQ) index	Wear debris, shock/overloading conditions, metallic contamination, dirty filters	Replace worn parts, inspect/replace filters, inspect/clean reservoir magnets, evaluate operating conditions
Ultracentrifuge (UC) rating high	High operating temperature, overloading condition, overextended oil drain, improper oil in service	Evaluate operating conditions, shorten oil drain intervals, evaluate equipment use versus design, use oil with oxidation inhibitor additives, flush system
Water and coolant	Low operating temperature, defective seals, oil cooler/heat exchanger leak, new oil contamination, coolant leak, improper storage, condensation	Tighten head bolts, check head gasket, inspect heat exchanger/oil cooler, evaluate operating conditions, pressure check cooling system, check for external sources of contamination

The following is an example of this concept. Oil analysis for a gearbox has identified an increasing trend of iron in the lubricating oil. The gearbox vibration analysis results reveals an input shaft bearing outer raceway defect, confirming the need to take action based on the oil analysis. In addition, the vibration analysis reveals high angular vibration (one times running speed) combined with higher temperatures at the input gearbox bearing and input motor bearing. This would indicate that angular misalignment may be the cause of bearing wear and the presence of excessive iron in the oil.

F. SUSTAINING A PROGRAM THROUGH DOCUMENTED PROCEDURES, METRICS, AND ASSESSMENTS

Having a well-documented in-service lubricant analysis program with regular assessments will help ensure the program will sustain optimal performance.

1. Procedures

Written procedures help drive consistency, effectiveness, and efficiency in an in-service lubricant analysis program. In addition, as people change the necessary program knowledge and capability do

In-Service Lubricant Analysis

not suffer. The procedures need to remain active via regular use by the practitioners and should be subjected to periodic review to ensure effectiveness. Some of the key procedures include:

- Overall program description with defined roles and responsibilities
- Sampling techniques (safety, method, frequency, location, etc.)—best practice includes photographic or video instructions
- Data analysis techniques
- Training or certification requirements
- Metrics to be collected and reviewed

2. Regular Key Performance Indicators or Metrics

As stated in the beginning of this chapter, meaningful metrics help steer the direction of the program, drive continuous improvement, and align the organization to commit to the program's importance. Some good metrics include:

- Number of samples taken on time per schedule
- Percentage of equipment meeting target lubricant cleanliness codes
- Number of sample alerts
- Types of alerts
- Sample turnaround time
- Practitioner competency levels and certifications
- Estimated value of equipment repair and downtime costs avoided

3. Assessments

Conducting regular assessments or evaluations of an in-service lubricant analysis program can help optimize the program's effectiveness. Internal reviews performed by knowledgeable individuals will lead to continuous program improvement. In addition, an assessment by a capable external assessor with "cold eyes" can shed light on how to improve a program. These assessments should evaluate the following areas:

- Management understanding, commitment, and leadership of the program
- Safety practices
- Procedures (both written and actual observed execution) to include sampling techniques, data analysis, and action taken based on analysis
- Training program and practitioners' competencies
- Metrics used and recorded
- Evidence of continuous improvement

Index

Page numbers followed by f and t indicate figures and tables, respectively.

A

Abnormal contamination conditions
    typical causes and remedies, 556
Abnormal lubricant conditions
    typical causes and remedies, 555
Accumulators, 127, 128f
ACEA European Oil Sequences, 254, 257t
Activated clay-type filter, 529
Active suspension system, 303
Actuators, 128–129
    hydraulic cylinders, 128–129, 129f
    rotary fluid motors, 129, 130f
Additives, 35–42
    antiwear, 41
    chemical, 4
    corrosion inhibitors, 38–40
    defoamants, 36–37
    detergents, 40–41
    dispersants, 40–41
    extreme pressure (EP), 41–42
    in greases, 71
    impact on lubricants properties, 14–15, 14t
    oxidation inhibitors, 37
    pour point depressants, 35–36
    rust inhibitors, 38–40
    steam turbine oil, 350–351
        entrained air release, 353
        load-carrying ability, 351–352
        oxidation stability, 352–353
        viscosity, 351
    sulfur- or aromatic-containing, 395
    VI improvers, 36
Additive technology, 4
Adequate cleaning capabilities, 490
Adiabatic compression, 404
Aeroderivative gas turbines, 331–333, 332f
AGMA Standards
    flexible couplings, 202–203, 203t, 204t
    for industrial gear lubricants, 195–196, 196t
    for open gear lubrication, 199
Air entrainment, 138
    in oil, hydraulic systems and, 134
    tractor fluids, 297–298
Air line oilers, 215–216, 216f
Air-operated grease pump/hoist, 502
Air-operated oil pump, 501
Air over oil (air–oil) systems, 517
Air spray application, 216, 217f
Alkalinity
    engine oil, 247
Alkanes, 103
Alkenes, 103
Alkyl aromatics, 9
Alkynes, 103
Allison Transmission, 295–296
All-loss methods, 211–218
    grease application, 216–218, 218f
    oiling devices, 211–216
Alphaolefin technology, 5
Alternate processing for Group III + quality, 31, 31f
Alternative materials, use of, 6
American Gear Manufacturers Association (AGMA), 195, 376
American Journal of Science, 2
American Natural Curiosities, 2
American Petroleum Institute (API), 9
American Petroleum Institute (API) Service, 415
American Society for Testing and Materials (ASTM), 380, 546
    test methods, 103
    viscosity–temperature charts, 170f, 171
Angular contact ball bearings, 173f
Antifoam quality, of oil, 134
Antifriction bearings, 171
Antilock braking systems (ABSs), 309, 310f, 311
Antioxidants, conventional, 397
Antiwear additives, 41
Antiwear agents, 393
Antiwear fluids, 133
Antiwear properties
    tractor fluids, 299
API 1509, 18
API base stock classification, 9–10
    Group I, 10
    Group II, 11
    Group II+, 11
    Group III, 11
    Group III+, 11
    Group IV, 11
    Group V, 11–12
API classification system, 249–252, 251t–252t, 253t–254t
API engine oil service categories, 249, 250t
API GL-1, 289
API GL-2, 289
API GL-3, 289
API GL-4, 289
API GL-5, 289
API Lubricant Service Designations, 289–290
API MT-1, 289
Apparent viscosity
    of greases, 83, 85
Aquatic toxicity tests (OECD), 102, 103t
Aromatic amines, 5
Ashless oils, 263
Asphalt, 9
Association Technique de I'Industrie Europeenne des Lubrifiants (ATIEL) Code of Practice, 15, 18
ASTM D130 test, 381
ASTM D665 test method, 381

ATFs, *see* Automatic transmission fluids (ATFs)
Atkinson cycle, 468
Automated manual transmission (AMT), 468
Automatic extraction turbines, 338
Automatic transmission fluids (ATFs), 67, 293–296
    Allison Transmission, 295–296
    Caterpillar, 296
    Chrysler ATFs, 295
    for European vehicles, 295
    Ford ATFs, 294–295
    General Motors, 293–294
    for Japanese vehicles, 295
    multipurpose, 295
    transmission fluids for, 295
Automative transmission, 273–275
    lubrication requirements, 280–281
Automobiles, 4
Automotive applications
    recommended used oil analysis test, 548
Automotive engine lubricants
    esters as, 5
Automotive equipment, 269
    typical sources of wear, 554
Automotive gear lubricants, 65–66, 288–292
    API Lubricant Service Designations, 289–290
    channeling characteristics, 290–291
    chemical activity, 291
    corrosion, 291
    foaming, 291
    frictional properties, 292
    identification, 292
    load-carrying capacity, 288–289
    oxidation resistance, 291
    rust protection, 291
    seal compatibility, 291–292
    storage stability, 291
    U.S. Military Specifications, 292
    viscosity, 290
Auxiliary groove, 168f
Auxiliary transmissions, 287
Axial distribution groove, 162, 163f
Axial flow compressor, 430
    lubricant recommendations, 431
    lubricated parts, 431
Axial groove, 162, 162f
Axial piston motors, 129
Axial piston pumps, 122, 123f, 130
    volumetric efficiency, 465
Axial spoke couplings, 202

B

Babbitt-lined guide, 365
Back-pressure, 337
Ball bearings
    terminology, 172f
    types of, 172, 173f
Ball thrust ball bearings, 173f
Base Oil Interchange (BOI) guidelines, 16
    example, 16–17, 17f
Base stocks, 9–33
    API classification system, 9–10
        Group I, 10
        Group II, 11
        Group II+, 11
        Group III, 11
        Group III+, 11
        Group IV, 11
        Group V, 11–12
    applications, 14–15, 16t
    blend properties, 13, 14f
    composition, 32–33, 32f
    conventional solvent processing, 25, 26f
        dewaxing, 27–29, 27f, 28f
        extraction, 26–27, 26f
        hydrofinishing, 28
    conversion processing, 29–30, 29f
        alternate processing for Group III + quality, 31, 31f
        catalytic dewaxing, 30–31
        GTL process, via FT synthesis, 32
        hydrocracking, 30
        hydroprocessing, 29–30, 29t
    crude oil role in manufacture of, 18–22
        chemistry of, 18–20
        selection of, 20–21
    for EALs
        options, 108–110
        synthetic, 109
    environmental characteristics of, 107–110, 108t
    global paraffinic demand forecast, 12–13, 12f
    impact on lubricants properties, 14–15, 14t
    marketer's responsibility, 18
    property comparison, 13, 13t
    refinery processing (separation *vs.* conversion)
        atmospheric distillation, 22–23
        propane deasphalting, 25
        vacuum distillation, 24
    selection, 12–13
    specifications, 18
    test methods, 13
Base stock slate, 15–18
    defined, 15
Base stock technology, 9
Batch filtration, 138
Batch purification
    settling tanks for, 526
Bath oiling, 225
Bearing housings, circulation system, 221–222
Bearing lubrication, of air compressors, 415
Bearings
    clearance, 159
    flat, 180–181
    grooving of, 161–169
    length/diameter ratio, 159
    materials, 160
    plain, 152–171; *see also* Plain bearings
    projected area, 159
    from Riffel test, 385
    rolling element, 171–179; *see also* Rolling element bearings
    surface finish, 161
Below-freezing temperatures, 498
Beta ratios (β), 137
Bevel gears, 183, 184f
Bin, for grease/oil, 482
Bioaccumulation
    EALs and, 101, 102, 103–105, 105t
    molecular weight and, 105

Biodegradability
  EALs, 101, 102–103, 104t
  of lubricant base oils, 107, 108t
  requirements, 107
Biodegradable, defined, 106
Biodegradation
  of PAGs, 108
Biomass energy, 371
Bio-mimicking tribology, 472
Bissel, George, 3
Bitumen, 1
Blade bearings, 384, 385
"Blank" experiments, 5
Bleeding, 497
Blend properties, of base stocks, 13, 14f
Blue Angel Ecolabel, 105
Boiling water reactor, 388
  characteristics of, 389
Bottle oilers, 212, 213f
Boundary friction
  elastohydrodynamic lubrication (EHL) contact, 450
  Raimondi and Boyd, 453
Boundary lubrication, 150, 151
Brake systems, 309–313
  antilock braking systems (ABSs), 309, 310f, 311
  conventional systems, 309, 310f
  fluid characteristics, 311–313
  power brake, 311
  wet brakes, 311
Buggies, 138
Bulb turbines, 355–356, 363
  installation, 363, 364
Bus, engine oil recommendations, 259
Bypass filters, 137

C

Cage design, 460
Cam followers, 208
Cams, 208
Carbon residue, of lubricating oils, 43
Case-hardened gears, 376
Cast/forged link chains, 206
Castor oil, 4
  history of, 4
Catalytic dewaxing, 11, 30–31
Caterpillar Inc., 296
Cavitation, 138
Centralized application systems, 226–231
  central lubrication systems, 226–229
  mist oiling systems, 229–231
Central lubrication systems, 226–229, 227f, 368
  direct systems, 226
  drum mounted pump and control unit, 226, 227f
  indirect systems, 226
  parallel systems, 226, 228f
  series manifolded system, 228
  series reversing flow system, 228–229, 229f
  single-line spring return, 227–228, 229f
  two-line system, 227, 228f
Centrifugal compressors, 428, 429
  lubricant recommendations, 430
  lubricated parts, 429
CFC refrigerants, 435

Chain couplings, 201, 201f
  lubrication of, 203
Chain oiling, 225
Chains, drive, *see* Drive chains
Channeling characteristics, automotive gear lubricants, 290–291
Charlevois, 2
Check valve, 124f
Chemical additives, 4
China
  hydraulic turbines, 355
Chrysler ATFs, 295
Circulation systems, 219–224, 219f
  applications, 219
  bearing housings, 221–222
  metallurgy composition, 222
  monitoring parameters, 223–224
  oil coolers, 223
  oil filtration, 222–223
  oil heating, 223
  oil reservoirs, 221
  paper mill dryer, 219, 220f
  pressure feed, 219, 219f
  pump suction, 221
  return oil piping, 222
Circumferential groove, 162, 163f, 165, 166f
Cleanliness, 492
Clean Water Act, 106
Clearance, bearings, 159
Clutches, 269–270, 270f
Coatings, 6
Cold Crank Simulator test method, 13
Collar oiling, 225
Color
  hydraulic fluids, 139–140
  lubricating oils, 43
Combined-cycle operation
  gas turbines in, 327–328, 327f
Combined cycle plant, 338
Combined filtration, 530
Combustion cycle, 233–236
  four-stroke cycle, 233–234, 234f
  two-stroke cycle, 234–236, 235f
Combustor, gas turbines, 322–323
Commercial lubrication cart, 510
Commercial vehicle engine test, 470
Compatibility
  EALs, 114–115
  grease, 83, 84t
  of oil, hydraulic systems and, 134–135
Compatibility testing
  steam turbine oil, 353
Compound steam turbines, 339
Compressed Air Institute, 413
Compressing natural gas, 417
Compressors
  centrifugal and axial flow compressors, 403
  centrifugal and axial flow deliver oil-free air, 403
  classification, 403
  crankcase, 407
  diaphragm compressors, 427
    lubricant recommendations, 427
    lubricated parts, 427
  dynamic compressors, 427

axial flow compressor, 430
centrifugal compressors, 428
gas, *see* Gas compressors
gas turbines, 321–322, 322f
helical lobe, 403
industrial single- and two-stage, 408
liquid piston compressors, 426–427
   lubricant recommendations, 427
   lubricated parts, 427
lubricant application methods
   bearing/running gear, 408
   cylinder lubrication, 406–407
multistage reciprocating compressors, 414
   factors affecting lubrication, 414–415
   lubricating oil recommendations, 415–418
reciprocating, *see* Reciprocating air compressors
refrigeration/air conditioning compressors, 431–437
   compressor affecting lubrication, factors, 433
refrigeration and air conditioning applications, 403
refrigeration system factors affecting lubrication, 434
   ammonia (R-717), 435–436
   carbon dioxide (R-744), 435–436
   fluorocarbons, 435
   hydrocarbon refrigerants, 436
   lubricating oil recommendations, 436–437
   sulfur dioxide (R-764), 436
rotary compressors
   lubricating oil recommendations, 420
   straight lobed, 418–420
   straight-lobed
      principal parts of, 420
rotary lobe compressors
   lubricated parts, 420–421
   lubricating oil recommendations, 421
rotary screw compressors, 421–425
   dry screw compressors, lubricated parts, 424
   lubricant recommendations, 424–425
single- and two-stage, 408, 414
   factors affecting cylinder lubrication, 408–414
   factors affecting running gear lubrication, 414
single- and two-stage trunk piston, 415
sliding vane compressors, 425
   lubricant recommendations, 426
   lubricated parts, 426
vertical single-acting, 407
Condensing, 337
Cone penetration
   of greases, 75–77, 75t, 76f
Conformal/nonconformal contacts, 447, 448
Conforming, 142
Consistency
   greases, 75–77
Constant pressure
   with flow control valve, 156
Constant pressure system, with flow restrictor, 155–156, 156f
Constant velocity (CV) joints, 282
Constant volume system, 155
Construction minimizes sealing requirements, 365
Contact points, 443
Contact pressure, 462
Contaminants, 403
Contamination
   rolling element bearings and, 177
   steam turbines lubrication and, 349–350

Contamination prevention, of foreign matter, 520
Continuous bypass purification, 522
Continuous full-flow purification, 522
Continuous independent purification, 523
Continuously variable automatic transmission (CVT), 276, 439
   transmission fluids for, 295
Control systems
   steam turbines, 339–342, 340f
Conventional antioxidant additives, 394
Conventional base stock, *see* Group I base stock
Conventional brake systems, 309, 310f
Conventional lube oils, 394
Conventional processing, 9
Conventional solvent processing, base stock, 25, 26f
   dewaxing, 27–29, 27f, 28f
   extraction, 26–27, 26f
   hydrofinishing, 28
Conversion processing, base stock, 29–30, 29f
   alternate processing for Group III + quality, 31, 31f
   catalytic dewaxing, 30–31
   GTL process, via FT synthesis, 32
   hydrocracking, 30
   hydroprocessing, 29–30, 29t
Cooling
   coils, 367
   internal combustion engines, 241
Co-ordinating European Council (CEC)
   test methods, 103
Corporate Average Fuel Economy (CAFE) standard, 439
Corrosion
   automotive gear lubricants, 291
   wire ropes, 208
Corrosion inhibitors
   lubricating oils as, 38–40
Corrosion protection
   engine oil, 247–248
   tractor fluids, 298
Corrosion protection test
   lubricating oils, 64
Cost/performance assessment
   base stock selection, 12, 12f
Coulomb's law, 392, 445
Couplings, flexible, *see* Flexible couplings
Cross compound steam turbines, 339
Cross-type universal joint, 281–282, 282f
Crude oil, 4; *see also* Hydrocarbons
   chemistry of, 18–20
   history of, 4
   hydrocarbons in, 18–19, 18f
   organic material, 18
   as raw material, 9
   role in base stock manufacture, 18–22
Curved chamfer, 168f
CVT (continuously variable automatic transmission), 276
   transmission fluids for, 295
Cylinders, 199–200
   hydraulic, 128–129, 129f
   lubrication of, 199–200
Cylindrical shell, 367

D

D'Allion, Joseph de la Roche, 2
Deasphalted oil (DAO), 11

Index

Defoamants, 36–37
   silicone polymers as, 5
Degradability, 101
Denison HF-6 Approval, 111
Denison Hydraulics HF-0, 1, and 2 approvals, 110–111
Density
   of lubricating oils, 43–44, 44f
Deposits, internal combustion engines, 242–243, 243f
Deriaz turbines, 356, 358, 362
   cross section of, 359
Detachable link chains, 206
Detergency
   engine oil, 246–247
Detergents, 40–41
Deterioration, of lubricants, 475
Deutsches Institut für Normung (DIN), 101, 375
Dewaxing, solvent, 27–29, 27f, 28f
DEXRON® Fluid (General Motors), 293
DEXRON HP (General Motors), 294
DEXRON IId, IIe, and III (General Motors), 294
DEXRON II Fluid (General Motors), 293
DEXRON VI (General Motors), 294
Diamond-like carbon (DLC) coatings, 473
Diaphragm compressors, 427
   lubricant recommendations, 427
   lubricated parts, 427
Dibasic acid esters
   history of, 5
Didodecyl selenide, 394
Diesel engines, 239–240
   in industrial and marine applications, 261–262
Diester oils, 5
DIN 51517-3 standard, 379
Direct drive turbines
   main shaft bearing lubrication, 385–386
Directional control valves, 125, 126f
Dirt
   open gear lubrication and, 198
Discharge pressure gauge, 404
Discharge temperature, 404
The Discoveries of Guiana, 2
Dispensing nozzles, 510
Dispensing oils
   basic requirements, 499
   central dispensing systems, 510–511
   closed system dispensing, 510
   faucets, 500
   grease gun fillers, 502–503
   highboys, 503–504
   industrial manufacturing/process plants, 499
   maintenance/services, 511
   oil house to machine, 504
      lubrication carts/wagons, 509–510
      oil dispensing containers, 504–505
      portable grease equipment, 507–509
      portable oil dispensers, 505–507
   rocker type drum rack, 500
   transfer pumps, 500–502
Dispersancy
   engine oil, 246–247
Dispersants, 40–41
Dispersion, mechanical, 102
Disposability, 101
Disposal, lubrication system, 536–537

Distillation process
   history of, 2, 4
   improvement, 4
Distribution grooves, in two-part bearing, 164, 164f
Double-acting machines, 404
Double row deep groove ball bearings, 173f
Drake, E. L., 3
Drive axle, 282–283, 283f
Drive chains, 205–207
   cast or forged link chains, 206
   silent and roller chains, 205–206
   viscosity selection, 206–207, 206t
Drive shafts, 281–286
   differential action, 283–284, 284f
   drive axle, 282–283, 283f
   limited-slip differentials, 284, 285f
   lubrication, 282
      factors affecting, 285–286
Drop feed and wick feed cups, 211, 212f
Drop feed cups, 407
Dropping point
   greases, 77, 78f
Drum dolly, 506
Drum faucets, 500
Drum handler, manual, 481
Drum pump
   with return, 501
Drum shelter, outdoor, 486
Drum stacker, 488
Dry lubrication, 418
Dry screw compressors, 425
Dual-clutch transmission (DCT), 468
Dual fuel engines, 265–266
Dust
   open gear lubrication and, 198
Dye pigments, 398
Dynamic compressors, 427
   axial flow compressor, 430
      lubricant recommendations, 431
      lubricated parts, 431
   centrifugal compressors, 428
      lubricant recommendations, 430
      lubricated parts, 429
Dynamic oxidation tests
   greases, 80

E

EALs, *see* Environmentally acceptable lubricants (EALs)
Edge-type oil filter, 527
EHC 45, 17
EHC 50, 16
EHC slate of Group II base stocks, 16–17
Elastohydrodynamic film, 142
Elastohydrodynamic lubrication (EHL) contact, 448, 449, 459
   common lubricant classes, 451
   configuration of ball, 450
   traction curves for a polyalphaolefin (PAO) fluid, 451
   traction measurement, 449
Elastomer testing, 382
Electrical machinery, 4
Electric power generation
   gas turbines in, 326–327
Electric power gun, 509

Electrohydraulic servo valve, 124f
Electrostatic precipitation filters, 532
Emergency shutdown, 368
Emission control systems
   protection of, 64–65
Emissions, 64–65
Emulsion and demulsibility tests
   lubricating oils, 59–60
Enclosed lubrication systems, 517
End product
   performance and quality, 15
Energy demand, 439, 440
Energy efficiency, lubricant contribution, 439
   friction loss mechanisms, 440
      boundary friction, 441–445
      in concentrated contacts, 447–452
      friction modifiers, 445–447
      viscosity-shear losses, 441
   gears/gearboxes, friction
      churning losses, 463–464
      gear teeth, 461–463
   hydraulic systems, friction losses, 464
      fluid selection, 465–466
      sources of friction, 464–466
   hydrodynamic fluid films, 452
      bearings friction in industrial applications, 452–454
      measuring bearing friction, 454–456
      rolling element bearing friction, 456–458
   vehicle/internal combustion engine efficiency, 466
      automotive design impacting efficiency, 467
      energy uses, 466–467
      engine friction reduction, 469–471
      materials and surfaces uses to improve efficiency, 472–473
      measuring fuel economy, 471–472
      trends, 469
Energy losses, 441
Energy sources
   development of, 7
Engine oil, 244–258
   alkalinity, 247
   antiwear, 247
   detergency, 246–247
   dispersancy, 246–247
   foam resistance, 248
   gasoline engine octane number requirement, 248–249
   identification and classification systems, 249–258
      ACEA European Oil Sequences, 254, 257t
      API classification system, 249–252, 251t–252t, 253t–254t
      API engine oil service categories, 249, 250t
      ILSAC performance specifications, 252–254, 255t
      manufacturer specifications, 257–258
      U.S. Military Specifications, 254–257
   low temperature fluidity, 245
   oxidation stability, 245–246
   recommendations by application
      aviation, 260, 260t
      diesel engines, in industrial and marine applications, 261–262
      farm machinery, 260
      motorcycles, 267
      natural gas fired engines, 263–266
      outboard marine engines, 266
      passenger car, 258–259
      railroad engines, 266
      truck and bus, 259
   rust and corrosion protection, 247–248
   thermal stability, 246
   viscosity, 244–245
   viscosity index, 244–245
Engine Oil Licensing and Certification System, 15
Engine oil licensing programs, 16
Engine tests, 472
   for lubricating oils performance, 60–65, 61t
Entrained air in oil, hydraulic systems and, 134
Entrained air release
   steam turbine oil, 353
Environmental acceptability, 101
   EALs selection and, 113
Environmental considerations, 101
Environmental criteria, 105–107
Environmentally acceptable lubricants (EALs), 101–115; *see also* Vegetable-based EALs
   and application data, 112t
   availability and performance, 110–113
   base stocks, characteristics, 107–110, 108t
   bioaccumulation, 103–105, 105t
   biodegradability, 102–103, 104t
   compatibility of, 114–115
   conversion to, 114–115
   cost issues, 110
   definitions, 102
   Denison HF-6 Approval, 111
   Denison Hydraulics HF-0, 1, and 2 approvals, 110–111
   environmental acceptability, 113
   environmental considerations, 101
   environmental criteria, 105–107
   equipment builder approvals, 113–114
   international labeling programs, 106–107
   maintenance practices, 114
   national labeling programs, 105–106
   operating conditions, 114
   overview, 101
   performance characteristics, comparison of, 108t, 109
   proven field performance, 114
   selection process, 113–114
   specifications, 113
   supplier reliability, 114
   synthetic base stocks, 109
   test procedures, 102–105
   toxicity, 102, 103t
   vegetable-based, 109
   Vickers 35VQ Pump Test (ASTM D6973), 110
Environmental sensitivity issues, 110
Environmental test criteria, 102
Equipment builder approvals
   for EALs selection, 113–114
Equipment commissioning, lubrication system, 536
Erosion, 1
Esters, 9
   as automotive engine lubricants, 5
   history of, 5
European Ecolabel, 105, 107

Index

European vehicles, ATFs for, 295
Evaluation
    greases, 78–85
    lubricating oils, 54–60
Eveleth, J. D., 3
Extraction, solvent, 26–27, 26f
Extreme pressure (EP), 393
    additives, 41–42, 151
Extreme pressure and wear prevention tests
    greases, 82, 83f
ExxonMobil Corporation, 4, 16, 514, 515
ExxonMobil's lubricants, 475

F

FAG Schaeffler Wind Turbine Bearing Test, 380
Fail load stage (FLS), 377
Farm machinery
    engine oil recommendations, 260
Fast breeder reactor (FBR), characteristics of, 389–390
Fatigue
    wire ropes, 208
Fatty oils, 4
Ferrocene derivatives, 5
Field performance, EALs selection and, 113–114
Filling slot ball bearings, 173f
Films, lubricating
    fluid, 141–149
    solid/dry, 151–152
    thin surface, 149–151
Filter carts, 138
Filtration, hydraulic systems, 136–138, 139
    batch filtration, 138
    Beta ratios (β), 137
    bypass filters, 137
    full flow, 136–137
    polishing filters, 137
    portable filters, 138
Final disposal, lubrication system, 536–537
Final drives, 288
Finished product sector
    lubricants demands by, 15, 15f
Fire extinguishers
    hand-operated, 493
Fire points
    lubricating oils, 44–45, 45f
Fischer Tropsch (FT) synthesis, GTL process via, 32
Fixed blade propeller turbines, 360, 361
Flammability, 483
Flammable liquids, 484
Flash point
    lubricating oils, 44–45, 45f
Flat bearings, 180–181
Flat surface cam followers, 208
Flexible couplings, 200–205
    AGMA Standards, 202–203, 203t, 204t
    grease-lubricated, 202–203, 203t
    lubrication of, 202–205
    lubrication techniques, 205
    steam turbines, 347–348
    types of, 201–202
Flexible member couplings
    lubrication of, 205
Flexing member couplings, 202

Flood-lubricated screw compressors, 424
    lubrication system for, 423
Flow control valves, 127
    constant pressure with, 156
Flow-of-stream river applications, 355
Flow restrictor
    constant pressure system with, 155–156, 156f
Fluid components
    in greases, 69–70
Fluid films, 141–149
Fluoride esters, 5
Fluorinated elastomers, 382
Flushing, lubrication system, 536
Foaming, 138
    automotive gear lubricants, 291
    tractor fluids, 297–298
Foam resistance
    engine oil, 248
    steam turbine oil, 352
Foam tests
    lubricating oils, 57–59
Force feed lubricators, 214–215, 214f–215f
Ford ATFs, 294–295
    MERCON ATF, 294
    MERCON LV, 295
    MERCON SP, 295
    MERCON V, 294–295
    Type CJ and Type H ATF, 294
    Type F ATF, 294
Forschungsstelle fuer Zahnraeder und Getriebebau (FZG), 377, 379
    FVA 54 micropitting test, 378
    weight loss, 379
Four-stroke combustion cycle, 233–234, 234f
Four-stroke high-speed gas engines, 265
Four-stroke low- to medium-speed gas engines, 264
Four-stroke motorcycle oils, 267
Francis turbine, 356
    assembly of, 360
    runner, 358
    spiral casing, 359
Frictional characteristics
    tractor fluids, 298
Frictional properties
    automotive gear lubricants, 292
Friction losses
    mechanical efficiency, 464
    sources of, 461
Friction modifiers, 445
    molybdenum friction modifiers, 446
    polyalphaolefin (PAO), 446
Front wheel suspension system, 301, 302f–303f
Fuel, 9
Fuel and combustion considerations
    internal combustion engines, 238–240
        diesel engines, 239–240
        gaseous fueled engines, 240
        gasoline engines, 238–239
Fuel economy
    lubricating oils, 65
Fuel efficiency, 470
Fuel energy
    distribution of, 467
Fuller's earth, 526

Full flow filtration, 136–137
Fundamental hydraulic systems, 118
Furfural, 10
FVA 54 test, 378
FZG, *see* Forschungsstelle fuer Zahnraeder und Getriebebau (FZG)

G

Galvanized containers, 504
Gas, *see* Gas compressors
Gas compressors and reciprocating air, 404, 406–408, 413–418
   methods of lubricant application
      bearing (running gear) lubrication, 408
      cylinder lubrication, 406–407
   multistage reciprocating compressors
      factors affecting lubrication, 414–415
      lubricating oil recommendations, 415–418
   single- and two-stage compressors
      factors affecting cylinder lubrication, 408, 413–414
      factors affecting running gear lubrication, 414
Gas-cooled reactors (GCR), 390
Gas engine oil, selection, 265
Gaseous fueled engines, 240
Gasoline
   history of, 4
   silicone polymers in, 5
Gasoline direct injection (GDI), 439
Gasoline engines, 238–239
   octane number requirement, 248–249
Gas-to-liquids (GTL) process, 9
   base stocks, 11
   via FT synthesis, 32
Gas turbines
   aeroderivative, 331–333, 332f
   applications, 326–329
   in combined-cycle operation, 327–328, 327f
   combustor, 322–323
   compressor, 321–322, 322f
   in electric power generation, 326–327
   industrial, features of, 325–326, 326f
   intercooling, 320–321, 321f
   jet engines, 324–326, 324f
   large industrial, 329–331, 329f–330f
   lubrication of, 329–333
   in marine propulsion, 328
   microturbines, 328–329
   overview, 317
   in pipeline transmission, 327
   principles of, 317–324
   in process operations, 327
   regenerative cycle, open system, 320, 320f
   reheating, 320–321, 321f
   simple cycle, open system, 318, 318f–320f
   theory of, 317
   in total energy, 328, 328f
   turbine, 323–324
GDI engines, 468
Gearboxes, 375, 462
Gear drives
   steam turbines, 347
Gear face profile deviation, 378

Gear lubrication
   full-pressure circulation system, 408
Gear oils
   condition monitoring, 382–383
   industry standards/builder specifications, 375–376
   interfacial properties of, 380–382
   viscosity/low temperature requirements, 376
Gear pumps, 119, 120f
Gears, 183–199
   action between gear teeth, 186–189, 187f, 188f
   AGMA Standard for, 195–196, 196t
   classification, 183
   convergent zone, 189, 190f
   film formation, 189, 191f
   lubricant characteristics for enclosed, 195–196
   lubrication of enclosed
      application method and, 194
      drive type and, 194
      factors, 189–195
      gear speed and, 192
      gear type and, 191–192
      lubricant leakage and, 194–195
      operating and startup temperatures and, 192–193
      reduction ratio, 192
      surface finish and, 193
      transmitted power and load, 193
      water contamination and, 194
   open, lubrication of, 196–199
   overview, 183
Gear type couplings, 201, 201f
   lubrication of, 203
General Motors ATFs, 293–294
   DEXRON® Fluid, 293
   DEXRON HP, 294
   DEXRON IId, IIe, and III, 294
   DEXRON II Fluid, 293
   DEXRON VI, 294
   Type A Fluid, 293
Genetic engineering, 109
Geothermal energy, 371
German bearing builder, 384
GF6 engine oil, 471
GHS pictograms, 477
Global paraffinic base stock demand forecast, 12–13, 12f
Graphite, 152
Gravity
   lubricating oils, 43–44, 44f
Grease, 4, 69–85
   additives in, 71
   apparent viscosity, 83, 85
   application devices, 216–218, 218f
   cartridge guns, 502
   characteristics, 75–79
      by thickener type, 71, 72t–73t
   compatibility, 83, 84t
   composition of, 69–71
   cone penetration of, 75–77, 75t, 76f
   consistency, 75–77
   dropping point, 77, 78f
   dynamic oxidation tests, 80
   effect of radiation, 397
   evaluation and performance tests, 78–85
   extreme pressure and wear prevention tests, 82, 83f
   flow test kit, 509

Index

fluid components in, 69–70
grooving to, 168
gun cartridges, 478
guns, 508
manufacture of, 74–75, 74f
mechanical stability tests, 78–79, 79f
need for, 69
NLGI grade numbers, 77, 77t
oil separation tests, 80–81
overview, 69
in plain bearings, 154–155
    selection for, 171
in rolling element bearing, 154
    selection of, 179
rust protection tests, 82
slides, guides, and ways, 154–155
static oxidation test, 79–80
structural stability tests, 78–79, 79f
thickeners in, 70–71, 72t–73t
water resistance tests, 81–82, 81f
Grease-lubricated bearings, 154–155, 370, 460
Grease-lubricated LMGs, 183
Grease polymerization
    effect of thickener, 398
Greece
    hydraulic turbines, 355
Grooving, of bearings, 161–169
    to grease, 168, 169
    for oil, 162–169
    for vertical bearing, 164–165, 165f
Group I base stock, 10, 13, 15
Group II base stock, 11, 13
Group II+ base stock, 11
Group III base stock, 11, 13
Group III+ base stock, 11, 13
Group IV base stock, 11
Group V base stock, 11–12, 15
Guides, 180–181
    grease lubrication, 180–181
    lubricant characteristics, 181

H

Handling, lubricants
    bulk products, 480–483
        special bulk grease vehicles, 483
        tank cars/tank wagons, 483
        unloading, 483
    defined, 478
    packaged products, 478
        moving to storage, 478–480
Hand oiling, 211; *see also* Oiling devices
Hand-operated fire extinguishers, 493
Hand-operated grease gun filler, 503
Hazardous chemical labeling, for lubricants, 476
Head hydraulic turbines, 361
Heat, from fission, 389
Heat, from reactor core, 390
Heat-insulating deposits, 434
Heavier hydrocarbons, 417
Heavy diesel engine
    guidelines for, 16
Helical gear, 183, 185f
Herodotus, 1
Herringbone gear, 183, 185f
Hetero-atoms, 18
Heterocyclenes, 5
Hexafluorobenzenes, 5
High ash oils, 264
Highboys, 503
High-performance synthetic lubricants, 516
High-pressure steam turbines, 4, 338
High-quality circulation oils, 169
High-speed engines, 261
High-temperature, gas-cooled reactor (HTGR), 390
Horizontal axis wind turbine (HAWT), 371
Hot pressurized carbon dioxide, 397
Hybridization, 469
Hybrid vehicles, 469
Hydraulic control systems
    steam turbines, 345–346
Hydraulic cylinders, 128–129, 129f
Hydraulic drives, 129–130
Hydraulic fluids
    types of, 135, 136t
Hydraulic lift gate, 480
Hydraulic oils, 110, 383
Hydraulic pumps
    axial piston pumps, 122, 123f
    gear pumps, 119, 120f
    in hydrostatic system, 130
    piston pumps, 120
    radial piston pumps, 120–122, 121f
    selection criteria, 122, 123t
    vane pumps, 120, 121f
Hydraulic retarders, 311
Hydraulics, 117–140
    actuators, 128–129
    advantages, 117
    basic principles, 117–118, 118f
    defined, 117
    fluid types, 135
    fundamental systems, 118
    hydraulic drives, 129–130
    hydromechanics, laws of, 117–118
    oil qualities, 132–135
    oil reservoir, 130–132
    pressure and flow, controlling, 122–128
    system components, 119–122
    system maintenance, 135–140
Hydraulic systems, 370
    accumulators, 127, 128f
    actuators, 128–129
    basic principles, 117–118
    dense media bypass filtration, 528
    filtration, 136–138
    fundamental, 118
    maintenance, 135–140
    oil qualities required by, 132–135, 132f
        air entrainment, 134
        antifoam, 134
        antiwear (wear protection), 133
        compatibility, 134–135
        demulsibility (water separating ability), 134
        oxidation stability, 133–134
        rust protection, 134
        viscosity, 133
        viscosity index, 133

oil reservoir, 130, 132, 132f
periodic oil analysis, 139
pumps, 119–122
reservoir oil levels maintenance, 138–139
routine inspections, 139–140
temperatures control, 138
transmit force, 464
valves, 122–128
Hydraulic turbines, 355, 356
    application of, 355
    hydraulic systems, 368
    impulse turbine, 356
    reaction turbines, 356–364
        bulb turbines, 363
        diagonal flow turbines, 359–360
        fixed blade propeller turbines, 360–361
        Francis turbines, 357–359
        Kaplan turbine, 361–362
        s-turbines, 364
    types of, 356
Hydrocarbon fluids
    physical and chemical properties of, 392
Hydrocarbon PAO, synthesized, 6
Hydrocarbons; *see also* Crude oil
    configurations, 18
    in crude
        structures, 18–19, 18f
        types, 18–19
    defined, 18
    history of, 5
    molecular composition in base oil, 13
    radiolysis processes, 392
Hydrochlorofluorocarbons (HCFCs), 435
Hydrocracking, 30
Hydrodynamic bearings, 152–155, 441
    pressure distribution, 154
Hydrodynamic bearing test machine
    schematic, 455
Hydrodynamic drive, 130
Hydrodynamic films, 142
Hydrodynamics, 118
Hydrodynamic theory, 452
Hydroelectric energy sources, 371
Hydroelectric installations, 355
Hydrofinishing, 28
Hydrogen-cooled generators, oil shaft seals for, 346–347
Hydrokinetic drives, 130
Hydrokinetics, 118
Hydrolytic stability
    vegetable-based EALs, 112–113
Hydromechanics, 117–118
Hydropower, 355
Hydroprocessing, 11, 29–30, 29t
Hydrostatic bearing
    applications, 156–157
    constant pressure system with flow restrictor, 155–156, 156f
    constant pressure with flow control valve, 156
    constant volume system, 155
Hydrostatic drives, 130, 131f
Hydrostatic films, 142, 148, 148f
Hydrostatic force, 118
Hydrostatic lubrication
    plain bearings for, 155–157
Hydrostatics, 118
Hydrostatic transmission, 277–279, 278f, 279f
    lubrication requirements, 281
Hydroviscous drive, 130
Hypoid gears, 183, 187f
Hypoid-type drive axle, 282–283, 283f

I

IEC 61400-4 document, 375, 376
ILSAC performance specifications, 252–254, 255t
Impeller rotary screw compressor, 422
Individual locker, 491
Industrial applications
    diesel engines in, 261–262
    recommended used oil analysis test, 549
Industrial equipment
    typical sources of wear, 554
Industrial gas turbines
    features, 325–326, 326f
    jet engines for, 324–326, 324f
Industry standards, 135, 135t
Inert gases, 416
Infinitely variable transmissions (IVT), 469
Inherent biodegradability, 102–103
In-plant lubricant handling, overview of, 514–515
In-service handling, on lubricant, 517
    contamination, elimination of, 519
        central reservoir maintenance, 519–521
        cross contamination, 521
        proper system operation, 521
    leaks/spills/drips, prevention of, 517–519
    reuse *vs.* all-loss systems, 517
In-service lubricant analysis, 539, 555
    abnormal oil analysis results, cause diagnosing, 553–555
    conducting regular assessments, 557
    corrective action recommendations
        to abnormal oil analysis results, 555
        correlating to other predictive maintenance techniques, 555–556
    documented procedures, 556–557
    elements of, 540
    grease analysis, 546
    implementation of program, 539
    interpretation, of used oil analysis results, 551
        condemning limits, used to evaluate sample result condition, 551–553
        overall sample result condition, 551
    oil analysis provider, 540–541
    oil analysis tests, common used, 546
    oil analysis test slates, application, 546–549
    precision of test, 549–551
    regular key performance indicators/metrics, 557
    testing, 546
In-service lubricant purification, 521
    continuous bypass purification, 522
    continuous full-flow purification, 522–523
    continuous independent purification, 523
    full-flow and bypass, 523–524
    periodic batch purification, 523
Instrumentation, 454

Index

Intercooling, gas turbines, 320–321, 321f
Internal combustion engines, 233–267
    combustion cycle, 233–236
    design and construction considerations, 233–238
    engine oil characteristics, 244–258
        alkalinity, 247
        antiwear, 247
        detergency, 246–247
        dispersancy, 246–247
        foam resistance, 248
        gasoline engine octane number requirement, 248–249
        identification and classification systems, 249–258
        low temperature fluidity, 245
        oxidation stability, 245–246
        rust and corrosion protection, 247–248
        thermal stability, 246
        viscosity, 244–245
        viscosity index, 244–245
    fuel and combustion considerations, 238–240
        diesel engines, 239–240
        gaseous fueled engines, 240
        gasoline engines, 238–239
    lubricant application methods, 237–238
    maintenance considerations, 243–244
    mechanical construction, 236–237
    oil recommendations by application
        aviation, 260, 260t
        diesel engines, in industrial and marine applications, 261–262
        farm machinery, 260
        motorcycles, 267
        natural gas fired engines, 263–266
        outboard marine engines, 266
        passenger car, 258–259
        railroad engines, 266
        truck and bus, 259
    operating considerations, 240–243
        cooling, 241
        deposits, 242–243, 243f
        sealing, 241–242
        wear, 240–241
    overview, 233
    supercharging, 237
International Civil Aviation Organization, 439
International Electrotechnical Commission (IEC), 375
International labeling programs, 106–107
    European Ecolabel, 107
    Nordic Swan, 106
International Standardization Organization (ISO), 101, 135, 369, 375, 546
    viscosity, 455
        base oil, 385
    Viscosity Grade Selection General Guidelines, 197t
Iron, 18
Isobutane (R-600A) gas, 436
ISO 14635-1 procedure, 379
ISO VG 68, 427
ISO VG 100, 417
ISO VG 220, 420
ISO VG 460 oil, 385
ISO VG 32 synthetic, 430
ISO 100 viscosity base oil, 385

J

Japanese vehicles, ATFs for, 295
Jet engines, 4
Joints
    constant velocity, 282
    lubrication, 282
    universal, 281–286
Josephus, 1
Journal bearings, steam turbines, 342–343

K

Kaplan turbine, 361, 362, 368
    bulb turbine, 363
    comparison of, 363
    cross section of, 362
    individual servomotors, 360
Ketone, 10
Kier, Samuel M., 3

L

Labeling programs
    international, 106–107
    national, 105–106
Laboratory oil analysis, 139
Landfill gas oils, 264
Lard oil, 4
Large industrial gas turbines, 329–331, 329f–330f
Large oil containers, 505
Leaf-type filter elements, 529
Leakage, 139
    of oil, 518
Length/diameter (L/D) ratio, of bearings, 159
Less flammable fluids
    steam turbines, 353
Lever-operated bucket pump, 506
Lever-operated grease gun, 507
Lift truck, for handling drums, 480
Limited-slip differentials, 284, 285f
Linear motion guides (LMGs), 181–183, 181f–182f
    grease lubrication, 183
    lubricant selection, 182–183
    oil lubrication, 183
Liquid piston compressors, 426–427
    lubricant recommendations, 427
    lubricated parts, 427
    oil-lubricated bearings of, 427
Liquid refrigerant flows, 431
LMGs, see Linear motion guides (LMGs)
Load
    gear lubrication and, 193
    open gear lubrication and, 199
    rolling element bearings and, 176
Load-carrying ability/capacity
    automotive gear lubricants, 288–289
    steam turbine oil, 351–352
Long-life synthetics, 101
Long-term storage, 498
Low ash oils, 264
Low-pressure steam turbines, 338
Low-speed engines, 262

Low temperature fluidity
  engine oil, 245
Low temperature performance
  of base stock, 13
  vegetable-based EALs, 111–112
Lubricant application methods
  bearing/running gear, 408
  cylinder lubrication, 406–407
Lubricants
  categories, 107
  churning, 460
  containers, 479
  demands of, 6
    by finished product sector, 15, 15f
    modern refining technology and, 9
  development of, 3–4
  EHL traction curves, 461
  future prospects, 6–7
  new techniques development, 4
  properties, base stock and additive impact on, 14–15, 14t
  recommendations, 369, 370
  substitutes, 3–4
  synthetic, history of, 4–6
  viscosity index of, 376
Lubricated contact, 445
Lubricating films
  fluid films, 141–149
  overview, 141
  solid/dry films, 151–152
  thin surface films, 149–151
  types of, 141–152
Lubricating oils, 35–67, 523, 530
  additives, 35–42
  antiwear additives, 41
  automatic transmission fluids (ATFs), 67
  automotive gear lubricants, 65–66
  carbon residue, 43
  color, 43
  corrosion inhibitors, 38–40
  defoamants, 36–37
  density of, 43–44, 44f
  detergents, 40–41
  dispersants, 40–41
  emissions and protection of emission control systems, 64–65
  emulsion and demulsibility tests, 59–60
  engine tests for performance, 60–65, 61t
  evaluation and performance tests, 54–60
  extreme pressure (EP) additives, 41–42
  fire points, 44–45, 45f
  flash point, 44–45, 45f
  foam tests, 57–59
  fuel economy, 65
  gravity, 43–44, 44f
  multicylinder high temperature engine tests, 62–63, 63t
  multicylinder low temperature tests, 63
  neutralization number, 45–46
  oxidation inhibitors, 37
  oxidation stability and bearing corrosion protection, 62
  oxidation tests, 54–56
  physical and chemical characteristics, 42–54
  pour point, 46–47
  pour point depressants, 35–36
  rust and corrosion protection tests, 64
  rust inhibitors, 38–40
  rust protection tests, 56
  single cylinder high temperature tests, 62, 62t
  sulfated ash, 47
  thermal stability, 56
  total acid number, 46
  total base number (TBN), 46
  viscosity index (VI), 54
  viscosity index (VI) improvers, 36
  viscosity, 47–53, 48f, 49f
Lubrication
  carts/wagons, 509–510
  hydraulic turbines, 364
  instructions, 509
  turbine/generator bearings, 364
    compressors, 369
    control valves, 368, 369
    governor/control systems, 368
    guide vanes, 368
    journal/guide bearings, 365
    lubricant application, methods of, 367–368
    thrust bearings, 365–367
Lubrication Recommendations, 430
Lubrication recommendations, in nuclear industry, 399
  general requirements, 399–401
  selection of lubricants, 401
Lungs
  petroleum hydrocarbons, 417

M

Maintenance
  hydraulic systems, 135–140
  practices, EALs, 114
Marine applications
  diesel engines in, 261–262
  jet engines for gas turbines, 324–326, 324f
Marine propulsion
  gas turbine applications, 328
Massachusetts Magazine, 2
Mechanical dispersion, 102
Mechanical handling equipment, 487
Mechanical stability tests
  greases, 78–79, 79f
Mechanical transmission, 271–272, 272f
  lubrication requirements, 279–280
Medium- and heavy-duty trucks, 313–315
  lubricant characteristics, 315
  lubrication requirements, 314–315
  Pitman arm, 313
  pneumatic braking systems, 313
  rigid front axles, 313, 313f
  single rear axle, 313, 314f
  steering system, 313
Medium-speed engines, 261–262
MERCON ATF (Ford), 294
MERCON LV (Ford), 295
MERCON SP (Ford), 295
MERCON V (Ford), 294–295
Metal drum covers, 486
  correct method of blocking, 487
Metallurgy composition, circulation system, 222

Metalworking fluids, 536
Metering control nozzle, 511
Micropitting damage
    extensive, 377
    minor, 377
Micropitting failure, 376
Micropitting protection, 378
Microturbines, 328–329; *see also* Gas turbines
Mineral oils, 3, 4, 5
Minimally toxic, defined, 106
Mist oiling systems, 229–231, 230f
Mixed lubrication film, 150, 151
Mixed-pressure steam turbines, 338
Mobil Oil Corporation, 4
    research at, 5–6
Mobil 1™ synthetic engine oil, 6
Modern refining technology
    demands for lubricants and, 9
Mo friction modifier, 447
Moisture, 414
    due to expansion and contraction, 485
Moisture condensed
    in intercoolers and aftercoolers, 409
Molecular weight
    bioaccumulation and, 105
Molybdenum (Mo), 446
Molybdenum disulfide, 152
Monitoring parameters, circulation system, 223–224
Motorcycles
    engine oil recommendations, 267
Multicylinder high temperature engine tests
    lubricating oils, 62–63, 63t
Multicylinder low temperature tests
    lubricating oils, 63
Multimachine collection system, 535
Multipurpose ATFs, 295
Multipurpose tractor fluids, 296–299
    EP and antiwear properties, 299
    foam and air entrainment control, 297–298
    frictional characteristics, 298
    oxidation and thermal stability, 298
    rust and corrosion protection, 298
    viscosity, 297
    viscosity index, 297
Multistage reciprocating compressors, 405, 414, 433
    factors affecting lubrication, 414–415
    lubricating oil recommendations, 415
        air compressors, 415–416
        chemically active gas compressors, 418
        hydrocarbon gas compressors, 417–418
        inert gas compressors, 416–417

N

Naphthenic base stocks, 11
Naphthenic crude, 11
National labeling programs, 105–106
    Blue Angel Ecolabel, 105
    Swedish Ecolabel, 105–106
    U.S. Department of Agriculture BioPreferred program, 106
    U.S. VGP (U.S. VGP 2013), 106
National Lubricating Grease Institute (NLGI), 370, 385
    grease grade numbers, 77, 77t

National Pollutant Discharge Elimination System, 106
Natural gas, 1
Natural gas fired engines
    engine oil recommendations, 263–266
Natural lubricants
    history of, 4
NBR rubber, dynamic tests, 382
Neutralization number
    lubricating oils, 45–46
Nickel, 18
Nitrile rubber seals, 382
Nitrogen, 18
Nitrogen compounds, 19
N-methyl-pyrrolidone (NMP), 10
Noack test method, 13
Noise levels, 139
Noncondensing, 337
Nonmagnetic sediment, 545
Nontoxic EALs, 102
Nordic Swan, 106
North America
    petroleum in, 2–3
Not bioaccumulative, defined, 106
Nuclear power generation, 387
    lubrication recommendations, 399
        general requirements, 399–401
        selection of lubricants, 401
    petroleum products, radiation effects, 390
        chemical changes, in irradiated materials, 392–396
        grease irradiation, 396
        grease thickeners, radiation stability of, 396–398
        radiation damage, mechanism of, 391–392
        turbine oil irradiation, 396
    reactor types, 387
        basic reactor systems, 387–388
        BWR, characteristics of, 389
        FBR, characteristics of, 389–390
        gas-cooled reactors (GCR), 390
        HTGC, characteristics of, 390
        PWR reactor, 389
Nuclear power plants, 355
Nuclear power reactor, 387
Nuclear radiation, 392
Nuclear reactor schematics, 388
Numerically Controlled (NC) machine tools, 136

O

Octane number requirement (ONR), 248–249
Odor
    hydraulic fluids, 139–140
OEM approvals, 135, 135t
OEM tractor fluid specifications, 296, 296t
Oil; *see also* Lubricating oils
    cart, 507
    conditioners, 533
    containers, large, 505
    coolers, circulation system, 223
    dispensing containers, 504–505
    film seals, 429
    filtration, circulation system, 222–223
    heating, circulation system, 223
    leakage, 139
    leaks, losses by, 518

mist lubrication systems, 517
oxidation, 408
in plain bearings
    selection of, 169–171
in rolling element bearings
    selection of, 177–178, 178f–179f
sampling, visible contaminants in, 545
sampling intervals, guidelines, 543
separation tests, greases, 80–81
shaft seals, for hydrogen-cooled generators, 346–347
viscosity chart, 170f, 171
Oil analysis
    for gearbox, 556
    interpretation of, 551
    predefined absolute and trend analysis oil limits, 553
    used oil analysis sample, 552
Oil analysis program, startup recommendations, 541
    recommended sampling procedure, 543–544
    sample intervals, 542–543
    sample point selection, 542
    taking a representative, 543
    visual inspection, of sample, 544–545
Oil analysis provider, 540
    guidelines, 541
Oil Creek, 2
Oil house, 488
    chain block and trolley, 489
    control of static electricity, 495
    layout, 492
    lockers of steel construction, 490
    lubricant labeling, 494
    storage racks, 489
Oiling devices, 211–216
    air line oilers, 215–216, 216f
    air spray application, 216, 217f
    bottle oilers, 212, 213f
    drop feed and wick feed cups, 211, 212f
    mechanical force feed lubricators, 214–215, 214f–215f
    wick and pad oilers, 213, 213f–214f
Oiling methods
    bath oiling, 225
    chain oiling, 225
    circulation systems, 219–224
    collar oiling, 225
    ring oiling, 225, 226f
    splash oiling, 224, 224f–225f
Oil-lubricated LMGs, 183
Oil reservoirs
    circulation system, 221
    hydraulic system, 130, 132, 132f
Olive oil, as lubricant
    history of, 4
Open gear lubrication, 196–199
    AGMA specifications for, 199
    dust and dirt and, 198
    load characteristics, 199
    method of application and, 198–199, 199t
    temperature and, 198
    viscosity recommendation, 200f
    water and, 198
O-phosphoric acid esters
    applications, 5
    history of, 5
Optimum film thickness, 443

Orderliness, 492
Organization for Economic Cooperation and Development (OECD)
    aquatic toxicity tests, 102, 103t
Original equipment manufacturers (OEMs), 399, 540
Otto cycle, 468
Outboard marine engines, 266
    engine oil recommendations, 266
Overdrive, 287
Overshot feed groove and chamfers, 164, 164f
Oxidation inhibitors
    lubricating oils as, 37
Oxidation resistance
    automotive gear lubricants, 291
Oxidation stability
    engine oil, 245–246
    oil, hydraulic systems and, 133–134
    steam turbine oil, 352–353
    tractor fluids, 298
    vegetable-based EALs, 111
Oxidation tests
    lubricating oils, 54–56
Oxygen, petroleum oils, 418

P

Packaged plant air units, 431
"Packed for life" bearings, 217
Packing design, mechanical rod, 406
Pad oilers, 213, 213f–214f
PAGs, see Polyalkylene glycols (PAGs)
PAOs, see Polyalphaolefins (PAOs)
Paper mill dryer circulation systems, 219, 220f
Parallel shaft gears, 462
Pascal's law, 117, 118f
Passenger car
    engine oil recommendations, 258–259
Passenger car engine
    guidelines for, 16
Pedanius, Dioscorides, 1
Pelton turbines, 356, 364
    cross-sectional components, 357
    horizontal shaft machine, 356
    vertical shaft, 356
Pennsylvania Rock Oil Company, 3
Performance tests
    greases, 78–85
    lubricating oils, 54–60
Periodic oil analysis, 139
Petroleum
    existence of, 1
    hydrocarbons, 417
    in North America, 2–3
    oils, 418
    overview, 1
    premodern history of, 1
    use of, 1
Petroleum products, in nuclear industry
    radiation effects, 390
        chemical changes, in irradiated materials, 392–396
        grease irradiation, 396
        grease thickeners, radiation stability of, 396–398
        radiation damage, mechanism of, 391–392
        turbine oil irradiation, 396

Phenol, 10
Pipeline transmission
    gas turbines in, 327
Pistol oiler, 505
Piston compressor
    horizontal balanced, 411
Pistons, 200
    pumps, 120
    rings, 200
Pitch gearbox applications, 376
Plain bearings, 152–171
    grease in, 154–155
    hydrodynamic lubrication, 152–155
    hydrostatic lubrication, 155–157
    lubricant selection, 169–171
        grease selection, 171
        oil selection, 169–171
    mechanical factors, 158–169
    shear rate, 154
    thin film lubrication, 157–158
Planetary gears, 274–275, 275f
Plastic containers, of oils, 504
Plinius, 4
Pliny, 1
Plutarch, 1
Pneumatic braking systems, 313
Point sources, 106
Polishing filters, 137
Pollutant source reduction, 101
Polo, Marco, 1
Polyalkylene glycols (PAGs), 103
    biodegradation of, 108
    history of, 4–5
Polyalphaolefin (PAO)-based synthetic lubricants, 424
Polyalphaolefins (PAOs), 5, 9, 11, 449
    as base stock for EALs, 109
    invention of, 5–6
    lubricant, 450
    synthesized hydrocarbons, 436
Polyethylene, 152
Polyethylene glycol fluids
    FKM elastomers used, 382
Polyglycols, 9
Polyol esters, 5
Polyphenyl ethers, 5
Polytetrafluoroethylene (PTFE), 152
Portable filters, 138
Portable grease equipment, 507–509
Portable grease gun filler, 508
Portable oil dispensers, 505–507
Pour point
    depressants, 35–36
    lubricating oils, 46–47
Power brake, 311
Power shift gearbox, 275
Precision insert bearings, 160, 161f
Pressure compounding, 335
Pressure feed circulation systems, 219, 219f
Pressurized water reactor (PWR), 388, 389, 401
    typical lubricant applications, 400
Process oils, 15
Process operations
    gas turbines in, 327
Product selection, on lubricant, 515

compatibility with other products, 516
ease of disposal, 516
long service life, 515–516
value as by-product, 516
Projected area, of bearings, 159
Propane (R-218) gases, 436
^{239}Pu (plutonium 239), 387
Pulmonary disease
    petroleum hydrocarbons, 417
Pump efficiency vs. fluid viscosity, 465
Pumps, hydraulic, see Hydraulic pumps
Pump suction, circulation systems, 221
Purification methods
    contaminants and additives, 525
    effectiveness of, 524
Purification methods, in industrial plants, 524
    filtration/straining, 525–526
    settling, 525
    size filtration, 526
        centrifugal oil filters (separators), 530–531
        centrifugation, 530
        clay filtration, 528–529
        dense media filters, 528
        depth-type filters, 526–528
        multipurpose purifiers, 529–530
        sludge/varnish removal, 531–533
PWR reactor, see Pressurized water reactor (PWR)

R

Radial piston pumps, 120–122, 121f
Radiation damage, 390
    penetration and base oil viscosity, 396
Radiation dosage, 391
Radiation effects, 390
    chemical changes, in irradiated materials, 392–396
    grease irradiation, 396
    grease thickeners, radiation stability of, 396–398
    radiation damage, mechanism of, 391–392
    turbine oil irradiation, 396
Radiation energy, 391
Radiation stability
    vs. sulfur and aromatic content, 393
    vs. temperature, 395
Radiolysis processes, in hydrocarbons, 392
Railroad engines
    engine oil recommendations, 266
Raleigh, Walter, Sir, 2
Rape seed oil, as lubricant
    history of, 4
Reactor, basic systems, 387–388
Reactor, types, 387
Readily biodegradable products, 102–103; see also Environmentally acceptable lubricants (EALs)
Rear suspension system, 303
Reciprocating air compressors, 404, 416, 433
    lubrication guide, 413
Reclamation, of lubricating oils, 533
    reclamation units, 533
Recycling, 101
Reduced floor space requirements, 481
Reduced residual waste, 481
Reduction ratio
    gears, 192

Refinery processing (separation *vs.* conversion), base stock
    atmospheric distillation, 22–23
    propane deasphalting, 25
    vacuum distillation, 24
Refining process, 11, 18, 19
Refrigerants
    lubricating oil recommendations, 436
Refrigeration/air conditioning applications, 403
Refrigeration/air conditioning compressors, 431–437
    compressor affecting lubrication, factors, 433
        bearing system conditions, 434
        cylinder conditions, 433
        oxidation, 434
Refrigeration system
    basic single-stage compression, 432
Refrigeration system factors affecting lubrication, 434
    ammonia (R-717), 435–436
    carbon dioxide (R-744), 435–436
    fluorocarbons, 435
    hydrocarbon refrigerants, 436
    lubricating oil recommendations, 436–437
    sulfur dioxide (R-764), 436
Regenerative cycle, open system gas turbines, 320, 320f
Reheating, gas turbines, 320–321, 321f, 338
Relative volatility, of base stock, 13
Relief valves, 125, 125f
Renewability requirements, 107
Renewable resources, 101
Reservoir oil levels, maintenance of, 135–140
Reservoirs, for circulating lubricating systems, 519
Residual gear compounds, 199
Resin filtration, 532
Resistance to flow, 441
Resource conservation, 101
Return oil piping, circulation system, 222
Reuse methods, 218–224
    bath oiling, 225
    chain oiling, 225
    circulation systems, 219–224
    collar oiling, 225
    ring oiling, 225, 226f
    splash oiling, 224, 224f–225f
Reversible pump/turbine, 365
Reynolds equation, 452, 453
Riffel test, 385
    rig, 384
Rigid front axles, medium- and heavy-duty trucks, 313, 313f
Ring-oiled bearings, 167, 167f, 168f
Ring oiling, 225, 226f
Ring reversal areas, 200
Rippling test, 384
Rocker type drum rack, 500
Roller cam followers, 208
Roller chains, 205–206
Roller-raceway contact points
    in cylindrical roller bearing, 460
Rolling element bearing manufacturers, 459
Rolling element bearings, 171–179, 172f, 457; *see also* Ball bearings
    contamination, 177
    EHL conditions, 458
    factors, 175–177
    friction generated, 458
    grease in, 154
    grease selection, 179
    load, effect of, 176
    lubricant selection, 177–179
    need for lubrication, 173–174
    oil selection, 177–178, 178f–179f
    sliding areas, 174
    speed, effect of, 175, 176t
    temperature, effect of, 176, 177f
    types of, 174f
Roman Empire
    hydraulic turbines, 355
Rotary compressors
    lubricating oil recommendations, 420
    straight lobed, 418–420
    straight-lobed
        principal parts of, 420
Rotary fluid motors, 129, 130f
Rotary liquid piston compressor, 426
Rotary lobe compressors, 420
    lubricated parts, 420–421
    lubricating oil recommendations, 421
Rotary screw compressors, 421–425
    dry screw compressors, lubricated parts, 424
    lubricant recommendations, 424–425
Rotary sliding vane compressor, 425
Rotary straight lobe compressor, 419
Rotating Vessel Pressure Oxidation Test, 396
Routine inspections, of operating systems, 139–140
Routine used oil analysis, 547
Routing, 534–536
Rreclamation units, 138
Rusting, 134
Rust inhibitors
    lubricating oils as, 38–40
Rust protection
    automotive gear lubricants, 291
    engine oil, 247–248
    of oil, hydraulic systems and, 134
    steam turbine oil, 352
    tractor fluids, 298
Rust protection tests
    greases, 82
    lubricating oils, 56, 64

S

SAE 30, heavy-duty engine oils of, 426
Safety container, 496
Safety Data Sheets (SDSs), 476
Schaeffler FE8 rig, 380
Screw compressor, 422
Scuffing protection, 379
Seal compatibility
    automotive gear lubricants, 291–292
Sealing
    internal combustion engines, 241–242
Segment-type guide bearings, 367
Self-aligning ball bearings, 173f
Self-contained grease lubrication devices, 217
Self-lubricated bearings, 367
Semiautomatic transmission, 276–277
    lubrication requirements, 281
Seneca Oil Company, 2, 3

Index   575

Sequence control valve, 127, 127f
Series manifolded system, 228
Series reversing flow system, 228–229, 229f
Servomotor, 383
    pitch and yaw, 383
Settling purifier, 530
Shaft gearbox, 461
Shear rate
    plain bearings, 154
Shear stress, 441
Shock loads, 139
Shockwaves, 118
Shorter storage life products, typical shelf life, 498
"Sicilian oil," 1
Siculus, Diodorus, 1
Siemens/Flender foam test rig, 381
Silent chains, 205–206
Silicone polymers
    applications, 5
    as defoamants, 5
    history of, 5
    use during World War II, 5
Silliman, Benjamin, Dr., 2
Simple cycle, open system gas turbines, 318, 318f–320f
Single- and two-stage compressors, 408
    factors affecting cylinder lubrication, 408–414
    factors affecting running gear lubrication, 414
Single cylinder high temperature tests
    lubricating oils, 62, 62t
Single-cylinder steam turbines, 337–339, 338f, 339f
Single-line spring return, 227–228, 229f
Single rear axle, medium- and heavy-duty trucks, 313, 314f
Single-row deep-groove ball bearings, 172, 173f
Single-screw compressor, 421, 424
Single-stage centrifugal compressor, 428
Six-stage axial flow compressor, 430
Slide-roll ratio (SRR), 450
Slides, 180–181
    grease lubrication, 180–181
    lubricant characteristics, 181
Sliding block couplings, 202
    lubrication of, 204
Sliding disk couplings, 202
    lubrication of, 204
Sliding motion, junction, 444
Sliding vane compressors, 425
    lubricant recommendations, 426
    lubricated parts, 426
Slipper-type couplings, 202
    lubrication of, 205
Sludge deposits, 531, 532
Small gas turbines, 325–326, 326f, 333
Smith, W. A., 3
Society of Automotive Engineers (SAE), 15
Solar energy, 371
Solid/dry films, 151–152
Solid-state power converter, 373
Solvent-refined materials, 15
Solvent refining technique, 10
Speed
    gear, 192
    rolling element bearings and, 175, 176t
Speed governors, steam turbines, 340–342, 341f
Spherical cam followers, 208

Spherical valve assembly, 369
Spinner II® centrifugal oil filter, 531
Spiral bevel gears, 183, 186f
Splash oiling, 224, 224f–225f
Spring-laced couplings, 201, 202f
    lubrication of, 204
Spring return cylinders, 128
Spur gears, 183, 184f
Squeeze films, 103, 149, 149f
Start-stop cycle guarantees, 452
Static oxidation test
    greases, 79–80
Stationary gas turbines, 317–333
Steam engine, discovery of, 3
Steam turbines, 335–354
    automatic extraction, 338
    combined cycle plant, 338
    compatibility testing of oil, 353
    compound, 338–339
    control systems, 339–342, 340f
    cross compound, 339
    high-pressure, 338
    less flammable fluids, 353
    low-pressure, 338
    lubricant application, 348
    lubricated parts
        flexible couplings, 347–348
        gear drives, 347
        hydraulic control systems, 345–346
        journal bearings, 342–343
        oil shaft seals for hydrogen-cooled generators, 346–347
        thrust bearings, 343–345
        turning gear, 348
    lubrication
        contamination, 349–350
        factors, 348–354
        requirements, 342–354
        temperature of oil and, 349
    maintenance strategies, 353–354
    mixed-pressure, 338
    oil additives, 350–351
        entrained air release, 353
        load-carrying ability, 351–352
        oxidation stability, 352–353
        viscosity, 351
    operation, 336–339
    overview, 335–336, 336f
    pressure compounding, 335
    regenerative heating, 338
    single-cylinder, 337–339, 338f, 339f
    speed governors, 340–342, 341f
    tandem compound, 339
    uncontrolled extraction, 338
    velocity compounding, 335, 336f
Steel worm
    worm over wheel configuration, 463
Steering gear, 306–308, 306f–307f
    lubricant characteristics, 308
    lubrication requirements, 308
Steering systems, 304–305, 304f
    lubricant characteristics, 305–306
    lubrication requirements, 305
Stick-slip effects, 180

Stop-start engine protection, 469
Storage stability
    automotive gear lubricants, 291
Storing, of lubricants, 483
    lubricant deterioration, 496
        high-temperature deterioration, 497
        long-term storage, 498–499
        low-temperature deterioration, 498
        water contamination, 497
    oil house, 488
        facilities, 490
        function of, 488–490
        housekeeping, 492–493
        manpower, optimum utilization of, 491–492
        safety/fire prevention, 493–496
        security, 496
        size and arrangement, 490–491
    packaged products, 484
        outdoor storage, 484–487
        warehouse storage, 487–488
Strabo, 1
Strainer, lubrication system, 536
Stribeck curve, 442, 471
Stribeck number, 152
Structural stability tests
    greases, 78–79, 79f
Stuffing box, 236
S-turbines, 364
Sulfated ash, 47
Sulfur, 18
Sulfur dioxide, petroleum oils, 418
Sulfur dioxide gas, 436
Supercharging, 237
Super tractor oil universal (STOUs), 297
Supplier reliability
    EALs selection and, 114
Surface finish, bearings, 161
Suspension system
    active, 303
    front wheel, 301, 302f–303f
    lubricant characteristics, 305–306
    lubrication requirements, 305
    overview, 301
    rear suspension, 303
Swedish Ecolabel, 105–106
Syngas, 431
Synthesized hydrocarbon PAO, 6
    history of, 6
Synthetic aromatic additives
    radiation protection of, 394
Synthetic esters, 103
    as base stock for EALs, 109
Synthetic lubricants, 87–99
    advantages, 87
    alkylated aromatics, 92–93, 93f
    applications, 87, 88t–89t
    classification, 87, 90, 90t
    comparative temperature limits of mineral oil vs., 87, 89f
    cycloaliphatics, 93–94, 94f
    defined, 4
    dibasic acid esters, 94–95
    halogenated fluids, 99
    history of, 4–6
    organic esters, 94–96

    overview, 87
    PAOs (olefin oligomers), 91–92, 92f
    phosphate esters, 97, 97f
    polybutenes, 93
    polyglycols, 96–97
    polyol ester, 95–96, 95f
    polyphenyl ethers, 98–99
    recommendations, 87, 88t–89t
    SHFs, 90
    silicate esters, 98, 99f
    silicones, 98, 98f
System leakage control, 518
System shock, 127

T

Tallow, 4
Tandem compound steam turbines, 339
Tandem cylinders, 129
Technical requirements, 107
Tehut-Hetep, 4
Telescoping cylinders, 128
Temperature
    gear, operating and startup, 192–193
    of oil, steam turbines lubrication and, 349
    open gear lubrication and, 198
    rolling element bearings and, 176, 177f
Temperatures control, hydraulic systems, 138, 139
Tetraalkylsilanes, 5
Thermal power plants, 355
Thermal stability
    engine oil, 246
    lubricating oils, 56
    tractor fluids, 298
Thickeners
    in greases, 70–71, 72t–73t
        characteristics by type of, 71, 72t–73t
Thick hydrodynamic films, 142–145, 143f–145f
Thin elastohydrodynamic (EHL) films, 146–148, 146f, 147f
Thin film bearing, 157–158
    wearing in of, 158, 158f
Thin surface films, 149–151
    nature of surfaces, 150
    surface contact, 150–151
Thrust bearings, steam turbines, 343–345
Tilting pad thrust bearings, 365, 456
    spring supported, 368
Torque converter fluids, 292–293
Torque converters, 130, 273–274, 273f, 274f
Total acid number (TAN), 548
    lubricating oils, 46
Total energy
    gas turbine applications, 328, 328f
Toxicity
    EALs and, 101, 102, 103t
Transaxles, 286, 286f
    lubrication requirements, 286
Transfer case, 287
Transmission fluids, for CVTS and ATFs, 295
Transmissions, 271–281, 276f
    automotive, 273–275
    auxiliary, 287
    CVT, 276
    functions, 271

Index

hydrostatic, 277–279, 278f, 279f
lubrication requirements, 279–281
mechanical, 271–272, 272f
semiautomatic, 276–277
Transmitted power
gear lubrication and, 193
Transport-related emission standards, 440
Triphenyl phosphate
history of, 5
Truck, engine oil recommendations, 259
Tubular centrifuges, 530
Tubular turbines, *see* Bulb turbines; S-turbines
Turbines; *see also* Hydraulic turbines; Wind turbines
blades, 371
compressors, 369
control valves, 368, 369
diagonal flow (Deriaz), 356
governor/control systems, 368
guide vanes, 368
inward flow (Francis), 356
journal/guide bearings, 365
lubricant application, methods of, 367–368
lubrication, generator bearings, 364
quality oil, 430
semiumbrella type, 365
thrust bearings, 365–367
Turning gear, steam turbines, 348
Two-line system, 227, 228f
Two-stroke combustion cycle, 234–236, 235f
Two-stroke gas engines, 264
Two-stroke motorcycle oils, 267
Type A Fluid (General Motors), 293
Type CJ ATF (Ford), 294
Type F ATF (Ford), 294
Type H ATF (Ford), 294
Typical depth-type filter, 527

U

^{233}U (uranium 233), 387
^{235}U (uranium 235), 387
Ultimate biodegradation, 102
Umbrella, 365
type bearing construction, 366
Uncontrolled extraction turbines, 338
Universal joints, 281–286
cross-type, 281–282, 282f
Universal tractor transmission oils (UTTOs), 297
Unloading valves, 125, 126f
Uranium oxide (UO$_2$), 387
U.S. Department of Agriculture BioPreferred program, 106
U.S. EPA Vessel General Permit (VGP) 2013, 106
U.S. Military Specifications, 254–257
Automotive gear lubricants, 292
U.S. Occupational Safety and Health Administration (OSHA), 476
Used oil analysis, 539

V

Vacuum distillation
history of, 4
Vacuum Oil Company of Rochester, 4
Vacuum pump, to extract used oil sample, 544
Valves, hydraulic systems, 122–128
check valve, 124f
directional control valves, 125, 126f
electrohydraulic servo valve, 124f
flow control valves, 127
relief valves, 125, 125f
sequence control valve, 127, 127f
unloading valves, 125, 126f
Vanadium, 18
Vane pumps, 120, 121f
testing, 466
Variable valve lift (VVL), 468
Varnish deposits, 531, 532
Vegetable-based EALs, 109; *see also* Environmentally acceptable lubricants (EALs)
hydrolytic stability, 112–113
low temperature performance, 111–112
oxidation stability, 111
performance concerns, 111–113
Vegetable oils, 103
as base stock for EALs, 108, 109
characteristics, 109
performance limitations, 109
Vehicle fuel economy, 466
Velocity compounding, 335, 336f
V6 engine configuration, 468
Ventilation equipment, 495
Vertical axis wind turbine (VAWT) machine, 371
Vertical double acting compressor, 410
Vertical shaft turbine
cylindrical bearing for, 367
VI, *see* Viscosity index (VI)
Vibration, 139
Vickers 35VQ Pump Test (ASTM D6973), 110
VI improvers, 36
Viscosity
automotive gear lubricants, 290
engine oil, 244–245
greases, 83, 85
lubricating oils, 47–53, 48f, 49f
oil, hydraulic systems and, 133
selection, drive chains, 206–207, 206t
steam turbine oil, 351
tractor fluids, 297
Viscosity Grade Read Across (VGRA) guidelines, 16
Viscosity index (VI), 5, 10, 54
engine oil, 244–245
lubricant, 455
multipurpose tractor fluids, 297
oil, hydraulic systems and, 133
Viscosity index improvers (VIIs), 133
Viscosity lubricants, 458
Viscosity selection chart
for rolling element bearings, 178f–179f
Viscosity *versus* temperature curves, 455
Viscous-related energy losses, 441
"V" type, 408, 411

W

Waste oil collection, 534–536
Waste oil disposal, 537
Water

jackets, 426
open gear lubrication and, 198
solubility, 103
Water-containing fluids, 135
Water contamination
and gear lubrication, 194
Water resistance tests
greases, 81–82, 81f
Water separating ability
of oil, hydraulic systems and, 134
Water-separating ability, 352
Water-stabilized calcium soap grease, 497
Watt, James, 3
Wax, 1, 9, 10
Waxy materials, 435
Ways, 180–181
grease lubrication, 180–181
lubricant characteristics, 181
Wear, internal combustion engines, 240–241
Wear protection
oil, hydraulic systems and, 133
Wedge-shaped fluid film, 143–144, 143f, 144f
Wet brakes, 311
Wheel bearings, 309
lubricant characteristics, 309
White oil quality, 15
Wick oilers, 213, 213f–214f
Wind energy, 371
Wind power, 7
Wind sensors, 371
Wind Swept Area, 374
Wind turbines, 371
AC-type, 373
antiwear performance, 376
bearing protection, 379–380
micropitting protection, 376–379
scuffing protection, 379
blade, 374
blades, 373–374
design, 371–373
direct current (DC), 373
direct drive, 385
FAG Schaeffler, 380
foam and air release characteristics, 380
gearboxes, long oil drain intervals, 375
gear oils, 379, 383
condition monitoring, 382–383

industry standards/builder specifications, 375–376
interfacial properties of, 380–382
viscosity/low temperature requirements, 376
generator, 374–375
generator lubrication, 385
horizontal axis, 373
main gearbox lubrication, 375
main shaft bearing lubrication, 385–386
material compatibility
seals/paint, 382
performance, 374
permanent magnet horizontal axis, 372
rotor blade lubrication
oil/grease, 383–385
synthetic hydrocarbon gear oils, 380
vertical axis, 372
Wire cloth strainers, 526
Wire ropes, 207–209
core protection, 208
corrosion, 208
fatigue, 208
lubrication
characteristics, 209
during manufacture, 208
need for, 208
in service, 208–209
wear, 208
World light vehicles test procedure (WLTP), 471
World War I, 4
World War II, 4, 317
crude oil shortage during, 4
demands during, 4
silicone polymers use during, 5
synthetic lubricants development during, 4, 5
Worm gearbox, 463
Worm gears, 183, 186f, 462
"W" type, 408, 412

Z

Zinc, in gearbox, 382
Zinc dialkyl dithiophosphate (ZDDP), 416
Zinc dithiophosphate (ZnDTP), 473
antiwear additive, 473
ZN/P factor, 152
ZN/P value, 153